ROMAN AQUEDUCTS & WATER SUPPLY

To
FRED HODGE
et, comme toujours,
à ma
COCO

Roman Aqueducts & Water Supply

A. Trevor Hodge

Duckworth

Second edition 2002
First published in 1992 by
Gerald Duckworth & Co. Ltd.
61 Frith Street, London W1D 3JL
Tel: 020 7434 4242
Fax: 020 7434 4420
inquiries@duckworth-publishers.co.uk
www.ducknet.co.uk

A catalogue record for this book is available from the British Library

ISBN 0 7156 3171 3

Typeset by Derek Doyle & Associates, Liverpool
Printed and bound in Great Britain by
Bookcraft (Bath) Ltd, Midsomer Norton, Avon

Contents

Preface

My aim in this book has been to answer a simple question, and one that springs naturally to the mind – how did an aqueduct really work? And how did this fit into the whole context of water use? Another question soon presents itself: who is going to be asking? What type of reader is the book intended for? The general reader? Students? Scholars? The range is not as wide as one might expect, for given the fact that the subject has relatively seldom been approached from this angle, the number of experienced specialists is limited, and the others are not all that far apart in level. Traces will yet be seen of an attempt to accommodate a varying clientele: Latin is usually translated, for example. I have also enough experience in research to know that this book will seldom be read through, continuously. It will usually be used for reference, in a more limited search for enlightenment on specific topics, following the index and chapter headings. There is consequently a good deal of cross-reference, a certain amount even of repetition, as matter in one chapter may also have a relevance elsewhere and would otherwise be missed.

To acknowledge the efforts of all those who have assisted me is a task both pleasurable and alarming, for there have been so many that I am bound to sin by inadvertent omission. First must come Mr Robert Noyes, of Noyes Press, who first suggested to me that I write such a book and he publish it. This was precisely the stimulus needed, and the book owes its existence to it. When Mr Noyes ceased to publish in the classical field, the project was welcomed and taken over by Mr Colin Haycraft, of Duckworth, to whom I am indebted for an understanding patience in awaiting the manuscript that goes far beyond the exemplary and approaches the frankly unreasonable. My various travels in search of relevant material in Spain, France, Italy, Turkey, Israel and Germany were financially aided and abetted by the Social Sciences and Humanities Research Council of Canada, with ready and effective further support from the Dean of Arts and the Dean of Graduate Studies and Research at Carleton University. To Wolfson College, Cambridge, I am grateful for their warm hospitality during the writing up process of the manuscript.

On a more individual basis, how much I owe to the various members of the Frontinus Gesellschaft will be fully appreciated only by those who have themselves had recourse to it. My greatest debt is perhaps to Dr

Henning Fahlbusch, of Lübeck, who proved to be a kind of personified 'Open Sesame' to current work on aqueducts and their hydraulics. Dr Klaus Grewe, of the Landesmuseum, Bonn, whose own work on the planning and surveying of aqueducts has opened up a whole new field of study, was invaluable during my work in the Rhineland, especially in tracing the Eifel aqueduct; to the weighty and knowledgeable authority of Dr Gunther Garbrecht I am indebted for ready and invaluable co-operation. In Turkey I was able to draw on the unrivalled local knowledge of Dr Unal Ozis, of Izmir, as on the great geological and urbanistic skills of Dr Dora Crouch. In Israel I owe a particular debt to Yehuda Peleg, who not only took me to see almost every ancient aqueduct in the country – most of which I would never have found for myself – but at Hippos, in the Golan Heights, safely guided my steps through the minefield surrounding the aqueduct; how I would have fared without him is something I have never yet cared to consider.

At Ottawa I have been able to lean on the hydraulic insight of Professor D.R. Townsend, of the University of Ottawa, a specialist in open channel flow; I will not be so presumptuous as to say that he has kept me straight, but my book would certainly have been a lot worse without him. Bev Hall, secretary of the Department of Classics at Carleton University, has typed and retyped my manuscript so often that I am convinced she now knows more about Roman aqueducts than I do, on the accepted principle that the onlooker sees more of the game; and for producing the bibliography the credit must go to Kathyryn MacKinnon, who has wrought upon her Apple Macintosh word-processor the requisite technical marvels of parturition.

Lastly, there remains the dedication to this, the Second Edition. Normally a dedication is not explained by the author, who leaves it in some enigmatic form, such as 'To B.L.J.', but I would wish to make an exception. Authors' wives always have a lot to put up with while 'the book' is being written, and I am glad to dedicate this book to mine. The other name is my father's. A fine man, but of simple education and standing – after serving in France in 1914-1918, he spent the rest of his life as a motor mechanic in a Belfast garage – he was such that I am proud to honour him in any way that I can. He died without ever having seen a book that I had written, and before my marriage. My wife has often said how she wishes she could, just once, have met him. So let them both share the dedication page of this book. It is the best I can do.

Department of Classics
Carleton University, Ottawa

A.T.H.

1
Introduction

> I ask you! Just compare with the vast monuments of this vital aqueduct network those useless Pyramids, or the good-for-nothing tourist attractions of the Greeks!
>
> Frontinus, *De Aquaeductu*, I 16

My translation is not, perhaps, a literal, word-for-word replica of the original, but I can guarantee that the spirit of it would be happily authenticated not only by Sextus Julius Frontinus himself, but by just about all Romans everywhere. Culture, art and the like might be all very well in their place, but an aqueduct, now, that was something on a completely different level: something solid, grand, practical, civilised – in a word, distinctively Roman. The aqueducts went wherever Rome went, an outward symbol of all that Rome stood for and all that Rome had to offer – in Gaul alone there were 300 of them.[1] One may perhaps argue that an even more widely accepted symbol of ancient Rome is the Colosseum, but, whatever a Roman himself would have felt about it, we, in modern times, would surely feel happier according our recognition to a structure devoted to human comfort and advancement rather than one consecrated to mass human slaughter for the fun of it.

The typical aqueduct

But what, in fact, was a Roman aqueduct like? Here we meet a weakness in the organisation of this book, for, though the reader may expect to have a pretty good idea of the Roman aqueduct by the time he finishes it, this could, in the early stages, cause trouble. A full appreciation of some element may not be possible without at least a rough knowledge of the overall context. I therefore think it would be useful at this point to give in summary form a brief outline of the typical Roman aqueduct and some of the issues it raises.

Most aqueducts did not draw their water from a reservoir. They drew it from a river or a spring that fed water into the system as fast as it was used at the other end. Like an electric grid, the system worked on the principle of constant throughput, with no provision anywhere for storage worth mentioning. The water was supplied as needed and used as supplied – immediately. It was normally carried in an open conduit – open in the sense that, although the channel was covered over on top, the cover was not always watertight and the water was not intended to fill the channel right up, so that it could be considered hydraulically as flowing in an open channel rather than a closed pipe. The actual size of the conduit was governed by the need for human access, for maintenance, and usually the water came no more than one-third of the way up the sides. The depth varied seasonally, according to rainfall; it helps to consider the aqueduct almost as an artificial river rather than as a water main.

The normal principle was gravity flow, which in effect meant that the engineer had to lay out a line avoiding both too steep a gradient (a fast flow would erode or damage the conduit) and one too shallow (to avoid the current stagnating and letting sediment settle on the bottom to clog the channel). Bridges and viaducts, sometimes very large, might be needed to cross intervening valleys, though this was often done instead by an inverted siphon. The oft-repeated objection that the ancients could not make siphon pipes strong enough to contain the pressures generated has no truth in it, but the siphons do present other problems not as yet adequately solved. The siphon pipes were usually made of lead, but often, in the East, of stone, and arrangements for releasing air pockets are particularly contentious. Mountain ridges were pierced by tunnels, usually dug simultaneously from both ends.

Settling tanks were provided along the way, to permit sediment in suspension to sink to the bottom, but a more difficult problem was posed by the calcium carbonate incrustation (often called by its German name, *sinter*) which separated out chemically from the water and built up along the sides and bottom of the channel. It had to be removed manually by chipping; this made the aqueducts costly to maintain, to say nothing of the expensive repairs necessitated by leaks, frost damage, bridge foundations settling, and the like. A further problem was the unauthorised manipulation of supplies by incompetent or corrupt workmen, and by dishonest customers.

Since many cities both had an acropolis to be served, and also themselves had to be approached across a plain (like the Roman Campagna), in the home stretch the aqueduct often arrived carried on top of an endless arcade of arches, to maintain water level. On arrival at the city it would pour into a *castellum* (*château d'eau*), a distribution tank from which supplies would branch off to other parts of the city, carried either by open conduits (sometimes on top of the city walls) or in lead pipes. There is no truth in the common belief that this caused widespread lead poisoning (though the Romans may indeed have suffered from it from other causes), for the inside of the pipes rapidly acquired an incrusted calcium carbonate coating separating the lead and the water, which was in any case in constant flow and never long enough in the pipe to take any harm from it. The oft-quoted description of the *castellum* in Vitruvius, and how its output was divided among three different systems, arranged to overflow into each other, is also sufficiently vague, confusing, and at variance with any actual remains discovered, for me to suggest that, at this stage, it is best and safest forgotten.

From the *castellum* the water proceeded to the baths, usually its most important destination. Other branches led, either directly or via small secondary distribution tanks, to the public street fountains and to a limited number of small businesses and private houses. Running water in the home was not to be taken for granted. Local regulations would vary, but it looks as if the householder either had to know somebody with the authority to issue the necessary permit, or pay an extra tax – very likely both. There being no meters, the maximum possible consumption was regulated by using a supply pipe (the *calix*) of an officially controlled size. This in turn led to taps being a rarity, and most outlets, private and public alike, were kept running 24 hours a day. To our thinking this is wasteful, but as long as the aqueduct kept constantly bringing in more water there was nothing else to be done with it, and any attempt to save water by turning off the aqueduct would only have made the water overflow from the conduits and flood the town. A Roman aqueduct could no more be shut off when not needed (except perhaps at its source) than can a river, not unless you provided somewhere else for the water to go. It follows, offtake being constant, that there were no peak hours or slack hours. This also had the advantage that the sewers were kept constantly flushed with a process of continuous dilution of solid sewage that no doubt played its part in helping the Romans to live in large cities wholly innocent of any sewage treatment.

Modern attitudes

In all ages the Roman aqueducts – or at least their most prominently visible manifestation, the great bridges and arcades – have excited envy and admiration, even if at the same time they were often being pillaged for ready-cut stone by the local farmers. Thus the great array of arches carrying the aqueduct of Carthage across the plains of the Oued Miliane were for long admired by Arab writers as one of the wonders of Africa, while the brightly multi-coloured piers of the aqueduct of Merida, in Spain, still retain locally their traditional name, *Los Milagros*, 'The Miracles'.[2] To the classical scholars of the nineteenth century, when the fledgling discipline of archaeology was just beginning to assert its independence from a field hitherto resolutely philological and literary, it was again the monumentality of the structures that appealed. There was also a second group who contributed to the formation of archaeology. This was the local antiquarians, a term we may extend to cover such interested amateurs as travellers, army officers posted overseas, and the like. To them the same argument applies. It was impossible for anyone with a historical turn of mind to live alongside the Pont du Gard and not to study it. But there is a proviso. The towering arches might command attention, but it was a different story for all the miles of subterranean conduit leading to them, weaving in and out the valleys, contouring around the hillsides, and all safely buried from sight under a metre or so of earth. There were no great attractions here, so attention was lavished on the chief monuments and the aqueduct as a whole was neglected.

Traces of this bias have continued down to the most modern times. Since the study of the aqueducts has been largely in the hands of the archaeologists, it is hardly surprising that they have been studied archaeologically. Much time and care has been expended on the study of brick stamps and the dating of different phases in their construction, much less on the water supply system as a whole. In a word, the aqueducts have generally been studied rather as archaeological monuments than as functioning machines. The major engineering works have too often been considered in isolation from the system of which they formed part, often even in isolation from each other, and have been treated as if they were temples or colonnades – to be admired for their appearance rather than expected actually to *do* anything. In this attitude traditional classical scholarship is getting increasingly out of touch with modern classical students, one of whose first questions, not unreasonably, tends to be 'How did it all work?'

This is a question to which modern scholarship also is turning its attention, though the impetus still comes mainly from outside the classical field. The establishment in many institutions of archaeology as a separate discipline in its own right, often linked to anthropology and the social sciences, has helped, as has the establishment as recognised sub-disciplines of industrial archaeology and the history of technology. There can surely be no aspect more central to the study of water supply than the technical, and until recently it has plainly suffered from being left in the hands of traditional classical scholars, who regarded it at the worst as a banausic skeleton best left in some basement closet of 'The Grandeur that was Rome', at the best as some arcane ju-ju beyond the comprehension of normal men. The end result was the same. The technology of the aqueducts was neglected, and if the reader seeking an answer to the key question that this book seeks to answer, 'What was the Roman water supply like, and how did it work?', looks through the professional literature of the last few decades, he will be largely disappointed. True, he will find abundant articles and reports describing, often in great detail, individual monuments and sometimes whole aqueducts, but the description will be largely factual, listing such details as the height and width of the conduit, the precise location of the remains, dating by epigraphical evidence, and so forth. He will end up with the dimensions of everything but only a sketchy idea of how the water flowed. Such studies he will find of value as evidence and source material, but not as a direct or even partial answer to his question.

The first real complaint against this attitude was voiced in 1899 by Clemens Herschel, an American hydraulic engineer, who, as he himself tells us, published his technically-oriented commentary to Frontinus because he got tired waiting for the orthodox classicists to do the job.[3] Yet attitudes change but slowly, and it is only now, at the time of writing, that serious attention is being directed to the technical aspects of the aqueducts. In this field the Germans are the acknowledged leaders, and though in this book I have done my best to address the problem, the reader must be warned that it is far from a definitive treatment of it. The subject is still too new and fresh for that, requiring much more discussion, analysis, and collation of material from disparate sources. The future will tell.

One point, however, may be disposed of here and now. As the mainstream of scholarly thought in ancient history has moved further away from conventional tradition and veered more toward an emphasis on social history, one has sometimes seen the aqueducts appraised in a new light and lauded for a new virtue. What other early civilisation, we are asked, set such store by public hygiene, by abundant pure drinking water, by the very essentials of health and life? On this basis, do not the aqueducts rank as Rome's greatest and proudest achievement? To the social historian the argument is irresistible, so we may well declare clearly that it is almost wholly false.

One could argue that the eastern Hellenistic kingdoms also had aqueducts, so Rome is not the first; but this is a case I would not wish to urge, for there is no question that Rome stands first in the sheer scale of the enterprise. The real argument comes from the fact that the Roman aqueducts were not built to provide drinking water, nor to promote hygiene. Nearly all Roman cities grew up depending for their water on wells or cisterns in the individual houses, and some cities (such as London) got through their entire history without ever having had an aqueduct at all. In most, when the aqueduct arrived, it came belatedly and only as the result of imperial or other munificence (or a concerted municipal effort), long after the city had grown up and already existed without one for many decades, even centuries, in apparent health and prosperity.[4] Of course, once the aqueduct was there and good water readily available in great abundance, people naturally drank it. Why not? But the wells were still there as a supplement, and in some cities probably even remained the major source of supply, if there were areas that the aqueduct could not reach, such as the tops of hills. The aqueducts, then, were not built to fill a basic human need.

They were in fact a luxury. The normal reason an aqueduct was built was to supply the baths. Of course, the water was also used for other purposes, ranging from domestic supplies to garden irrigation, aquatic shows (*naumachiae*), flour mills, decorative fountains, and what Frontinus shudderingly calls 'uses too foul to mention' (presumably public toilets); but the voracious appetite of the Roman baths for constant and vast supplies of fresh water made them a prime consumer. It is not unknown for a bath to function, on a small scale, without any connection to an aqueduct, but for any serious enterprise water was needed in quantities that only an aqueduct could supply. Usually, then, the construction of a large bath complex was the chief reason for building an aqueduct; sometimes it was the only one.

Ancient attitudes

Were the baths, then, and their concomitant aqueducts, a luxury? The question is relative, for one man's luxury is another man's necessity. In the Roman world, the baths were popular places of mass resort. They may perhaps be likened to the sidewalk cafés that today form so characteristic an element of continental European urban life, for their function was largely social, but that must not lead us to underestimate the bodily comforts that they offered. In the blazing sun of the Mediterranean summer a wash was always welcome, and in the grim chill of the North European winter, given the inadequacies of domestic heating in the poorer homes, the baths at least offered somewhere one could look in to get warm. The baths, in short, were where one went not just to meet one's friends but also to find facilities that we regard as standard equipment in our homes but which in antiquity often did not form a normal part of domestic accommodation. This, then, explains the importance of baths in the Roman world, and hence the aqueducts serving them. It is probably the greatest single reason (though not the only one) why the Roman aqueducts were built. To the 'barbarians', provincials, and others as yet unaccustomed to the Roman way of doing things, it no doubt seemed an effete extravagance, if not a downright criminal waste of resources, both physical and financial. To the Roman legionaries, to whom a bath-house in their camp was as standard a piece of equipment as a refrigerator in a North American apartment, to retire as veterans to a city devoid of baths and aqueduct no doubt seemed quite unthinkable. Whether this then ranks as luxury or necessity, the reader must himself judge.

But there was also a second reason for building an aqueduct, at least as viewed through Roman eyes. This was civic pride. It manifested itself in various ways. For one thing, aqueduct-building was incredibly, even ruinously, expensive, and whole cities had been known to go bankrupt over building even quite a small one.[5] To possess an aqueduct was thus an outward mark of prestige and prosperity that none could mistake. To possess several, as did Ephesos, Lyon or Pergamon, marked a city as being indeed in the major league. It was something like, in modern times, having a large jet airport, or perhaps we should go back to the nineteenth century: the communal feelings aroused in so many towns by the arrival of the local railway line, at last linking the burghers to the great world outside, and celebrated by official dinners, speeches, beanfeasts and other festivities, must have been reflected in many ancient cities upon the arrival of the water through the conduit. Architecturally, moreover, this municipal pride took various forms. The great bridges, such as the Pont du Gard, and the long arcades crossing the Roman Campagna, made their own highly visible contribution. Given the great cost, I would not suggest that they were ever deliberately built purely for effect where in engineering terms they were unnecessary, but it seems equally improbable that, when a bridge had to be built anyway and a spectacular effect produced, the Romans were insensible to it. In a later age, after all, one of the reasons for building the enormous bridge at Roquefavour, near Aix-en-Provence, on the modern aqueduct serving Marseille, was a conscious and deliberate desire both to imitate and to outdo the Pont du Gard. Moreover, the treatment, architectural and otherwise, of the aqueduct and water system within the city often illuminate local attitudes to it.

Nowadays, although we are very conscious of the crucial importance of water, especially in conferences on the underdeveloped countries and the like, we do not consider water supplies as something particularly glamorous. Water mains and pumping stations do not appear on picture postcards, and sewage farms are fit only for the attentions of music-hall comedians. Nowhere is there a sense of the public greatness of the enterprise, with one exception. The exception is dams. The sheer bulk of a large dam cannot but impress, particularly when allied to the feeling of the vast latent power of the water contained behind it. The public admires and visits dams, and the fact is recognised by the provision of car parks, viewing points, and often a good deal of architectural decoration: when the Greeks covered the modern Marathon dam with a veneer of Pentelic marble and then built at the bottom of it a replica of the Athenian Treasury at Delphi, they were perhaps being a bit extreme, but there is nothing inherently uncommon about the attitude itself.

One does not find this spirit extended to hydraulic arrangements within the city, or at least no longer so. It was not always thus. In quite a few European cities our forebears lavished upon such things as the distribution *castellum* on the Peyrou esplanade at Montpellier or the *Wasserturm* in downtown Mannheim (Figs 1, 2) a whole array of columns, pediments, and other architectural embellishments which their descendants today would certainly deny to something now seen as a mere 'public utility'; and in the seventeenth century the Spanish successors to the *conquistadores* built in Mexico a whole range of aqueduct arcades imitating the Roman aqueducts of their native Spain (notably Segovia).[6] But, dams apart, hydraulic arrangements are not considered fit subjects

1. Montpellier: eighteenth-century *château d'eau* on the Promenade de Peyrou.

for public celebration. True, the Romans often felt this way too. Nothing could be more utilitarian than the treatment of the water supply at Pompeii, with its brick-piered watertowers and barely adorned street fountains. No marble columns here! But one sees a different attitude in the *castellum* of Nîmes.[7] Nothing could be more bleakly unattractive than this site today, but in antiquity it was fitted out as a centre of social resort, featuring something like a small shrine, murals of aquatic themes, and a bronze balustrade on which idlers could lean and watch the water swirl by. The important thing here is the indication that not only were the ancient Nîmois proud of their water supply: they were evidently expected to demonstrate that pride by going there and looking at it. And though unpretentious street basins on the Pompeii model were common enough, so also were nymphaea, exedrae and decorative fountains of all sorts. In one single year the Roman administrator Agrippa installed 300 statues on the fountains of Rome. These are often treated simply as architectural or sculptural adornments, but the flowing water lent another dimension. In normal sculpture a nymph was just a nymph, but a nymph pouring out a constant stream of water from a cornucopia was actually *doing* something, and something that moved. It was probably the closest the ancient world ever got to op-art. It was also, of course, a celebration of water. This celebration achieved its ultimate expression in the practice, especially common in the cities of Turkey during the second and third centuries AD, of terminating the aqueduct upon its arrival in the city in a massive decorative façade some two or three storeys high, heavily embellished with sculpture and with fountains spouting and cascading in

2. Mannheim, Germany: the *Wasserturm*.

all directions;[8] the best examples are the terminal nymphaea (or fountain complexes) at Miletos, Aspendos, Side and, especially, Perge, where the water continued on from the exedra fountains through a series of cascades spaced out along an open water channel running down the middle of the main street, all the way across the city (Figs 3, 4, 5).

Extravagances such as this surely reflect more than just a pride in the water system. They reflect the particular form that took: an ostentatious insistence on abundance, or, as we would now call it, conspicuous consumption. Water was not in particularly short supply at Miletos or Aspendos, not at least as compared with some herdsmen's hamlet on the edge of the Sahara, but everyone knew its vital importance, and 'shortage', like 'luxury', is a relative term; even where it was abundant, water was heavy and carrying it from the source or spring was troublesome. Indeed, it is not so long ago that one can remember, in rental notices, cottages, villas and holiday boarding houses listing among their various attractions 'running water' as a facility worth mentioning. In the ancient Mediterranean a good deal of the peasant mentality rubbed off even on to the city dwellers, and for simple men living a hard life with not much of anything to hand, what counts is quantity, not quality. Gourmet dining is the prerogative of the sophisticated and well-off, who have never gone short, while at a rural feast what makes the impression is the groaning board, the feeling of limitless plenty beyond

3. Miletos: terminal exedra of aqueduct, restored.

4. Aspendos: terminal exedra of aqueduct on the acropolis.

5. Perge: main street, with water channel (cascades) running down the middle.

all dreams. The same thing surely often influenced ancient attitudes to aqueducts. Even if, without an aqueduct, the citizens usually nevertheless had their own wells, the traditional ways of thought still held good. The particular feature of aqueduct water that should be celebrated above all others was its profusion; hence the emphasis on cascades and fountain jets. Water not carried in jars, not hauled up, pot by pot, out of wells, but endlessly tumbling, night and day, out of a great spout right in the city centre – that was the true magic of the aqueduct. It might have been built for the baths, but here was its most striking and outwardly visible manifestation, and the message it sent was unmistakable: 'Water? Why, we've got heaps of it, we just throw it away. That's what an aqueduct is – and our city's *got* one!'

Possible lines of study

How, then, are we to approach our subject? Several plans of study suggest themselves. One is chronological, to write a history of Roman aqueduct development. This approach naturally commends itself to the archaeologist, in whose mind, very properly, the dating of any artefact or monument bulks very large. It has the general advantage, in common with other possible approaches, of automatically providing a coherent framework for our own exposition, and the particular advantage that a survey of aqueducts in chronological order makes it easy to tie in their building with the political and historical context, to correlate with each other the construction of series of aqueducts and so clarify which was built first, and why and when others were needed, and, finally, to identify chronological trends in the development of new techniques and hydraulic practices.

All of these are valid reasons, yet I do not feel that any of them are strong enough to lead me to adopt this approach. The truth of it is that Roman aqueducts basically did not develop, and the latest ones were built and run essentially in the same way as their predecessors several centuries earlier. Like all such simplifications, this is not wholly true. The coming of concrete as a standard building material around the Augustan age made a great difference in bridge and arcade construction, as it did in architecture generally, and there were other differences as well. But there were no great breakthroughs, such as those that marked the Industrial Revolution, nothing that would enable us to divide up Roman aqueducts into an early, middle and late period. One aqueduct might be bigger and better than another but, within limits, they all operated on the same principles. I have therefore felt that this would not offer the best framework for a general book such as this. The reader must be warned that in what follows he may find a lack of emphasis on dates and periods, and this may well constitute a weakness in the book.

Another possible approach is the geographic. Could not the various aqueducts be grouped geographically, with chapters on North Africa, Turkey, Gaul, Italy outside Rome, and so forth? This again would have the advantage of readily identifying local trends and regional specialities, developed in response to local conditions. But this approach, though it has been used successfully for books on Roman architecture, would, I think, tend to turn this present book into something like a gazetteer, a work of reference to be consulted when the reader needs information on one particular aqueduct, but not to be read through from one end to the other for an answer to what I see as our key question, 'How did the Roman aqueducts and water supply actually work?'

There are other possible approaches as well: aqueducts as viewed through the literary sources, from the legal standpoint, and so on.[9] But the approach I have in fact adopted is to follow the same sequence as the water passing through the aqueduct, from catchment to its final dispersal by drains and sewers, and within this framework to see what problems confronted the typical Roman hydraulic engineer and what solutions be typically found for them.[10] When exceptions arise – such as the prevalence of wooden pipes in Northern Europe, dams in the Middle East, or stone pipelines in Turkey, and the African habit of running the city end of aqueducts not into fountains but into storage reservoirs – they will be noted as they become relevant.[11]

Ancient sources

Inscriptions

What, then, is the evidence? Archaeologically, our main ancient source, apart from the remains of the aqueducts themselves (which we will deal with in the rest of this book), is inscriptions. These are numerous, for, aqueduct-building being so expensive and prestigious an undertaking, whoever undertook it did not want his name overlooked, and dedicatory inscriptions recording either the construction or renovation of aqueducts are common. Usually it is the reigning emperor that is so commemorated, or if not he will at least be mentioned in the text, thus facilitating the dating of the aqueduct. The inscriptions are thus often vital for establishing a chronology and fixing the building of the aqueduct in the general narrative of historical development. It has also been pointed out that the almost equally frequent inscriptions recording repairs and refurbishing help to illustrate, by their very frequency, a sad truth often forgotten – that, however impressive they were and are as monuments, in pure engineering terms the Roman aqueducts were a very unsatisfactory way of carrying water, and needed constant work to keep them going.[12] Apart from dedicatory inscriptions such as these, the principal inscriptions bearing on our topic are the Lamasba inscription (Algeria), for details of agricultural irrigation; the Crabra inscription (Rome), for the same thing; the 'Pierre de Chagnon' (Lyon), for laws on encroachment on the aqueduct right-of-way; and the Nonius Datus inscription (Algeria), for the tribulations of an unfortunate Roman army surveyor in charge of bringing an aqueduct tunnel through a mountain.[13]

Vitruvius

Our literary sources include a wide range of chance references in such authors as Juvenal, for current attitudes to aqueducts and a few references to topographical features (such as the fact that at Rome the Porta Capena had the aqueduct running on top of it, and it leaked), and Pliny, in whose *Natural History* one is liable to find assorted trivia on almost anything one cares to name; but for all serious purposes they boil down to two names, Vitruvius and Frontinus.

Vitruvius Pollio comes first chronologically, having lived in the early first century AD. Paradoxically, he is in modern times much the more famous of the two, while in antiquity it was Frontinus who was better known: he was a very senior public servant with a distinguished army record, while Vitruvius is not known ever to have done anything except write his book *De Architectura*. But since this work is the only ancient treatise on architecture to have survived, it has received the concentrated respect and attention of artists and architects from the Renaissance onwards, sometimes a respect deeper and more exaggerated than it in fact deserves.

One of the prime weaknesses of the ancient world was the doctrine of *banausis*. Originating with the Greek thinkers – especially Plato – this decried practical, manual pursuits and extolled the virtues of abstract thought, on the logically impeccable grounds that if performing technical labour got your hands dirty, then even thinking about it probably got your brains dirty. Though the hard-headed Romans, particularly their engineers, were never quite as convinced of this as were the Greek intellectuals, the doctrine left its mark, and many professional men sought to upgrade the prestige of their profession by either stressing or inventing a connection with philosophy, and downplaying the strictly practical. As applied to Vitruvius, this resulted in his portraying the architect as a sort of intellectual superman who had to be conversant with geology, meteorology, hydraulics, mathematics, history, art and, in short, just about the whole span of human experience. One is left wondering how any mere mortal can measure up to so exacting a prescription, except, of course, that one assumes that Vitruvius himself did. No doubt this is precisely what Vitruvius wanted his readers to think, particularly the emperor of the day, who had building commissions in his gift; significantly, the one subject Vitruvius never mentions, though most architects consider it highly relevant one way or another, is money.

This has led to a common misapprehension of Vitruvius' *De Architectura*. The first mistake is to consider it as a sort of architectural/technical encyclopaedia. This is not what it is, but it is what the modern scholar often needs and Vitruvius is the closest thing to it there is, so, the wish being father to the thought, the identification is often, wrongly, made. It is wrong because Vitruvius' technical understanding and reliability is, at the best, variable. A good deal of his work does genuinely deal with the details of building construction, and when he dilates further into allied topics and scientific principles, he is often beyond reproach: his account of water wheels, for example, is both clear and correct. But Book VIII, devoted to water supply and aqueducts, is not Vitruvius at his best. His explanations are often confused and confusing, the account of siphons being particularly obscure and contentious. Technical terms are used without any explanation of what they mean. Scholars have speculated on whether some of his sources were in Greek, and how good his Greek was. Others, daringly hazarding their reputation, have wondered whether he had the remotest idea of what he was talking about. Others yet again, quailing before a step so desperate, have publicly berated their own and their colleagues' denseness in being unable to make sense of him, maintaining that whatever is wrong, it's all their own fault.[14]

As a professor, I may perhaps be forgiven for detecting in his siphon account something that, far from being confusing, I find depressingly familiar. To me, it reads very much like a poor undergraduate essay, hastily thrown together in the library the night before the class deadline for submission, and consisting largely of passages clipped from two or three standard works, chosen partly for their impressive terminology but otherwise more or less at random, and strung end to end to form an account which is all factually correct if the reader already knows the subject, but which does not show any real understanding on the part of the author. One would dearly love to call Vitruvius into the office and give him an oral examination on his book. The trouble is that, as with the essay, most of the information is probably correct if only one could sort it out, and it is hard not to think of Theseus' theatre review of Quince in *A Midsummer Night's Dream*: 'His speech was like a tangled chain; nothing impaired, but all disordered.' But there is a further complication. As always, the only really valuable pieces of information are those that we do not know already from other sources, and, by definition, this usually means that, when we do come across them, there is no way of verifying that what we have got is reliable. Yet, when all is said and done, we have no reason for giving up on Vitruvius. For most of his material, Vitruvius is the only source we have, and we should be grateful for that rather than harp on his deficiencies.

The second mistake is to think that Vitruvius has written an account of how the Romans built. This is not so. What he has written is based on this, true, but in reality his book is rather a series of recommendations on how, in his opinion, they ought to build. His work is something like a building code, but without the force of law behind it, whereas what would be useful to us would be a description of actual practice, something that explained to us the monuments as they are and as we find them. An example makes the point clear. Vitruvius was well aware of the poisonous qualities of lead, and recommends strongly and clearly against the use of lead water pipes. If all we had by way of evidence was Vitruvius' text, we would therefore confidently assert that the Romans never used them, for the inherent dangers were well known. In fact, as archaeology shows us, lead pipes were used, despite Vitruvius, everywhere, at every period, and in vast quantities; and the criticisms of Vitruvius are, as we shall see, invalid.

A further weakness resides not in Vitruvius but in the use we make of him. His reputation is great, and, in many areas, deserved. And the traditionally trained classicists, in whose hands rested until recently so much of the study of aqueducts, had themselves usually had a literary rather than a technical training. This sometimes led to an overuse or even misuse of Vitruvius. The temptation, when a problem arose, was always to reach for the volume of Vitruvius, conveniently at hand, rather than go to the site, often inconveniently distant, to look at the actual remains, much less, to cross the quad to that, psychologically at least, even more distant *ultima Thule*, the Faculty of Engineering. This has led to cases where the remains have been interpreted in the light of Vitruvius rather than the other way round, even to the extent of denying what plainly exists if Vitruvius says otherwise.[15] One lesson the modern scholar must quickly learn is that Vitruvius is valuable, but he is not gospel.

Frontinus

Sextus Julius Frontinus is a very different proposition. His work *De Aquaeductu* must qualify as one of the driest ever written, and is wholly devoid of any literary pretensions or elegance whatever. Unlike Vitruvius, he is completely reliable, a specialist writing on his own field. His weakness, other than his indigestibility, is that sometimes his expositions of hydraulics are confused or inaccurate, because they reflect a Roman technical knowledge that was itself incomplete or erroneous, and also that his treatment of the subject has major gaps. Frontinus was a bureaucrat, not an engineer, and what we get in him is the view from head office rather than the job as seen by the section maintenance foreman or the surveyor planning the line.

Born probably around AD 35 and dying in 103 or 104, he served in the army and as governor of Britain (where he pacified the Welsh), held the consulship three times (in 73, 97 and 100), and in 97 was appointed by Trajan *curator aquarum*, Water Commissioner, at Rome. The importance of this office was in Roman eyes immense, for, prestige apart, upon it depended not only the health, comfort and sanitation of the city, but even its safety; Rome was plagued with fires and the memory of the Great Fire of Nero was only forty years old. The *De Aquaeductu* was the fruit of his tenure of this post.

Although highly technical in content, the book was inspired by political motives. Rome was at the time facing a dilemma familiar in the modern commercial world: is management essentially a profession in itself, so that a competent manager can run any enterprise, irrespective of its product, or, especially in technical fields, is first-hand experience of the trade essential to running the business properly? Hitherto, under the Julio-Claudian and Flavian emperors, Rome had preferred the second answer. Senior technical positions went to imperial freedmen, the personal representatives of the emperor, with a long training behind them in the particular business they served. But now Nerva and Trajan changed the policy, appointing distinguished public servants from the senatorial ranks, much as a modern Prime Minister reshuffles his cabinet ministers, each of whom is deemed capable of taking charge of whatever department is allotted to him. Frontinus was one of the first representatives of this new order of things. Vested interests are always present, and no doubt there were many at Rome who looked forward with both pleasure and confidence to watching Frontinus, out of his depth in this highly technical quagmire, foundering ignominiously and dragging down with him into obloquy the whole new official policy. But Frontinus did master his hydraulics, did reform the Water Office (which sorely needed it), and what better way to prove the point – and so defend Trajan, the emperor who appointed him – than to publish a detailed and technical treatise on the Roman waterworks? True, he offers the excuse that he compiled these notes for his own guidance, but only some motive such as the above can account for their publication and circulation.[16]

The *De Aquaeductu* throws a great deal of light upon Frontinus' own character and the high standards of public service that he both set and himself lived up to. Fraud is anathema, and incompetence not much better. The worst incompetence of all is that of the minister who, out of ignorance of his own department, perforce rubber-stamps the decisions of his own subordinates and ends up as a puppet in their hands; and the only remedy is for him to take the trouble to learn the business for which he has undertaken responsibility.[17] Granted, then, that Frontinus' book is based on these attitudes and this experience, what, more precisely, are its strengths and weaknesses? One weakness is that, just because his experience is limited to metropolitan Rome, he is obviously a poor authority for Roman aqueduct practice in general throughout the empire, and standard features which happen not to occur often on the Rome network – siphons are a case in point – are passed over in silence. The day-to-day work of the maintenance crews is underplayed, and the never-ending cleaning of the aqueducts and chipping away the calcium carbonate deposits is never mentioned, though it must have been the single greatest task of the workmen. And since Frontinus' job was running the aqueducts, not building them, we hear nothing of the work of the surveyors, even though we know him to have been expert in that field – he wrote a book on it.[18] He pays little attention to the major engineering works along the Rome aqueducts, and one has the impression that he could write a full account of the aqueduct of Nîmes without ever mentioning the Pont du Gard. His knowledge of hydraulic theory is spotty and, worse, conveyed only in a disjointed series of asides, appended as footnotes to individual topics and difficulties. There is nowhere any attempt at a coherent and comprehensive general statement. Such a corpus of hydraulic knowledge and its application to engineering, though in places imperfect, yet certainly did exist by Frontinus' time, and Vitruvius, though he occasionally got things wrong, was well aware of it; he quotes, for example, the extremely sophisticated dictum of Archimedes that all water on the earth forms, potentially, part of one uniform sphere, with the same centre as the earth itself.[19] There is nothing like this in Frontinus.

His strengths are based on his personal experience as Water Commissioner. He excels in the bureaucratic overview, comprehensively grasping the aqueduct system as a whole. One feels that he carried in his head a map of the entire network and could visualise, almost pictorially, the traffic it was carrying – what volume of water was coming from where and to where, where additional supplies were flowing into the mainstream and where they were branching off, and what particular kind and quality of water was in each. He is always conscious of the needs of scheduling shutdowns for repairs, of critical path programming to minimise interference and inconvenience, taking into account varying demand, availability of a surplus for diversion to points of need and other relevant factors. Facts and, even more, figures, are meat and drink to him. The stirring quotation at the head of this chapter comes at the end of several pages of the driest statistics on aqueduct lengths and pipe discharges that the mind can conceive, but there is no indication that Frontinus thought they were a necessary evil, something to apologise for. On the contrary, he glories in them. It is that very recital of endless figures that fires his spirit, and it is what the ringing declaration refers to.

Administrative techniques are his *forte*, and engineering is really of interest to him only in so far as head office had something to do with it. When I was a boy, our school library was full of improving works with titles such as *The Romance of Engineering*. That is a concept which, when you get right down to it, Frontinus would not have understood; but he would very likely have substituted for it something even more alien to our own attitudes – *The Romance of the Filing Cabinet*. Thus he always pays a great deal of attention to the techniques of measuring the quantity of water (even though we are not sure exactly how these worked), and a great deal also to aqueduct law, and the rights and duties both of consumers and of the landowners across whose estates the aqueduct passed.

This hard-headed practicality has led many scholars to acclaim him as an ideal public servant,[20] reinforced by his reportedly forbidding the erection of any grave monument to himself ('Money wasted! My memory will survive if the achievements of my life deserve it').[21] Such has long been the accepted evaluation, including in the First Edition of this book, but it may have to be seriously questioned; much of the praise of Frontinus apparently derives only from his own unsupported word, so that his image is self-created. For a wider discussion I would now refer the reader to the chapter 'Reappraising Frontinus' in my book *Frontinus' Legacy* (see Supplementary Bibliography, below, p. 358). Nevertheless, he remains our prime source for Roman aqueducts and for that is to be respected. That is an honour nothing can take away from him.

2
The Predecessors of Rome

Rome stands founded on the men and manners of old.

Ennius, *Annals*, lib. inc., 37

When that stalwart conservative, the Roman poet Ennius, insisted that the Roman state was founded on the achievements of the men of old, he meant, of course, the early Romans. He was not thinking of Greeks or Etruscans. But in aqueducts it is different. One cannot embark on a work such as this without first taking a sideways glance at what the Romans owed to their non-Roman predecessors. Another equally valid but less obvious distinction is that between local and transported water. By local I mean water that is tapped and used at its source, as opposed to water that is transported long distances: it is thus a distinction between, on the one hand, wells, cisterns, fountain houses and other installations designed to give access to water already locally present, and, on the other hand, aqueducts. Both systems were in widespread and often complementary use in antiquity, but we will find it convenient here to concentrate only on transported water in pre-Roman times, leaving local water, Roman and pre-Roman alike, till the next chapter.

The eastern empires

Aqueducts and canals, usually for irrigation rather than urban supply, are found in several of the early eastern empires. Armenia (at this period known as Urartu) was an early leader in complex irrigation networks,[1] followed closely by Assyria, where at a slightly later date Sennacherib (705–681 BC) canalised the waters of the Atrush and Kohsr rivers. The scheme was executed in three phases spread over a thirteen-year period (703–690), and the final phase was the most spectacular. This involved building a dam on the Atrush at Bavian and diverting its waters by canal to Nineveh, some 55 km to the south. The canal followed a winding course through the foothills, crossing several valleys by embankments and masonry structures rather after the fashion of a Roman aqueduct, the most imposing being at Jerwan; here the valley is spanned by a 300 m long dyke built of limestone blocks and carrying a conduit no less than 12 m wide.[2] Another of Sennacherib's schemes was a 20 km aqueduct, largely in tunnel, to supply the town of Arbela, later to achieve fame as the site of possibly the greatest victory of Alexander. Like the Bavian project, it drew on water impounded by damming a river. At approximately the same date, in Judah, Hezekiah (727–669 BC) brought water supplies by an underground conduit 537 m long from the spring of Siloah to within the city of Jerusalem.[3]

The qanat

In their essentials the aqueducts of Assyria do not perhaps differ too much from their classical Roman successors, but there was another hydraulic installation that was both highly characteristic of eastern water supply and unique in its principles. This was the qanat. Found throughout the Middle East, but particularly common in Iran, the qanat is a tunnel driven into a hillside to tap an aquiferous stratum deep inside it. The tunnel has just enough of a downward slope for the water tapped to run down it and into the open air by gravity, and is punctuated at intervals of 20 m or so by vertical shafts to the surface (Fig. 6).[4] The qanat is important to our studies for three reasons. First, although not originally a Roman form of water supply or of Roman inspiration, a great many qanats remained in operation in lands occupied by the Romans, so that the qanat was in fact a common water source in the Roman empire. Second, although the natural home of the qanat is Iran, its use expanded at an early date far beyond the Iranian borders. It is possible that the Romans came in contact with it indirectly (through, say, the Etruscans, whose *cuniculi* may have been a form of qanats) and that they were influenced by it in the formative years of their own hydraulic skills. Third, the construction techniques of the qanat engineers may throw light upon the techniques of the Romans, particularly in tunnelling and surveying. The argument here is that for the Romans' techniques, and instruments used in such tasks as levelling and determining the aqueduct gradient, we depend almost entirely on literary sources. They describe the *chorobates*, and *dioptra*, and the *groma*,[5] and these accordingly bulk large in our accounts of Roman aqueduct engineering. But there is also a range of simpler techniques employed by the qanat engineers, using simpler instruments, to achieve the same ends, which we know about not from written sources but because the same techniques are still in use building qanats in Iran today; they will be considered below, when we deal in detail with the engineering of aqueducts.[6] Though these techniques apparently date back to antiquity, there is no evidence as to whether the Romans did or did not use them. But, given that they were actually used by hydraulic engineers in the Mediterranean basin at the same time that the Romans were beginning to build aqueducts, one

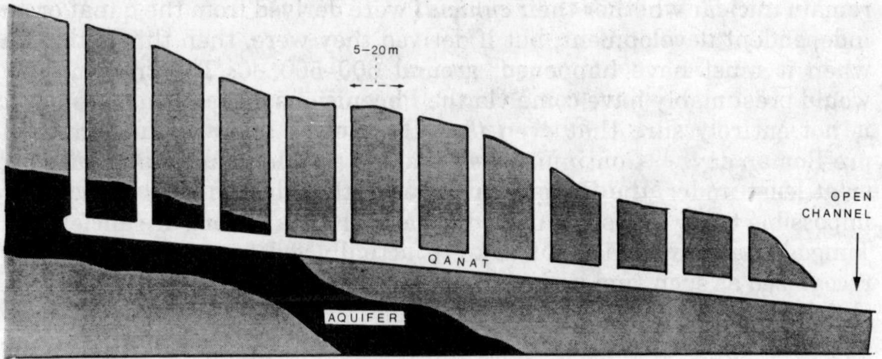

6. A typical qanat, in section. In reality they are usually much longer in proportion, and the vertical shafts more numerous.

plainly cannot rule out the possibility that sometimes the Romans did the same thing the same way, even if it is not mentioned in the written sources.

The name qanat is derived from the Akkadian *qanu*, 'reed', and is cognate with the Greek *kanna* and Latin *canna*, from which in turn is derived *cannalis*, 'shaped like a reed, i.e. a pipe, thence canal'. The spelling qanat seems to be the commonest, but one also finds 'kanat', 'ghanats', and 'quanate', to say nothing of the Biblical city of Qanatha,[7] named after its qanats. Other and quite different terms are also employed for qanats. Sometimes, especially in North Africa, the name is 'foggara'. 'Kariz', often anglicised to 'kareez', is also found, as is 'khettara': 'madjira' means the same thing, and, from the profusion of Saracen qanats, gave its name to the city of Madrid.[8]

Geographically, qanats were and are most at home on the Iranian plateau, but at an early date they spread beyond its limits. Interpretation here seems normally to follow the diffusionist principle of archaeological evaluation – that is, where qanats are found outside Iran, it is assumed that this represents a transmission of ideas and not independent invention; an exception is made for Northern Europe, where qanat-like structures may have evolved locally. The dating of the various stages of transmission is difficult, but it would seem certain that the qanat was already common in Iran by the eighth century BC. They are thought to have spread into Arabia in the sixth, and may be so recorded in Herodotus, though this is not clear.[9] From Arabia they spread into Egypt, introduced by the Persians and probably by Darius I (around 500 BC). The best-known site is the oasis of Kargah, where some of the qanats are of very impressive size: one has no less than 700 vertical shafts. In Egypt at least there seems to be no doubt that the qanat network was both maintained and expanded by the Romans. As for the Etruscans, it must remain unclear whether their *cuniculi* were derived from the qanat or an independent development, but if derived they were, then this is the era when it must have happened, around 600–500 BC. The transmission would presumably have come via the Phoenicians of Carthage, though it is not entirely sure that even they themselves knew about qanats in pre-Roman days.[10] Continuing west, qanats existed in antiquity, built by, or at least under, the Romans in Algeria, though it is often difficult or impossible today to distinguish which actual installations are ancient. At Timgad the Roman aqueducts were actually fed by qanats, often not recognised as such, and it may also be that many of the tunnels in Algeria classified as drainage works should more accurately be categorised as qanats.[11]

This is apparently as far west as the qanat went in antiquity. One naturally wonders about Morocco, where a large number of qanats of mediaeval, or Saracen, date have been found. If the idea had got along the North African coast as far as Algeria, is it impossible that it made the last step and got to Morocco too? The idea is tempting but our chief authority regretfully rejects it, suggesting that it is more likely that the qanat was brought by Islam to Spain, where it enjoyed wide popularity, and was thence transferred south to Morocco only around the eleventh century AD. In Spain it flourished, and Madrid, as noted, had a large number of them, some of which are still in use, serving fountains. It will be fitting to close this geographical and historical survey by noting that when the *conquistadores* came from Spain to the New World, they brought with them the knowledge of this hydraulic technique and employed it in circumstances deemed suitable. Thus it is that it was qanats that provided the first water supply to – of all unlikely places – the city of Los Angeles.[12]

The invention of the qanat is almost certainly to be seen as a by-product of mining, rather than agriculture or irrigation. The easiest way of draining a mine, if the strata and topography are such as to permit it, is to cut a gallery from the inundated area running slightly downhill to the open air, so drawing off the water. Naturally this will only work in special circumstances, such as when the mine has been driven into a mountainside, but it is commonly enough found in ancient workings.[13] Moreover, mining produced another by-product useful to the qanat engineer: a study of strata and underground topography. It is not without significance that the oldest known drawn plan is not a plan of an architectural monument, but of a mine.[14] For this combination of skills to produce a qanat all that was needed was an inversion of priorities. One dug a tunnel that, instead of getting rid of unwanted underground water and disposing of it no matter where, tapped underground sources and delivered the water where it was needed. The mines themselves go back to a very early date in Iran. The qanat was derived from them at some later period, but although the first qanats actually attested date only from the eighth century BC, one presumes that, given the early date of the mines, the derivation took place well before that.[15]

The qanat is generally located in the foothills of a sizeable ridge or mountain, for two reasons. First, the sloping land surface makes possible a tunnel that sinks deeper underground while actually running uphill from the entrance. Second, such a ridge is liable to feed the aquiferous strata by forcing higher the rain-bearing clouds and inducing precipitation. In theory all that is then required is a sloping tunnel driven into the hillside to tap the water and bring it running by gravity to the outflow, to form an artificial spring. The slope of the tunnel is carefully surveyed to be constant throughout – unlike aqueducts, where this gradient often varies – and is at the shallowest gradient that will keep the water flowing: the greater the gradient, the greater the danger of damage from erosion and the stream eating into its bed. The gradient normally used is 0.5‰, or 0.5 m in 1 km.[16] As compared with most Roman aqueducts, this gradient is slight, though it is in the same range as those of Nîmes, Arles or the Aqua Appia at Rome. The water is usually carried in a gutter or channel in the floor of the tunnel, the tunnel itself being of dimensions determined by the need for human access, and usually about 0.6 × 1.2 m;[17] we will again encounter this practice in Greek and Roman aqueduct tunnelling. The tunnels were normally unlined, but, in modern work at least, are often reinforced when passing through strata where there is a danger of collapse by the insertion of large terracotta rings (*kaval*) of about 1.2 m diameter, which interlock with each other after the fashion of classical terracotta pipes, to form a continuous tube.[18] It does not seem to be known whether this technique was used in antiquity.

However, although it is the tunnel that is the heart and *raison d'être* of the qanat, it is the vertical shafts that are its most outward and visible manifestation. These were dug chiefly for the evacuation of excavated spoil during the construction of the tunnel, and the spoil was then heaped up in a ring around the mouth of each shaft, forming a concave or saucer-shaped depression with the shaft in the middle. The course of the qanat is thus characteristically marked by a string of these rings across the land, like a line of stitching.[19] In most handbooks the qanat is shown diagrammatically, as in my Fig. 6, so the number of shafts is small, only four or five. In reality, qanats can be very long indeed. An underground tunnel of 10 or 15 km, getting all the time gradually deeper below the surface, is not at all unusual, and in the largest known qanat, at Gonabad, the tunnel runs for no less than 35 km. Over this distance the shafts will be spaced a mere 5–20 m apart – a wider spacing reflecting a greater skill on the part of the engineers[20] – which means that even an ordinary average-sized qanat is liable to be marked by a string of over 500 of them. We may note that in aqueduct tunnels the superior skill of the Roman engineers enabled them to space their shafts much further apart

– Vitruvius recommends a spacing of an *actus*(?) or 35 m.[21] In depth, the shafts increased progressively as one got in further from the outflow, the last one (*le puits-mère* or, in Arabic, *gamaneh*) being the deepest. It was often 100 or 150 m deep, though in the giant qanats of Gonabad it was a full 300 m, and had to be broken up into three successive, staggered shafts of 100 m each, with a landing between each pair. As well as for evacuating spoil, the shafts had other uses. Their close spacing made for rapidity of construction of the tunnel, through simultaneous access to a multiplicity of work-faces.[22] They also provided for ventilation of the tunnel and perhaps, to some extent, aerated the water as it flowed. Finally, they provided a convenient means of evacuating the silt, mud and other accumulated debris during cleaning operations.[23]

On the volume of water produced by a qanat it is obviously impossible to generalise, any more than one can set an average for a spring. In modern Iran the overall average per qanat works out at somewhere around 1,300 m^3 per day (15 l/sec), or about half the discharge of the smallest of Roman aqueducts. The largest, on the other hand, at Sarud, 400 km east of Teheran, produces 77,760 m^3 (900 l/sec), about the total of all four Lyon aqueducts combined, though here the very high output may come from an underground stream intersected by the qanat rather than water seeping from the aquifer into the catchment tunnel in the ordinary way. At the other extreme, in the arid climate of the Sahara and Morocco, some of the smaller qanats yield only 432 m^3 (5 l/sec) daily, or even less, though this is still enough to provide a small village with a living.[24] It must be emphasised, however, that unlike the Roman aqueducts, which carried large volumes of water to a town already established, and that largely for luxury uses such as the baths, the water from qanats was used locally for irrigation at whatever point the surveyor found it convenient for the delivery tunnel to come to the surface. Some qanats might serve orthodox aqueducts, as at Timgad, or be built to supplement an oasis, but mostly the village, and its irrigated land, sprang up around the qanat and its outflow. The situation was therefore the reverse of what happened with aqueducts. With qanats, it was the qanat that came first and the settlement it served grew up afterwards, while with aqueducts it was the other way round. It remains only to be repeated that while this vast qanat network was largely outside the boundaries of the Roman empire, a sizeable portion of it lay within them and, particularly for rural irrigation in North Africa, was an important part of ancient Roman water supply.

The Greeks

Archaic and classical

Water supply in ancient Greece can most conveniently be divided into two periods, for the coming of the Hellenistic age, with its siphons and other ambitious schemes, effectively changed the whole picture. In archaic and classical Greece, one seldom finds anything that, especially in comparison with the more grandiose Roman projects, one can really call an aqueduct. Most cities had some sort of local spring at their centre, and probably grew up around it. In a major city they often achieved a symbolic fame that far exceeded their actual utility, and Glauke and Peirene at Corinth, the Clepsydra on the Athenian Acropolis,[25] to say nothing of Castalia at Delphi, were celebrated far and wide. Usually these urban springs were, at one time or another, artificially systematised by the provision on the site of a fountain house, stone basins for filling water jars, and various architectural adornments.[26] In particular, water works of this kind were a favourite project for tyrants. It might also be that, as the city grew, the local spring had to be supplemented with additional supplies piped in from a distance. This happened with the best-known fountain in Athens, the Enneakrounos ('Nine-spouter'), which was fed not only from the local spring Kallirhoe ('Fair-flowing') that preceded it on the site, but from a terracotta pipeline that brought supplies from somewhere in the direction of Hymettos, or possibly Pentelikos.[27]

The terracotta pipeline seems to have been the normal method of conveying water in classical Greece. Often the pipes might be laid along the bottom of a large channel or tunnel, provided to facilitate access for maintenance, as on Samos, but it will usually be found that this channel was not itself full of water, while in Roman aqueducts it usually was. It follows that Roman aqueducts normally carried a much greater volume of water than these Greek pipelines, though, as at Kuttolsheim (p. 116 below), they too sometimes used pipes.[28] The terracotta pipes, usually around 20–25 cm diameter, were designed to fit into each other, and sometimes had an opening in the top covered by a lid. Intended either for inspection and cleaning or to enable the workman to get his hand inside to plaster the joint during installation, these holes at least make it plain that the water inside was not running under pressure; in fact, on the Enneakrounos pipeline, 'inside the pipes is a hard lime deposit up to half an inch thick along the bottom and part way up the walls',[29] making it plain that normally the pipes were not even full, but instead operated hydraulically like a gravity-flow open conduit. This was not invariably so. Where the terrain made it advisable, the pipeline ran full, and up and down minor irregularities in ground level. On these stretches it was laid with pipe sections with no inspection holes, as the water ran under at least nominal pressure.[30] It is impossible to say whether such a system led to the inverted siphons that formed so prominent a feature of Hellenistic work, but it is undeniable that the principle is the same, and it is only a matter of degree.

To this simple system more sophisticated modifications could be added. One way was to attach to the fountain house carrying the delivery spouts a large storage reservoir. Its purpose can only be guessed but the most

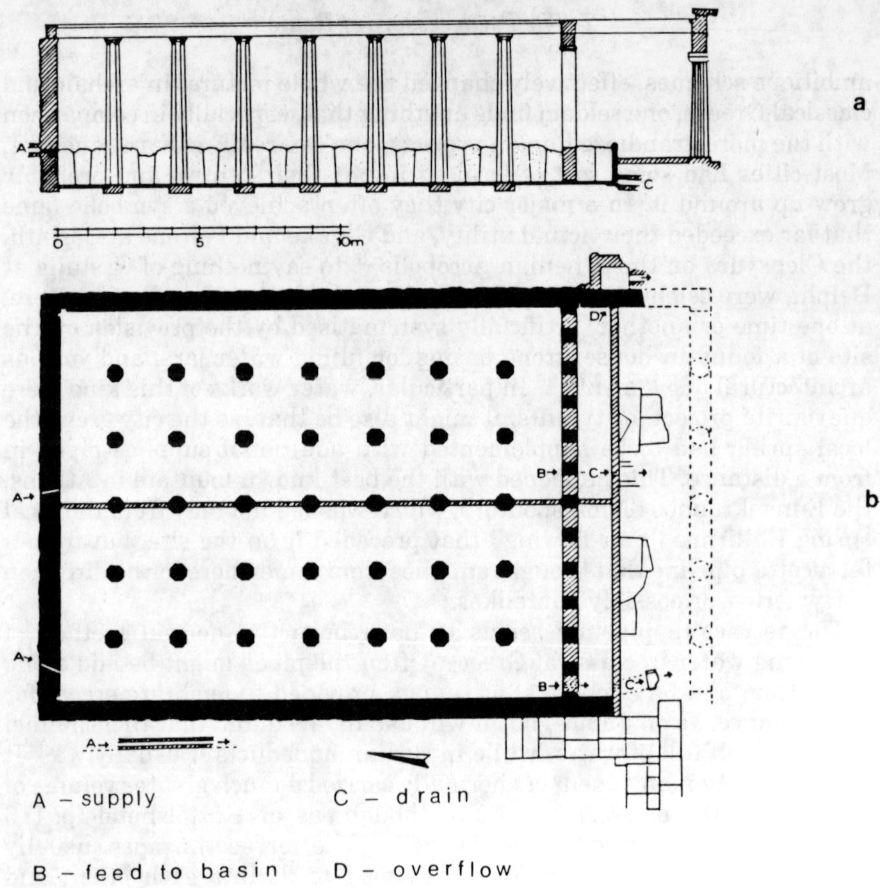

7. Megara: fountain house of Theagenes (sixth century BC): (a) section; (b) plan (Gruben, 1964).

obvious conjecture is that it was employed where daily needs were found to exceed the daily discharge by the spring or pipeline; a storage reservoir would enable surplus water to be accumulated and stored during the off-peak hours of darkness, and so help satisfy the increased demands of the coming day. The best-known is the so-called fountain house of Theagenes at Megara (Fig. 7), a late-sixth-century structure inaccurately ascribed to the tyrant by Pausanias.[31] This was an oblong tank some 18 × 13.5 m, covered with a flat roof carried on five rows of seven octagonal pillars each and directly running into a line of dip basins across one end, from which the public filled their amphorae. The capacity is around 380 m^3, and it has been calculated that, at a discharge of 15 l/sec, it would take the delivery pipe seven hours to fill it, which would confirm that the intention was for it to fill up by night and be used by day.[32]

A separate and parallel refinement was to build a similar reservoir at the other end of the pipeline, into which the springs discharged directly and from which the water then entered the pipes for conveyance to the eventual delivery point in the city. The pipeline could thus be equipped with two reservoirs, one at each end, or with either one independently. A good example of the reservoir at the springs is that on the sixth-century BC aqueduct built by Eupalinos on Samos. It is roughly triangular in shape, as dictated by the natural rock formation upon which it is built, and covered by a roof carried on fifteen square limestone piers. On one side two openings deliver the water from the springs in the rock; facing them a third is the take-off point for the pipeline. It is noteworthy that this is not at floor level but some distance above it. The pipeline could thus run dry while there was still water in the reservoir. The purpose of this arrangement was evidently that the reservoir served not only as a collector to feed the pipeline, but also as a settling tank, allowing particles of earth and other sediment carried in by the water to sink to the bottom. For cleaning out, a drain hole at floor level was provided below the offtake pipe.[33] It was probably not uncommon for such a settling tank to be provided on archaic Greek aqueducts at the point where the water left the spring and entered the pipeline, though not perhaps on the scale of the installations at Samos.

The fame of the Samos aqueduct, however, rests chiefly on another feature, itself perhaps the most striking single testimony to the hydraulic skill of the Greeks of the archaic age. This is the Tunnel of Eupalinos. We are fortunate in having the clear statement of Herodotus that during the tyranny of Polycrates, that is, in the late sixth century, Eupalinos, son of Naustrophos, of Megara, built a 'double mouthed' (*amphistomon*) tunnel on the Samos aqueduct; he also gives full dimensions, which are more or less accurate.[34] The fortune consists of the fact that without this firm dating we would undoubtedly have ascribed the tunnel to the Hellenistic age in the full confidence that the archaic Greeks could never have managed so advanced a feat of engineering. Indeed, almost the greatest value of the Eupalinos tunnel is the salutary lesson it offers to modern scholars and their estimates of the technical capacity – or incapacity – of the ancients at any given period. The tunnel (Figs 8, 9) was dug simultaneously from both ends and, running under the Kastro on Mt Ampelos, is just over 1 km long; in cross-section it is more or less square, 1.8 × 1.8 m.[35] More remarkable was the fact, established only by careful measurement in 1960 by Kastenbein, that the main tunnel is actually 'horizontal, with a slight sag in the centre'; in spite of this, almost all the handbooks print drawings showing it, reasonably but wrongly, with a downhill slope.[36] Along the floor of the main tunnel was dug a continuous trench, in the bottom of which ran the terracotta pipeline actually carrying the water. Instead of being in the centre of the tunnel, as was usual with such an arrangement, it ran along one side, close against the

8. Samos: tunnel of Eupalinos. The water was carried in a pipeline at the bottom of the trench alongside the left side of the floor of the main tunnel (photo: Deutsche Archäologisches Institut, Athens).

wall, and though the tunnel was horizontal, the bottom of the trench was not, getting progressively deeper below it and so giving the pipeline the necessary downward slope to carry the water – the gradient actually works out at 0.4%, dropping around some 5 m through the length of the tunnel. Even at the shallow end, this trench is remarkably deep, going down some 3.5 m below the floor, while at the lower, or southern, end it is 8.5 m, so that the pipes run at the bottom of a deep cleft or chasm. The top of the trench is often covered over with stone slabs, and it has been partly filled in; the fill, however, does not actually go right down to the bottom, so that the pipeline runs in an open space forming, in effect, a second tunnel underneath the main one (Fig. 9(b)). The most reasonable explanation for this approach is probably that the problems in underground measurement and levelling precluded the driving of a main tunnel at the steady downward slope required. Instead, a horizontal gallery was established as a base line from which the gradient of the pipeline could be set by frequent vertical measurement, after the fashion in which the slope of the draining gallery in a qanat was fixed by measurement down the vertical shafts (but see p. 212f. below). No doubt it was also used as a service gallery for construction and maintenance[37] and, since its southern mouth lay actually inside the walls of the ancient city, may have also provided a handy means of secret access or escape in times of military emergency.[38] However, we must also note that in Greek

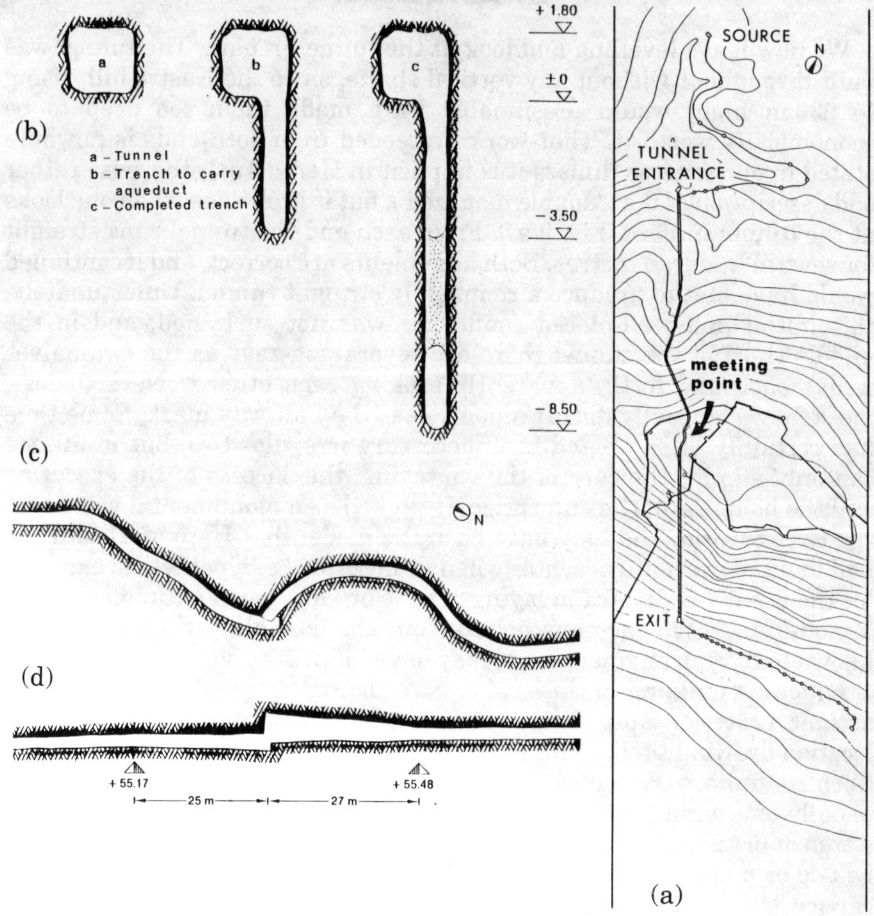

9. Samos: tunnel of Eupalinos (sixth century BC): (a) plan; (b) cross-section of tunnel, showing, left to right, preliminary bore, digging of trench to carry aqueduct, and deep trench at downstream end: the water ran in a terracotta pipeline laid along the bottom of the trench; (c) meeting point of the two halves of the tunnel in plan, and (d) vertical section (Kienast).

water supply there are other cities that built tunnels in pairs, one on top of the other and connected by regular vertical shafts. Such an approach – it may be observed at Athens, Syracuse and Acragas – certainly recalls the qanats, and may be of relevance to a proper understanding of Samos, though the true answer remains problematic.[39] The truth of it is that there is much here still to be understood, particularly on the whole question of levelling, how Eupalinos went about it, what his intentions were and, of the various features of the tunnel, which are accidental and which deliberate.

We now leave levelling and look at the tunnel in plan. The tunnel was built throughout without any vertical shafts, since the Kastro hill, rising to 230 m high, would presumably have made them too deep to be economically worth it. That work proceeded from both ends is nowhere stated in our sources (unless it is implicit in Herodotus' otherwise rather odd description of it as 'double mouthed'), but it is obvious when one looks at the tunnel in plan (Fig. 9(a)). From each end the tunnel runs straight for several hundred metres. Both alignments are correct, and if continued would have met to produce a completely straight tunnel. Unfortunately, this initial and well-placed confidence was not sustained, and in the middle third of the tunnel there are several zig-zags as the two halves weave back and forth, apparently seeking each other.[40] Nevertheless, meet they eventually did, at a point near the half-way mark. Some time had certainly been wasted in unnecessary irregularities, but in all one can only admire the care of the surveying, the success of the execution and the boldness of the enterprise. It ranked as a monumental wonder in the Greek world, and may have been the model that Hero of Alexandria had in mind six centuries later when he described a hypothetical example for the use of his dioptra in laying out the orientation of a tunnel through a mountain.[41] We can thus not rule out the possibility that it was also known to Roman engineers, and may have influenced some of them.

Aqueduct building in classical Greece showed no great difference from archaic practice. Most cities were served by fountains in some central location drawing their water from a local source, or supplied by a conduit. Such conduits were almost invariably of terracotta piping.[42] The pipes, usually of around 25 cm internal diameter, were laid along the bottom of a trench or tunnel, ensuring both access and protection, and there might be two or more pipes in parallel, depending on the volume of water to be carried.[43] The water supply was always carried underground, just as reservoirs were always roofed, and it has sometimes been suggested that this was done for military reasons, to protect the supply from enemies. This does not seem to be the whole answer, however. At Acragas the entire system was so protected even though it lay wholly within the city walls, and this must have been done as a protection from pollution. It was probably not just to prevent odd debris from falling into the channel. The Greek medical writers are very conscious of the importance of pure water, and it looks as if they were reluctant for it to be exposed even to the open air.[44]

The most remarkable and fully published work of classical Greek water supply is probably that of Perachora, but unfortunately this no longer seems as clear-cut a case as it did upon excavation, and the picture is now very confused. The water source and the point of delivery are only 200 m or so apart, and the most obvious anomaly is that while both feature elaborate arrangements that look as if they were meant to handle water in large quantities, they are connected only by a small surface runnel or gutter of insignificant dimensions (11 × 10 cm), that seems to be conceived on quite a different scale. Indeed, it was so small that it was covered with roof tiles, and ridge tiles at that.[45] The disparity in proportion is enough to make one wonder whether these three elements do, in fact, all belong together as parts of the same system, for at the other two the concept is positively grandiose. The fountain house has the usual row of dip basins with an imposing Ionic façade, served by a set of three dead-end tunnels running back 29 m into the rock-face. Although they look like catchment tunnels to collect seepage from an aquiferous stratum, the walls and even the ceiling were lined with waterproof plaster; no water could seep through in either direction, and they were therefore storage reservoirs pure and simple. Their capacity was 350 m^3, and it would have taken the runnel 97 hours to fill them.

The arrangements at the source were even more remarkable. The water had to be brought up from a depth of 30 m, and to do so there was sunk into the rock a complex of three deep 'shafts' served by an access staircase running down a long sloping tunnel to reach the shafts at their bottom. The 'shafts' might be more accurately described as long, deep slots, around 15.5 × 1.3 m, and the question is how the water was raised. The excavator proposed for each slot an endless bucket chain passing over a large wheel mounted in the slot and turned by oxen 'through a simple pair of cog-wheels engaging at 90°', the well-known Persian sakia apparatus. Unfortunately there were no scratch marks on the sides of the slots made by the turning wheels, such as one often finds with water-wheels, nor vertically, as might be expected from the buckets on their way up and down, and further excavation around the slots showed no signs of the necessary foundation for the proposed installations. It would also mean the sakia turning up in use long before what is generally thought to be its first appearance, and accordingly the most recent authority, J.P. Oleson, rejects the whole bucket-chain hypothesis. He cannot, however, suggest a reasonable sure alternative, and Perachora remains a mystery.[46] It is a notoriously arid region where no doubt special measures were called for to ensure a water supply, and the whole concept of a public fountain being fed, not by gravity from a spring, but by machines, is highly atypical of Greek hydraulic practice. For a general study such as this, we would therefore probably be wise to leave Perachora aside as uncharacteristic, however it worked.

Hellenistic

It is with the Hellenistic age that the great breakthrough comes. It sprang from two sources. The political and economic situation in the Greek world was changed with the growth of the Hellenistic kingdoms of the east, under the rule of the monarchs usually compendiously known as 'The Successors', leading to an upsurge in architectural development and urban beautification. It was natural that aqueducts should play a prominent part in the new movement. Second, the progress in Hellenistic science gave to the hydraulic engineer a whole new dimension of technical expertise with which to confront his new responsibilities. The result is that aqueducts now achieve a new, almost Roman, magnitude of conception. With some, indeed, the dating is disputed and it is uncertain whether we are dealing with a Hellenistic or Roman installation. Naturally, one expects to be able to date an aqueduct by its masonry or architectural elements (although current doubts about Patara make even that questionable); conversely, it would seem that in their absence uncertainty is to be expected, for there is little that is datable about a rock-cut channel. But that is not quite the point. When one compares, say, the Enneakrounos pipeline with the Anio Vetus it is plain that, quite apart from the archaeological datability of the individual component elements, we are dealing with two schemes of quite different design, scope, concept and execution. As one says nowadays, they are just not in the same league, and we can, in a general way, identify one league as typically Greek, and the other as essentially Roman. And Hellenistic waterworks, on the whole, belong in the Roman league. This is not to say that they are the same. For one thing, Hellenistic aqueducts still usually used pipes, instead of the Roman masonry conduit, to carry the water, and accordingly delivered it in much smaller quantity. For another, however ambitious the planning of the line, they lacked the Romans' practised skill in civil engineering, particularly in bridging and the use of the arch.

The Greeks did, of course, know how to construct an arch theoretically, and occasionally actually built one, as in the celebrated decorative arch at Priene,[47] but this was rare. Whether the Hellenistic aqueduct builders lacked the theoretical knowledge to build arches or were simply inexperienced in it, it came to the same result. They were not used. This meant that the engineer was in serious difficulty when he wanted to raise his conduit above ground level. It was not just that he had no equivalent to the great Roman bridges and arcades, though if the reader can conceive in his imagination a Pont du Gard built entirely on the principles of trabeated architecture, Doric columns and architraves, he will receive a striking illustration of just where the problem lies. But, apart from the great bridges, a large proportion of Roman aqueducts are carried on arcades that are quite low, where it is necessary for the conduit to run 3 m or so above ground over perhaps quite a long distance. Faced with such a situation, the Hellenistic engineer could have recourse only to the precedents of embankments and causeways. Bridges did exist in the Hellenic world, going right back to the viaduct of Cnossos, but they were always a rarity, even in Hellenistic times. The truth was that in a Hellenistic city the engineer's whole training and experience was architectural rather than in civil engineering, and when it came to laying out an aqueduct and countering the irregularities of the natural terrain, the tunnel was almost the only weapon in his armoury of structural techniques. Indeed, it may be that the traditional Greek practice of running the channel continuously underground with regular vertical access shafts was retained because this made it easier to maintain a constant channel level and to compensate for surface irregularities: if there was a surface depression to be crossed it was maybe easier to keep a subterranean channel level by digging shallower shafts than to elevate on an embankment one that was already at ground level. This, however, would reflect only the local picture. On the wider scale, Greek – and Hellenistic – aqueducts solved the problem by generally following the contours and so often managed to produce an entire aqueduct without any major engineering works.

But there is one striking exception to this generalisation. This is the inverted siphon. It will be discussed in detail when we come to consider Roman work (p. 147ff. below). Here let us note simply that it involves crossing a valley or depression by running the water through a closed pipe that goes down one side, across the bottom, and up the other side (Fig. 102 below), roughly in the form of a U. By convention, archaeologists refer to this as a siphon (as shall I also), although 'inverted siphon' is the more proper term; in American hydraulic engineering the term 'sag pipe' is recommended, precisely to avoid such confusion.[48] In the absence of bridges, the siphon was the only way the Hellenistic engineer had of carrying his aqueduct across any valley that could not be avoided or contoured around, and our surprise must be directed not to the fact that they used siphons, but that they did not use more of them. Statistically, one per aqueduct seems to be about the maximum frequency and there are many aqueducts with none; there may also, of course, be many siphons as yet undiscovered, particularly small ones. The dating is intrinsically difficult, so some of those listed may be in fact early Roman rather than Hellenistic, but the catalogue of sites in the east alone is nevertheless impressive: Ephesos, Methymna, Magnesia ad Sipylum, Philadelphia, Antioch on the Meander, Blaundos, Patara, Smyrna, Prymnessos, Tralleis, Trapezepolis, Antioch in Pisidia, Apamea Kibotos, Akmonia, Laodicea and Pergamon.[49]

There can be no doubt that these siphons provided a model for later Roman work, but they did have certain features that were generally typical and yet were not copied. First, they were generally single-pipe siphons, in distinction to the multi-pipe installations at such places as Lyon, where one finds batteries of nine pipes side by side. This was reasonable enough, since they were built on ordinary pipelines, where the volume of water was nothing like that handled by a Roman masonry conduit. Second, the pipes were not usually of lead, the normal Roman material, but of terracotta or stone. Stone pipes were of a highly characteristic and easily recognised form for they normally consisted of a

10. Ephesos: block from stone pressure pipe, showing recessed male-female joint. Also to be noted are the two small holes, one on top and one in the side of the block; their purpose is uncertain.

11. Ephesos: blocks from a stone pressure pipe, piled in the agora; some (e.g. top centre) carry in the side a vent (?) hole.

12. Patara: siphon, formed by stone pipe blocks (photo: K. Grewe, Bonn).

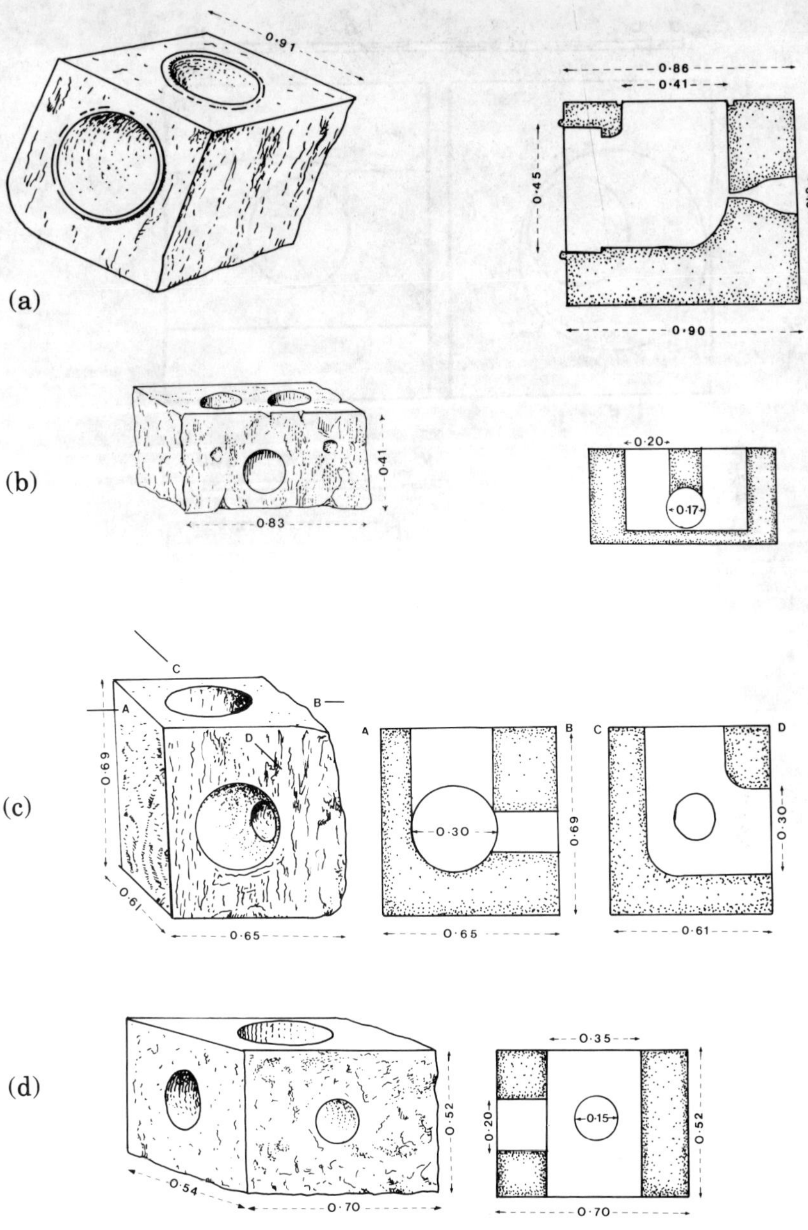

13. (a) Laodicea: angle block from stone pipeline, showing (section), possible air vent hole; (b), (c), (d) Smyrna: various angle and junction blocks from stone siphon piping (Weber).

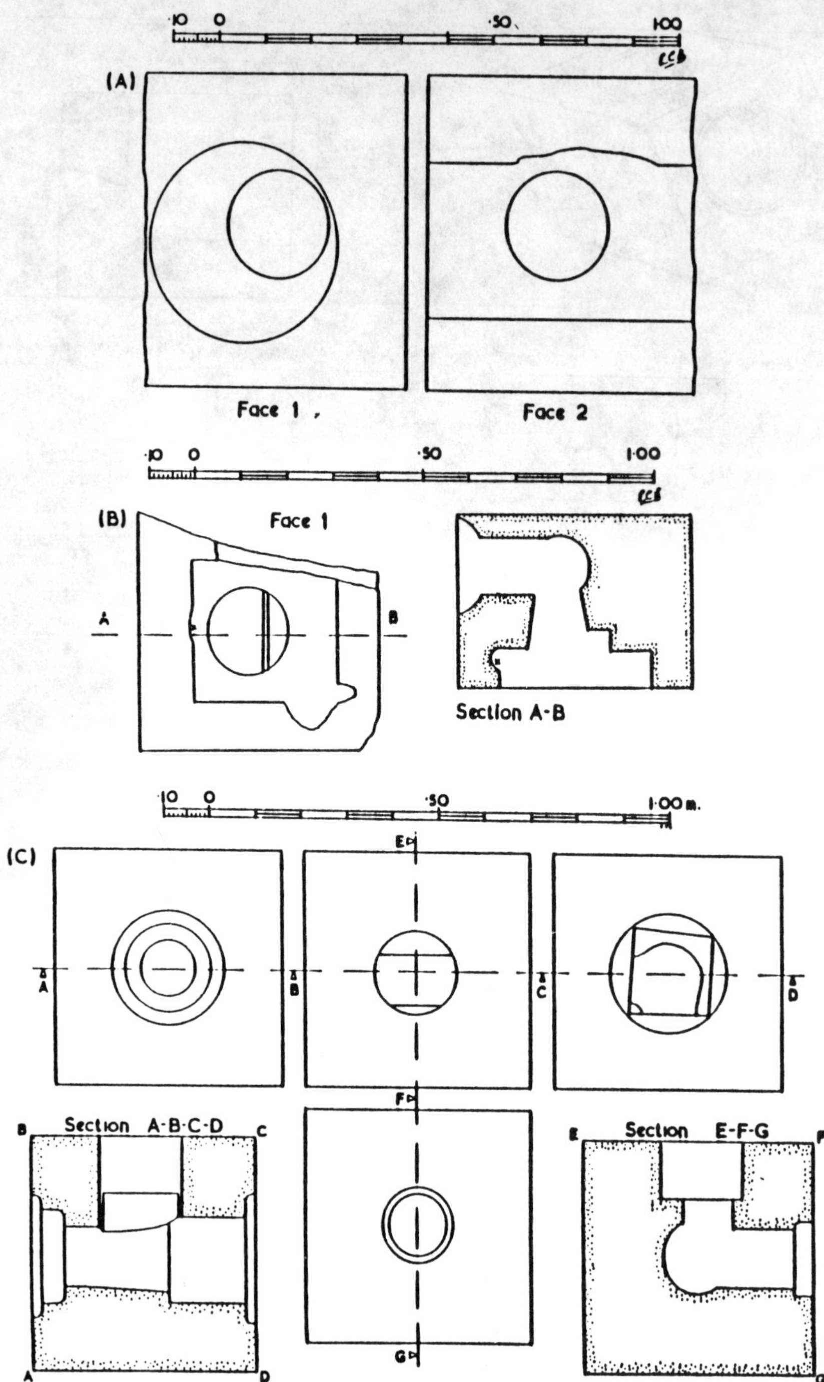

14. Oinoanda: various angle and junction blocks from stone pipeline (J.J. Coulton).

series of stone cubes, each perforated by a cylindrical hole which, when the blocks were laid end to end, like a continuous course of ashlar masonry, formed in effect a pipe inside it. The blocks also fitted together in a male/female joint, one contact face carrying a projecting circular flange around the end of the hole, and the other a circular recessed sinking to receive it (Figs 10, 11, 12, 16). As well as what we may call ordinary blocks, with the pipe (or hole) running straight through, several special ones have been found featuring elbow bends, vent holes, T-junctions, or even more complicated arrangements the interpretation of which challenges or even baffles our imaginative skill. Smyrna and Oinoanda offer a particularly rich assortment of these; some are shown in Figs 13 and 14.[50] In this respect also the Hellenistic siphons significantly differ from the Roman ones, and throw light upon them. It is eminently probable that the average siphon did not consist simply of a plain pipe throughout its length, but must have occasionally been provided with fittings of some sort – no matter what. In Roman siphons the pipes, being of lead, have entirely disappeared, so that any special features, such as Vitruvius' enigmatic *colliviaria*, have to be imagined. The stone pipes of many Hellenistic siphons, more durable and less liable to looting, have often survived, and provide our best evidence for such features (as with those found at Smyrna), though interpreting it is a different matter. It must also be repeated that although most of these works have traditionally been dated to the Hellenistic period, a dating I respect to the extent of including them in this chapter, there is a very serious possibility that they are really Roman, and deserve a place instead in Chapter 6.

Nevertheless, one can clearly identify one major feature of ancient siphons upon which the Hellenistic stone pipes do provide evidence. One of the major problems to be faced in running a siphon is, or at least may be, air pockets. In theory there should be none. A leading technical authority has insisted that in practice, during normal operation, there were none either (in filling and draining the siphon conditions are quite different); and in a long article published in 1983 I followed this view.[51] I am now not so sure. Modern hydraulic engineers, in a very extensive correspondence, have stressed to me that in siphons today, for they are still used in modern installations, air getting into the pipes is, one way or another, a real and common problem.[52]

A number of these stone pipes are fitted with small, round vertical holes, leading from the pipe inside to the upper surface of the stone block. One from Aspendos is published by Lanckoronski; it is actually on the joint between two blocks (Fig 15(a)), and being cylindrical in bore (16 cm in diameter) could perhaps be interpreted as a T-junction supplying a pipe joined on to it.[53] At Laodicea, however, we encounter the same phenomenon, except that here the holes are funnel shaped (Fig 15(b)). These must have been meant to be some kind of a vent, the tapered sides being probably to facilitate plugging. Indeed, Weber even mentions that

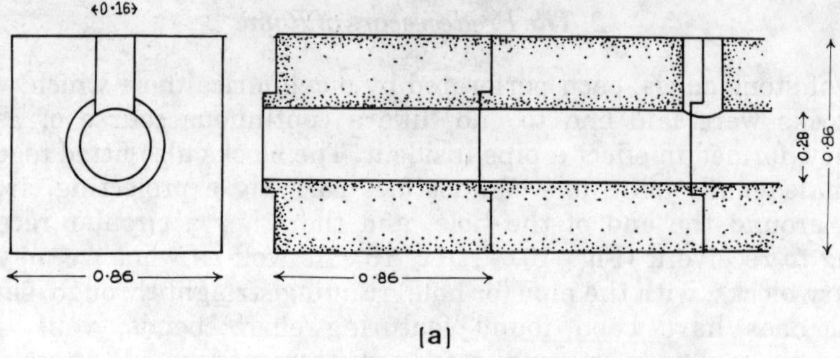

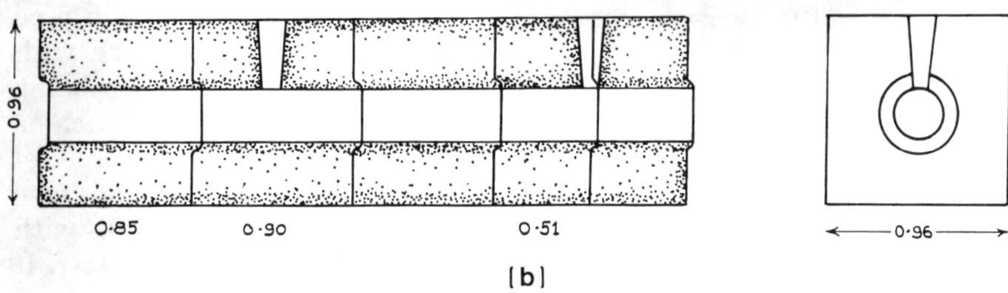

15. Stone pipeline blocks showing (vent?) holes, from (a) Aspendos; (b) Laodicea (Lanckoronski, Weber).

at Pergamon a similar hole was found still plugged with a round stone, sealed in place with plaster.[54] Other similar holes have been found in a stone siphon at Susita (Israel) (Figs 16, 17).[55]

There are various possibilities for the purpose of these holes. They may have been intended for cleaning out the pipe, perhaps by some sort of pull-through device, a common way of cleaning out modern pipes.[56] Alternatively, they may have been connected with the release of air pressure. This might have been done by periodically opening them to allow the escape of air that had accumulated in pockets during normal operation. It might have been intended to get rid of air trapped in the siphon when it was being filled up from dry, in which case they could be left open and then closed off, one by one, to keep pace with the mounting level of the water in both sides of the U as it filled up from the bottom; conversely, when the siphon was being emptied, the holes would be opened to admit air and let the water drain out. A further possibility is that the holes were stopped up with some sealing designed to blow off under a set pressure, in which case they are to be seen as an automatic protection, like safety valves, against water hammer or excessive internal

16. Hippos (Susita), Israel: stone siphon pipe *in situ*. The stone blocks have been left only roughly dressed, but again regularly carry holes on top, giving access to the pressure channel within (for air venting? maintenance?) (see Fig. 17).

static pressure. We must also note (Fig. 13(a)) that at both Laodicea and Smyrna blocks have been found carrying a small hole, much smaller than those hitherto considered, which from the narrowness of the aperture (1 cm in diameter), must surely have been intended for the passage of air, not water; though whether the purpose was to let air out to relieve excessive pressure within, or to let air in to equalise pressure by destroying an internal vacuum, I would not like to say.[57] Again, the holes may be connected with efforts to clear the siphon pipe when it got blocked by an obstruction. In this case, they would have two possible functions. The first is to locate the obstruction. This would be done by opening the hole and seeing whether the pipe was at this point empty or full of water. Either set of holes, large or small, would be suitable for this. The second job is, having located the obstruction, to destroy or remove it. This means getting inside the pipe, and only the larger holes would give suitable access.[58]

Whatever their purpose – vent holes, access holes, *colliviaria*, safety valves, or anything else – these holes in the Hellenistic siphon pipes must surely offer valuable evidence for their Roman successors where the pipes have now disappeared, even if we cannot interpret it with certainty.

The size of these Hellenistic siphons is variable, but in depth they are usually smaller than their Roman equivalents, which only begin at a

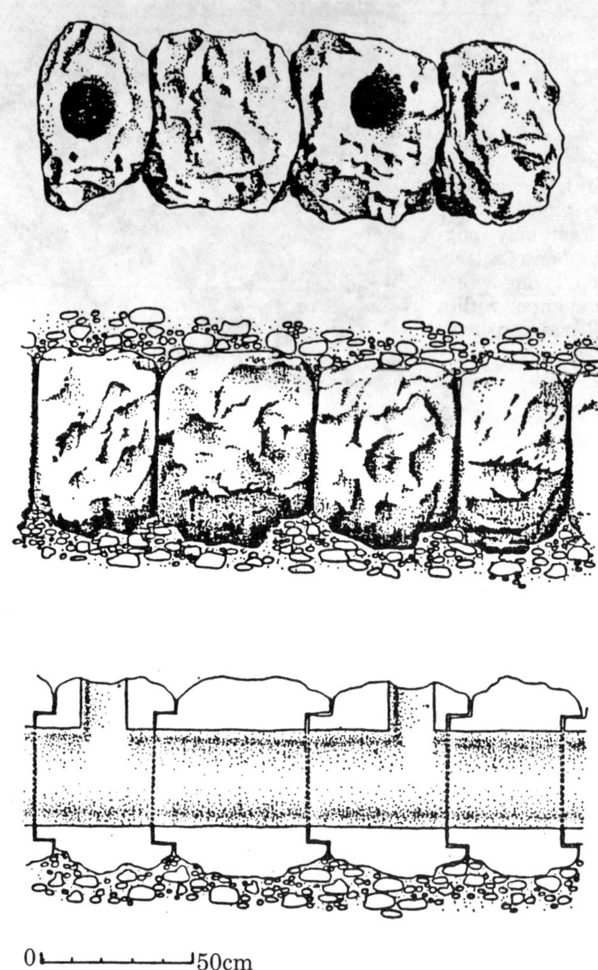

17. Hippos (Susita), Israel: stone siphon piping, in plan, elevation and section, showing (vent?) holes (Y. Peleg).

depth of around 50 m. To list those already mentioned, the depths, where known, are: Magnesia ad Sipylum, 30 m; Philadelphia, 20 m; Antioch on the Meander, 15 m; Blaundos, 15 m; Patara, 20 m; Smyrna, 158 m; Prymnessos, 40 m; Tralleis, 75 m; Trapezopolis, 40 m; Antioch in Pisidia, 28 m; Apamea Kibotos, 28 m; Akmonia, 25 m; and Laodicea, 25 m. In many of these locations a Roman engineer would instead have used a bridge. Hellenistic siphons are relatively small because their engineers could not build bridges and when they came to a valley, even a small one, a siphon was the only way of crossing it, other than by contouring around, whereas the Romans had to fall back on siphons only when the valley was too big to be bridged.

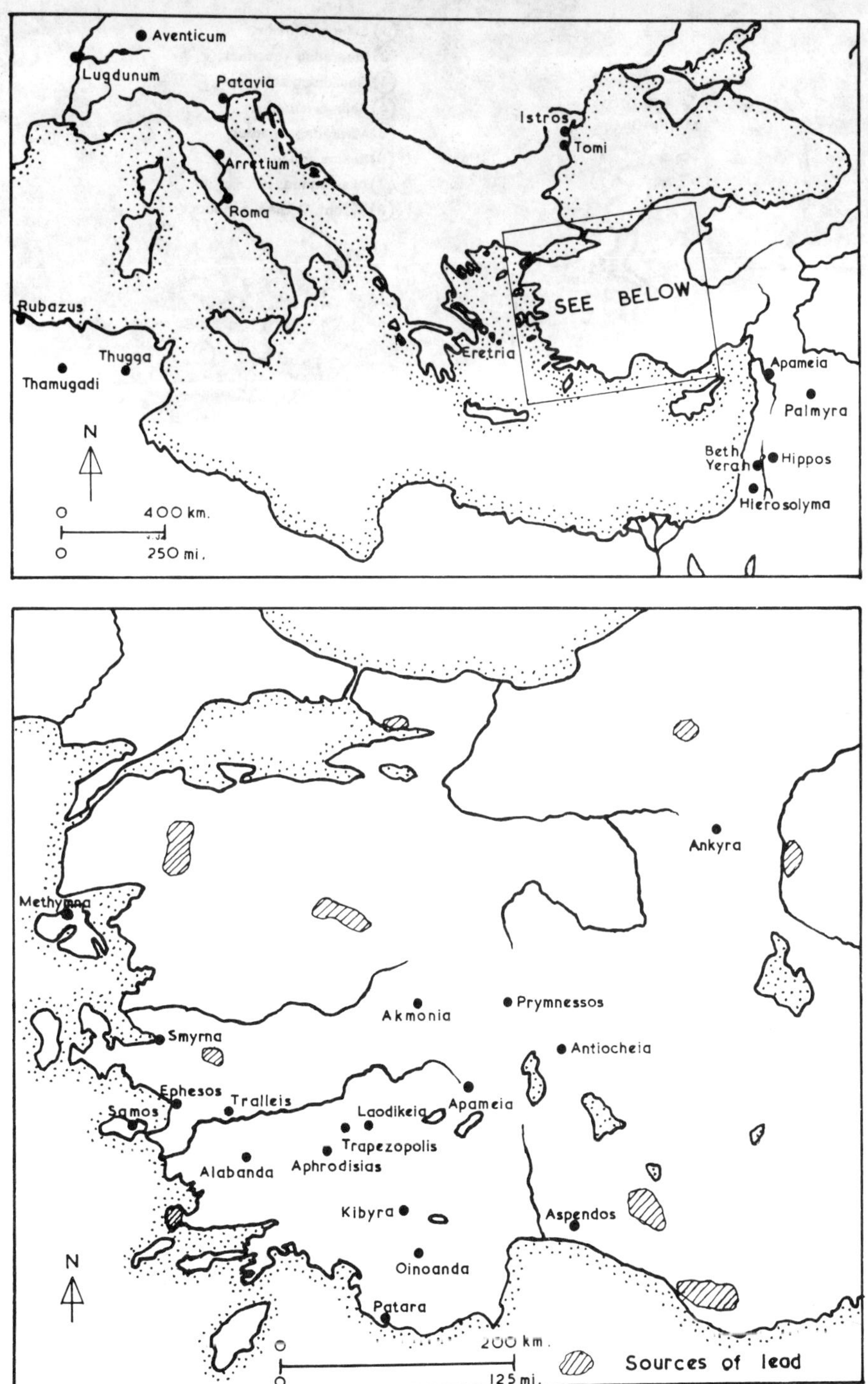

18. Distribution map of stone pipeline blocks (J.J. Coulton).

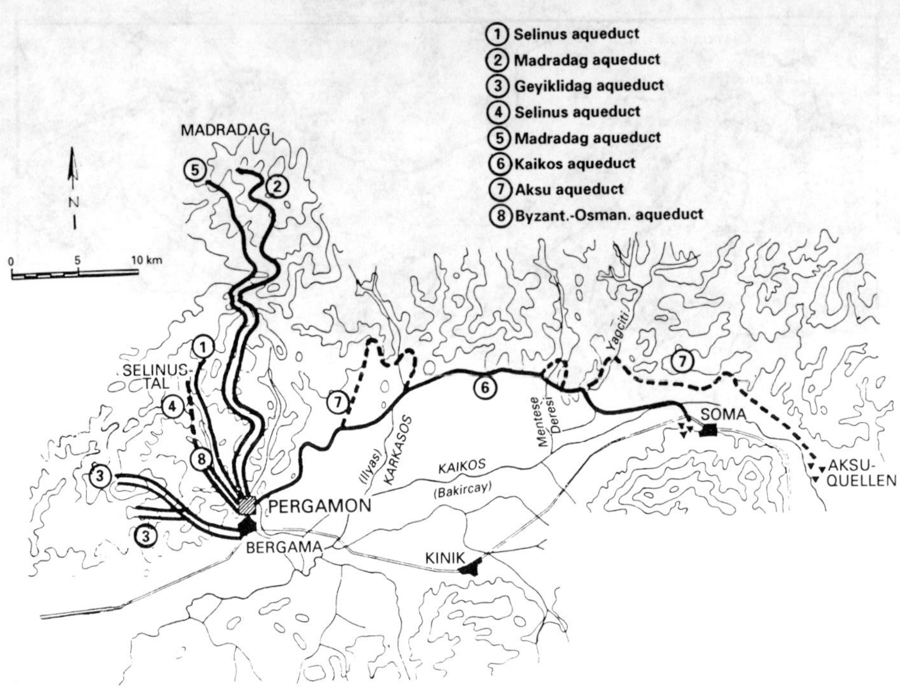

19. The aqueducts of Pergamon: the principal lines are the Madradag and the Kaikos (Garbrecht).

But there is one further siphon that stands apart from all others, Greek and Roman alike, by reason of its gigantic size. This is at Pergamon. Pergamon was remarkably well off for aqueducts, having a total of no less than eight of them (Fig. 19), but it is the Madradag aqueduct, apparently built by Eumenes II (197–159 BC), that here concerns us.[59] For most of its course the aqueduct is relatively orthodox, running in a triple pipeline of three terracotta pipes side by side with a total daily discharge of 4,000 m^3. For the last part of its run into the city, however, it ran in a siphon of dimensions without parallel in antiquity, either before or since: 3.2 km long and about 200 m deep (Fig. 20). The loss of height from beginning to end of the siphon is 41 m, giving an overall hydraulic gradient (see Fig. 102) of 13.3 m per km, or 1.3%;[60] this is four times as steep as the best known Roman siphon, at Beaunant, on the Gier aqueduct at Lyon. The steepness of the drop in the hydraulic gradient accelerated the flow of water through the siphon, enabling its single pipe to carry the discharge of the three pipes of the aqueduct serving it.[61] Unlike most siphons – unlike, indeed, the pipes of most aqueducts, including the earlier, gravity-flow piping of this aqueduct itself – the pipes were above ground. They were originally thought to have been of

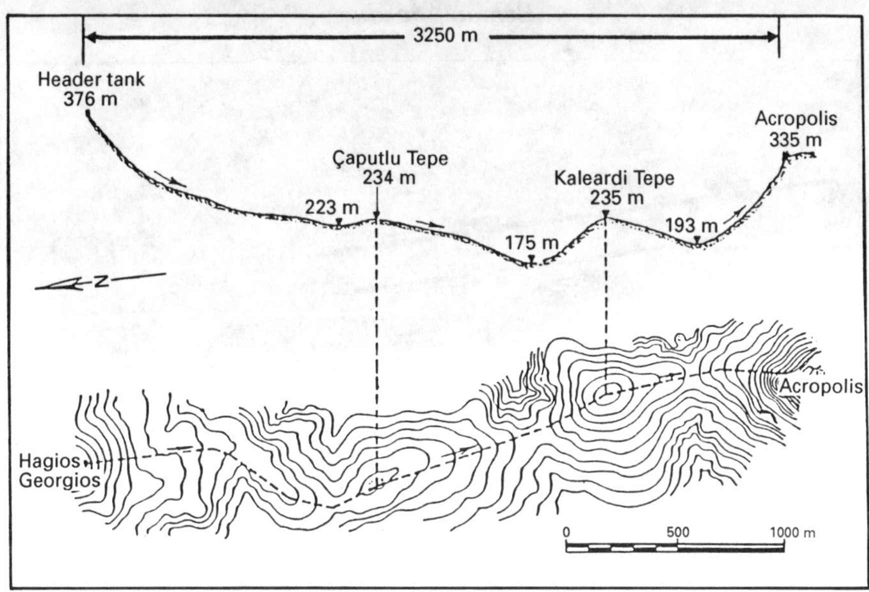

20. Pergamon: Madradag aqueduct: profile (above) and plan of siphon bringing water to the Acropolis (second century BC).

bronze (so R.J. Forbes) but since 1976 are now known to have been of lead, following chemical analysis of the soil along the line of the siphon which showed an average lead content some 56 times higher than samples taken a mere 8 m away to one side.[62] The pipes apparently came in lengths of 1.2–1.8 m, of known external diameter around 30 cm and hence estimated internal diameter of 17.5 cm. They were presumably joined by collars or sleeves of metal, and were carried in round holes bored through a series of stone slabs set upright on the sloping hillside. Many of these slabs, of a local stone known as trachyte, are still in situ, and the preserved holes give the pipe diameter (Fig. 21).

The profile of the siphon is both interesting and impressive. The great pressures generated at the bottom of the valley are enough to impress even modern engineers; they came to around 18–20 atmospheres of 240 lb/in^2 (18.5 kg/cm^2), and were evidently successfully contained by those same ancient pipes that modern commentators so often maintain could not have handled any serious pressure.[63] No doubt there was a lot of leaking, but that does not obscure the main point, that this was a project of monumental magnitude, recognised by its contemporaries as a wonder, and that in practice it did actually work. But there was also a peculiarity about the profile that was, perhaps, highly significant. Normally, ancient siphons, Greek and Roman alike, are of a U-shape. However, the profile of the Madradag siphon at Pergamon shows (Fig. 20) two bumps in the middle, where the line first climbs over a small eminence known as

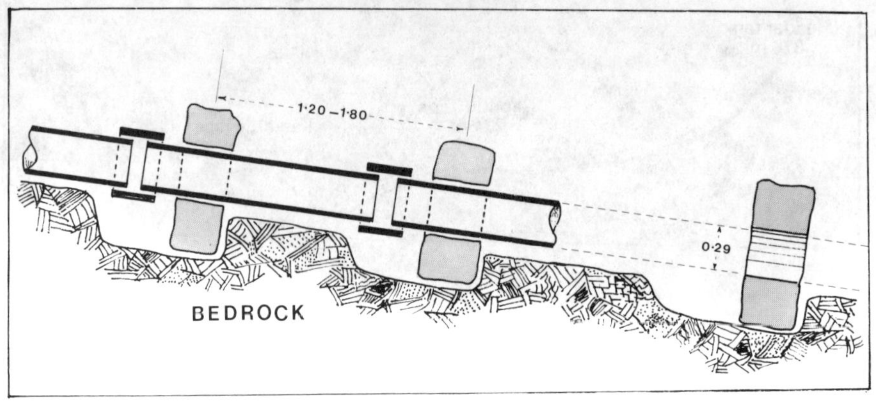

21. Pergamon: Madradag aqueduct: lead pressure pipes of siphon resting in stone supports.

Çaputlu Tepe, descends again to the lowest point reached on the profile, climbs again to a second peak (Kaleardi Tepe) and runs down the far side of that before eventually climbing again to the end of the siphon in the city of Pergamon. This gives the siphon an irregular W-shape profile that is not only without parallel in the ancient world, but productive of serious hydraulic problems. These occur at the two peaks, Çaputlu Tepe and Kaleardi Tepe, though in more acute form at the second. At each the pipe has a bend (*geniculus*) in the vertical plane, where the inertial thrust of the water would exert considerable pressure on the upper surface of the pipe as it rounded the bend. Moreover, if there was any question of air in the pipeline, from whatever source, this is precisely where it would collect in a troublesome and potentially dangerous air pocket. It was probably to avoid exactly such a situation that at Les Tourillons, on the La Craponne aqueduct at Lyon, an open tank was inserted between two valleys, thereby avoiding a peak in the pressure line and, in effect, breaking the W into two halves.[64] In point of fact, the inertial thrust at Pergamon was supposedly contained by embedding the pipe at the critical points in heavy masses of masonry, as recommended by Vitruvius;[65] the air pockets, if any, may have been accommodated by some kind of release vent, but there can of course be no evidence in the total absence of the pipes.

However, the true significance of the two bumps or peaks only becomes apparent when one compares the siphon profile with the contours of the relief map. Why was the line of the siphon so laid out as to include these two peaks? Forbes echoes a common opinion when he says that it went 'over a ridge'; K.D. White calls it 'a saddle'.[66] This is what one would expect, for a ridge could not be avoided and the hydraulic inconvenience would simply have to be accepted. But the map (Fig. 20) makes it plain that both peaks were more or less conical mounds and could easily be avoided by contouring around the base. The smaller peak, Çaputlu Tepe, indeed, would be very easily eliminated. Close study shows that things are even worse. At both peaks, there is slight but perceptible deviation from the natural existing line of the aqueduct to ensure that it went right up to the top of the peak. Far from any attempt to avoid these potentially difficult summits, the line has deliberately been laid out to include them and even maximise their effect. The irregular W-shape of the Pergamon siphon is the product of intentional choice. The only reason for it that I can think of is another hydraulic factor, static pressure. Static pressure is generated by the vertical column of water supported, so that when a siphon pipe runs up over a bump or peak, the static pressure is less there because, in effect, it is not so far below the natural water level. Running up to the summit of Kaleardi Tepe and down the other side, instead of circling around its base, did thus reduce the static pressure on a long section of the pipeline, in much the same way that a *venter* bridge does on an orthodox, U-shaped siphon (Fig. 102), but it did so at the price of introducing a danger point vulnerable to inertial thrust and air pockets; and, of course, even so it could do nothing to reduce the static pressure at the point where it was greatest of all, crossing the saddle between the two peaks at an altitude of 175 m. The inference surely must be that the builders of the Pergamon siphon were more worried about static pressure than inertial thrust on *geniculi*, and that the alignment, in plan, that they chose reflects their relative scale of priorities.

The Etruscans

It is widely believed that the Romans inherited many characteristic features of their civilisation from this neighbouring people. In temple design, gladiatorial shows and religious practices, Etruscan influence has been clearly identified, and though their waterworks had nothing near as sophisticated as the Hellenistic siphons, they may yet in some ways have inspired Roman practice, being that much closer to home. In fact, the Etruscans seem to have been noted for drainage rather than water supply, and the form in which they excelled was the underground tunnel.

Much of Etruria was formed geologically of tufa rock, so soft that the simplest and easiest way of conveying water was by digging an underground conduit. It is sometimes suggested that the Cloaca Maxima at Rome owed something to their influence, and we have already noted the possibility that their tunnels may have been derived from the qanat.[67] This is questionable, though the resemblances are considerable. The Etruscan tunnels resemble qanats in that they have frequent vertical shafts, but they are usually intended to remove unwanted water rather than provide a supply. Known as *cuniculi* ('rabbit burrows'), and dating from the fourth century BC, they have been found in particular abundance in the areas around Veii, to the north of Rome, and around Ardea and

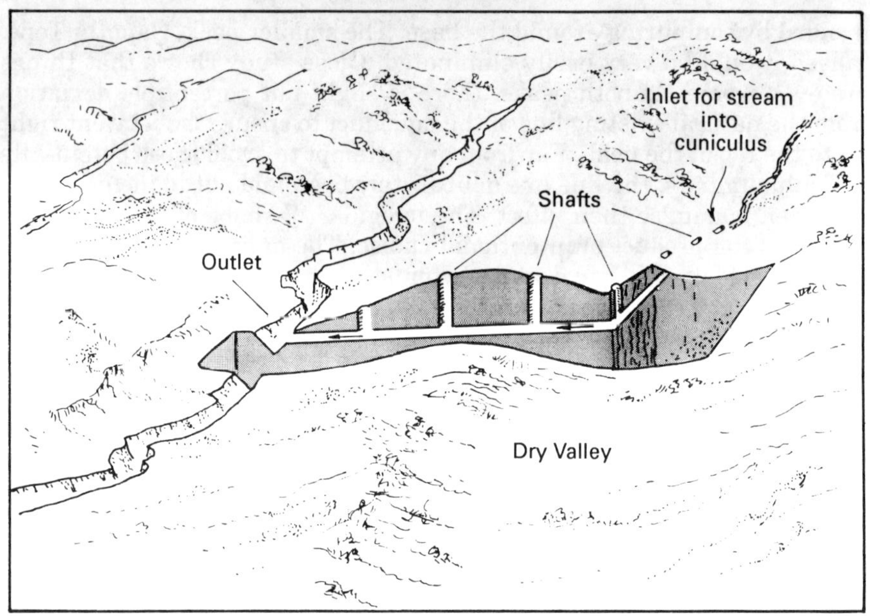

22. Etruscan *cuniculus*: diagrammatic view of drainage *cuniculus* leading from one valley into another (Judson and Kahane).

Velletri south of it.[68] The tunnel is usually around 1.75 m high and 0.5 m or so wide, and 'they are just large enough to allow the upright passage of a small man', though since most still have water running through them they have been enlarged by erosion and no longer retain their original shape.[69] The vertical shafts are usually about 33–34 m apart, and are rectangular in section, around 1.2/1.6 m by 55/75 cm. The depth is usually four to six metres, for the tunnel normally runs not far below ground level, but where it turns to tunnel under a ridge the shafts may become up to 30 m deep. Because the tunnel is laid out to maintain a constant level below the surface of the ground, the gradient in it reflects that of the ground itself; usually the tunnels are built running down along valleys, and the gradient in them varies from 1.2%, or 12 m/km, to 2.6%, or 26 m/km. In other words, they are considerably steeper than Roman aqueducts. The longest one known is 4.5 km.

Cuniculi are found in various locations and for various purposes, but there is no doubt what their principal function is. The typical *cuniculus* is found out in the open country, running lengthways along a shallow valley. It is slightly off the central axis of the valley but parallel to it, usually on its right-hand side. In this position it runs down to the mouth of the valley, where there is an outlet; sometimes, however, it turns sideways, burrows under a ridge and discharges into the next valley

(Fig. 22). Its function is to drain land otherwise too waterlogged for agriculture, and it was highly effective. Usually such a valley will have a stream running along the bottom of it. This is diverted into the *cuniculus*, leaving the valley dry. A second effect is to lower the water table in the subsoil, which now drains into the subterranean gallery running through it, and it thus drains more effectively than would a surface channel. A third result is that, since the stream is diverted underground, surface erosion is virtually eliminated: it has even been noted that 'the contours of the whole area today are conspicuously different from those of the adjoining area that did not receive the same treatment'.[70] It should be noted that quite a number of these galleries are still in operation.

We may also note a further and quite separate function of the *cuniculus*, which is as a road bridge. For a people who evidently found tunnelling relatively easy and bridge building difficult, the easiest way of bridging a stream was often to divert it through the tip of some conveniently projecting ridge via a *cuniculus*, and then run the road over the top of the tunnel.[71] Other possible uses are listed by the chief authorities, Judson and Kahane, such as domestic drainage, aqueducts, irrigation, water power (mills) and mining, but these are either rare or hypothetical, and bridges and agricultural drainage remain the *cuniculus*' main *raison d'être*.

Nevertheless, although there is no evidence of Etruscan aqueducts or water-supply systems of the type we have hitherto been studying, the skill and experience of their engineers at tunnelling must often have been a convenient and welcome resource for the Romans to draw upon. After all, the earliest aqueduct at Rome, The Aqua Appia, was itself entirely underground and in engineering, if not in purpose or function, can have differed but little from an Etruscan *cuniculus*. 'In hydraulic engineering, as in road building, it seems that Rome began where Etruria left off ...; the Roman systems of drainage and of water supply were founded on a solid basis of practical Etruscan experience.'[72]

3
Wells and Cisterns

> For 441 years after the city was founded, the Romans were content with water supplies drawn from wells, springs, or the river Tiber.
>
> Frontinus, *De Aquaeductu* I, 4

It is now time to turn to Rome. When we do so, it is natural that we should think of the aqueducts and their monumental arcades. A Roman would think of them too. Frontinus, the director of the Roman Metropolitan Waterworks, was assuredly speaking for all his countrymen in his proud boast: 'With such an array of indispensable structures carrying so many aqueducts, compare, if you please, the Pyramids, or the famous but useless works of the Greeks.'[1] Modern scholars, too, have often surveyed those mighty ranges of arches and, impressed by their unspoken but compelling indictment of the bathless Middle Ages, when sanitation was unheard of and even washing suspect, have declared them one of the greatest manifestations of Roman civilisation. While nobody would dispute the main point, there are qualifications to be borne in mind. As we shall later see, the arches, for all their prominence, are not at all typical of the aqueducts, most of which ran at ground level or a metre or so underground. More to the point, an ordinary Roman town was not supplied with water solely by its aqueduct: it also drew copious supplies from a great multiplicity of private wells and cisterns.

There were, in fact, two quite separate water systems operating in parallel and independently, fulfilling different purposes and observing different rules. Nor should we think of the aqueduct as necessarily the senior partner of the two, and the wells a kind of auxiliary or supplement. In a way, it was the wells that provided essential services and the aqueduct that was a luxury. In a city's earliest days, drinking water was the personal and individual responsibility of its inhabitants, and they met it by sinking wells or cisterns, usually inside their houses, though local springs or fountains might also help. This normally provided enough water for essential human needs; some cities never had any other supply, and got by without any aqueduct right to the end of the Roman empire – Ampurias and London are only two of many such. Many others were founded, flourished, and reached an advanced date in their history before any aqueduct was built. A city as large and as close to Rome as Ostia acquired one only in the early empire.[2]

What usually made the difference was baths. Baths were notoriously heavy users of water, and though the small, local street-corner baths might perhaps get by without an aqueduct to supply them, any serious establishment needed one. However, even though the aqueduct might be built principally for the sake of the baths, there were plenty of other users for its abundant supplies once it was installed. Thus the whole pattern of life in the city was liable to be changed – in Pompeii the coming of the aqueduct changed even the layout of the gardens and the flowers grown in them[3] – and the wells and cisterns relegated to a lesser role, much as piped water must have reduced the impluvium in the Pompeian house to a function largely traditional and decorative. One man's necessity is another man's luxury, and whether the baths and the aqueduct serving them are to be considered as essential services, depends on one's point of view. The Roman legionaries who founded and built many of the colonial cities had become accustomed, during their military service, to a relatively high standard of living, certainly one that would be considered so by many of the provincials and natives. On Hadrian's Wall, all the forts have baths as standard equipment,[4] and when the soldiers were discharged they were no doubt often a potent force in any city where they settled in demanding, as a matter of course, baths and an aqueduct as a regular feature of ordinary civilised living.

Yet the large number of wells found makes it plain that, even in an aqueduct-served city, their importance must not be underestimated. Indeed, it is as important not to underestimate it as it is difficult to estimate it accurately. We will return to this later in considering urban water distribution. Here, however, we may note three further differences. First, the aqueducts were, and had to be, a public and collective enterprise. The wells and cisterns, on the other hand, were a strictly private and individual one. Second, the aqueducts served only the cities. Even though they might run for great distances through the countryside, the rural population in no way benefited from the water passing in such volume through their own farms and lands. Today, settlement follows communications. When a new main highway is opened, residential and commercial construction rapidly forms a ribbon of development along it; in North London, the population patterns were formed in the early 1900s as the suburbs progressively marched north, following behind the ever-extending lines of the underground railway, reaching out to the then-virgin lands of Edgware and Stanmore. The same thing could have happened with Roman aqueducts, but it did not. No doubt there did occasionally exist short branches off the aqueduct to serve users out in the country that it passed through. For example, we know of one, off the Aqua Marcia, at Tivoli; perhaps typically, it was to serve a rich man's villa, not a rural community.[5] This, however, was rare – certainly rare enough that the aqueducts did not significantly affect country life. Even though located in the country, they were an urban phenomenon, underlying the urban orientation of Roman civilisation.

There is no evidence that country-dwellers resented this. In other societies, in other ages, complaints might have arisen of exploitation, of unfair preferential treatment at the expense of the farmers, but we hear nothing of such feelings. Of course, it may be that they existed but are not recorded in our literary sources. It is no secret that while ancient authors vie with each other in adulation of that stock figure of Roman historical mythology, the stout-hearted yeoman farmer, literature, too, is an urban creation, in which the voice of the authentic, working farmer goes mute and unheard.[6] Even Cato, for all his practical familiarity with agriculture, could hardly claim to be representative of all the hosts of ordinary countrymen, nor of their views either. There is therefore always the possibility that, given the urban slant of our sources, countrymen did complain about aqueducts passing them by, but that it has not been recorded. Be that as it may, the fact remains: aqueducts were for cities, wells and cisterns were for city and country dwellers alike.

From this springs the third distinction. Surprisingly, it is racial – even political. The assimilation of native peoples was a Roman policy, and on the whole a successful one. But in some provinces the natives were more assimilated than in others, and even in those wholly integrated, it was not so at all periods in their history. Where, or when, there were two separate peoples in a given region, Romans and natives, it was natural for the Romans to live in the towns, which often did not even exist before their arrival. As that is also where the aqueducts were, it is thus possible to interpret Roman water-supply policy as aqueducts for the Romans, wells for the natives. Having got so far, it is hard not to embark on a moral judgment, which in turn is liable to assume political overtones in the context of nineteenth-century colonialism and its twentieth-century aftermath.

The point is perhaps most clearly illustrated by North Africa. When the French colonised the North African littoral from Morocco to Tunisia, they came in the conscious guise of latter-day successors of the civilising Roman legions.[7] It followed that the French, at this period, admired the Roman aqueducts, appositely quoting Gibbon's encomium of them as 'among the noblest monuments of Roman genius and power', and their engineers, logically, often sought to provide water by getting the Roman aqueducts working again. Further experience frequently showed that this initial optimism was misplaced, and something quite different had to be built to satisfy modern needs and technical criteria. More recently, the boot is on the other foot. The argument now seems to be that, under Roman rule, it was yet the rural African peasants who formed the economic backbone of the region, and it was the cisterns and wells that supplied them. Cisterns and wells thus formed the most important element in the area's water supply, and the aqueducts become a useless and expensive luxury that positively depressed the general standard of living by exploiting the native population for the benefit of a colonial elite. This viewpoint is essentially similar to the putative countrymen's complaints for which I could find no evidence in antiquity. The stand of the French excavators of Cherchel on the matter is expressly and explicitly socio-political: the study of the Cherchel aqueduct 'nous a aidés à prendre conscience de la réalité matérielle de l'exploitation d'une société (rurale et africaine) par une autre (urbaine et romaine), c'est-à-dire du fait colonial romain', a judgment warmly espoused by the local African authorities.[8] It is not my wish to enter into a discussion of these matters, though one can only regret that Frontinus is no longer here to speak in reply. He would certainly have some pungent comments to offer that would be worth hearing.

Wells

As Forbes remarks, 'Unfortunately most authors do not properly distinguish between the natural "spring" and the man-made "well". The word "well" should be confined to man's attempts "to obtain water from the earth, vertically below the spot where it is required, when it is not obviously present at the surface". Many of the so-called "holy wells" are in reality enclosed springs, often deepened much later.'[9] The techniques for choosing a good spot to dig a well were thus essentially the same as those for finding a spring to serve an aqueduct, and will be considered later in greater detail (see p. 74f. below). There is no evidence of any real or systematic geological understanding, and in spite of the approaches to water-divining outlined hopefully by Vitruvius, one cannot but feel that much of the true picture has been accurately caught by Smith's verdict that 'in the final analysis one is bound to suppose that well-sinking was wholly speculative. People must have dug for water where they needed it, with no guarantee of a successful outcome, and many a disappointment.'

In the circumstances it was lucky that archaic man often believed that the Earth floated on a sea of underground water; the well-digger must have needed such a faith to hold on to. The closest thing to this ancient belief actually existing is probably the artesian well, but though the ancients may have occasionally hit on one by accident, they had no concept of the principle, and it was not until AD 1126 that it was systematically exploited.[10] Ironically, because unknown to the Romans, there was an area in Italy itself rich in artesian possibilities, the territory around Bologna and, more particularly, Modena, but this only came into use in the sixteenth century. In antiquity, at least in the present state of our knowledge, it seems that artesian wells remained a wholly untapped resource.[11]

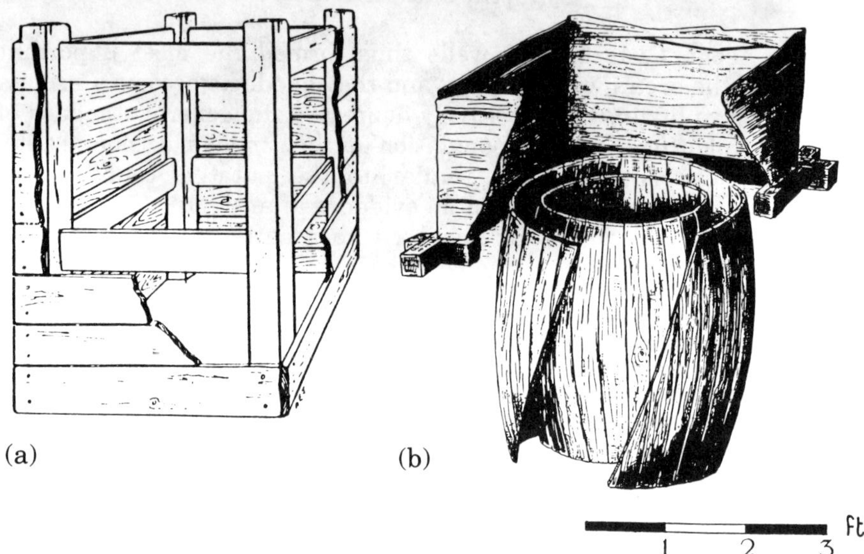

23. Wood-lined wells: (a) from Krefeld-Gellep (Germany) (Anne Johnson); (b) from London, Lime Street: wood-lined well changing from square to round (barrels) construction (Merrifield).

The ordinary well, as we generally understand the term, goes back far in history. It was always dug, not drilled in modern fashion, was usually round, and was lined, or 'steined', with stone, brick or wood. Not surprisingly, wood was most commonly used in the heavily afforested regions of northern Europe (we will later make the same observation on the use of wooden water pipes), some of the best-known examples coming from the Rhineland and Britain.[12] As one would expect, wood-lined wells were often square in cross-section, the exigencies of boards and carpentry making it easier to build a square frame than a round one. The result was 'a square box-like lining of planks'[13] (see Fig. 23(a)), sometimes with small internal diagonal braces across the corners. For round wells, a simple prefabricated lining was sometimes formed by using old barrels stacked one on top of the other with the bottoms knocked out. It was not unusual for the two systems to be combined, resulting (Fig. 23(b)) in a well that starts out square and changes to round half way down.[14] Unlined wells, though unusual, did exist. They depended on the stability and cohesion of the rock or subsoil and have been found dug in shale, clay and chalk.[15]

For most of the empire, however, the commonest form of well was that lined with masonry, either with mortared joints or of dry-stone. Occasionally rings of terracotta, like the *kaval* of the qanats (p. 23 above), were employed.[16] The well was usually also provided with a well-head, or low parapet, to stop objects and people falling in, and often also, as today, with a movable cover or lid, to be lifted off when the well

(a)

(b)

24. (a) Relief from the Palace of Sennacherib at Nineveh (Assyrian, seventh century BC): two banks of shadufs raising water from a river for irrigation, in successive lifts. Note the conical form of the buckets; (b) Assyrian relief (eighth century BC) in the British Museum, showing earliest known use of the pulley: a besieged city is about to have its water supply cut off.

was in use. In North Africa the well-head sometimes had another and remarkable characteristic. It was occasionally necessary, for hydrological reasons, to locate the well in an area that, though normally dry, was liable to become the bed of a torrent after cloudbursts. The well-head thus had to be strongly and carefully built to withstand the surface water swirling around it during these periodic inundations.[17]

In size and shape the wells varied almost as much as their modern counterparts. Most are around 0.5–2.0 m in diameter, and their depth depends entirely on the water table. Most ranged from between 3–4 to 25–30 m deep, but it is impossible to generalise. Assisted by the development of mining techniques, really deep shafts became a possibility at quite an early date. In the Nementchas region of the North Sahara a depth of 60 m is common (and dug through exceptionally hard rock). Forbes mentions one at Lachish, Palestine, 80 m deep, and another of Hellenistic date at Cairo, of 93 m. For comparison we may note that vertical shafts up to 120 m in the Laurion silver mines, near Athens and of fifth-century date, are fairly common, and the 80 m Gallo-Roman well near Poitiers was doubtless not unique.[18]

In fact, with most of these very deep wells the governing factor was probably not the engineering difficulty in sinking a well to that depth, but the practical problems of lifting the water so far to the top once the well was in operation. We should note that of all the various forms of water supply considered in this book, the well is the only one where this problem exists. In qanats, aqueducts and piped supplies, when the water has to be moved from its source to the point of use, it moves by itself, under gravity. In springs and fountains it is carried away in jars, by hand. Only in the well (and, to a much lesser degree, in the cistern) does water have to be lifted; and, in so far as this often means a resort to mechanical means, the well is the only aspect of ancient urban water supply involving the use of machinery.[19] In its simplest form, this involved no more than letting down a bucket on a rope. Few have survived since they were normally made of leather, but from bas-reliefs it is plain that they were conical in shape (Fig. 24(a)). This is much more efficient than the modern form, as can be testified by anyone who has let down a bucket only to find it obstinately floating on the water surface: a conical one automatically tips over immediately and fills up.[20] Provided the well was not too deep, the process could be assisted by the use of a shaduf or swipe, a swing-beam like a see-saw with the bucket permanently suspended from one end and the other end counter-weighted, enabling the operator to lift the full bucket with minimal effort. Though making for rapid and easy operation it will only work with a fairly shallow well, since the possible lift is limited by how high the tip of the beam can swing, usually not more than 4 m or so. The shaduf was thus best suited to continuously raising a moderate supply of water a short distance, and was much used in the east for rural irrigation, where it was employed, for example, to raise water from a river or ditch to the surrounding fields. Sometimes shadufs were there employed in batteries, either in parallel (to increase volume) or in a series (to increase the height the water was raised, by successive lifts), or both combined. But they were also regularly used for wells and may still sometimes be found so employed.[21]

For deeper wells, no doubt often in early days, as Forbes suggests, 'relays of men and women down to the water's surface passed the water-jar on to each other'. Wells, like mine shafts, were often equipped either with wooden ladders or at least with cut toe-holds by which a man could descend while holding on to a rope. The vertical shafts of Etruscan *cuniculi* regularly had them, and the square box-like wood-lined wells of London and elsewhere were often fitted with small wooden slats as a diagonal bracing across the internal corners; as well as reinforcing the

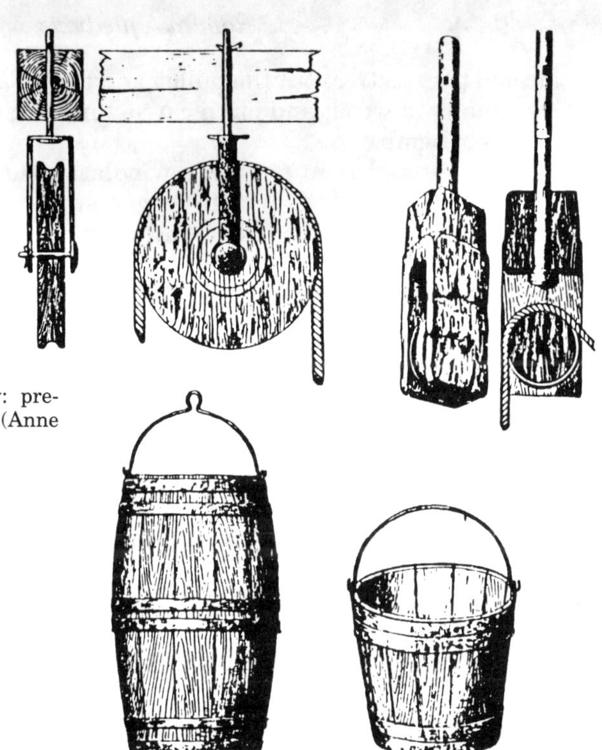

25. The Saalburg, Germany: preserved buckets and pulleys (Anne Johnson).

construction, these served as hand-holds or a ladder for human access. Once the ladders were there, there was, of course, in theory no limit to how far down one could go, and one modern scholar, Henri Goblot, recounts how in a lead mine in Sardinia he personally climbed down them to a depth of 180 m, negotiating sixty ladders on the way.[22] In fact the obvious solution for deep wells was to use a bucket on a rope, and this was what was normally done. At the mouth of the well the rope was as a rule simply led in over the side, where its repeated friction has often left marks in the form of worn grooves on the stonework of the coping or parapet.[23] This is particularly true of decorative well-heads in marble, found in many houses, though there they usually give access to a built cistern under the floor rather than an actual well shaft sunk into the ground (see below).

The next technical improvement was to run the rope over a pulley suspended above the middle of the shaft. This facilitated matters in two ways: the bucket was held out clear of the sides of the well-shaft, so that it did not bump against them and possibly catch on any projections; and the operator could exert more strength since he was now pulling down on the rope instead of lifting up. The pulley seems to have come into use around the eighth century BC, when it is depicted in an Assyrian relief (Fig. 24(b)). A lot of artistic licence has been taken with the depiction of both scale and perspective, but the pulley and its use are clear enough. Pulleys were usually of wood, and often carried in a wooden block (Fig. 25). The rope was of hemp.[24]

The next step was to provide a windlass, and this too was often done where the well was deep enough to warrant it. As the well got deeper, so did the weight to be lifted. The bucketful of water, naturally, remained constant. What increased was the length, and therefore the weight, of the rope. In the ordinary garden well this is of course minimal and in the ordinary way one never thinks of it, but in deep wells it can become significant. Indeed, we may note that in strict theory it is the governing factor limiting the depth of the well, because while digging or climbing ladders are processes that can be continued indefinitely, there comes a point at which a longer rope becomes impossible: not because a longer one cannot be manufactured, but because the rope becomes incapable of sustaining even its own weight.[25] Though ancient wells were not deep enough for this factor to become serious, there were two other difficulties. One was, given a very long rope, the weight of the rope to be lifted, the other the time taken to lift it. The first made the task more laborious, the second reduced the productivity of the well, particularly if, as in the case of irrigation, continuous operation was desired, rather than fetching the occasional bucketful for domestic needs. One simple way of reducing both problems is to have two buckets mounted on the same windlass, one descending as the other mounts (after the fashion of the cars on most funicular railways). This halves the time between bucketfuls, and to some extent the rope being paid out on one side of the windlass counterbalances that being pulled up on the other. There is little evidence that the ancients ever did this,[26] though it has been proposed for some of the copper mines of early China and is a simple enough procedure.

Another possibility with a very deep shaft is to break it up into two or three shorter ones, each with its own winding gear, staggered or offset in plan and connected by short horizontal landings. This was done in the Chinese copper mine of Tonglushan, and on the 300 m deep vertical shaft of the thirteenth-century (AD) qanats of Gonabad in Iran, but again there is little classical evidence for it.[27] Another even simpler answer was to use animal power, and raise the bucket by hitching a donkey, mule or even an ox to the other end of the rope and having it walk away from the well. Not only did the extra power permit an exceptionally large bucket, but in India at least it could incorporate a simple self-tipping mechanism for when it arrived at the top.[28]

In further developments, machines of all kinds could be and were used for lifting water in antiquity. There were the continuous bucket chain, the noria, the tympanum, the screw, the force-pump, powered by various means – manual, animal and water power – applied and delivered via a lever, the capstan, the windlass, the treadmill and the sakia gear (basically two large wooden gear-wheels engaging at right angles and permitting a vertical bucket chain to be operated by an animal walking in circles around the well). Some of these, particularly the bucket chain, could be used in wells and were so employed. Others, such as the compartmented wheel (tympanum) or noria, required space and are usually found outdoors, though whole nests of them were installed in caverns excavated underground to raise water from the Rio Tinto mines in Spain.[29] Others again were used in what we might call well-like situations, such as the Stabian baths at Pompeii, where a treadmill-driven bucket-chain operating in a rectangular masonry shaft forming part of the building, lifted the water from ground level up to a reservoir on the roof, in order that the supply into the baths should be delivered under sufficient head, pressure and continuity.[30] Indeed, the question of water-lifting machinery opens up what would be a whole new study in itself which could long occupy us. We will pass over it without further attention, for two reasons. First, these machines were used mostly in two circumstances: for rural irrigation; and to serve the limited and particular needs of individual establishments, such as baths. They therefore did not play a major part in Roman urban water supply, which is the chief focus of this book. Second, this very topic has now been so comprehensively treated and the machines so clearly and exhaustively described in the monumental work by J.P. Oleson[31] that any exegesis here would be pointless duplication.

Wells are extremely numerous. Within the fort of the Saalburg no less than ninety-nine have been found, though they were presumably not all in use at once, and in the Athenian Agora excavations, some 230. One reason for a multiplicity of wells is the fact that during the digging or while in use they sometimes collapsed, particularly if unlined. Moreover, sometimes the source dried up. Again, most towns only acquired an aqueduct after they had already become an established community, and so houses were normally built each with an individual well (or cistern, as we will see below) as a matter of course. Apartment blocks often had a communal one in the garden or central courtyard. In addition to these private wells, we also find public ones at strategic locations in the city streets or in the forum. Sometimes these were in large groups, and it has been acutely suggested that their public status is both indicated and confirmed by the unusually large amount of junk and debris thrown down them; private owners were presumably more careful about their own property.[32] In the country, villas and farmhouses naturally often had wells, either for drinking water or for irrigation. Irrigation wells were usually fitted with some kind of lifting machinery to ensure a constant supply, but even so the area of land irrigable from one single well was limited to about 0.25–0.5 hectares (= ½–1 acre) under ideal conditions, and accordingly well irrigation was normally restricted to the orchard or market-garden. Pliny recommends a stream passing through the garden as the most convenient way of watering it, and a well as the next best thing. Columella adds the practical detail that the well ought to be dug in September, when the water level is at its lowest after the summer drought; this will ensure a perennial supply.[33]

Cisterns

While a well taps an underground spring or aquiferous stratum, the cistern is essentially a masonry tank, either built at ground level or excavated a little below it. It is fed from above and is normally used to collect and store rainwater, either directly from the roofs of associated buildings or from the surface run-off from the ground. Sometimes it may be tall, round and narrow, in which case the excavator may only with difficulty be able to distinguish it from a well; frequently the mouth is narrower than the rest, and the whole is then called a 'bottle cistern'. More often the cistern is simply a circular or oblong tank, often under the floor of the house it serves and fitted with a well-head through which a bucket can be lowered to get water. This is standard practice in the houses of Delos; it is usually under the peristyle, with the well-head supplemented by a stairway to give access for cleaning. Another site rich in domestic cisterns is Ampurias, though there they are a good deal smaller and their characteristic form is long and narrow, with rounded ends. The *impluvium* traditionally found in the centre of the atrium of the Roman (or at least Pompeian) house must also count as a cistern of sorts. This was an uncovered shallow, oblong tank set into the floor and fed by rainwater from a square hole in the roof above, the entire house roof being sloped inwards from all sides towards the hole ('compluviate roof') to ensure an adequate catchment area. Presumably the original purpose was water supply. Presumably also this was superseded when the aqueduct brought fresh running water to the fountain on the street corner, if not into the house itself, reducing the *impluvium* to a traditional and decorative function. Whatever their original use, it is hard to believe that by AD 79 many of the *impluvia* in Pompeii were in fact anything but ornamental pools, though their overflow was often put to practical use by being diverted into a cistern under the floor.

For rural agricultural or industrial use the cistern was likewise common, usually a simple reservoir that today looks like a low-walled enclosure, which in antiquity would often have been roofed over with a wooden cover to prevent pollution and evaporation (but see below). Inside, all cisterns were lined with waterproof cement, usually pinkish in colour, and often, when they are in fragmentary state, it is the presence of this hydraulic sealing that tells us that it is a cistern we are dealing with. Stone access stairways again are common. Some of the most numerous and best preserved cisterns are to be found in the hills around Thorikos and Laurion, in Attica (Fig. 26) where they collected from surface rainfall the large amounts of water needed for washing the ore from the Laurion

26. Laurion, Greece: circular cistern for storage of surface runoff, to be used in washing silver ore from the mines.

silver mines. They are also (Fig. 28) common in farms in the arid regions of North Africa.[34]

The only place where calculations have actually been done on the capacity of urban cisterns, the amount of rain water available to be stored in them, and its adequacy for domestic needs, is the Hellenistic site of Morgantina in Sicily, but presumably the figures would also be roughly relevant for much of the Roman empire. At Morgantina the average annual rainfall is 70 cm and the average cistern capacity 30 m^3; to fill it would require the run-off from a roof area of 45 m^2. Of course, this calculation is based upon the cistern being filled only once, and would require the inhabitants to live for a year on one cistern-full of water. In practice the supply would be augmented by other considerations. Run-off from the ground could also be channelled into the cistern, and a roof area of 45 m^2 is provided by a very small house, a 9 × 5 m plan, or the equivalent in other proportions. Most ancient houses are around twice this size, with a correspondingly larger roof catchment area. One could easily deduce from this that the average cistern, at least at Morgantina, was filled by rainfall around three times per year, but this would only give a deceptively precise impression, there being so many variables. We are probably safer simply to stick with the generalisation that 'such a cistern, even without any special treatment to conserve water, could provide water for all but the driest months of the year',[35] and even then there were plenty of local springs and public fountains to fall back on. We may also note that on the basis of the 500 litres daily per person estimated by Eschebach as the average water consumption in Pompeii,[36] this means that a full cistern would maintain a single householder for two months, or a year if filled six times. Given that Eschebach's figure seems high, I would imagine that these statistics might well be roughly valid not for one single person but for the entire household living in the house, cooking water and the like being shared. I need scarcely emphasise the vagueness and unreliability of this estimate, but on balance I think it may have some value as an indication at least of the order of magnitude with which we are dealing. By comparison, the capacities of the four industrial cisterns at Laurion listed by Ardaillon (n. 18 above), are 160, 239, 421 and 579 m^3, putting them into a completely different class from the 30 m^3 domestic cisterns at Morgantina.

In general, wells were preferred to cisterns, as the water from them was fresher and, presumably, more abundant. Again, however, much depended on local circumstances. The local well water might be unpalatable, and if cisterns were what they were used to, a conservatively minded population might cling to them. This has been known to happen in modern times, particularly in Provence, where a large number of villages in the late nineteenth century not only depended entirely on rain-fed cisterns, but vigorously resisted official attempts to change the system to something more modern.[37] Another local problem might be the level of the water table and its ease or difficulty of access. This did not always remain constant. The best documented case of this seems to be Athens, where until the fourth century BC wells, usually in the courtyard of houses, formed the main source of private water supply (in addition, that is, to public fountains such as the Enneakrounos). In the fourth century, however, comes a change. The bottle cistern makes its appearance and soon largely replaces the well, being again located in the courtyards; in the residential districts around the Agora, 140 of them have come to light. This represents a major change in urban water supply: instead of tapping the underground supply, the inhabitants now relied on collecting and saving the run-off from their roofs. Since the cistern was in no way a superior technical development, this can only mean that the water table had sunk so low that they could no longer get at it. Whatever the explanation (the two possible causes seem to be depletion by over-use and climatic drought), the condition was temporary. As early as the third century BC the water table evidently began to rise again, and in Roman Athens the well had made a come-back, now outnumbering cisterns by a factor of three or four to one. Indeed, several cisterns were converted into wells, the well-shaft being dug through the floor as the inhabitants once more drew on the underground supply.[38]

A special case must be made for cisterns in arid countries, such as the Maghreb (North Africa). One of the chief difficulties here is classification.

27. Ampurias: urban cistern. Ampurias having no aqueduct, there are many such cisterns throughout the city.

The cisterns themselves perhaps may most usefully be divided into covered and uncovered, but it is not so easy to categorise them by function, at least on the lines along which this book is organised. Sometimes a cistern may be, in effect, a large city reservoir, aqueduct-fed. Cisterns of this type more properly belong to a study of aqueducts, and they will be described in Chapter 10 below (pp. 279-80). Sometimes a smaller cistern will, in effect, perform the function of a *castellum divisorium*, or perhaps a settling tank.[39] In urban areas individual cisterns served private houses and workshops. Where there was no aqueduct (as at Ampurias, Fig. 27), the run-off, possibly from quite a large area, would be channelled into a cistern and possibly used to serve public fountains. At Tiddis even the baths were served in this way, though for an establishment as profligate in water use as baths to depend purely on rainfall is highly unusual.[40] In rural areas the cistern might also be used for serving hamlets or farms. The water to fill it might be conveyed there through a short channel, perhaps likewise from cistern to farm; whether or not one dignifies such a channel with the title of aqueduct is a matter of semantics, but we are plainly dealing with a concept very different from the large city aqueducts of later chapters, and so this material will be included here.

The number of these rural cisterns in the Maghreb is very great, and, contrary to what one might expect, a lot of them have been published. The reason is that in the nineteenth century the French colonial authorities instituted several such surveys of Roman hydraulic works with the express purpose of seeing how they could be repaired and put back into service. The *Archives des missions scientifiques et littéraires* are full of such reports; in particular, Gauckler's study of Tunisia and Gsell's of Algeria were both prompted by this desire, and the Italians did the same thing in Libya.[41] The result was a sometimes undiscriminating catalogue in which, lacking a disciplined archaeological approach, almost any hydraulic ruins were liable to be listed as 'Roman', the whole being compendiously damned by a modern critic as 'boring catalogues of countless wells, storage basins and aqueducts ... lacking in any firm methodology or direction, and as arid as any desert in their *monotonie désespérante*'.[42] However, though the interpretation or analysis of these works may be lacking, the actual description of the monuments (perhaps apart from dating) presumably is in general reliable, and Saladin's outline of the system seems fair: 'Par le barrage des petites vallées, on emmagasine dans de petits reservoirs nombreux des quantités d'eau que l'on dirige ensuite par les aqueducs ... jusque dans de grands reservoirs ou *fesguia*.'[43]

Covered cisterns may in theory be covered by anything, such as flat stone slabs, but in practice the commonest covering is the traditional Roman barrel-vault. In plan, the commonest shapes were square, rectangular and cruciform. Sometimes they are found in small groups of two or three, close together and interconnecting. At Cherchel some twelve cisterns have been identified within the urban area; five of those have dimensions sufficiently complete to enable an estimate to be made of capacity and they range from 30–130 m^3. Their collective total capacity, indeed, was quite enough to serve the needs of the city, the Cherchel aqueduct being something of a luxury.[44] Usually they are from 2.5–3 m deep, and when they were full the water usually came up to the beginning of the vault. When the cistern was too large to be covered by a single vault without undue trouble, it was either divided up into a series of separate compartments with holes in the intervening walls to allow the water to flow through freely, or subdivided internally by rows of piers dividing up the cistern into aisles, as was commonly done in the large aqueduct storage cisterns in Africa and elsewhere; these were much larger than the small domestic ones and run from 6,000 and 9,000 m^3 for the two cisterns at Dougga, up to 25,000 m^3 at Bordj Djedid (Carthage).[45] Practice varied from one city to another, the single large cistern, which was normally used when there was an aqueduct to feed it, being replaced by a number of medium-sized ones, the size of each depending on the needs of the district it served and the extent of the catchment area available to fill it with its run-off.[46] The small cisterns serving individual houses, where they existed, formed a third category.

Uncovered cisterns are, surprisingly, particularly numerous in the Maghreb. One is surprised, because one would expect that the potential loss by evaporation from an uncovered water surface would stop them being used in such a hot and arid climate. Perhaps we overestimate this: Goblot is at pains to point out that in uncovered conduits, at least, evaporation loss is minimal even in hot weather, as a protective layer of very humid air forms immediately above the water surface, effectively insulating it.[47] This is certainly not at all what one would naturally expect. Conduits, of course, are a very different thing from the open ponds or cisterns we are here considering, but for us too Goblot's note of caution may be a salutary warning against taking too much for granted. Another explanation for the multiplicity of open cisterns in the Maghreb may be that in fact no more existed there, but more have been found because the efforts of colonial administrators seeking Roman hydraulic works capable of rehabilitation resulted in these regions being much more intensively surveyed (particularly in rural areas) than other comparable parts of the empire. Be that as it may, open cisterns in the Maghreb are commonly found. They are usually in rural areas, serving isolated farms or hamlets, and have a characteristic and unexpected form. They are usually broad, flat and shallow, rather like artificial ponds, around 1–1.5 m deep and, one would have thought, but for a hesitant reticence induced by reading Goblot, exceptionally subject to evaporation loss. In plan, they can be round, semi-circular, oblong or of any other convenient configuration (Fig. 28). The side walls are often strengthened with buttresses, sometimes internal, sometimes external. One would of course expect to find buttresses round the outside, to contain the pressure of the water within, and internal buttresses at first come as something of a surprise. The reason usually seems to be that these reservoirs were sunk into the earth, rather than built above ground level (though this is often not clear from the published plans),[48] and the internal support is needed to hold back the surrounding earth when the water level in the reservoir is low. Later, we shall encounter exactly the same problem in dams (p. 85 below).

As for the unexpectedly broad and shallow proportions, this may often be caused by the local topography and the exigencies of gravity flow. A deep reservoir presents serious difficulties in a terrain that is flattish, or only gently sloping. If it is sunk below ground level, then once the water has run down into it there is no way of getting it to run out again, short of pumping or raising it in buckets. If it is built above ground level, the water can be drawn off easily enough by pipes or channel, but how can it be raised to get it in, over the top of the reservoir wall, in the first place? The only answer is to have it arrive already running at a high level, which means an aqueduct, like those crossing the Campagna at Rome. For the average Maghreb farm or village this would be a prohibitive expense. The only convenient way to store water arriving at ground level and to be delivered for use at ground level, is to store it at ground level too; and if

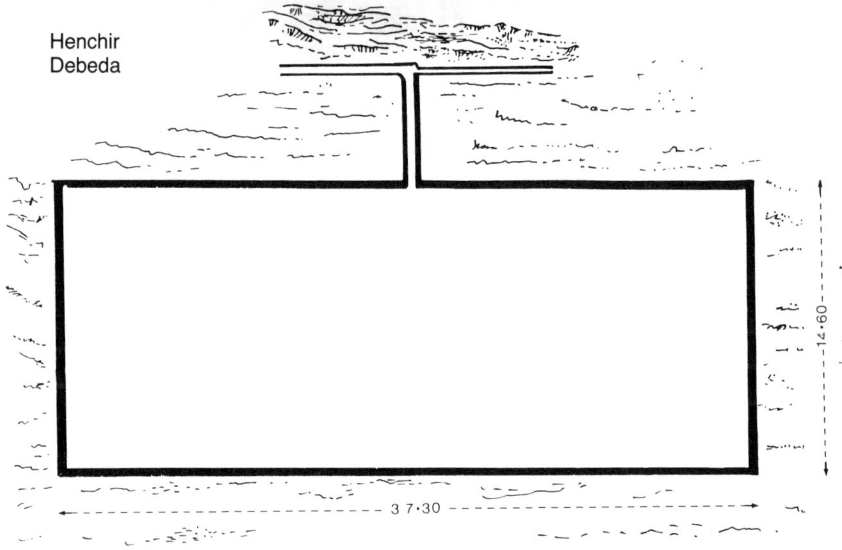

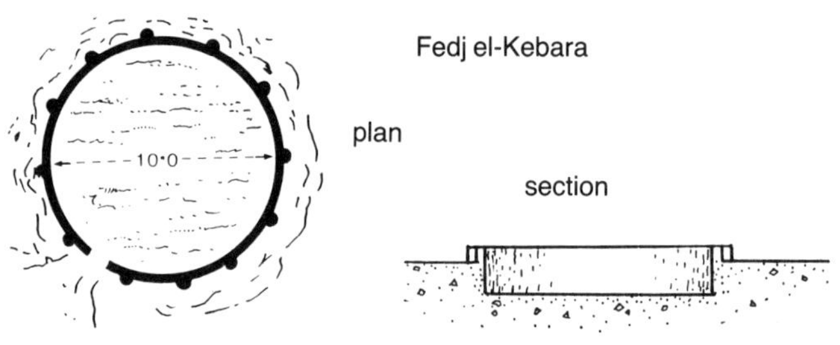

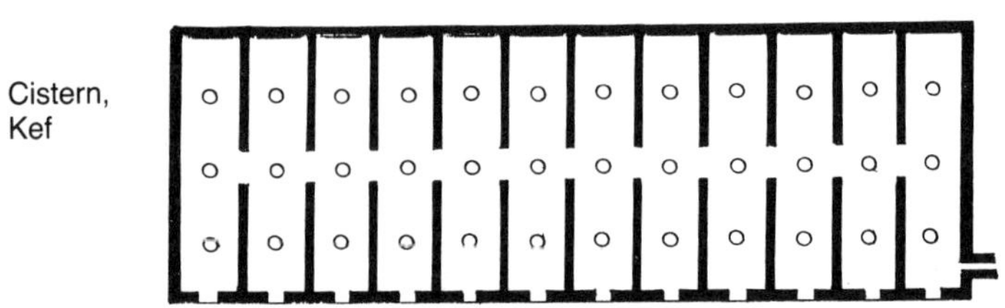

28. The Maghreb: plans of various cisterns (Birebent).

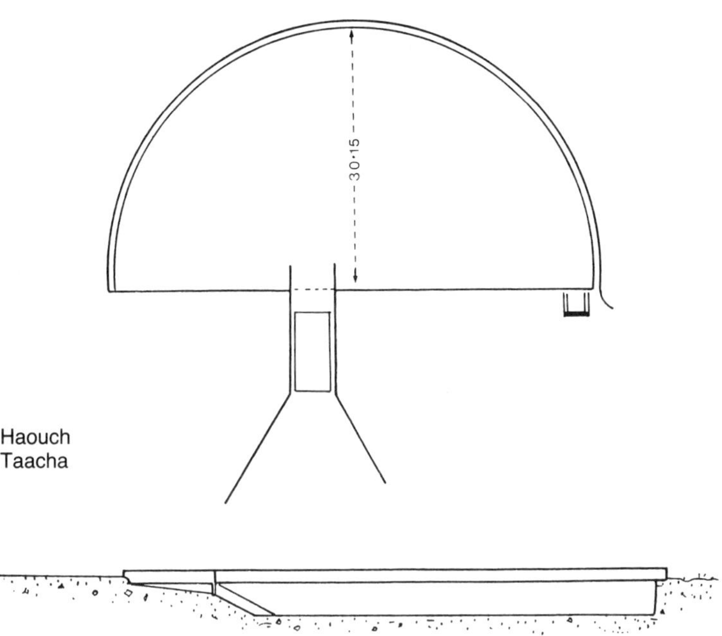

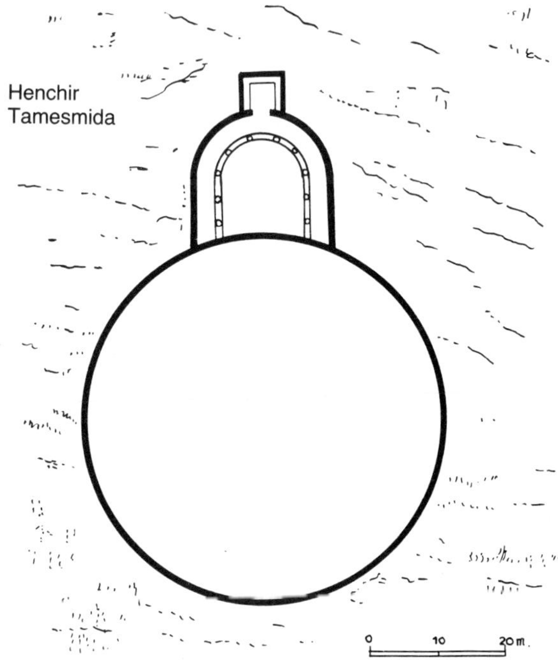

Henchir
Tamesmida

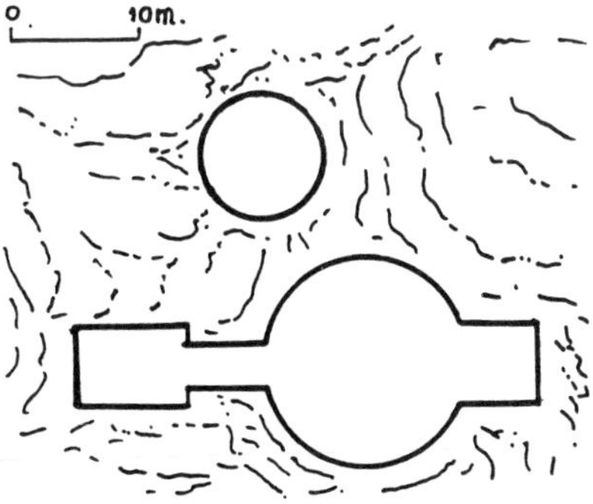

0 10m.

Henchir Mzira

0 10 20m.

the amount to be stored is at all considerable, then this means extending the storage horizontally, resulting in a wide, shallow tank. The width will then in turn often make roofing or covering impractical. It is tempting to conclude that, given the arid climate and the desperate need for water, the evaporation problem then often solved itself: the water, being constantly used and replenished, was simply not in the reservoir long enough for much of it to have any chance of evaporating. Of course, this theory is no more than speculation, but I would not be surprised if, in some cases at least, it were true.

4
The Source

O spring of Bandusia, clearer than glass!

Horace, *Odes* III, 13

The Roman water supply system is best considered by following the course that the water followed through it. All fresh water begins as rainfall, or at least precipitation. On falling to earth, its subsequent behaviour and the factors acting upon it form the science of hydrogeology, which may in its essentials be defined as the study of the interaction between the geologic framework and the fresh water. A modern authority explains further of what this interaction chiefly consists: 'In the hydrologic cycle, water modifies rocks by erosion, solution and deposition. The rocks contribute mineral constituents to, and alter the heat content of, the water and thus materially change the character of the latter.'[1] The cycle ends when the water is returned to the atmosphere, either by evaporation (especially from the surface of the sea) or by transpiration (from the leaves of trees and plants).

In its intermediate phase, after falling as rain, but while still present on earth in liquid form and before evaporation, the water is stored in three separate ways: as surface water, soil water and ground water. Surface water is self-explanatory, including lakes, streams and, for that matter, aqueducts. Soil water is water retained in the soil, dampening it and enabling plants to grow; it is drawn from the soil by their roots and returned to the atmosphere by transpiring through the leaves.[2] Ground water is what the layman would probably call 'underground water', or perhaps the 'water table'. Both terms are, superficially at least, rather misleading: we are not here dealing with underground rivers (which would be included in this category, but are rare), nor is the water surface of the 'table' necessarily level. Basically, ground water is found underground, contained in a geologic stratum of some porous material (e.g. limestone), saturated with water, and known to the geologist as an aquifer (Fig. 29(a)). The upper level of this water forms the water table; water above the table is classified as soil water, below it as ground water. The table may or may not coincide with the upper limit of the permeable

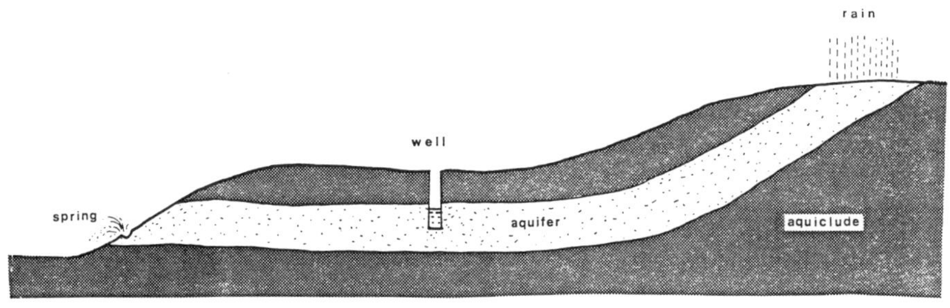

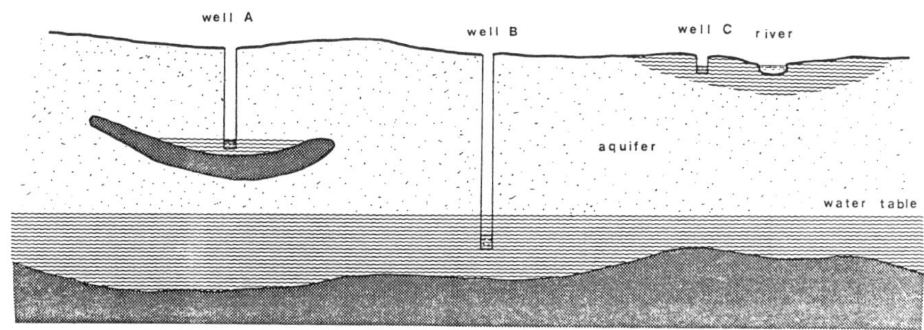

29. Diagram illustrating (above) the functions of aquiclude and aquifer; (below) how three wells, close together, can tap water from different levels.

stratum, depending on whether the stratum is, so to speak, full up to the top. The water may be stationary, so that the stratum forms a reservoir, or it may be moving, slowly percolating, so that the stratum forms a conduit. If, as often happens, the stratum is sandwiched horizontally between layers of dense, impermeable material, such as clay, then these layers form in effect the limits of the conduit.[3] If the stratum is not level, the water will percolate through it in a downhill direction forming, in effect, a solid, very slow-moving, underground river, of the thickness of the stratum (which will vary) and of a width limited only by the stratum's horizontal extent (which may be almost indefinite). When, or if, the stratum surfaces, the water emerges as springs.[4] It may, of course, be tapped before then vertically by wells (see Fig. 29(a)). Given the complex and undulating profile of many of the strata, it is also quite possible to find isolated pools of water, perhaps at a quite different level from the general water table. Fig. 29(b) shows a typical example, where three wells sunk quite close together have to go to very different depths to strike water because each is tapping a different source. Well C finds it quite close to the surface, water that is in effect seepage from a nearby river. Well B, not far away, is beyond the limits of the seepage, and so has to go

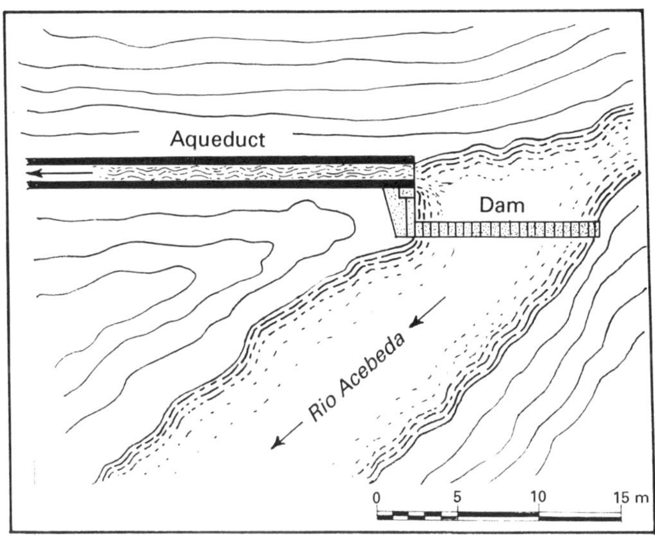

30. Segovia: weir across Acebeda stream to divert water into the aqueduct conduit.

right down to the overall water table; in the Poitou area of France, near Poitiers, a Gallo-Roman well has been identified that went down 80 m to strike the table.[5] And Well A has been lucky enough to strike a pool of water contained on top of an isolated and concave section of aquaclude. One way or another, whether by wells, local springs, or by distant springs tapped and brought to town by aqueducts, it was this ground water that filled by far the greatest part of the ancients' needs.

Surface water, whether from lakes or rivers, was used less often than might be expected. At Rome some of the aqueducts, notably the Anio Vetus and Anio Novus, took their water (as their names indicate) from the Anio river. The Segovia aqueduct drew its water from the Rio Acebeda, a small stream some 12 km south of the city, the water being diverted into it by a weir set at an angle across its course (Fig. 30). At Aix-en-Provence likewise a small weir across a brook at St Antonin directed its waters into the St Antonin aqueduct, and on the Traconnade aqueduct, also at Aix-en-Provence, the stream of Les Pinchinats, in the northern suburbs of Aix, was diverted into the aqueduct by a dam.[6] In south Turkey, the aqueduct of Side was fed from the river Manavgat by an arrangement both unique and ingenious (Fig. 31). The chosen point was the confluence of the Manavgat and the very copious waters of the Dumanli spring,[7] which come sweeping into the waters of the mainstream at right angles with great rapidity and volume. It was the Dumanli that the engineers really wanted to tap, since its water was of superior quality. The trouble was that it was on the wrong side of the river for the aqueduct; how then to get its waters across without the

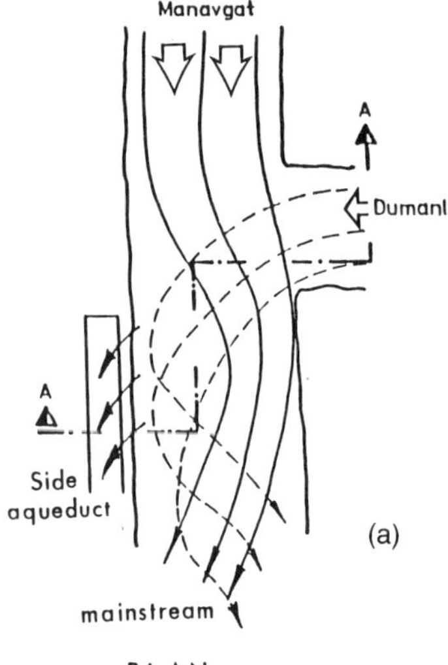

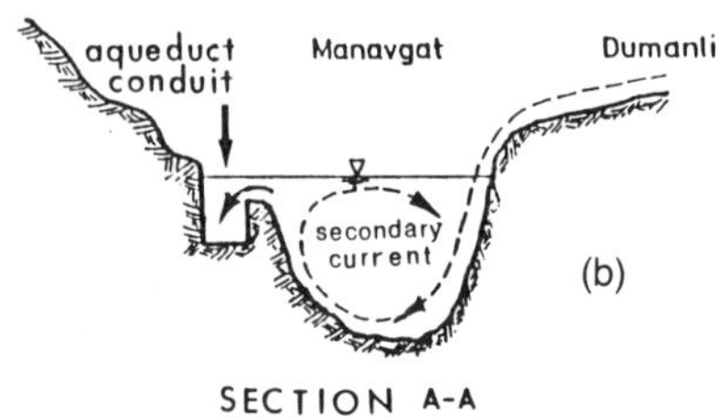

31. Side: diagram showing how the channel of the Side aqueduct (left in plan, (a)) was fed by water from the Dumanli spring spilling in (cross-section, (b)) across the intervening stream of the Manavgat river (second century AD).

difficulty and expense of building an aqueduct bridge across the Manavgat? The secret lay in taking advantage of the natural currents, for the Dumanli enters the mainstream with such force (it is possibly the biggest spring of its type in the world) that it goes swirling over to the opposite bank as an independent and cohesive body of water, and it is only further downstream that the Dumanli and the Manavgat really mix together. The aqueduct channel was therefore located parallel to the river, slightly downstream from the confluence and on the opposite bank, so that as the turbulent waters spilled over into it what it received was almost entirely Dumanli water – a good example of the Romans observing river currents and profiting by them.[8] Again, in Spain, North Africa and the Middle East, as we shall see, they often drew water from reservoirs created by damming, but this was otherwise rare.

Cities, even when built upon major rivers for communications and commerce, seldom used the river water flowing by their doors. For one thing it was too polluted, and for another, given gravity flow, a river can hardly directly serve a town built on its banks. The water has to come from a higher source, particularly if, as so often, the city is itself built on high ground: Fourvière, the ancient Lugdunum (Lyon), high above the Saône, well illustrates the point. True, Frontinus assures us that for the first four centuries of its life Rome drew its water from the Tiber, but this only emphasises its primitive state, and even at this stage he makes it clear that the river was not the only source, but was supplemented by springs and wells.[9] What the wells were producing was of course river water that had percolated sideways, but in its passage through the soil it must have been at least partly filtered. For most Roman urban supplies, therefore, rivers are to be ruled out (after all, that's where they dumped the sewage), and we return to springs, and the geological principles underlying them.

The way these geological principles apply in particular to the Mediterranean area has been clearly expressed: 'The geological configuration prevalent in most of the ancient Greek lands on the rim and the islands of the Aegean is a permeable limestone cap superimposed on an older impermeable layer. Consequently, there is a general shortage of surface water. The water seeps through the limestone and collects at its base, where the limestone meets the impermeable strata. This seam is also where most springs and trickles of water are found, often issuing from a natural cave or fissure.'[10] But how far were the ancients aware of the geology governing distribution of the ground water, on which they were so dependent? Plainly, a lot must have been learned empirically, particularly by miners. Already in the fifth century BC in the Laurion silver mines, vertical access shafts were being driven straight down, cutting through all the strata to a depth of 120 m or more, and showing a considerable command of technique. Miners, above all, who were always facing water leaking into their workings, must have developed a good knowledge of practical hydrogeology.

Theory was a different matter. Philosophers might speculate on the role of water and earth in the creation of matter, cataloguers such as Aristotle might enumerate and classify different kinds of rocks, usually with particular attention to their potential as sources of minerals, or even gems, but that was usually as far as ancient geology went. It has been well said that 'their approach was, throughout, descriptive rather than analytical', listing various geological phenomena as examples of them occurred, but without trying, or at least succeeding, to infer from them a systematic and coherent general picture.[11] The idea that water could be conveyed considerable distances through soil or geologic strata is inherent in the statement of Plato that as an irrigation technique dams could be built to pond run-off surface water, which would then be absorbed into the soil and surface elsewhere as springs;[12] and underground rivers were a known phenomenon. Indeed, they proved so attractive to poets, particularly when running under the sea bed, that a whole range of firmly established classical legends grew up attributing to well-known springs and rivers sources in impossibly distant (but exotic and poetically attractive) localities overseas, an underground channel forming the link; the best known was the spring of Arethusa at Syracuse (Sicily), which was supposedly fed by the River Alpheus, or part of it, going underground near Olympia (Greece) and running for 600 km under the sea bed.[13] In short, the ancients seem to have seen the earth as a positive honeycomb of underground caverns, channels and galleries, but to have had no concept of hydrogeologic principles that actually produced, bubbling up at the springs, the water on which they depended.

Springs and spring-water

Springs were the commonest source tapped by an aqueduct. In output, springs vary enormously, from the romantic trickle the mind conventionally associates with the term, to great floods before which it simply boggles. A large spring located in a natural declivity, as many of them are, often simply fills it up until at some point it overflows, creating a river running out of a spring-fed lake, which means that no one ever sees the spring itself; it is perhaps because the larger springs are often hidden in this way that the popular concept of one is of a modest source, and one wonders how a spring could aspire to keeping filled a full-size city aqueduct.[14] In point of fact, this is no problem.

What then was this water like as it emerged into the air and became available?[15] We are accustomed to thinking of spring water as pure compared to other sources, but this is deceptive. A simplified generalisation will make the point. All water is by nature (acid rain apart) entirely pure when it falls as rain. It then becomes adulterated in two ways. Surface water (rivers) remains chemically pure but is physically polluted; ground water (spring water) remains physically pure but is chemically adulterated. The surface water acquires mud, sand, sediment in suspension (and, of course, any pollutant that gets into the river). This makes it distasteful, even unhealthy, but since the polluting matter does not physically form part of the composition of the water, it can, in theory, be separated from it by physical means, such as filters and settling tanks. Even though these did not always work, they were at least within the scope of Roman technology.

Ground water, on the other hand, is delivered at the spring physically pure – clear, sparkling water with no sediment – but in seeping through the limestone it has become hard through the chemical absorbtion of calcium (sometimes also magnesium, and, to a lesser extent, iron and other minerals depending upon what deposits it has touched). Since it is a chemical condition, it is unaffected by physical means – filters and so forth – and once hard, the Romans could not even try to make the water soft again. Because it was hard it then formed the incrustation inside pipes and conduits that was such a regular feature of Roman waterworks.[16] Mud and sediment from surface sources could of course clog the channels too but were more easily removed. Mud could often be flushed or washed out and in closed pipes could be cleared out by some pull-through device; pipes heavily incrusted might simply have to be replaced.

The generalisation thus stands as follows: faced with a choice between soft and dirty water that could in theory be cleaned, and clear but hard water that could not be softened and would, in time, wreck the whole system (through untreated incrustation), the Romans chose the latter. Looking at the muddy waters of a river and a clear bubbling spring, one can understand their decision; and the above is, in any case, no more than a thumb-nail simplification.[17] There were presumably regions in the Roman empire where the catchment area was not limestone, and which did not suffer from hard water, but Roman aqueducts everywhere *were* famous for incrustation, showing that hard water was a general complaint, and presumably also the limestone that caused it.[18]

The ancients themselves were well aware of the different qualities of water, though not always accurately or correctly so. Everyone was agreed on the purity of rainwater. Hippocrates actually says that water coming from springs in the rock is hard, using the Greek word for it (*skleros*). It is not quite clear what he means by that since he relates it to the hard texture of the rock and seems to be saying that water coming from a hard rock naturally has to be hard. The accepted modern touchstone of hardness in water, the ease or difficulty with which it raises a lather in soap, was of course, unavailable to an age when soap did not exist, and Hippocrates' chief objection to what he calls hard water is that it inhibits urination and causes constipation.[19] His recommendations for picking a spring to be tapped are that it should be high on a mountain (for health, not level of the supply, Hippocrates being a doctor rather than an engineer), in an earthy, not rocky, terrain, and facing east. Vitruvius, who devotes much of his space to catchment and the varying qualities of water, recommends tufa and 'black earth' as productive of pure springs, preferably on a mountain, or just at its base, where they are mountain-fed; he also likes them to be facing north, on the sound principle that northern slopes are shaded from the sun and so more likely to be wooded, assisting the retention of water, but unfortunately he adds the distinctly unscientific argument that springs facing north are also to be preferred since that is the direction the biggest and best rivers come from.[20]

Testing for the purity of the water was usually no more than a matter of common sense. A simple precaution was to examine the local inhabitants who regularly drank from it and see if they *looked* healthy, and the same thing could be done with animals. The Roman practice of augury is often ridiculed – how *could* one expect to presage good or bad fortune by inspecting the entrails of pigs and sheep?! – but it probably had a reasonable and rational origin in some such internal examination as part of the routine procedure in locating and founding a new settlement; one had to try and determine whether the water was good. Other simple tests include boiling water to evaporation, to see if any residue of impurities is left in the vessel, and cooking vegetables in it, and seeing whether they are properly cooked.[21]

When we read Roman authors, we get the impression that, especially in the city, 'drinking water' was hardly ever actually drunk, at least in its natural state. It was normally mixed with something else, usually wine, and was frequently impressed to hold at bay the climatic rigours of the chill northern provinces, in the welcome form of hot toddics and herbal tisanes. But this is probably misleading. The poorer classes presumably drank a lot of water straight, and Romans affecting a simple or austere lifestyle followed suit. Cato made a point of it when on military service.[22] Certainly the local water and its characteristics were in antiquity a topic of engrossing popular interest (as in small Greek villages it still often is).[23] Nothing else could justify Vitruvius' endless (and, for an architect, rather pointless) catalogue (VIII, 3, 7–25), a kind of Ripley's believe-it-or-not tour of the hydrological curiosities of the entire Mediterranean basin.[24]

The next step was to determine where the spring was to be tapped. The profession of water-diviner or water-seeker was recognised in antiquity (Gk. *hydroskopos*, Lat. *aquilex*),[25] but they have themselves, of course, left no writings and for their activities we are dependent on references in more general literary sources. Thus, if no natural spring was visible at ground level, it might be necessary to sink a shaft to tap the water-bearing stratum (aquifer), and it is again Vitruvius who outlines for us simple techniques for finding water, techniques that are, of course, equally valid for sinking wells. Metal vases can be buried overnight, and next day examined for traces of water collecting in them. An unfired clay vase can be buried and if there is any moisture in the earth will absorb it and itself begin to disintegrate. Surface vegetation may offer a clue to the

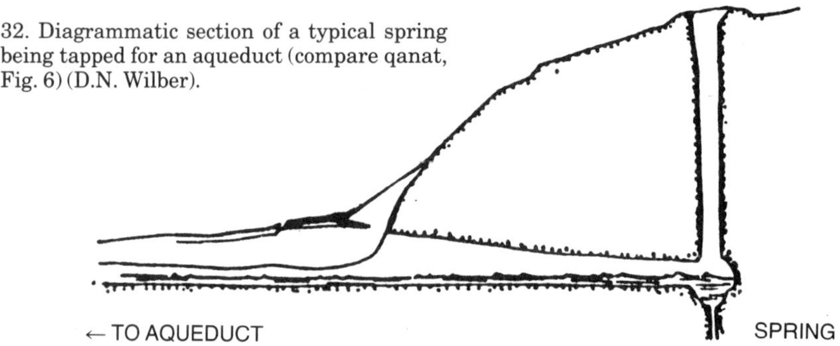

32. Diagrammatic section of a typical spring being tapped for an aqueduct (compare qanat, Fig. 6) (D.N. Wilber).

existence of water below, and one can lie flat on the ground at dawn observing where, in the early light, humid vapours are to be seen arising from the earth.[26] As to the method most popularly celebrated, water-divining or dowsing by a diviner's twig, there is no evidence for this in antiquity. The twig, even the forked twig, was known and used in divination, but only in a general context of augury and auspices. There is no reason to believe it was ever used in the search for water, and indeed we do not find it in water-divination before the sixteenth century AD.[27]

In the event that the terrain was a sloping hillside, with a water-impregnated aquifer some distance below the surface (a not uncommon situation), the aquifer would be tapped by an almost-horizontal adit, with a slight downhill slope to bring the tapped water out to the surface by gravity. To help locate the aquifer, and hence the adit tapping it, a vertical shaft would often first be sunk (Fig. 32). If the ground slope was anything less than 45° this shaft would be a good deal shorter than the adit, and hence require less work to repeat in case the diggers came up with a 'dry hole'. Once the aquifer was intersected, the angle and position could then be calculated to dig the adit, very much on the same principles employed in digging a qanat.

However, it was probably much more common for the engineers simply to pick an abundant spring already gushing forth on a convenient hillside. It might happen that this was already located conveniently within the city it was to serve (or, to put it another way, the city had originally been built there because of it). If so, some kind of fountain or fountain-house would be built on the spot, and nothing more was required: every city had its own such local fountain, often the setting for primitive myths and legends, and the fountains of Juturna and the Camenae[28] in Rome were merely two of many. More often, however, it was not, and its waters had to be brought to town by an aqueduct. In such a case, the spring, though in some distant and perhaps inaccessible locality up in the hills, might yet be celebrated architecturally with columns and niches after the fashion of a nymphaeum – the best example is at Zaghouan, the source of the aqueduct of Carthage (Figs 33, 34). More often, the water from the spring could be collected straight into a masonry basin with the intake conduit for the aqueduct running directly out of the far side, the best preserved example of this probably being the Kallmuth spring on the Eifel aqueduct at Cologne (Fig. 35: compare Sens, Fig. 36). However, it was more usual first to seek to augment the supply by driving one or more tunnels or adits back into the hillside to tap other branches of the spring. The Grüne Pütz spring (Fig. 37) also on the Eifel line at Cologne, is a good example: an open tank, 1.93 m × 1.86 m × 2.0 m deep, with the aqueduct running out of one side in a rectangular, slab-roofed channel, and the water delivered at the other by a smaller vaulted tunnel some 80 m long, collecting the water from a marsh at the base of a hillside.[29] More often several tunnels were used; the Marcia at

33. Carthage: at Zaghouan the source of the aqueduct was enclosed within a semicircular decorative colonnade (photo: M. Goodfellow).

34. Carthage: at Zaghouan the water collected in a figure-of-eight shaped basin (photo: M. Goodfellow).

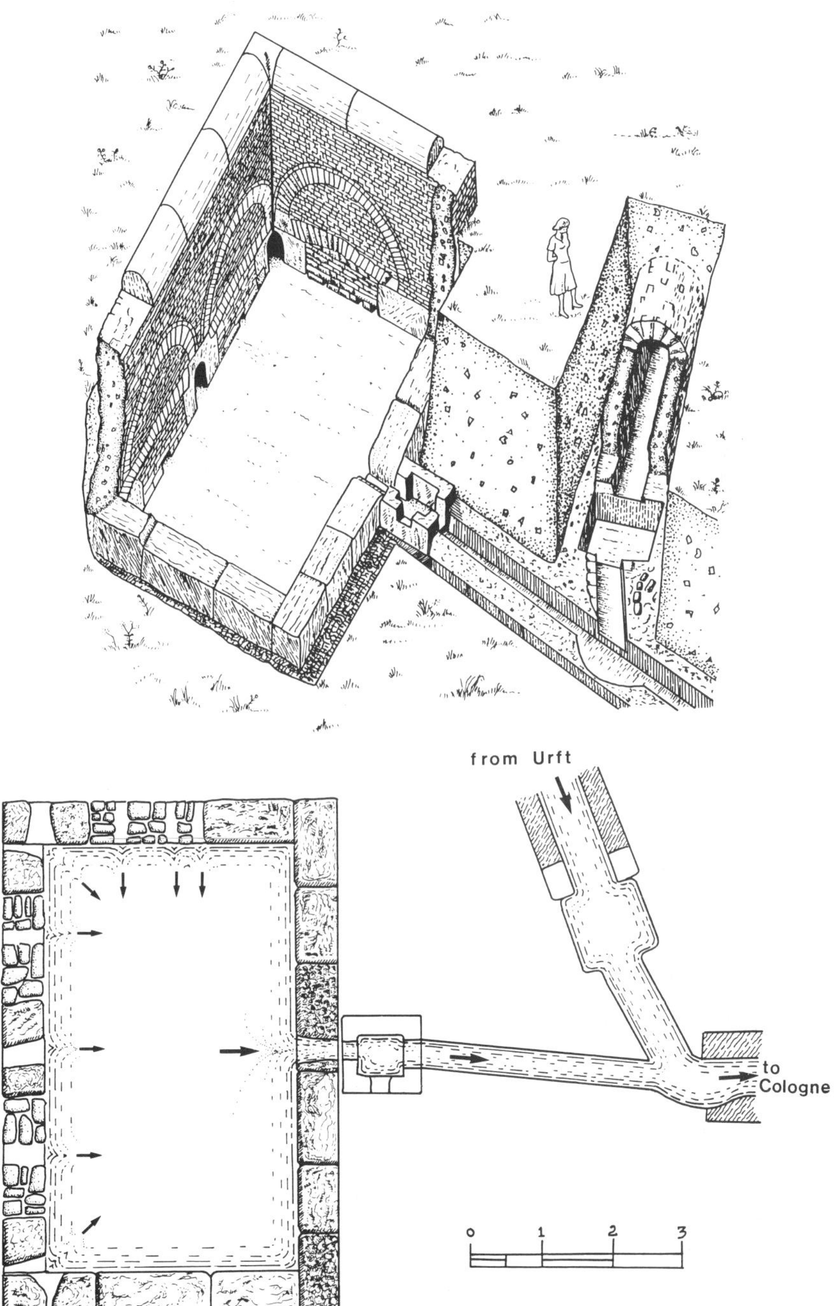

35. Kallmuth, Germany: spring serving the Eifel aqueduct (Cologne). The water from the spring entered the rectangular basin through a series of small holes along the base of the walls; outside the basin is the junction with the aqueduct from Urft. Above: reconstruction; below: plan (W. Haberey).

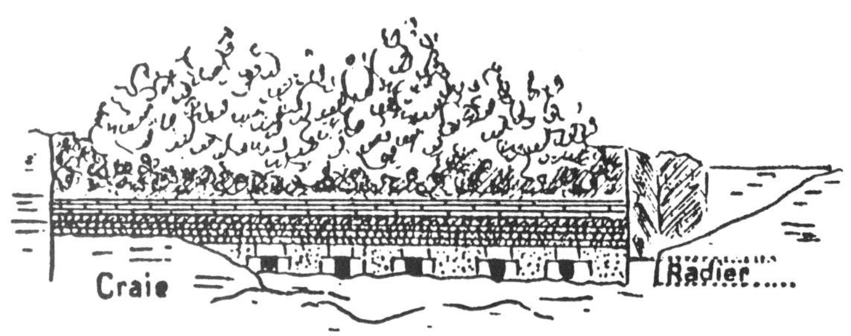

36. Sens, France: springs at Noé (Sens aqueduct), showing holes by which water entered the basin (*MAGR*).

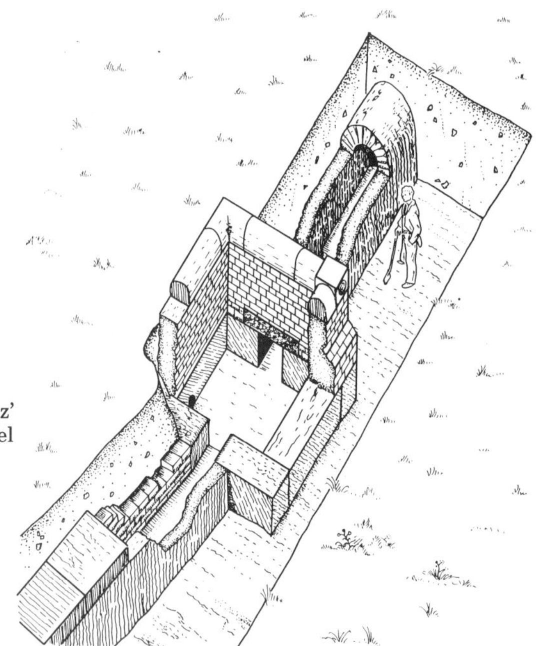

37. Cologne, Germany: 'Grüne Pütz' spring and settling tank on the Eifel aqueduct (W. Haberey).

Rome was served by a whole network of such feeders.[30] Vitruvius recommends something of the sort with his specification for digging wells – if water is struck then this main well should be supplemented by tunnels radiating from the bottom of it in various directions to bring in additional water to the centre. Interestingly, precisely the same technique is recommended by a modern handbook on well-drilling, and in almost the same terms. It should be understood that the purpose of these tunnels is not to gain access to other springs. If the rock in which the main spring is found is aquiferous then it is, in effect, itself impregnated with water which will seep out through its surface wherever that is exposed. The primary purpose of the tunnel is thus to increase the exposed surface area of rock, rather than to act as a conduit or means of conveyance from some spring at the far end of it, and it taps the water rather by having it percolate through the tunnel walls and roof all the way along.[31]

With springs located on a hillside, more often tunnels would be driven parallel, side by side into the hill. As with all ancient tunnels, they could not be bored or drilled in the modern sense but had to be cut by hand. Human access thus being the governing factor, the resultant tunnel was often much larger than needed for tapping the water. Inside, it must have looked like a particularly wet mine gallery, with water constantly trickling down the walls and dripping from the roof. Sometimes these tunnels could be filled, or allowed to fill, with water and be used as reservoirs; the best known examples being Peirene (Corinth) and Perachora (? – see pp. 30-1 above); though Greek, not Roman, work and though both immediately serving a fountain house at the end of the tunnel instead of an aqueduct, they well illustrate the principle.[32]

It was also possible to collect surface run-off water and direct it into the conduit, either at the source to supplement the spring water, or at some intermediate point along the aqueduct. This procedure either was rarely adopted or has rarely been identified, but it has been attested near Fontvieille on the aqueduct serving Arles. A whole fan-shaped network of small, 25 cm wide, rock-cut channels gathered the surface water from an area of some 2–3 km^2 and funnelled it into the aqueduct.[33]

Dams and reservoirs

Springs, as described above, were by far the commonest source for Roman aqueducts. Rivers, diverted directly into the conduit, were used occasionally, lakes, surprisingly, almost never.[34] That leaves us with reservoirs, artificially formed by damming, and the picture there is more obscure. Small earthen or concrete dams a few metres long built, for example, to divert the waters at a spring in the direction of the aqueduct, must have been common enough, though rarely thought sufficiently important to rate extensive publication.[35] Large dams are a different matter. The modern idea of creating by damming a large storage reservoir from which water can be drawn as needed was at variance with the traditional Roman aqueduct principle of constant offtake. The whole concept of water storage and use only as required represents a step towards a different way of thinking, a conscious attempt to economise and to get the most out of limited resources. Perhaps naturally, the Romans first applied this thinking at the delivery end of the aqueduct, in the construction of cisterns and underground reservoirs such as the Piscina Mirabilis at Misenum and Bordj Djedid at Carthage. Impounding and storing the water at source, and that in a reservoir formed mainly by the natural contours of the terrain instead of a built cistern, and of many times a cistern's capacity, required a new attitude and approach. None the less, necessity, in the shape of limited water supplies, often required that it be done, and it was. Unfortunately, these dams and the reservoirs they formed are located chiefly in three of the more arid regions of the empire – North Africa, the Levant and Spain – and so have been little studied. The result is twofold. An exaggerated impression of the rarity of dams has arisen, and the dam itself as a form of Roman civil engineering has been almost completely neglected.[36]

Dams are of three different types: gravity dams, arch dams and arched dams. The difference between the first two is easily defined. An arch dam is curved, with its convex face to the water, and in essence acts as an arch laid on its side. It contains the water by transmitting the stress horizontally to the abutments or rock-face on either side, in the same way that in architecture an arch transmits the superimposed weight to the two piers on which it stands. Many modern dams are of this type (the Kariba Dam in Zambia/Zimbabwe is a good example), and from our familiarity with it springs the layman's common expectation for a dam to look curved. Moreover, the outer and visible face (the air face) is very steep, often almost perpendicular, so that the structure looks like a curved wall (which in fact can be quite thin).

The gravity dam operates by a very different structural principle. Essentially, it is made so massive and heavy that the water pressure pushing it on one side is unable to move it, and insofar as the pressure is transmitted anywhere, it is transmitted not outwards to the ends, as in an arch dam, but downwards to the base. Because it is not acting as an arch, a gravity dam can be made straight (though it does not have to be). Because it has to be heavy and wide-based, its outer (and sometimes inner) face is no more than a gentle slope, and the structure resembles an embankment rather than a wall. Moreover, it need not be of masonry. Earth and rock, or earth alone, can often be quite adequate, even for a very large dam. The High Aswan Dam is of this type, and so, though not strictly a dam, is the 25 km-long dyke closing off the mouth of the Zuyder Zee from the North Sea in Holland. Gravity dams can then be subdivided into two kinds. One is earth dams. These are composed of earth alone, though sometimes with a thin stone facing to retard erosion. In cross-section, both sides of the dam slope, that is, the water face, as well as the air face, are sloping, not vertical (e.g. Cornalvo, Figs 45, 40(a)). In the other kind, the dam is basically a straight masonry wall, reinforced on the downstream side (the air face) either by masonry buttresses or by earth heaped against it in a gentle slope. This gives a cross-section of very different profile, one side of the dam being sloped and the other (the water, or inside, face) being vertical. In Roman work this face was often not truly vertical but merely very steep, being formed by successive masonry courses each projecting a little to form a series of steps. This can be seen in the dam of Proserpina (Figs 40(c), 41; compare Alcantarilla,

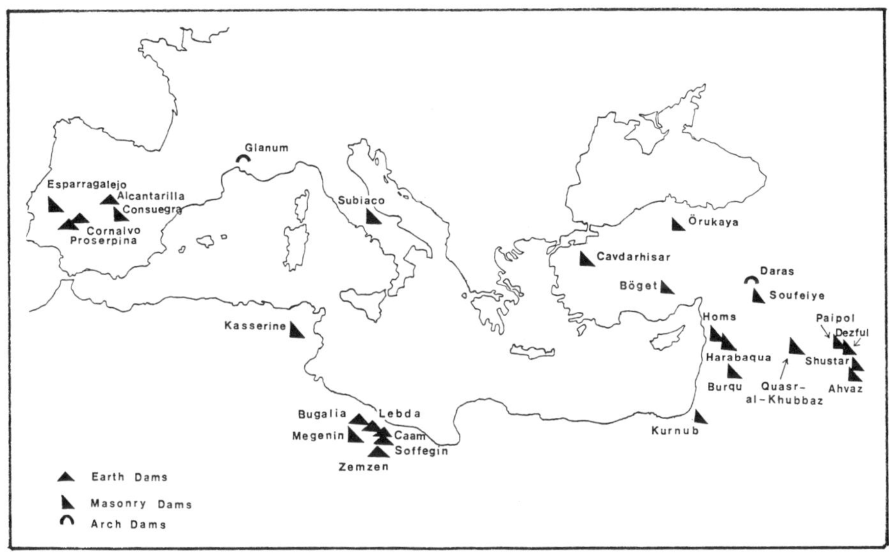

38. Principal known Roman dams in the Mediterranean world (after Schnitter).

Fig. 40(b)). Dams of this type we may refer to as masonry dams.

An arched dam is a combination of these two, a dam 'which in plan view is curved, generally to a circular shape but not necessarily so, but which at the same time resists the water-pressure primarily by its weight. In short, then, an arched dam is a curved gravity dam.'[37] A good modern example is the Hoover Dam in the USA.

The rule as it affects our studies is simple. 'In antiquity the vast majority of dams were gravity dams, a few were arched dams, and no example of an arch dam is known.'[38] A possible exception is the Roman dam at Glanum. It was certainly curved, and looks as if it was too thin (3.5 m), and hence light, successfully to resist the water pressure by gravity. It may therefore have been an arch dam, but if so it was unique, doubly so because of its early date (first century BC). Unfortunately, the remains are no longer to be seen, obliterated in 1891 by a modern dam built on the same site.[39] This inability, or at least reluctance, of the Romans to build arch dams is reflected in their siting. Gravity dams, being basically embankments, are long and low, arch dams tend to be short and high. They fit well into narrow gorges with a rock face on either side for ends of the dam to brace themselves against, while gravity dams are suited to broad shallow valleys. Faced with a river to be dammed, therefore, a Roman engineer often chose the second option where a modern engineer might pick the first. The dam would probably be long, low and straight, but might incorporate a curve or angle (Fig. 43); when it did, the reason was not, as is sometimes suggested, better resistance to

Name	*Earth or masonry*	*Max. height (m)*	*Max. length (m)*	*Purpose*	*Date*	*Source*
Cornalvo (Mérida)	E	15	194	aqueduct	Hadrianic	SR 22; HD 47
Proserpina (Mérida)	E	12	427	aqueduct	Trajanic	SR 23; HD 47
Alcantarilla (Toledo)	M/E	14	550	aqueduct	2C AD	SR 23; HD 48
Consuegra (Toledo)	M	5	664			SR 23
Esparragalejo (Mérida)	M	5	330			SR 23
Subiaco (Rome)	M	50	*c.*70?	pleasure lake; aqueduct	Nero	SR 22; HD 26-32
Glanum (St Remy)	M	12	18	aqueduct	1C BC	SR 25; HD 33-5
Kasserine (Tunisia)	M	10	150	irrigation	2C AD	HD 35
Wadi Caam (Leptis Magna)	E	7?	900	diversion from aqueduct	Septimus Severus?	SR 24; HD 37
Homs (Syria)	M	7	2,000	irrigation/ city supply	284 AD	SR 25; HD 39-41
Harabaqua (Palmyra)	M	21	365	irrigation	132 AD	SR 24; HD 39
Orukaya (Turkey)	M	16	40	irrigation		SR 25
Cavdarhisar (Turkey)	M	7	80	irrigation		SR 25; Schnitter 144

SR = N. Schnitter Reinhardt, 'Les Barrages Romains', in *Dossiers d'Archéológie* 38 (Oct-Nov. 1979), 20-5.
HD = Norman A.F. Smith, *History of Dams* (London, 1971).
Schnitter = N. Schnitter, 'A Short History of Dam Engineering', *Water Power* (1967), 142-8.

39. Table of dam dimensions.

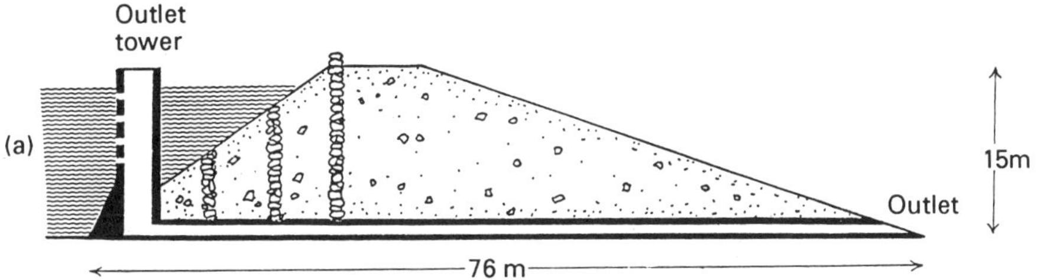

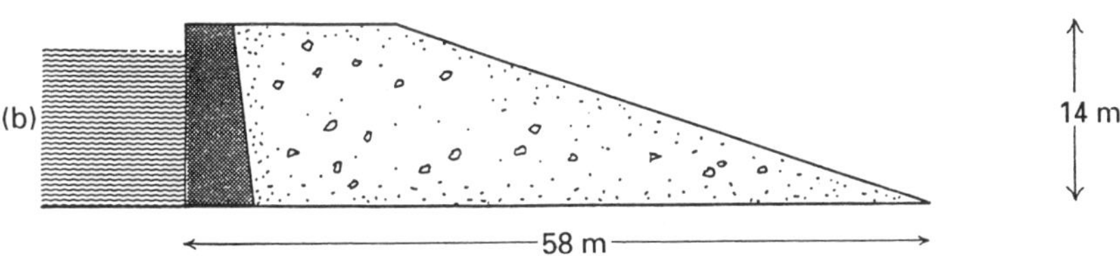

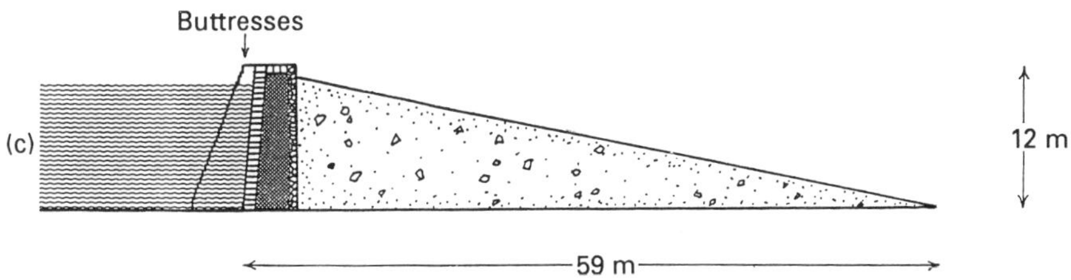

40. Roman dams, in cross-section: (a) Cornalvo; (b) Alcantarilla; (c) Proserpina.

water pressure, but more likely the availability of good foundations.[40] The dam would be either of earth (and would look like an embankment), or of masonry. As Smith puts it, 'All masonry dams which have been found follow the same basic layout. A rubble-and-earth core, sometimes consolidated with concrete, is covered with carefully cut and fitted blocks whose joints are sealed with very strong hydraulic lime mortar. The water faces of the dams are specially sealed with *opus siginum*, a type of

41. Proserpina, Spain: dam serving the city of Mérida.

plaster made up of hydraulic lime mixed with crushed brick or pottery. The air faces of the dams are stepped to a greater or lesser degree and many of them feature buttresses on the downstream side for added strength. In steep and narrow watercourses where good rock foundations were available, this was the "standard" Roman dam.'[41] In shallow and broad watercourses, especially in North Africa, an earth dam was more usual.

A dammed reservoir is subject to three main hazards. The most obvious and spectacular is a failure of the dam. This seldom seems to have happened in antiquity, though our evidence is scanty. A dam will burst for various reasons. It may have been insufficiently waterproofed, or not properly keyed into its foundations and abutments, so that the water pressure either undermines it or even pushes the whole dam bodily in front of it; or the dam may be simply too weak so that it bursts under the pressure from behind. The second-century AD dam at Kasserine (Fig. 42; 200 km south-west of Tunis, and supplying the town of Cillium), a masonry gravity dam, can be shown by calculation to have been, uncharacteristically, only just heavy enough to work, but apparently it did, though with no margin in hand. The Proserpina and Cornalvo dams, twin earth dams (probably Trajanic) serving the city of Mérida (Spain) were each reinforced by a heavy earth embankment piled up on the downstream side, giving the dam an almost triangular cross-section. This seems to have been an early anticipation of the later, non-classical buttress dams, where masonry buttresses reinforced the outer face of the dam in the same way. But, although bursting outwards is the familiar form of dam failure, the opposite is also a real danger. A dam designed to

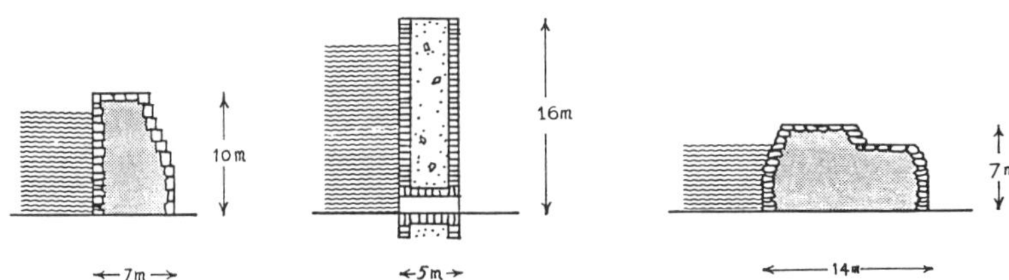

42. Roman dams at (left to right) Kasserine, Örükaya and Homs.

resist and contain water pressure may, if that pressure is removed because of low water level, become unstable and topple inwards into the empty reservoir. This actually happened to the second-century AD dam at Alcantarilla, near Toledo, though it is not known when, and it was presumably to avoid such a collapse that the engineers strengthened the inside face (i.e. the one under water) of the Proserpina dam with a series of buttresses (Figs 40(c), 44).[42]

The second risk to a dam is spillover. Because of heavy rains or other causes the water level in the reservoir may rise until it actually overflows, cascading over the crest of the dam and in effect turning it into a weir. If the dam has not been designed for this it may be seriously damaged; an earth dam may be partly washed away, perhaps enough to precipitate general failure. One of the very earliest dams known, the Sadd el Kafara dam in Egypt, failed from this cause shortly after completion.[43] Two precautions can be taken against spillover; in Roman dams they sometimes were, sometimes were not. One was the provision of spillways, usually one at each end of the dam; they took the form of a masonry escape-channel, set rather lower than the top of the dam, and normally closed by stop-logs set in slots which are often still preserved.[44] The provision of spillways in an earth dam, particularly susceptible to erosion from spillover, was quite crucial but was not always done.[45] The other precaution was to make the dam spillover-proof by cladding it, and particularly the crest, in a watertight masonry casing. This was done at Kasserine, in the evident expectation that the dam would regularly function as a weir once the water rose high enough. At Shustar in Iran, the river Karun was spanned by a third-century AD dam some 516 m long and 10–12 m thick, the Band-I-Mizan, which, being in a state of permanent overflow, was really a weir, intended to raise water level enough to serve the town. It was remarkable in that a road bridge across the river was then built on top of the dam, the water continuously cascading through its arches, and it is a signal tribute to Roman engineering that, with maintenance and renovations, this dam remained in service till 1885.[46]

The third risk is silting. It is not so much a hazard as a sure consequence that all reservoirs suffer, the only variable being the degree or the rapidity. A reservoir being in effect a giant settling tank, any sediment in suspension in the water as it enters is inevitably deposited on the bottom. In ancient reservoirs, indeed, this silt layer is often the evidence for how long the dam remained intact and the reservoir operational, and it brought with it various complications. The offtake sluices were normally located at the bottom of the dam, as they would have to be to exploit its full capacity, and were therefore usually the first thing to silt up. At Cornalvo we can see one possible way of correcting this. In the water close alongside the dam is built a 4.5 m square tower, or outlet well, with an internal stairway giving access to the bottom. From the bottom of the tower runs the outlet channel, through the dam. The sides of the tower are pierced with apertures at various levels, enabling water to be drawn from the reservoir and so reach the outlet even when the silt had progressively accumulated around the base of the tower. Moreover, the reservoir itself would slowly fill up with earth, reducing capacity and increasing the frequency of the water overflowing the dam. Eventually it would fill up completely, forming a flat tract of high-quality arable land with the dam acting as a retaining wall. Sometimes, in the wadis of North Africa and the Middle East, this was a welcome if unanticipated consolation for the disappearance of the reservoir; sometimes it was planned from the start, and the whole purpose of the dam was soil retention rather than impounding water, a technique particularly practised by the Nabataeans.[47] Silting also led to a further development. As the reservoir filled up, one way of retaining its original capacity was to heighten the dam, and many of the North African wadi dams bear additions of this kind. Further modification, such as lengthening, might also be required, for another hazard facing the North African dam builder in particular was that the wadi itself might change its course, either because the dam had somehow deflected it or because desert streams do tend to change course anyway.[48]

The purpose of the dams varied; often, indeed, we cannot even be sure which function they were intended primarily to fulfill. They might be intended for the provision of urban drinking water, conveyed by aqueduct. More often they were for irrigation, sometimes for flood control, and sometimes for soil retention. Often several of these functions could be combined, particularly in arid regions subject to sudden rainstorms: a dam across a wadi might normally have enough water in it to satisfy the needs of irrigation, while it would also have enough surplus capacity to accommodate a flash flood from a sudden rainstorm, a flood that would then both deposit in the dam any earth it was washing away and also be prevented from eroding any more on the downstream side. Frequently it is not even known exactly how the water was used – the Harabaqua dam, for example, certainly served Palmyra, but whether only for irrigation or whether some of the water was also channelled into the city remains uncertain. However, it does seem that the dams at Cornalvo, Proserpina, Alcantarilla and Wadi Caam (Leptis Magna) were built entirely as part of the city aqueduct system.

One can in general say that Roman hydraulic engineering showed little progressive development over the years. Details changed, and so did building materials, as ashlar gradually gave way to brick and concrete, but in the general principles of aqueduct layout and operation things were done very much in the same way in Augustan days as in the reign of Hadrian, or even Constantine, and the chief difference was the increased actual number and extent of the aqueducts. But dams were an exception. There was nothing revolutionary about the building of dams, and in the Middle East they long antedated the Roman empire. Within the empire, however, they really only begin to appear in any numbers around AD 100, and the Roman use of them to store drinking water at source (as opposed to irrigation or other agricultural dams) really does seem to have been a hydraulic revolution.

It will be fitting to close this account with notes on some of the more important and technically significant dams. First must come Subiaco, partly because it is the only one in Italy, partly because it is both early in date and exceptionally large. There were actually three dams at Subiaco, built by Nero across the river Anio so as to form along it three pleasure lakes as an adjunct to his villa there;[49] later, their waters were tapped by Trajan to serve the Anio Novus, apparently with some success since Frontinus speaks highly of their quality. All three are now gone, almost without trace, but it was the middle one that is chiefly of interest. It survived until 1305, and a painting of it (published by Smith) is still to be seen in the nearby monastery of Sacro Speco. From this and from the evidence of topography (e.g., the water level had to be kept high enough to flow into the Anio Novus) we can deduce that this dam must have been 40 m, perhaps even 50 m high. This not only makes it by far the highest dam in the world when built (as high as the Pont du Gard), but it still held that record for something like 1,500 years afterwards. The top of the dam is said to have been about 13.5 m thick and paved with tiles as a protection against overflow; it also dipped in the middle to provide a sort of primitive spillway. On top of the entire dam was then built a bridge giving access to Nero's villa.[50]

The best existing evidence is offered by three well-known dams in Spain – Alcantarilla, Proserpina and Cornalvo.[51] The first evidently provided drinking water for Roman Toledo, the other two for Mérida. A degree of progressive technical development is observable among the three. Alcantarilla[52] seems to be the earliest, datable to the early second century AD and of relatively crude construction, and may have been the first ever built to store water for urban supply. It was what we have called a masonry dam, formed by a concrete wall backed by a mound of earth on

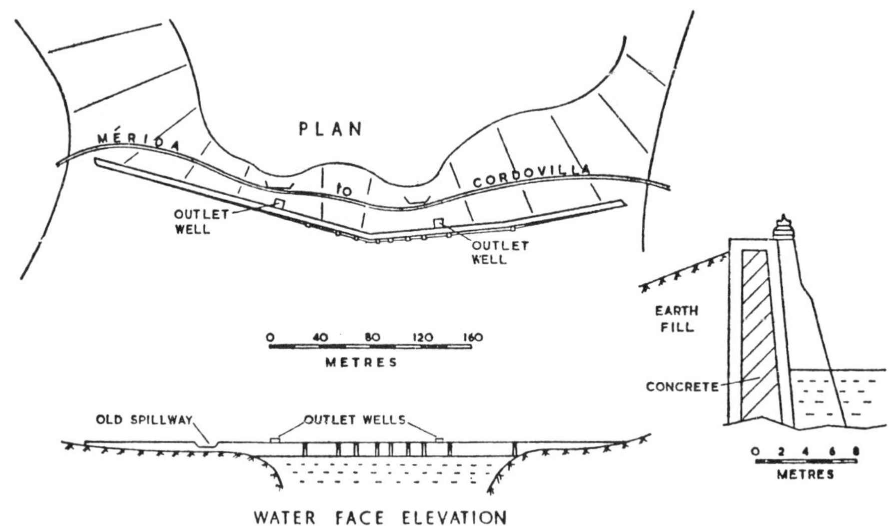

43. Proserpina dam: plan, elevation and section (N.A.F. Smith).

the downstream side (air face). The reinforcement it offered proved excessive, and at some time when there was no corresponding resistance because of low water level, the wall toppled over into the empty reservoir.

In the Proserpina dam[53] countermeasures were taken. This was a masonry dam, formed by a thick wall of stone-faced concrete with an earth mound again heaped against it on the downstream side to contain the thrust of the water. On the upstream side (water face), however, a repetition of the Alcantarilla debacle was prevented by a series of nine masonry buttresses propping up the dam face (Fig. 44). Although this does represent an advance in technique over Alcantarilla, one cannot of course say that it was directly inspired by it. Quite apart from uncertainty in the relative dating of the two dams, so that one cannot be sure that Alcantarilla came first, there is no telling at what date Alcantarilla collapsed, and it is only with the collapse that its lesson would become plain. The spacing of the buttresses is quite irregular, and in plan the dam itself is not straight, having two bends, one of 20° and one of 5°. This gives it the form of a very wide, obtuse-angled V, with the apex upstream, but although it has been suggested that this was done to resist water thrust,[54] the alignment was determined, as usual in such cases, by the availability of suitable foundations. Water was let out from the reservoir by two tunnels piercing the bottom of the dam. In each, the control sluice was set not at either end of the tunnel but at a point in the middle, access to it being by a vertical, stone-lined well (of section 5 × 4 m, and with internal stairway) sunk through the earth fill of the supporting embankment immediately next to the masonry dam wall. Of the two wells, one is 11 m deep and the other 17 m. Unfortunately, all this did

44. Proserpina, Spain: dam showing internal buttresses to prevent it toppling inwards into the reservoir when water level is low.

nothing to prevent the reservoir silting up and covering the water outlet tunnels, which it duly did to such effect that by the seventeenth century they were several metres underground. At this date new, higher outlets were built to clear the obstruction, and the dam continues in use today, though only for irrigation. The reservoir had a capacity of 10,000,000 m^3, making a striking comparison with the 25–30,000 m^3 of the Bordj Djedid cistern at Carthage or even the 325,000 m^3 of the reservoir 'of the 1,001 columns' at Istanbul.[55] Plainly, once the Romans began using outdoor natural reservoirs formed by damming instead of built cisterns in the city, their ability to store water and the whole concept of water storage and its role in the water supply system was changed radically. In fact, these new possibilities were only exploited from the second century onwards, from which date most dams, and then in certain areas clearly defined geographically: Spain, North Africa and the Middle East. Elsewhere the old system of continuous offtake, supplemented by the occasional storage of small amounts in cisterns, remained the standard practice, presumably because there rainfall, and hence springs, provided enough water to keep it functioning and there was no need to change.

The Cornalvo dam (Fig. 45) also served Mérida. It was built later than Proserpina and consequently further away, the nearer source being tapped first (the respective aqueduct lengths are 12 and 38 km). It is an earth dam with stone cladding on the water face; unlike Alcantarilla and Proserpina, it has no masonry dam wall. Accordingly both sides of the dam

45. Cornalvo, Spain: dam serving Mérida. The masonry tower in the water is ancient, and gave access to the outflow sluices.

slope, the air face at 1:3, the water face at 1:15. To bond the earth together a series of masonry crosswalls was built, intersecting to form a series of rectangular open compartments. The earth to form the dam was then piled around, into and on top of these compartments which thus held it together and keyed it firmly into position on the ground below. The earth slope on the water face fulfilled the function of the buttresses at Proserpina, preventing toppling at time of low water, and the silting problem was met by an outlet tower, as described above (p. 86). It is tempting to see Cornalvo as a direct development from Proserpina and an answer to the problems it raised.

Outside Spain, Roman dams are found chiefly in Tripolitania and the whole of the Middle East (including Turkey). In Tripolitania, apart from small earth dams across wadis to provide local irrigation, the most significant dams are grouped around Leptis Magna.[56] One of them, on the Wadi Lebda, was something highly unusual, a diversion dam, the purpose of which was not to store the run-off but to direct it into a drainage canal discharging into the sea, thereby preventing it from running straight down the wadi and silting up Leptis Harbour. The Wadi Caam, 17 km east of the city, provided the city's water supply and was also remarkable in that it took two dams to do it. Behind the lowest one accumulated a reservoir of drinking water, fed by local springs and running thence to the city by subterranean aqueduct. To preserve this clear water from pollution by the ordinary muddy run-off coming down the wadi, a second dam, no less than 900 m long, was built as a silt trap 2 km upstream, the water it collected being used for agriculture.

In the Middle East, dams had a long and distinguished history that went back far before the coming of the Romans, right to the ancient god Nimrod or Marduk, who was credited with having dammed the Tigris at Samarra in Iraq around 1750 BC. The earliest dams historically attested in the region were a series apparently built by the Assyrian monarch Sennacherib, who in 703 BC dammed the river Koshr, a tributary of the Tigris, to supply Nineveh; in 694 he added another two dams further upstream to increase supplies, and in 690 a fourth was added at Bavian on the river Atrush. All four were masonry dams or, more properly, weirs, since the water was intended to overflow, and had the peculiarity that they were not aligned straight across the river but at a very oblique angle, presumably in a search for good foundations. In later times we have already noted the extensive use of earth-retention dams by the Nabataeans to create arable land in the Negev, a technique evidently borrowed from them by the Romans for use in North Africa.[57]

The greatest of Roman dams in the Middle East is beyond question that on the Orontes at Homs[58] in Syria, built by Diocletian in 284 and, after various renovations, still in service. A masonry dam, formed by a rubble core faced with basalt ashlars, it was no less than 2 km long, and of height and thickness varying according to locality, to a maximum of around 6–7 m high and 7 m thick (at top), 20 m (at base). In alignment, it is built in the form of a very flat V with the point facing upstream. This has nothing to do with resistance to water pressure; the dam is founded upon a long ridge of basalt that itself follows this course. The reservoir thus impounded, known as the Lake of Homs, had an original capacity of 90 million m^3 (now 2,000 million, after an addition in 1938 to raise the level of the dam), used mostly for irrigation but perhaps also for urban supply.[59] Remarkably, there has been very little silting.

Other smaller but still notable Roman dams are to be found in the same area at Harabaqua (18 m high and 200 m long), and at Quasr al Khubbaz (serving the needs of a frontier garrison on or near the Euphrates). At both the reservoir is now completely silted up.[60]

Two further structural points remain to be noted. One is a peculiarity of Turkish dams. Three such exist, at Cavdarhisar (Aezani), Örükaya, and Böget; all are masonry dams, with a core of clay revetted in a casing of large stone ashlars. The peculiarity consists in the fact that sometimes molten lead was run in to seal the joints between the blocks and make the dam waterproof, in place of the cement or concrete more usually employed for this purpose in Roman work.[61] The dams were also quite unusually thin in cross-section. Örükaya (Fig. 42) indeed looks incapable of containing the water accumulated behind it and seems to have survived only by good luck, probably assisted by its two ends being very firmly keyed into the sides of the valley. Our second and final point is of much greater importance. In the reign of Justinian (AD 527–565 and therefore only marginally within our period), Chryses of Alexandria was employed to build a flood-control dam (apparently also used for urban water supply) at Daras, on the Turkish-Syrian border about 30 km north-west of Nusaybin. No trace of it survives, but there is a full description in Procopius,[62] who includes the vital phrase that it was not straight but built 'in the form of a crescent, in order that its arch which was turned against the stream of the water, might be better able to resist its violence', also noting that the ends of the dam were 'mortised into each of the two cliffs' between which the river flowed. This is a clear description of an arch dam with the two vital points clearly stated: the dam is curved specifically to resist water thrust, instead of following a more or less curved line simply because that is where the best foundations were, and, second, the thrust is contained by the two abutments, braced like feet against the side walls of the gorge, not by the immovable weight of the dam or the security with which it is fixed down to its foundations below. If we except the questionable and fragmentary evidence of Glanum, Daras is the first historically proven arch dam in the world and represents perhaps the greatest breakthrough in dam engineering ever achieved. The thrust of the accumulated water is now for the first time contained on scientific principles, whereas the gravity dam may be said to rely simply on plain brute force. The difference is akin to that marked in architecture by the replacement of the traditional Greek column-and-lintel construction by the Roman arch and, by an economy in the material used, makes possible dams that are much larger or cheaper. Given the Roman mastery of the arch, and indeed the concrete vault, there is nothing surprising about them adapting the principle to dam construction, and it would be rash to ascribe the invention to Chryses. There may have been many earlier such dams, now disappeared; indeed, had Procopius' text not survived, we would not know about Daras either.

In the absence of such evidence, however, we must content ourselves with recognising the superiority of the Roman dam-builders in other fields. Smith justly summarises: 'Works of this sort were made possible by the Romans' ability to plan and organise engineering construction on a grand scale. That they were able to build so many big dams, many of which have lasted such a long time, was also a result of their evolving better methods of construction based on better materials, especially hydraulic mortar and concrete. Moreover, they paid proper attention to hydraulic problems, being careful to ensure that the water could not percolate through their dams and that, when it overflowed them, spillways were provided to prevent erosion if the crest itself was not suitably protected. All in all, Roman dam-building shows a completeness which was not emulated until the nineteenth century.'[63]

5
The Aqueduct

It must be admitted, this is the greatest wonder the world has ever seen.

Pliny, *NH* XXXVI, 24, 123

The aqueduct channel

We have now tapped the water supply at source or spring, and turn to the means whereby it was transported to the city – the aqueduct. To most people, familiar with pictures of the great bridges and arcades, arches and aqueducts are largely synonymous. One forgets the rest of the aqueducts, for which Vitruvius[1] specifies three types of conduit: masonry channels, lead pipes and terracotta pipes. In fact, the arches, despite their prominence and fame, are always a rarity considered against the total length of the Roman aqueducts. And in spite of Vitruvius' offer of a triple choice of materials for the conduit (as we shall see, there were actually not three but six materials in common use), there was only one generally employed in the circumstances we have envisaged. We can thus produce, as it were, a portrait of a typical section of Roman aqueduct, and this reconstruction will be true of some 80 to 90 per cent of the total mileage of all Roman aqueducts everywhere in the empire.

The typical Roman aqueduct was a surface channel. By this I mean a conduit that closely followed the surface of the land, instead of being raised on arches or sunk deep beneath it in a tunnel. To protect the channel it was usually some 50 cm to 1 m below ground (as in Fig. 46), and on this basis is often, not unreasonably, described as a subterranean aqueduct. Today it is intermittently traceable in spots where the channel has collapsed or been interrupted by modern excavations, such as roadworks; or where it perforce emerged from underground in order to maintain its level in uneven terrain. Sometimes it was equipped with vertical inspection manholes at regular intervals which reveal the presence and course of the channel below. But such a structure does make the conduit hard to detect from a surface survey, and whole aqueducts of this type have remained untraced, though known to exist.[2]

Presumably, being only a metre or so underground, the channel was normally built on the 'cut and cover' principle – that is, by digging an

46. Cologne, Germany: a stretch of the Eifel aqueduct, showing vaulted roof and an inspection manhole; normally subterranean, this section was uncovered during road-building (photo: K. Grewe, Bonn).

open excavation, constructing the conduit, and then covering it over again, as opposed to tunnelling proper.[3] The channel was built of masonry, and since the great majority of aqueducts, especially in the provinces, were built during the empire and hence post-date the widespread adoption of concrete in Roman structures, concrete and rough stone construction is common. So is brick. Cut and fitted stonework, such as we see in the Pont du Gard or the arcades of the Aqua Marcia, is very rare in a surface, or subterranean, channel.[4]

In cross-section the conduit normally formed a tallish oblong: its size naturally varied, but the figures for the Marcia (90 cm wide by 2.40 m high) and the Brevenne (Lyon: 79 cm by 1.69 m) give the general proportions.[5] The roof was usually formed by a vault (Fig. 46), sometimes recessed back from the tops of the vertical side walls to form a ledge on which the wooden centering could rest during construction, sometimes flush with them. Less commonly the channel was roofed with a flat stone slab, or occasionally a pair of flat slabs tilted in to meet and form a pointed roof[6] (see Figs 48, 49).

The reason for these proportions must be clearly understood. The channel had to be accessible to a man for maintenance and cleaning, and

47. Pont du Gard: the conduit of the Nîmes aqueduct, a short distance to the south of the great bridge. The channel, here running at ground level, has lost its vaulted covering but clearly shows the thick *sinter* incrustation on its sidewalls.

it was this factor that governed its size, not the volume of water to be carried.[7] In fact, the channel was normally only half to two-thirds full of water (which has the incidental result that calculations of the aqueduct's discharge cannot be based on a simple cross-section of the conduit), and was never intended to carry more. Overflows in the side walls were occasionally provided to ensure that it did not, though sometimes this might happen by accident as the water blocked back before some bottleneck or obstruction.[8]

The floor and side walls were lined with waterproof cement, though sometimes not all the way up to the top but just as far as it was thought the water would actually rise. Sometimes in the two bottom corners of the

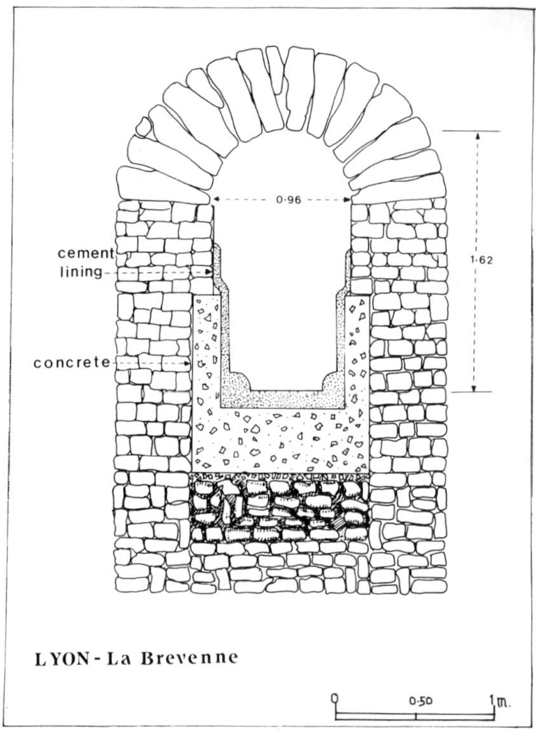

48. Lyon: (above) cross-section of typical aqueduct, showing vaulted roof, foundations and cement lining of channel (aqueduct of La Brevenne); (below) channel with corbelled roof. The top cover slab could be lifted to facilitate cleaning (aqueduct of Mont d'Or) (Cl. Lutrin).

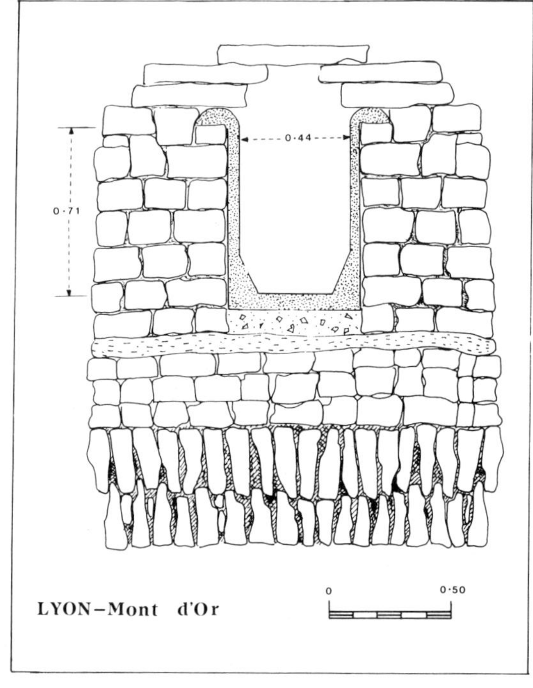

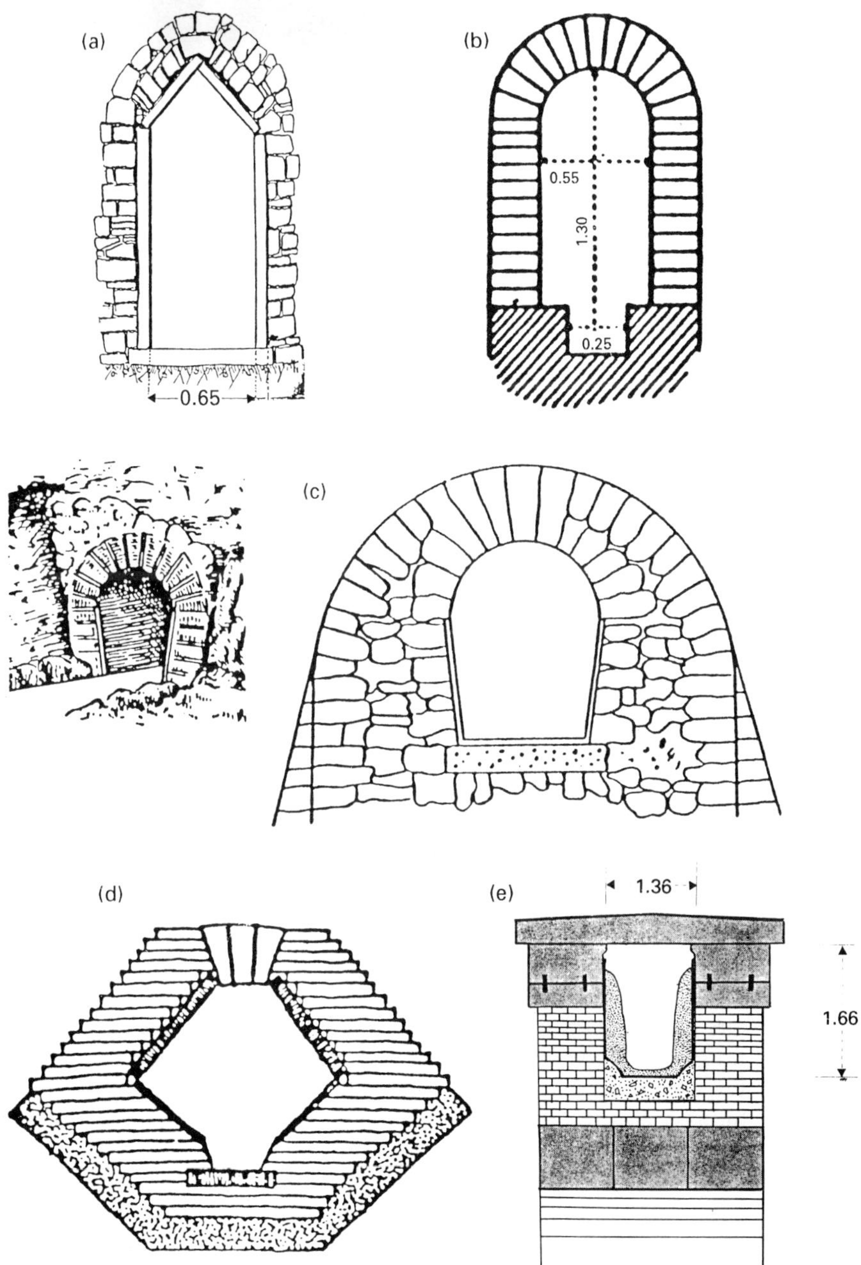

49. Aqueduct channels of various shapes: (a) Ain Ferhat, Algeria: gabled roof; (b) Serino aqueduct, Pompeii and Naples; vaulted roof, with sunken channel in floor to carry the water. The actual water channel is about 0.25 m^2, the rest of the tunnel being to provide access for maintenance; (c) Fréjus: horseshoe-shaped conduit; (d) Alteburg, Rhineland: hexagonal- or diamond-shaped conduit; (e) Pont du Gard, Nîmes: rectangular channel, showing also calcium carbonate deposit.

channel, the cement formed a continuous beading or 'quarter round', as a further reinforcement against cracks developing lengthwise at their vulnerable points.[9]

The function of the cement was threefold: to make the channel impervious to leaks or seepage; to provide a smooth, friction-free contact surface; and to make this surface continuous and uniform with no joints from one end of the aqueduct to the other. As well as performing these functions it also had to be strong and resistant to cracking, whether from heat expansion, freezing, or other causes. Both the composition of the cement and its installation were therefore complex and specialised tasks to which only recently has proper study been devoted. There are three vital features to be noted. First, the material itself is characteristic and easily recognised, looking like either an exceptionally fine concrete or a rather rough-grained cement. It is basically a mixture of quicklime and either sand or crushed brick, looking yellowish or reddish in colour depending on which is used. Second, it is laid in not one, but several layers, like plywood or coats of paint, with pauses between to let it harden; for the whole to set fully took about three months, and probably depended on the weather and season as well.[10]

The number of layers and their composition depended on local needs and circumstances. In the aqueduct of Caesarea (Israel), built around AD 100, Malinowski has identified and analysed no less than six separate layers, forming a total thickness of some 2.6–3.0 cm.[11]

Third, once the last layer has been laid it is carefully rubbed and polished 'to mirror-brightness'. The process as described by Vitruvius refers to stucco applied to the walls of buildings, but something like it seems to have been done to the cement lining of aqueducts. The purpose was threefold. First, as Malinowski points out, this ultra-smooth surface 'inhibits the formation of lime sediments on the walls of cisterns and aqueducts due to a better flow of water. The removal of such sediments during maintenance work is also easier' – that is, it is easier to split off the calcium carbonate incrustation from the channel walls if they are smooth and glossy. Second, this hardens the top layer, which thus acts as a protective skin for those below it. Third, in the rubbing and polishing process the particles of lime and marble assume a horizontal orientation which lends strength to the material and prevents cracking from shrinkage. The full elaboration of this process was not always necessary, because 'for closed canals and tunnels where there is continuous moisture and no danger of cracks caused by shrinkage, the problems were solved in an easier way without such careful polishing'.[12]

On top of the cement lining came the incrustation (Figs 49(e), 159, 160). Not all aqueducts carried this unwelcome addition, for it depended on the quality of the water. The water could carry two kinds of impurities. If it was rather dirty there might be sand, mud or other very small but solid bodies carried along in suspension in the water. The other kind was lime, chemically dissolved in the water, and separating out to form a calcium carbonate incrustation on the channels and pipes it flowed through. Water falling as rain is completely pure, so the presence of lime depended on whether it had afterwards percolated through limestone before eventually surfacing in the form of springs to feed the aqueduct. Usually it had, for Roman cities generally seem to have been built in areas where the catchment area was composed primarily of sedimentary rocks, and limestone is indeed retentive of water and hence productive of springs. No geological study of the Roman empire with reference to the hardness of its water has ever been carried out, but the truth is indicated by the end result – heavy incrustation is a standard feature of Roman hydraulics.[13]

This thick lime deposit is itself often built up of clearly discernible layers, not unlike those of the cement liner, except that this time the layering was fortuitous. Should the flow of the water at any time be interrupted and the channel left dry, the exposed surface of the lime deposit would, of course, acquire a skin of weathering. When the flow resumed so would the build-up of incrustation, but with the line between the layers clearly visible. Normally this incrustation dates from the mediaeval period when maintenance and cleaning was at a minimum, and the commonest reason for interruption of the supply was the upheaval of war and siege. It is possible to relate the sequence of layers to the actual historic events that caused them, almost on the principles of dendrochronology, though no one has yet tried to identify and equate a similar sequence of layers in two or more different aqueducts. They are particularly prominent on the aqueduct between the Pont du Gard and Nîmes (Fig. 50).[14]

Whether of layered structure or monolithic, the incrustation could become a serious impediment if allowed to build up long enough. For Aventicum (Switzerland) it has been calculated that with a 3 cm thick incrustation in a channel 24 by 40 cm, discharge was there cut by two-thirds, from 16,600 litres per minute to 5,600.[15] The aqueduct at Nîmes carries an incrustation up to 47 cm thick on the floor and each side of a channel only 1.20 m wide to start with, leaving little more than a narrow slit still clear down the middle. But such a drastic reduction in the aqueduct's capacity did not necessarily reduce the amount of water still coming down from the springs, with the result that at many points along the route it was in a state of copious and permanent overflow (compare Fig. 51 (Perge)).[16]

The build-up of this deposit could only be prevented by chipping it away as it formed, which accordingly became a constant and never-ending duty for the waterworks staff.[17] This is why the channel had to be large enough for human access. In closed lead pipes where no such access was possible this accretion may have been cleared by some sort of pull-through device operating between inspection holes in the pipe (this is commonly done in modern work); or the problem was perhaps simply given up as insoluble

50. Nîmes: edge-on view of the *sinter* (calcium carbonate deposit) accumulated on sidewall (left) of aqueduct conduit, near the Pont du Gard. The sandwich-like layered structure of the accretion is clearly visible.

51. Perge: north side of main street: mass of *sinter* formed by overflow from aqueduct. The vertically fluted appearance was caused by the installation of terracotta pipelines to tap the supply; the water, continuing to leak, then led to ridges building up between them, so that when the pipes eventually fell down or were removed (a short section is still visible *in situ*, bottom right), the tracks where they had been remained as open 'flutes'.

52. Trier: inspection shaft and manhole (A. Neyses).

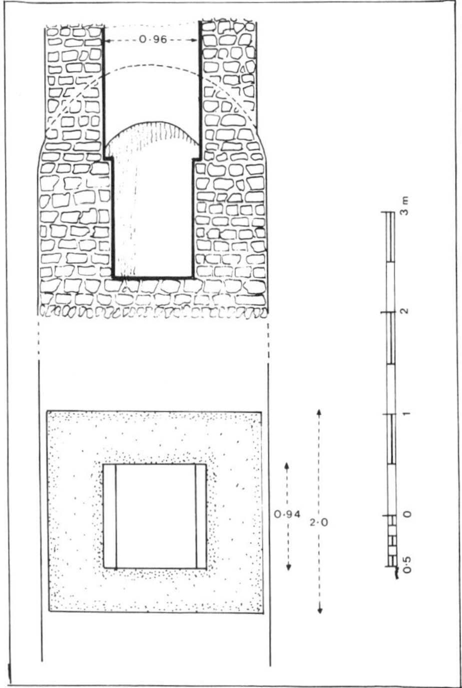

53. Cologne: Eifel aqueduct, inspection shaft at Buschhoven (A.T.H., after W. Haberey).

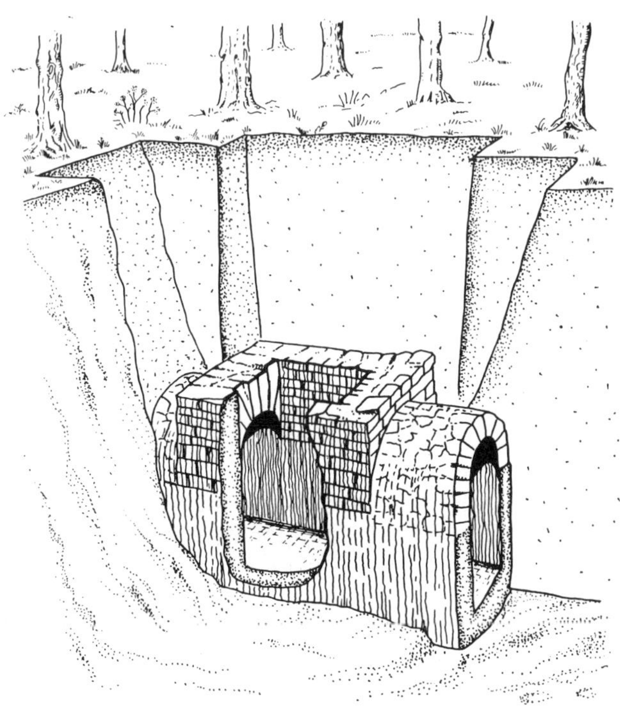

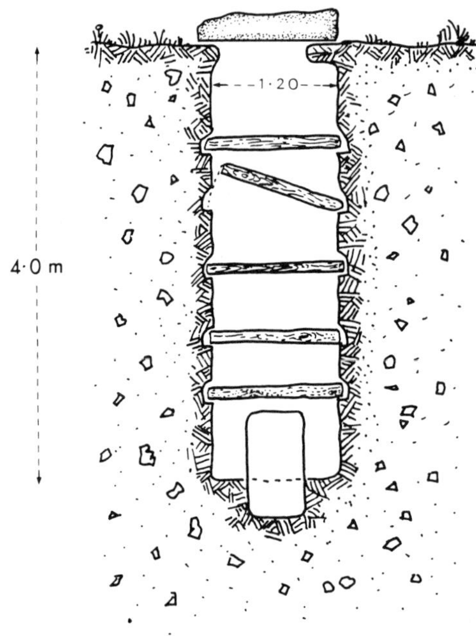

54. Arles: inspection shaft with crossbeams forming access 'ladder' (A.T.H., after Stübinger).

and the pipes completely replaced once they had become fully choked.[18]

When the channel was roofed by flat cover-slabs (see n. 6), these could simply be removed as required, continuously if necessary, to give access for cleaning. When, as was normal, the channel was vaulted on top, access had to be provided by manholes, or, if it was some distance underground, inspection shafts (Figs 46, 52, 53, 54). They are found on the channel of the Carthage aqueduct even where it is running raised above ground, prominently projecting some 70 cm up above the rounded top of the channel like the conning tower of a submarine. Understandably, there is no set distance apart for the manholes (though Vitruvius specifies an interval of one *actus* (= 35 m), VIII, 6, 3), but one usually finds them at intervals of 75 m or so. They often come closer together when there are special circumstances, such as the conduit being on a curve or sunk in a tunnel. Sometimes they were specifically located to facilitate needed repairs.[19]

The openings are usually round, sometimes square (particularly at the bottom of a rock-cut shaft when the channel is in a tunnel); it is not known whether the Romans were influenced by the one great advantage a round manhole has over a square one – it is impossible to drop the lid through the hole.[20]

The manhole was normally fitted with a lid, often in stone (as in Fig. 55). These were very heavy, weighing as much as 300-400 kg and frequently required a crane to lift them. Handles for ordinary manual

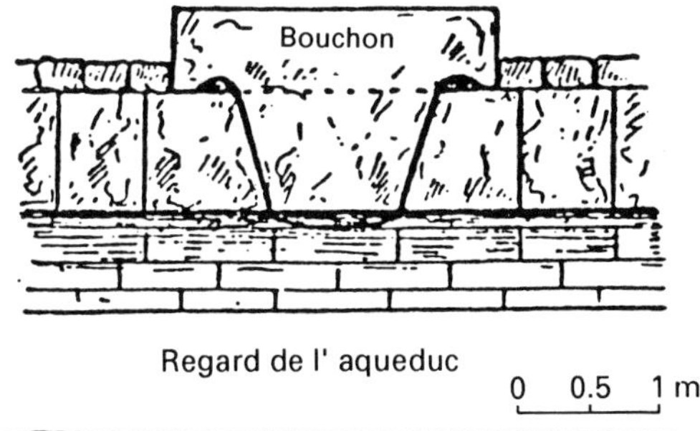

55. Geneva: stone manhole cover, showing tapered fit and recessed handgrips under the edge (L. Blondel).

lifting would have been pointless. At Carthage the manholes, now without covering, are reasonably supposed to have had wooden lids, and this may also have been common practice.[21]

The other kind of impurity, sand and the like in suspension in the water, was normally screened out by settling tanks arranged at intervals along the line, which do not concern us here. Sometimes, however, these were supplemented by small settling pits, recesses a foot deep or so and set in the floor of the ordinary channel. They could be round or square and do not seem to have been very common. Examples are preserved on the Brevenne (square; Fig. 57) and at Carthage (round),[22] where they have been explained on the ground that the aqueduct could maintain only a very shallow gradient (actually 1.51 m per km) as it crossed the plain to the city of Carthage, and sediment accordingly would be very prone to settle on the bottom of the conduit because of the low velocity of the water. The presence of such settling pits would, by itself, require close and convenient access by manhole for the removal of the accumulated sediments.

Alongside the aqueduct channel its course might be marked by *cippi*, usually about 71.3 m (=240 Roman feet) apart. A *cippus* (literally 'a gravestone') was a small stone marker set into the ground, and it performed two functions: where the channel ran underground, the *cippi* marked its location, and since they were numbered, like milestones, they gave the maintenance staff a convenient point of reference to any particular section of the line. They might also indicate the limits of the dedicated right-of-way through which the aqueduct ran, and upon which encroachments by neighbouring proprietors were prohibited.[23]

Cippi seem to be found only on the aqueducts of Rome, and not on all of them. They were apparently instituted by Augustus, who installed them on existing aqueducts such as the Marcia and the Anio Vetus, which had previously lacked them, as well as on new construction and renovation.[24]

However, though the *cippi* gave an easy framework of reference, it looks as if it was not actually much used. Frontinus, no doubt following the common practice in the water office, regularly describes any location on the aqueduct line, not by its *cippus* number, but by the corresponding

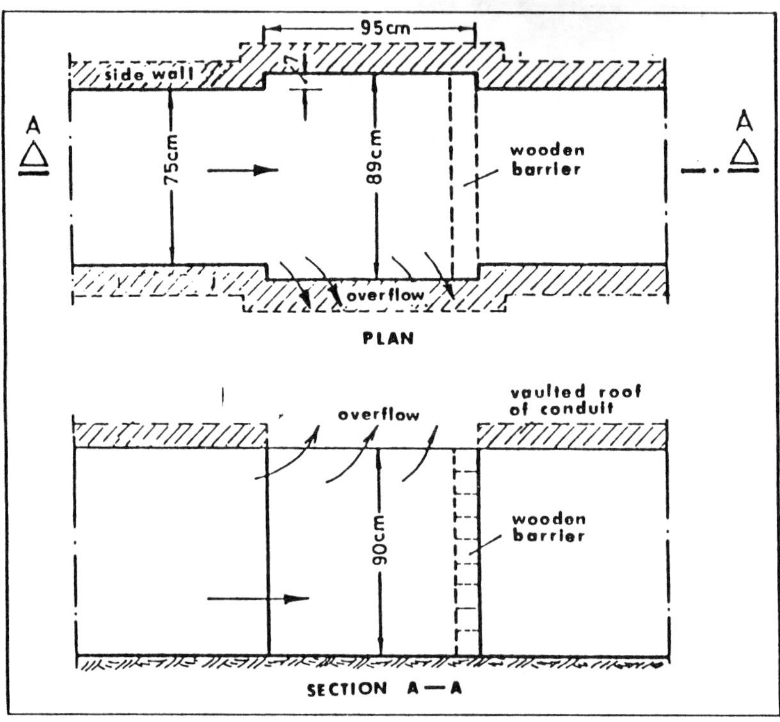

56. Lyon: overflow chamber on Gier aqueduct. When the ongoing channel (right) was shut off by the wooden barrier of stoplogs, the water was diverted to overflow through the opening in the roof and upper sidewall of the conduit (Jeancolas, 1978).

milestone on the nearest main road.[25] This was not as unlikely as it sounds, for roads and aqueducts often paralleled each other and the road was in any case the normal route of access for the maintenance crews and the milestones on it their accustomed point of reference. At Rome, moreover, roads and aqueducts had originally been under the charge of the same magistrates (the censors or aediles).[26] Where no road was convenient, a service road might be built alongside.[27]

The final point to be noted about our typical section of aqueduct is that it is probably running at right angles across a hillside. Aqueducts tend to be built through the hills, and not just because of the traditional purity of mountain springs. This is in fact true, but more important than purity is height. The water had to be brought in at a level high enough to serve all parts of the city. This was always liable to cause problems, particularly where the city had some kind of an acropolis, such as the Palatine and Capitol at Rome or the heights of Fourvière that formed the core of ancient Lyon, and it usually meant that water had to be sought from a source in the hills even if there was an adequate, but lower, one closer at hand. During its passage through the hills the line of the aqueduct

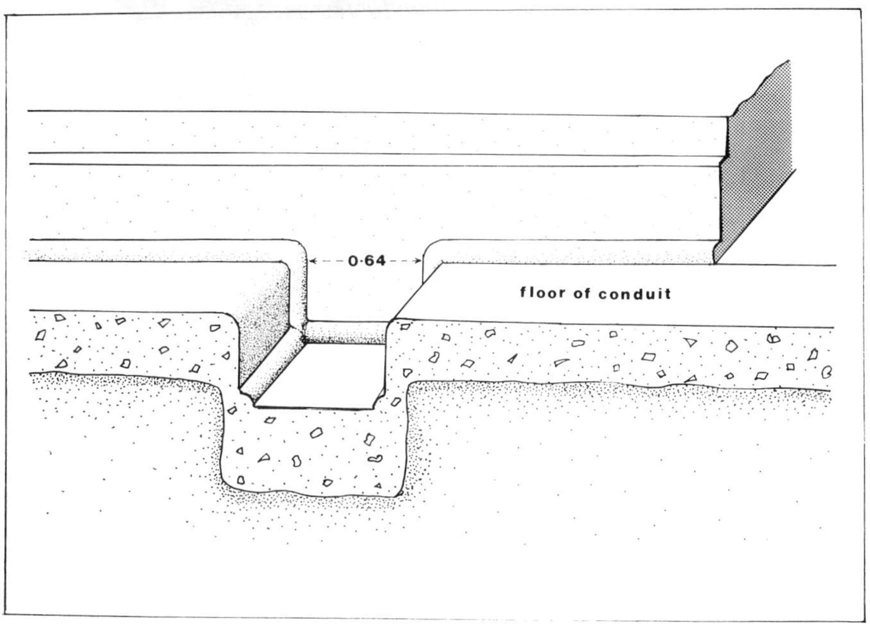

57. Lyon: settling basin set into floor of aqueduct channel at Sotizon, La Brevenne aqueduct (A.T.H., after Cl. Lutrin).

normally remained level so far as was possible (if we neglect the slight slope required to maintain gravity flow), like a canal or a contour line on a map. Therefore, just because it did more or less follow the contour lines, so did it usually have the land higher on one side than the other. Typically, this led to trouble, for the masonry conduit, submerged perhaps a metre or so below ground level, was liable to act as a kind of underground dam, a continuous concrete wall some two metres or so high, against which rainwater, percolating downhill through the topsoil, would accumulate. A conduit was therefore liable to be accompanied on its uphill side by a continuous band of heavily sodden earth, the water coming not from leakage out of the conduit but from surface water outside. In winter this was then liable to freeze (since, unlike the water in the aqueduct, it was not moving), cracking the conduit. The problem could be solved by providing adequate drainage (at Cologne there was a drainage canal running parallel to the aqueduct and slightly uphill from it), but this was not always done.[28]

Other forms of conduit

The open channel

The masonry conduit just described is by far the commonest form that

Roman aqueducts take, but there are five other possibilities that must be mentioned. The closest is the stone-cut channel and its earthen equivalent, the open clay-lined leet. The other four are pipes, of various materials, which therefore present rather different problems.

That the normal aqueduct channel should sometimes become rock-cut whenever rock was encountered in its path goes without saying. If underground, it then might or might not be also masonry-lined. Minor aqueducts, 'particularly in mountainous or sparsely inhabited country',[29] or those for agricultural or industrial (as opposed to urban) use, often followed this pattern, running in a rock cutting where necessary, and otherwise in a dug channel, like a stream, across the fields. Like streams, and unlike the upright, oblong cross-section of the masonry conduits just studied, they were usually squarish in section, or even wider than they were deep, and the water was liable to fill only part of the channel.[30] It seems unlikely (though not really known) that they were continuously covered over, but attempts could be made (apart from the usual settling tanks) to minimise sand, earth and other impurities getting into the stream by lining the banks with clay and the bottom with 'stretches of smooth laid stones'.[31]

On hillsides, if rocky, the channel could be formed by cutting back the rock on the uphill side and forming the downhill side of the channel by building a continuous masonry wall, or sometimes insetting vertical cut stone slabs (compare Figs 58, 59). On an earthen slope the same revetting effect could be achieved by shoring the channel up on one side with wooden planking. Streams or small gullies could be crossed by carrying the water across in a wooden trough supported by trestles, or even, if the supply was small enough, in wooden pipes made of hollowed-out trees.[32]

It may well be that on some of these small aqueducts (particularly in Mediterranean and North African lands where flash floods are common) stream or wadi crossings were often formed of some temporary wooden structure deliberately regarded as disposable and provided in the full expectation that it might be swept away by the spring or autumn rains and have to be replaced as a matter of course.[33]

Many of these industrial and irrigation aqueducts were no doubt small and minor in scope – some modern equivalents in the *falaj* of Oman are illustrated in Costa's article (n. 32 above) – but industrial needs could be quite extensive. The Dolaucothi aqueduct was in places 1.5 m or so wide, and carried a volume of water roughly equivalent to the Aqua Alsietina at Rome. Despite its industrial purpose, some importance was also attached to purity, hence the stone or clay lining and settling tanks.[34]

Pipes

Pipes can be made of terracotta (earthenware), lead, wood and stone. The use of all four is attested in Roman aqueducts, though in differing

58. Side, Turkey: rock-cut channel of the aqueduct (photo: H. Fahlbusch).

59. Side, Turkey: rock-cut channel carrying the Roman aqueduct (indicated by arrows) along the gorge of the river Manavgat, immediately below the modern dam.

60. Hammat Gader (Golan), Israel: stone pressure pipe in the Roman Baths.

61. Miletos, near the Agora: block from stone pipeline, with right-angle bend.

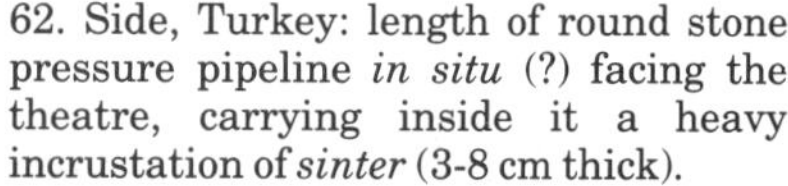

62. Side, Turkey: length of round stone pressure pipeline *in situ* (?) facing the theatre, carrying inside it a heavy incrustation of *sinter* (3-8 cm thick).

63. Segovia: re-used block from a stone pressure pipeline (arrow) embedded in the arches of the Roman aqueduct.

64. Mainz: typical length of wooden pipe (courtesy C. von Kaphengst).

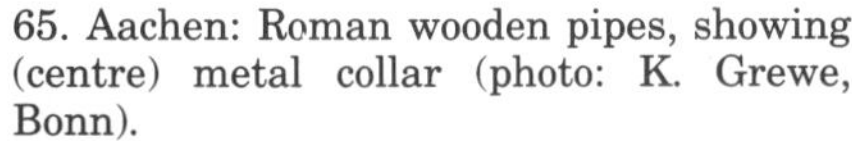

65. Aachen: Roman wooden pipes, showing (centre) metal collar (photo: K. Grewe, Bonn).

amounts and varying circumstances. There is no evidence for bronze pipes, other than the adjutages (*calices*) at the offtake points of urban domestic lines, and the long-standing suggestion that the Pergamon siphon was laid with bronze pipes, apparently based on nothing more than conjecture but still repeated in some works, is now known to be wrong: the pipes at Pergamon were of lead.[35] Despite Vitruvius' suggestions, on the country aqueducts lead pipes were used only for siphons, where metal pipes were needed to withstand the pressure generated. In the city distribution network, on the other hand, they were the standard thing and will be discussed in Chapter 10.

Stone pipes also occur, and they are not as rare as used to be thought. Usually they are of large size, for the main aqueduct rather than the various small-gauge branches of the distribution network, and normally consist of a series of stone cubes each bored through with a round hole, which fitted together so as to form a continuous tube, while externally looking like a course of squared ashlars; sometimes, however, these outer surfaces have been left rough (Figs 17, 60). The joints were shaped to fit together on the male/female principle, much like the ends of terracotta pipes to be discussed below, and once one knows what to look for, it is not uncommon to find individual blocks of this sort lying around sites where a stone pipeline has not been published or otherwise attested. Their commonest use seems to have been in siphons (for which see pp. 33-41 above), but they were also found in other positions (such as some of the internal city mains) where the water ran under pressure, even low or nominal pressure (Figs 11, 61). Special blocks incorporating a right-angle elbow joint or T-junction are relatively common, as are those featuring vertical 'vent-holes' – the inverted commas are advisable, as their actual function, to me at least, remains uncertain.[36] Round stone pipes (instead of the cubical format) are rare but do occur (Fig. 62), though one must here, if possible, be careful not to confuse them with the round stone sleeves ('*marmormuffen*') sometimes used, as at Ephesos, to accommodate the joints on a pipeline otherwise made of lead; similar isolated stone blocks sometimes occur at bends.[37]

These stone pipelines are particularly numerous in Turkey and the Middle East, though examples do occur in Italy, North Africa, and even as far west as Baelo (ancient Bolonia), near Cadiz on the Atlantic coast of Spain; and what looks like a re-used block formerly associated with such a pipeline is even to be found embedded in the arches of the aqueduct at Segovia (Fig. 63). The best known, and most accessible, such pipleine is the siphon at Aspendos (which ran in a single stone pipe), but the best preserved are probably those at Patara and Laodicea: the Patara pipeline exists almost in its entirety (Fig. 12), and at Laodicea there is a section of over 25 blocks still *in situ* and all joined together. As previously noted, the dating of these stone pipelines, whether in siphons or not, is unclear. Traditionally they have always been considered Hellenistic (at least the

66. Wooden pipe hollowed from tree-trunk, with square access hole on top (to assist cleaning?) (A. Neyses).

67. Terracotta pipes: various forms, from (1) Nemi; (2) Wiesbaden (mediaeval?); (3) Worms; (4) Strasbourg; (5) Jagsthausen; (6) Mainz (Samesreuther).

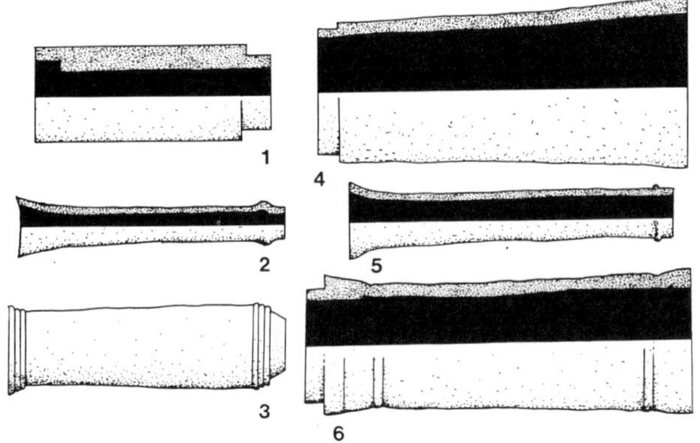

ones in the east), but a stone pipeline is in itself not intrinsically an easy thing to date, and it now appears that they may all be Roman work.[38]

Wooden pipes, which were again much more used for urban networks than the main aqueducts, but were common in small isolated systems individually supplying villas, forts and the like, were particularly numerous in northern Europe, especially Germany (Figs 64, 65). In Britain they replaced lead piping as the normal practice and at Caerwent were even used in a siphon, where several of them burst under the pressure and were repaired with small lead patches. In spite of a common belief that 'wood rots and splits', this is in fact incorrect. Wooden pipes do not decay as long as the wood remains constantly wet, and it is only an

68. Ephesos: a stack of terracotta pipe sections of different kinds; apart from size, differences include presence or absence of internal *sinter*, and varying thickness of sidewalls, depending on whether they were intended to be subjected to pressure.

alternation of wet and dry that causes rot to set in.[39] A wooden pipeline normally consisted of tree-trunks with a hole bored longitudinally down the middle (Figs 64, 66). This ensured an actual water channel of neat, uniform cylindrical bore, but an irregular exterior as the trunks were usually only roughly trimmed and always retained, from the natural profile of the tree, one large end and one small. The joints were normally reinforced with circular iron collars, which have often survived as the only evidence for the aqueduct. At Caerwent a series of collars was found *in situ*, alternately of three inches and four inches diameter, indicating that here the sections of piping were regularly laid in pairs, with the two large ends together, and the two small ones likewise. This would of course facilitate the jointing, and we may presume that such a sequence was common if not invariable practice. The pipes (or trunks) were most commonly made of oak (Pliny, *NH* XVI, 81, also recommends pine, pitch-pine, and alder),[40] and come in lengths varying from around 1.5 to 7.5 m; the hole for the water could be from 5 to 10 cm in diameter. The whole pipeline might then be embedded in packed clay, with supporting stone slabs set under the joints. Internal cleaning might be difficult, but we sometimes find wooden manholes or access shafts, so that a cleaning device could be pulled through the pipeline, or its contents and condition at least inspected.[41]

The advantage of wooden pipes was of course cheapness and in Northern Europe ready availability. Trees did not come in standard sizes the way lengths of lead piping did, nor did the bored holes always match

69. Ephesos: terracotta pipelines beside the Library of Celsus. The various sections are of differing size (note the small one to the left end of the upper pipe), showing that these pipelines are late work, made of re-used materials.

the various gauges of pipe so rigidly specified by Frontinus, but this was not a problem that affected small, self-contained aqueducts where the question of drawing a prescribed amount of water from a larger system did not arise.

Another advantage was the length of the sections, reducing labour costs on the joints (this must have been a considerable factor *vis-à-vis* the very short lengths used in most terracotta conduits, as we shall see). Of course, the use of long, straight sections also brings a loss of flexibility in laying out the line of the aqueduct, and to follow a serpentine course the smaller terracotta sections would presumably be more suitable. Samesreuther also maintains that a wood surface reduces friction on the water-flow, accelerating the velocity.[42]

Terracotta pipes (Figs 67, 68, 69), after a masonry channel, were the commonest form for an aqueduct to take. They are found both in some of the smaller main-line aqueducts and in local urban distribution systems, to say nothing of drains. Contrary to traditional belief, they were also sometimes used for pressure lines (identifiable when one notes that the pipe walls are exceptionally thick relative to its diameter). They were the one all-purpose form of conduit, and in a study such as this it is hard to know where they would most appropriately be considered.[43]

The individual sections are normally around 40-70 cm long and with an internal diameter of up to 10 or 15 cm. The length was presumably often limited by their having been thrown on a potter's wheel, nor were they usually completely cylindrical, for the joint was usually effected by having one end slightly narrower so that it could slide inside its larger neighbour (with an appropriate flange or groove to help seal the joint).

70. Mt Beuvray, France: pipeline made of re-used amphorae, with tops and bottoms removed so that they fit into each other. Photograph by the excavator, J. Délechette, in 1907) (photo © Beuvray).

Though this need not have affected the internal diameter, in fact it usually did, and a terracotta conduit seldom has a completely parallel internal bore. We may also in passing note (Fig. 70) a unique curiosity from the Gallo-Roman site of Bibracte (Mont Beuvray, near Autun, Burgundy), which boasted a pipeline made entirely of re-used wine amphorae, their tops and bottoms knocked off so that they fitted snugly into each other at the joints.[44] I know of no parallel to this. The short length of terracotta pipes meant a very large number of joints – the terracotta pipelines of the Pergamon aqueduct had 200,000 of them and used so many pipe-sections that they were evidently manufactured not only on the site but at several different locations along the aqueduct, as attested by differences in the clay.[45] Terracotta pipes are also often provided with an opening on top, covered by a removable lid. The purpose of this may be simply to facilitate cleaning; its effect is clearly to destroy any possible claim that this be considered a pressure pipeline: once the pipe ran full, plainly the water would begin overflowing through the holes along the top, so that a pipe constructed in this way must be thought of as operating on the same general principles as an open channel under gravity flow. Sometimes every single section of pipe has an opening and it is plain that in this case the intention was to enable the workman to get his hand into the pipe to plaster from the inside the joint with its neighbour.[46]

The joints were fitted into each other and sealed with a plaster, in composition somewhat like the waterproof cement lining of masonry channels. This has again been studied and reproduced by Malinowski. Its basic ingredient was quicklime, *calx viva*, from which was produced an ' "expanding rim" sealant'.[47]

Pipes: the line of the aqueduct

Although it was not common for the Romans to follow Vitruvius' suggestion and actually build an aqueduct carried in pipes throughout, it did occur. The use of pipes entailed a radically different approach to the routing of the aqueduct. Unlike the normal masonry channel, which was designed to run only half-full and under gravity flow, pipes, except where their integrity was destroyed by inspection openings on the top as just mentioned, were intended to run full. This in turn means that a pipeline would run under at least nominal pressure. We are not here speaking of the high pressures generated in a siphon or any of the other locations that have so worried modern commentators. All that is meant is that, unlike an open channel, a pipe need be laid only roughly level. Minor bumps and dips in the line are of no importance. The water will surmount them without trouble, so long as the general tendency of the line is gently downhill, and though this will inevitably cause some changes in pressure within the pipes, the changes will be so small and the pressure so low as to be insignificant and easily contained both by terracotta pipes and even their plastered joints. In theory, this means that a pipeline is probably easier to lay out than a conventional channel. For there is a much greater tolerance in establishing the level, nor are the results of an error so drastic; 50 cm too high could result in an aqueduct overflowing and the rest of the channel running dry, but could be accommodated in a pipeline. A pipeline would also certainly be shorter for it need not adhere so slavishly to the contour. It can cut corners, and its course need not be so serpentine – it is rather like the difference between building a railway and a canal. This does not mean, of course, that the pipeline followed a straight line with all the pressures and/or engineering works involved, only that it was more flexible than the orthodox channel. The disadvantage of a pipeline was its small capacity, as compared with a masonry channel. This meant two things – that, the size of pipes being limited, they were often laid in groups of two or three (on the analogy of lead siphon pipes, which were often laid in batteries of up to nine); and that their use was normally restricted to small aqueducts.

It must also be admitted that this issue is not nearly as clear-cut as it might appear, for discharge through a pipe depends on velocity, which itself varies with circumstances, and these circumstances varied much more with a pipeline, which might be running under a head of pressure varying at different points along it, than with a conventional channel running under more or less uniform gravity flow. The point is well

71. Caesarea, Israel: 'flat' siphon on north aqueduct, with three terracotta pipes laid in the bottom of the open conduit they replaced.

demonstrated by the Hellenistic aqueduct of Pergamon which, on the main section of its course from the springs at Madradag to the settling tank at Haghios Georgios, ran through an orthodox, unpressurised, set of three terracotta pipes (16-19 cm inner diameter). From the settling tank on, the aqueduct went into a siphon, in which one single pipe of similar dimensions (actually 22 cm) apparently handled all the water delivered by the other three. There is no sign of any other offtake or overflow at the tank, and theoretical calculations have confirmed that the flow-capacity of the triple-pipe conduit and the single pipe siphon are in fact the same.[48]

Probably the best example of a line entirely laid with terracotta pipes is the Kuttolsheim aqueduct, serving Strasbourg.[49] It was 20 km long and laid with a pair of terracotta pipes, 30 cm in diameter, running side by side and usually about 1 m deep in the ground, an adequate protection against frost in these northern regions. It is described as having 'une pente variable mais continue' and, where necessary, crosses depressions not on a built masonry *substructio* of the familiar type, but on an earthen embankment created by digging a pair of ditches parallel to the desired alignment, one on either side, and heaping up the excavated material in the middle. The 'pente variable mais continue' is exactly the kind of arrangement that can be accommodated by pipes but which might present difficulties in an open channel, making this a good example of a simple aqueduct, unambitious in scope but efficiently engineered.

Another possibility is the use of pipes to replace a section of aqueduct channel that, for one reason or another, had become unserviceable. This seems to have happened at Lyon and also at Caesarea Maritima in Israel, where at some time a stretch of the aqueduct, here running on arches, was replaced by three earthenware pipes (17 cm in diameter), running along the actual bottom of the open channel and embedded in a mixture of mortar and small stones (Fig. 71). They reduced the carrying capacity of the aqueduct by 90 per cent – from 38,400 m^3 per day to 3,600 – and it is suggested that the substitution was required because of subsidence: the aqueduct structure sank in the soft ground, leading the water to overflow and cut off supplies to the channel on beyond. The replacement by pipes turned this section, in effect, into a long and very shallow siphon, and no doubt 3,600 m^3 was much better than none at all. It puts it, after all, into the same general range as the Pergamon siphon. We may also note in passing that at at least two points this siphon was interrupted by being 'brought up for air', to a surface level tank (see Figs 170, 171 below) after the fashion of Les Tourillons, on the Craponne aqueduct.[50]

Curves, junctions and settling tanks

As it followed the contours through broken terrain, the line of the aqueduct often developed an extremely sinuous course, as sinuous as the contours themselves. The greatly increased distance made no difference to discharge or real efficiency in the aqueduct, for the longer time the water took in getting to its destination was plainly of no importance to anyone; but it made an enormous difference to maintenance costs, both of the perennial task of cleaning the channel free of incrustation, and of repairing cracks, leaks and even landslides. The increased length could also bring down the aqueduct gradient below acceptable levels, resulting in stagnation of the water. These provisos apart, Roman aqueducts often do meander (in striking contrast to Roman roads), entailing plenty of gentle sweeping curves.

Sudden abrupt curves, sharp angles and acute bends, one thinks, might be a different matter. Would it not assist the flow to have the course streamlined? In fact, abrupt bends are a rarity in piped supplies. Terracotta pipes in the form of a right-angled elbow bend have rarely been found, and the more common technique was to achieve the more gentle sweeping curves mentioned above by laying short, straight pipes, with the curve being taken up in the joints. Another procedure occasionally used to negotiate a corner was to instal at it a large, round *pithos* or other jar, with the terracotta pipeline emptying into it through a hole in one side, and leaving by another offtake pipe cemented into the jar at right angles. The jar thus served both to eliminate inertial thrust as the current rounded the bend, and as a silt trap, the accumulated debris being removed through its open mouth.[51] Another possibility was to

72. Miletos: stone angle block *in situ* at bend on terracotta pipeline.

introduce at a bend on a terracotta or lead pipeline a single, isolated stone block pierced with an L-shaped hole. The bend was thus contained entirely inside the block, which provided the requisite strength to resist the centrifugal thrust as the water negotiated it (Fig. 72).

But the conventional masonry or rock-cut channel was a different matter. There was no reluctance to build or cut a channel to form an abrupt angle – even right angles are not rare – and that often where it would have been quite simple to introduce a curve into the line, at least for a metre or so on either side, to ease the current into its change of direction.[52] One often finds them at one end (or both) of an aqueduct bridge where the conduit, running along one side of the valley, goes round a sharp corner to continue out on to the arches.[53]

Sometimes the bridge is a later addition to cut off a circuitous loop in the line; it is then liable to form a T-junction at each end where it leaves and rejoins the original line, and if half of the T (the abandoned loop) is then blocked up, one has a right-angled bend even if this was not in the original design. This happened at Cherchel, and one may see something like it at Saintes.[54]

It also sometimes became necessary to introduce a bend in the middle of an aqueduct bridge or arcade, and here again the bend is usually an abrupt one, though it would have been easy enough to build the arches in a curve. On such occasions the Romans may have been aware of the consequent sideways thrust on the masonry structure at this bend – at Aspendos they strengthened it with buttresses, at Segovia (Fig. 73), in similar circumstances (but with a very small conduit), they did not.[55]

Irrespective of the effect on the bridge structure, however, there cannot be much doubt about its effect on the water flow. These sharp angles, if

73. Segovia: angle in the arcade. There are no buttresses to counter centrifugal force at the bend.

the current was running fast, must have created considerable turbulence and acted as something of an obstruction. The Romans apparently did not mind, perhaps rightly. Uniformity of flow was never attained anyway in a Roman aqueduct, but will 'always be in a state of change, trying to adapt to the new conditions'.[56] To put it another way, obstructions in the way of changes of channel width and gradient abounded, so that the current was always either being already affected by the next one to come, or still recovering from the last one. The turbulence created by right-angled elbow bends may therefore be less important than one would think.

In the open country (as opposed to the city *castella*), junctions occur when additional channels join the main line[57] of the aqueduct as tributaries, sometimes added later to increase the discharge; and where branches leave it to serve other locations. To generalise, the first are most numerous near the beginning of the channel, in terrain where there are other springs to be tapped. The second usually occur on or shortly before arrival at the city, where there are other suburbs to be supplied. On the middle section of an aqueduct – one might say the middle 80 per cent of its length – junctions of either kind are rare. Once it has collected all possible water from local springs, and until it reaches the urban area, the aqueduct usually runs through as a single line, so junctions do not arise. When they do, they are of two different kinds. First, the two channels, whether masonry or piped, may join directly in a T or Y-junction. Second, without actually joining, they can be arranged both to empty

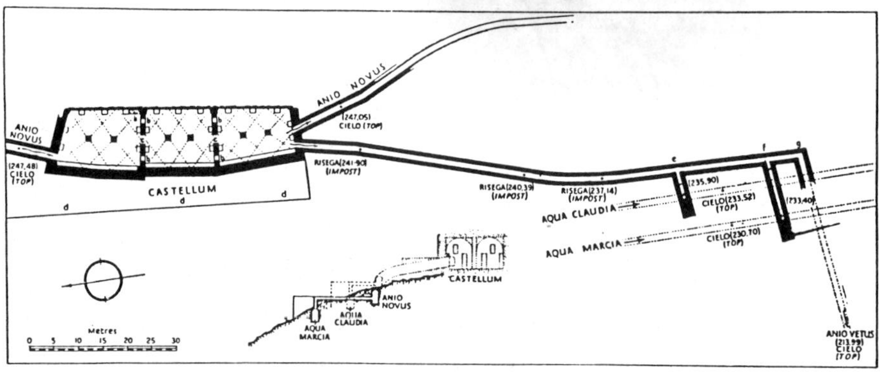

74. Rome: Grotte Sconce, a settling tank and junction chamber on the Anio Novus, showing crossover link for diverting water down into the Marcia and Claudia (Ashby).

independently into some large reservoir or receptacle, in the other side of which is the offtake for the continuing main line of the aqueduct.

To consider first junctions of the direct type, few if any of these are to be found on piped aqueducts. They are common enough on pipes of the street or urban distribution systems, and will be considered there. By direct junctions in conventional masonry channels, I refer to Y or T-junctions where one channel simply runs straight into the other, such as at Osteriola junction on the Anio Novus. At such a junction each of the incoming lines may also be fitted with a small clearing tank just before the junction point. Other more complicated forms of junctions are found when it is a question of putting in some sort of cross-link from one aqueduct to another, at a point where they run nearby, to enable water to be diverted. The complication comes from the fact that the two channels are hardly ever at the same level (which means that flow through the link is not reversible, and it is only the upper aqueduct that can feed the lower), resulting in arrangements such as we see at Grotte Sconce (Fig. 74) and S. Cosimato on the Rome system. At the first, a pair of cross-links from the Anio Novus supplied the lower-lying Claudia and Marcia, in each case apparently joining them by coming in vertically through the roof. At S. Cosimato, the Claudia passed directly above the Marcia and the two were joined by a vertical shaft, opening apparently out of the side of the Claudia and regulated there by a sluicegate, and falling straight down and into the Marcia through the roof. This shaft, or *puteus*, on the Marcia, which was built first, was 1.2 by 1.08 m, and some 9.2 m deep (Fig. 75); the water thundering down it must have been an impressive sight when it was in operation.[58]

Indirect junctions on piped aqueducts usually take the form of some sort of small box into which the ends of the pipes are introduced.[59] In orthodox channels the junction can take the form of a large reservoir,

75. Rome, Aqua Marcia: view directly up the vertical shaft at S. Cosimato, through the roof of the Marcia channel to the channel of the Claudia, 9.20 m above.

rather after the fashion of a *castellum divisorium*, with the various channels entering or leaving it as required. It may also double as a settling tank. Thus at Grotte Sconce the Anio Novus entered a large triple-chamber vaulted reservoir, from which were led off the cross-link channels to the Claudia and Marcia just mentioned:[60] from the opposite corner left the Anio Novus, continuing on its way, but dropping 0.43 m in level as it left. The water of the Anio Novus was notoriously sandy, and no doubt this reservoir constituted an attempt to get rid of as much of the impurities as possible before (when this was necessary) diverting them into the purer waters of the Marcia and Claudia, so that this constituted both a junction and a settling tank combined and conceived to serve a single purpose (as well as achieving a drop in level). An analogous arrangement can be seen at Grüngürtl on the earlier Cologne aqueduct. A junction here being necessary, the opportunity was seized to combine it with two large settling chambers, a complex array of regulatory sluices, and a drop in level of 2.1 m, this time only on the branch line.[61]

A simpler and no doubt commoner layout is to be seen, also on the Cologne system, at Eiserfey junction on the Eifel line (Fig. 76). This is a shallow, circular open tank of 3.05 m in diameter, its paved bottom 2 m below ground level. As restored, it is surrounded by a low guard wall with a sandstone coping, and much resembles the well known *castellum* at Nîmes. At floor level the main aqueduct from Urft, running in a vaulted masonry conduit just below ground, enters through an arched opening and leaves through a similar but larger one on the other side on its way to

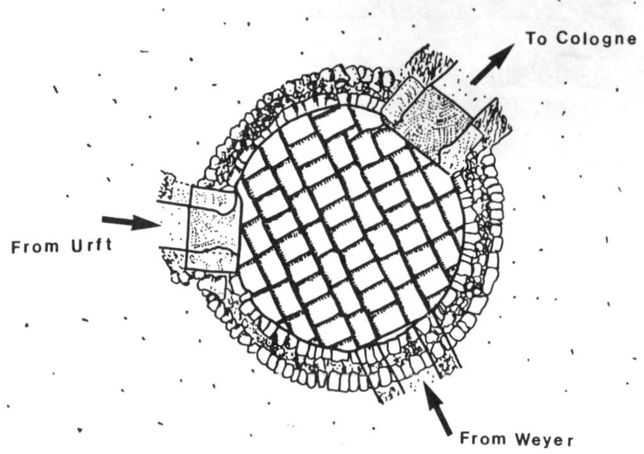

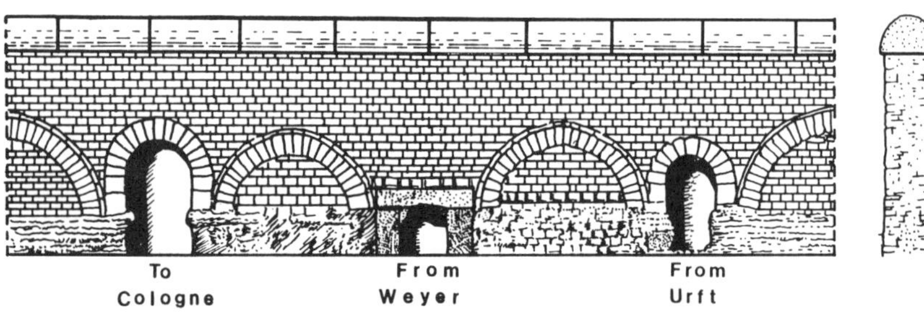

76. Cologne: Eifel aqueduct, Eiserfey junction chamber and settling tank. Above, plan: the main line of the aqueduct arrives (left) from Urft and continues on to Cologne (top right). At the bottom a branch line from Weyer, running in a much smaller tunnel, joins in. Below, restored elevation. The whole tank was open to the air (W. Haberey).

Cologne. Between them a short tributary branch from Weyer enters through a small square aperture. The three channels are spaced at equal distances around the circumference of the reservoir, and enter and leave it flush with floor level so that there is no question of this having been intended as a settling tank. Its function was presumably threefold: to effect a junction conveniently and in a manner avoiding, so far as possible, excessive turbulence at this confluence, which could harm the conduit lining; to give ready access for inspection and maintenance at a key location in the network; and to provide an impressive public display. This last hypothesis rests on the tank being open to the sky, and hence to view; the whole appearance suggests a place of public visit and resort (as was the Nîmes *castellum*), although it was of course in the open country and not an urban setting as at Nîmes. It may also have been used as a convenient place for drawing water.[62]

It may well be that something like this was the commonest form of indirect junction, where such was used, but the evidence is really very thin. Usually actual junctions on aqueducts are either not preserved or not identified and excavated, and their existence rests on purely topographical evidence: that is, enough is preserved of the two channels to give the alignment and it is known that they must have joined somewhere, but the location of the junction is conjectural; so, therefore, is its form.[63]

Settling tanks were also found along the main stretches of the aqueduct, though, like junctions, they occurred more often near the beginning, to clear the water on leaving the springs, or near the terminal *castellum* and the urban network.

On the short (8.5 km) aqueduct at Siga (Algeria) there was a series of at least twenty small basins (some oval, some round, and of about 1 m^3 capacity). Each was located not on the aqueduct line but close beside it. Water was brought into the basin by a short connecting channel, and left it by a second one, at right angles, back into the aqueduct, so that the basin was situated on a sort of loopline or by-pass.[64] Was this a settling basin or an access point from which local farmers could draw water? We cannot tell, and hopes must rest on finding a similar installation elsewhere that will throw light on the matter.

Large settling tanks are both rarer and more common – rare because any given aqueduct will have only a few of them as compared with the twenty basins at Siga, more common because plenty of aqueducts do have them, while the Siga arrangement remains unique. Again their function is not wholly clear. A good example of such a circular tank has been found at St Cyr, near Lyon; it is 3.5–4 m in diameter and 0.8 m deep.[65]

Often we find simply an enclosed vaulted reservoir, and it is hard to tell whether purification was part of its intended function, since on going through a tank designed for any function – such as a junction with a branch line arriving or leaving – the current *would* slow down and impurities settle, whether that was the chief purpose intended or not. Rarely do we find an open settling chamber, as at the 'Grüne Pütz' at Cologne (Fig. 37), where the underground channel leading from the springs into the aqueduct conduit is interrupted by a rectangular tank, open to the sky and surrounded by a decorative coping, on much the same principle as the junction reservoir already noted as Eiserfey and possibly again serving as a local point from which to draw water.[66]

On the other hand, Frontinus does specify that the measuring equipment for recording the discharge, whatever it is, is often installed and operated in a *piscina*, a small reservoir and hence presumably a settling tank. Indeed, he specifically identifies one such as being ('ubi … linum deponunt', *Aq.* I, 19), adding that this particular one is covered ('contectis piscinis'); he also adds that there is measuring equipment in it. In spite of this, I would be inclined to think that it would be so much easier and more satisfactory to manipulate the equipment in the open, where one could see what one was doing, than in the dark and awkward

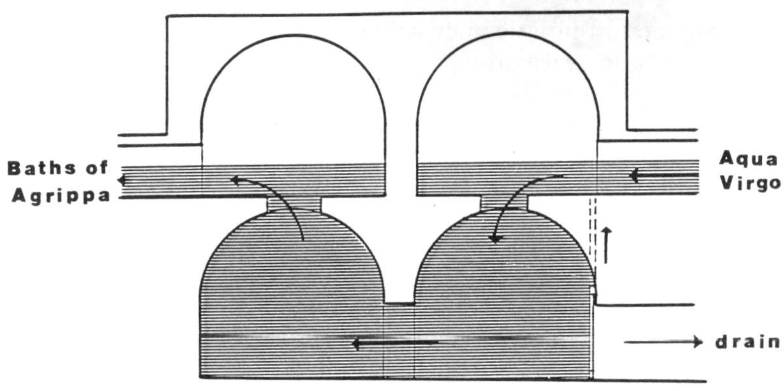

77. Rome: settling tank complex on the Aqua Virgo, near the Via Latina (J-P. Adam).

confines of a covered tank full of water, that the fact that the measuring equipment was often mounted at a settling tank[67] would indicate that many of them were uncovered. I must readily grant, though, that not many such have been found.

The normal form is a barrel – or cross-vaulted chamber or set of chambers – through which the water makes its way (Fig. 77). The importance of the settling may be gauged from the pebbles found in the settling tank of the Anio Novus (a notoriously impure stream) at the Villa Bertone, Capanelle, shortly before its entry into Rome. The pebbles are the size of a pea and completely round, having apparently made the journey all the way from Subiaco in the Sabine Hills along the bed of the Anio Novus, a distance of some 70 km. Even more impressive is the amount. Not only was the tank entirely full of them when discovered, but with them the modern owner of the property, a certain Cav. Bertone, 'has gravelled about a kilometre of avenues'. The mound upon which his villa stands also 'is artificial, and owes its origin to the refuse of the cleaning tank of the Anio Novus', from which was also forthcoming so much sand that 'Cav. Bertone was able to make plaster for six or seven buildings on his farm'.[68]

Nothing could speak louder for the vital role settling tanks played in the Roman aqueducts than this great heap of rubbish filtered out of the Anio Novus: for though not all streams were as impure as this, neither is there any way of telling how often the tank had already been cleared out and the pebbles carted away in antiquity before this enormous heap began accumulating in the period of neglect. We may also note that at its source the Anio Novus drew directly on the waters of the river Anio. This proved too muddy, and so Trajan realigned the intake to draw from the middle of three large pleasure lakes which Nero had created at Subiaco by damming the river, and which thus now acted as gigantic settling tanks for the aqueduct; the great mass of sand and pebbles that ended up at the Villa Bertone was thus only the final residue that escaped this screening process, from which we may with profit speculate on what would have been the result had there been none at all.

6
Engineering Works

PATIENCE – SKILL – HOPE

motto on the inscription of Nonius Datus, aqueduct engineer

Hitherto we have been considering the normal course of the aqueduct as it followed the natural contours of the land, on or just below the surface. This procedure was adhered to whenever possible, but sometimes engineering works could not be avoided. They fall into five categories: tunnels; surface works such as low embankments (*substructio*); bridges or viaducts; siphons; and continuous arcades across the plains.

Tunnels

One often finds references to a tunnel as going under a mountain. Strictly speaking this is never done. If there is a mountain in the way one goes round it. The place one normally finds tunnels is piercing ridges, watersheds, saddles between two mountains, and projecting spurs that it would be uneconomical to contour around.[1] The longest tunnel on a Roman aqueduct was probably that on the direct cut-off of the Anio Novus from Osteriola junction to Fosso di Ponte Terra via Valle Barberini. Running from the Valle S. Gregorio to Gericomio, it was at least 2.25 km long[2] but no trace of it survives and its existence is attested only by the presence of an otherwise impenetrable ridge of hills across the line of the aqueduct. Shorter tunnels – say from 50 up to 400 m – are not uncommon. If possible (as in the Halberg tunnel: Fig. 78), they were naturally constructed by sinking a number of vertical shafts (*putei*) and tunnelling in both directions from the bottom of each.[3] The chief advantage of this technique was that it provided ventilation to the workmen, and by multiplying the workfaces it accelerated progress, but it also meant that the shafts could afterwards serve as access and inspection shafts, while during construction the line of shaft-mouths at surface level was a useful check on alignment. Several of these shafts are still preserved and visible. Where they are not, the excavated spoil and deposit around the shaft heads often survives as a prominent series of patches of soil differently

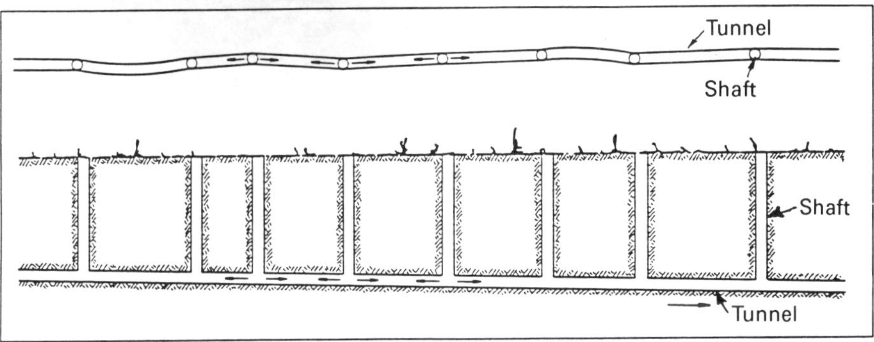

78. Halberg, Germany: tunnel on the Halberg aqueduct, showing correlation between vertical shafts and irregularities in tunnel orientation. Above, plan; below, section (arrows show direction of work) (Keller, 1965).

79. Arles: manhole at top of inspection shaft, sealed by stone cover slab (Stübinger).

coloured from the surrounding topsoil, providing a useful guide to the course of the tunnel underneath. The mouth of the shaft would usually be protected with a stone slab covering it, perhaps sealed down with a protective strip of cement round the edges, removable when the shaft was to be opened. It could also be protected by a waist-high wall around it, so that it looks like a well (Fig. 79). Vertical access in the shaft could be by footholes cut in the side walls or by wooden ladders.[4] All other things being equal, the tunnel normally ran in a straight line, the shortest route.

80. Lyon, Soucieu: typical section of *substructio*, here carrying the conduit of the Gier aqueduct, made of concrete and faced with *opus reticulatum*.

Sometimes, however, other things were *not* equal. The Mornant tunnel (500 m) on the Gier aqueduct was on a curve, and the Crucimèle tunnel (400 m) on the Nîmes aqueduct zig-zagged, apparently following the strata of rock easiest to cut.[5]

Moreover, it might be that the ridge to be tunnelled under rose too high, so that the vertical shafts would be too deep to be practical. In such a case, the multi-shaft approach would perforce be abandoned and the tunnel driven in one continuous bore. It seems to have been normal to begin at both ends simultaneously and meet in the middle. This cuts the working time in half, as two workfaces are available, but orientation is then critical to ensure that both halves do in fact meet. We have already seen how in the Greek tunnel on Samos they almost missed, and in the well-known inscription from Saldae in North Africa, a certain Nonius Datus, an army engineer, complains bitterly of a tunnel where the failure was complete. Although nominally in charge, his supervision seems to have been minimal: the two work crews missed each other altogether, and, since they kept on digging hopefully long after the half-way mark was passed, had almost ended up with two tunnels instead of one when Datus finally intervened to save the situation.[6] This complication would of course be avoided by tunnelling from one end only. Presumably the saving in time made the risk acceptable, particularly since Roman aqueduct tunnelling involved a great deal of unproductive labour in any case. This was because, just as the dimensions of the conduit were determined by the need for human access for maintenance, so, even if the aqueduct was a piped one, the tunnel had to be big enough for the men cutting it to work in.[7] Accordingly, the actual water channel was sometimes no more than a small conduit cut in the floor of the tunnel proper and the remaining tunnel space, excavated at such enormous time and cost, was, in a sense, wasted. (Compare the Serino channel, Fig. 49(b).)

Surface works

Little need be said under this heading, and it is included only because, though straightforward and requiring no exposition, they were fairly common. Frequently there was a need to carry the aqueduct a few feet above ground level, and, unless it was carried on an earthen embankment as at Kuttolsheim (p. 116 above), the normal answer was a *substructio*. This was a built masonry structure (Fig. 80) resembling a thick wall, with the aqueduct running along on top. It was often faced with brick or *opus reticulatum* and up to 1.5 – 2 m high; above that height an arched structure would be used.

Bridges

Bridges were used whenever there was a valley to be crossed and going all the way round was, for one reason or another, impractical. Bridges are therefore usually comparatively short and to be found in the hill country where the aqueduct has its source. They are essentially different in concept and purpose from the long lines of arches that often are to be seen crossing the plain as one approaches the city.[8]

In size, they range all the way from the meanest little culvert to monsters towering 50 m high, such as the Pont du Gard (Figs 81, 82), which have always been considered some of the greatest and most spectacular monuments of Roman civilisation. However, the 48.77 m height of the top of the Pont du Gard above the waters of the river Gard below does evidently represent the maximum the Romans thought attainable for bridges; above this, siphons were used instead.[9]

Originally, bridges, like other structures, were built of squared ashlar blocks. This was true down to around the end of the first century BC. All the bridges on the Marcia are so constructed, and so is the Pont du Gard. The actual stone was naturally sought as near by as possible, so that the various bridges along the length of an aqueduct might well be built of different material, depending on what was locally available. Thus on the Marcia the Ponte Lupo is built of reddish-brown tufa, while the Ponte San Pietro is of porous limestone. The quarries for the Pont du Gard and the Ponte Lupo are both still to be seen, in the side of the valley and only a few hundred metres from the bridge,[10] while at Tarragona there is an ancient, but smallish, quarry only 100 m north of the great bridge.

One is so used to admiring Roman work in concrete that it comes as something of a surprise to realise that as enormous a monument as the Pont du Gard was built entirely of cut stone blocks, not only free of mortar, but apparently nowhere even bonded by clamps. The blocks, some of which are of six tons, remain in place purely by gravity and the careful precision with which they have been shaped in a manner reminiscent of the best quality Greek work; to assist their placing, they often carry masons' marks.[11] Even more striking is that in the voussoirs of the great arches of the middle and bottom tiers there is no cross-bonding through the thickness of the bridge. Each arch is three voussoirs thick (Fig. 84; four on the bottom tier), but it is as if there were constructed three independent arches, side by side. This peculiar (but in this case unquestionably effective) construction has parallels in various Roman vaults and arches in Provence, but is otherwise rare.[12]

Towards the end of the first century BC concrete came to replace ashlar stone. It was a gradual process, as the practice grew of transforming mere inert rubble fill in the centre of a wall into a cohesive core by the greater use of pozzolana and lime, but was well established by the time most aqueduct bridges came to be constructed, and is the form they normally take[13] (*opus reticulatum* is also encountered, but less often). There was, however, a complication. Many of the bridges had to be rebuilt, strengthened or renovated, sometimes frequently. It might be because, as on many of the Rome bridges, extra aqueducts had been added on top of them that they were not originally intended to carry; it might be, as at Cherchel, that they stood on unstable ground. For whatever reason, the result, particularly on the Rome network, is often such a *mélange* of different styles, dates and materials that even the basic shape of the bridge is barely discernible behind a confused welter of buttresses, walled-up arches and other alterations. Aesthetically, the Ponte Lupo (Figs 93, 94, 95), which ought to be the splendid queen of the Rome aqueducts, looks a shapeless mess defying the analysis of all save an Ashby or a Van Deman, and one cannot look at it without regretting the clear, clean lines of the Pont du Gard.[14]

The design of the bridge was dictated by the terrain. This is more than a mere truism, for it was usually more a viaduct than a bridge proper. Strictly speaking, a bridge carries a route of some kind (e.g. a road or an aqueduct) across an obstacle such as a river or a gorge where intermediate support is difficult or impossible; a viaduct carries it across a dip in the land where almost continuous support can be provided and the purpose of the structure is to maintain the level of the route. In a bridge, therefore, the emphasis is on a wide, clear span. In a viaduct the span of the arches is almost unimportant, and the emphasis is on height. The design requirements are thus quite different, resulting in a few wide arches for bridges, and a row of high narrow ones for viaducts.

Traditionally, most studies use the term 'bridge' for such structures as, say, the Ponte Lupo, but in fact they more closely correspond to our description of a viaduct. When a Roman aqueduct goes on to a 'bridge', it is usually a question of reaching the other side of a valley while maintaining level, not of bridging a river or spanning a chasm. The Pont du Gard, which not only crosses a valley but also spans a major river, the Gard, is a rare exception to this rule, and usually Roman aqueducts are not laid out so as to have to cross major river valleys: the valleys they do cross are generally those of minor tributaries or depressions that, while they may have a small brook finding its way along the bottom, are not really to be thought of as river valleys at all. Accordingly, the typical Roman aqueduct 'bridge' (for this study we will retain the traditional term) is formed by the series of tall narrow arches familiar from countless illustrations, and there are no wide spans. In engineering terms, the emphasis on height rather than span means an escape from the problems of constructing a big arch, but only at the expense of adding to those of lateral stability. The widest spans known are those on the Trajanic aqueduct at Antioch (22 m), the Ponte S. Pietro on the Marcia (15.50 m), and the great centre span of the Pont du Gard (24.52 m): most bridges had their piers set at an interaxial interval of about only 7.5 m.[15]

The Romans' use of repeated arches – in effect a modular system of construction – meant that they could build an aqueduct bridge of indefinite length (compare Segovia, Figs 86, 96), which, indeed, is essentially what was done in the arcades across the plains. Length, therefore, was no problem. Height was, and in two ways. In its purest form, it theoretically caused no difficulties. The piers of an aqueduct, whether of concrete or cut stone, were strong in compression and could easily carry the direct vertical load. True, trouble might arise with the footing: if the ground was unstable and the piers ill-founded, then as the weight of each pier increased with its own increasing height, something might begin to shift.[16] But let us consider the stability of the structure as contained within itself, and setting aside the possibility of something giving way underneath it.

The first problem stemming from height was that if the piers were unduly tall and thin, they lacked both longitudinal and lateral stability, and individual piers might tend to buckle or tilt, either along the line of the bridge or crossways to it. One solution to this was to construct the bridge of two or three superimposed tiers of arches, thus effectively limiting the height of each arch and its supporting piers to manageable proportions. The classic example of this is the Pont du Gard, which is, in effect, three self-contained bridges built one on top of the other, with none of the piers particularly tall.[17]

The bridges at Segovia and Tarragona (Figs 85, 86) are built on the same principle, though there it is not so striking as there are only two, not three, tiers, and the arches are all of the same size, while in the Pont du

81. Pont du Gard, carrying the aqueduct of Roman Nîmes across the river Gard (Gardon). It was the largest of all Roman bridges – 48.77 m high by 275 m long, with a maximum free span of 24.52 m. The scale is convincingly given by the automobile crossing it on the adjacent roadway; it is just discernible in the first archway from the left in the middle tier, but is so small as to be almost invisible. For the mutilation to its piers which it sustained without collapse, compare Fig. 83. Compare also Fig. 135.

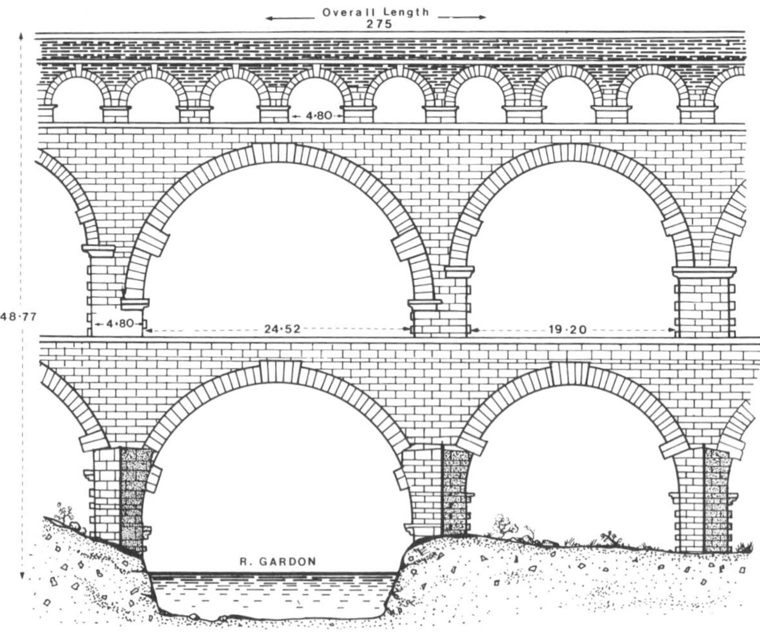

82. Pont du Gard, Nîmes: elevation of central section, showing 24.52 m span over the river Gard. On the bottom and middle tiers, the arches become smaller on each side (19.20 m span instead of 24.52), while on the top one they remain constant, with the thickness of the piers varying to take up the difference in spacing (Esperandieu, Léger).

83. Pont du Gard: the bridge in 1557. This somewhat imaginative drawing (it can be seldom that the river rose high enough to flow under *all* arches of the bottom tier) yet shows how the piers of the middle tier were cut back to half their thickness to facilitate road passage (Esperandieu, Poldo d'Albenas).

84. Pont du Gard: inner face of central archway, middle tier. It is constructed of three separate courses of voussoirs, not bonded to each other.

85. Tarragona: the bridge carrying the 35 km long aqueduct (attributed to Augustus) into the city. It is 249 m long and 26 m high.

86. Segovia: built under Claudius, the arcade is 728 m long, and has a maximum height of 28 m. It is built of granite, with no mortar between the blocks.

87. Ephesos: bridge on the aqueduct of S. Pollio, 5 km south of Selçuk.

88. Metz: bridge at Jouy-aux-Arches. This is the downstream extremity of a kilometre-long bridge crossing the river Moselle. Considerably restored.

89. Izmir: late Roman aqueduct serving ancient Smyrna.

90. Pont du Gard: the *specus* (water channel); the *sinter* on the sidewalls is very prominent.

91. Tarragona: water channel.

92. Selçuk, Turkey: the piers of the Roman aqueduct incorporated into the buildings of the modern town; on top of each pier is a stork's nest.

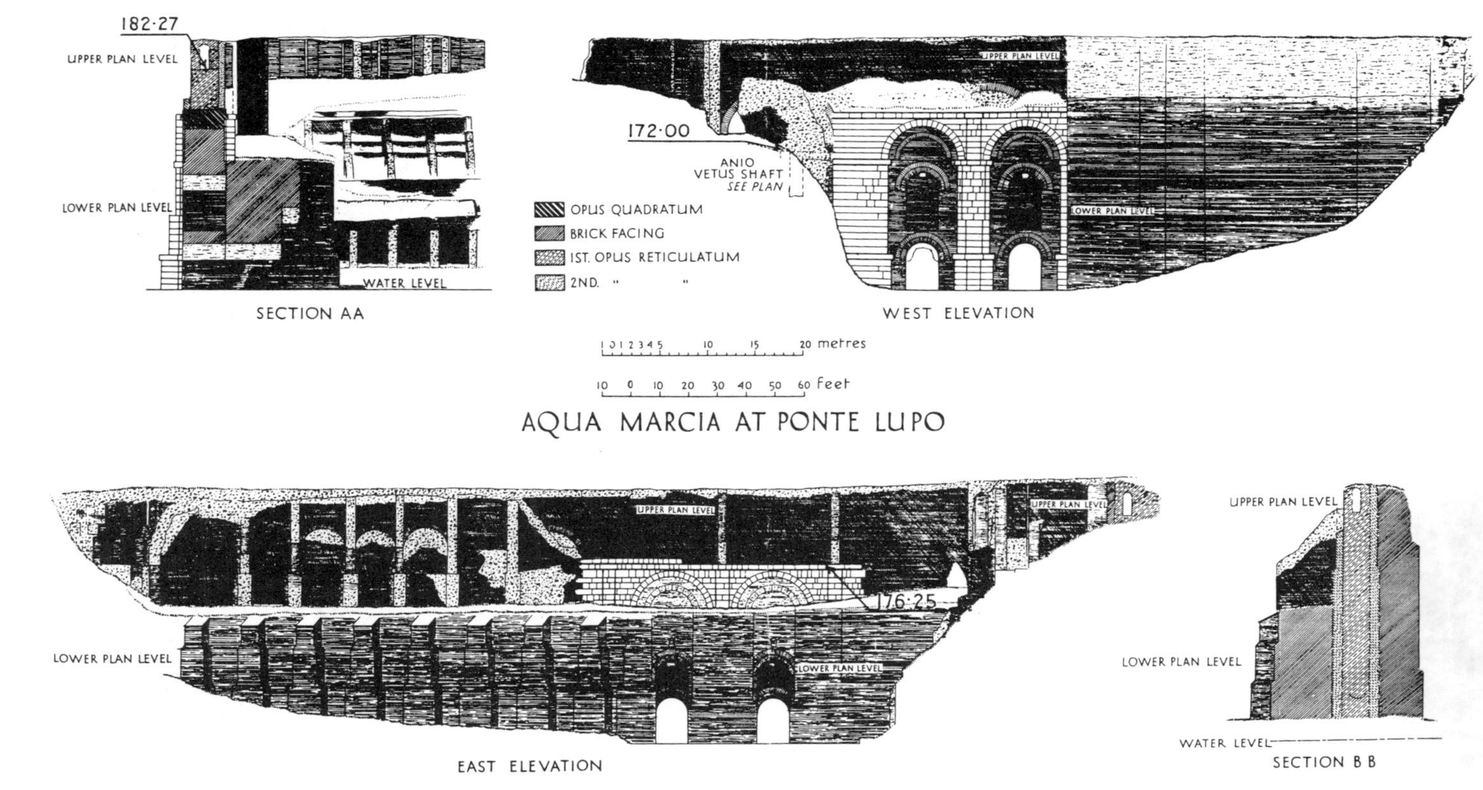

93. Rome: Ponte Lupo (Aqua Marcia), elevation (restored by Ashby).

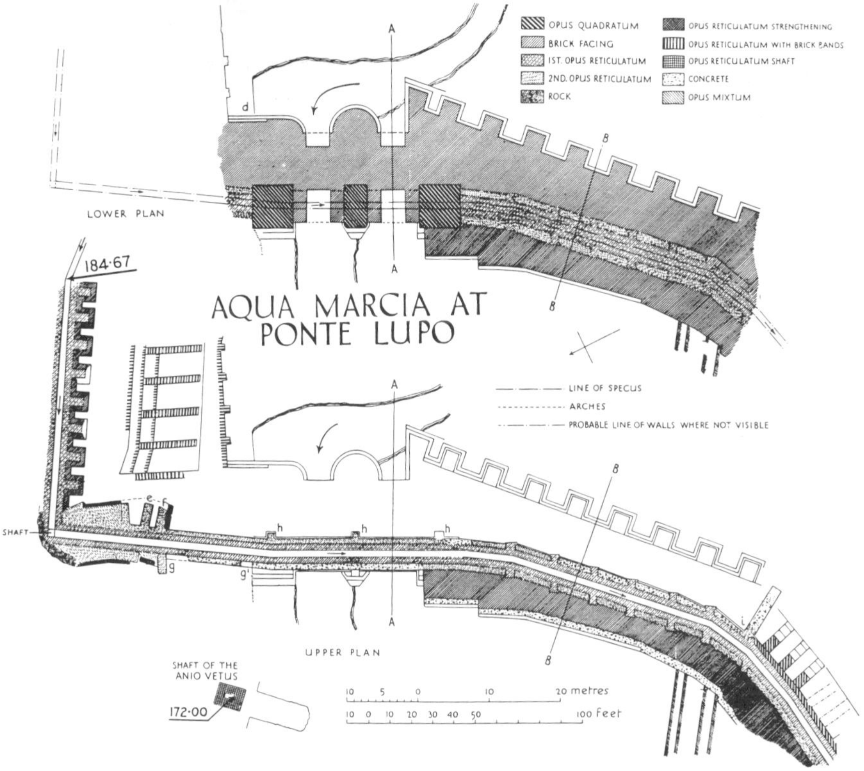

94. Rome: Ponte Lupo, plan (Ashby).

95. Rome: model of the Ponte Lupo (photo: Museo della Civiltà Romana, Rome).

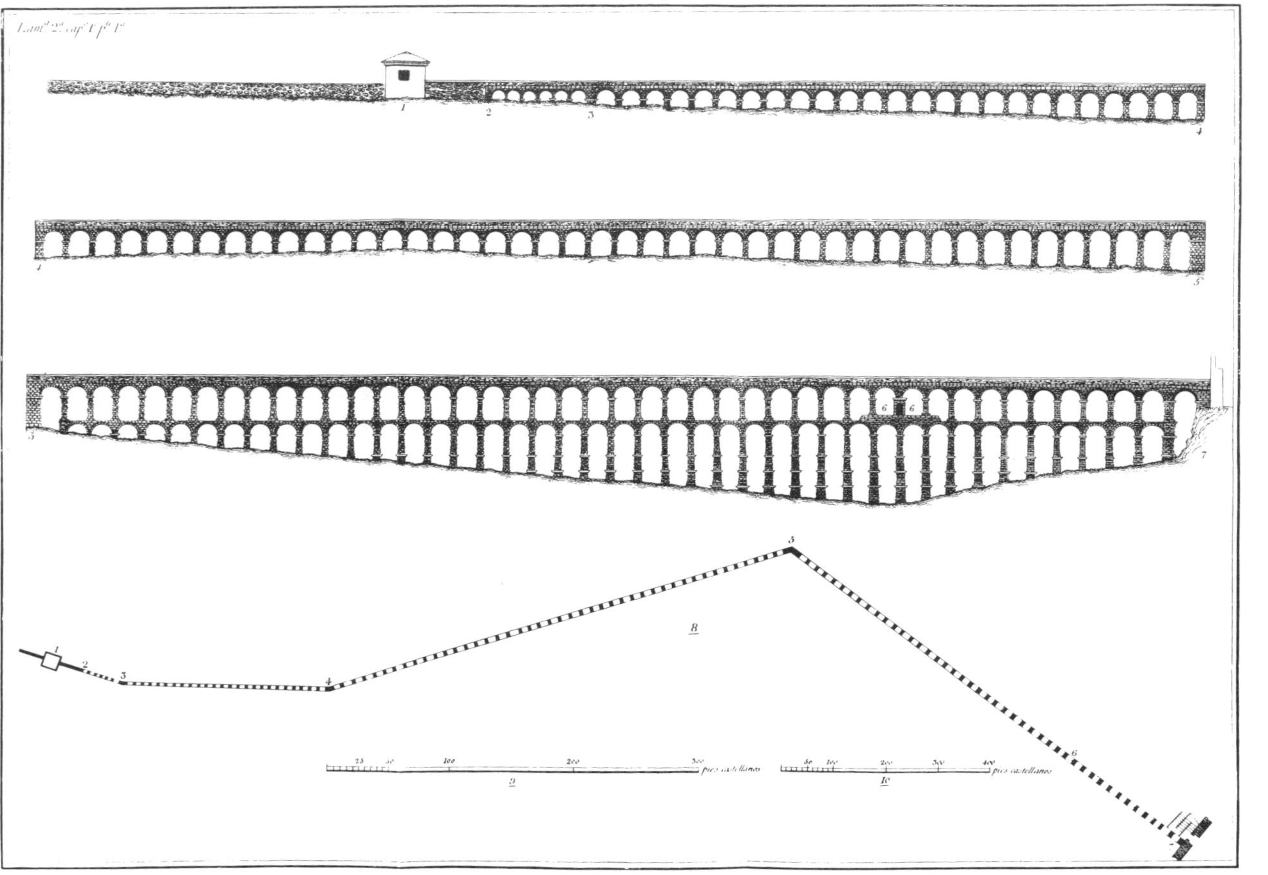

96. Segovia: elevation and plan of aqueduct bridge (1842).

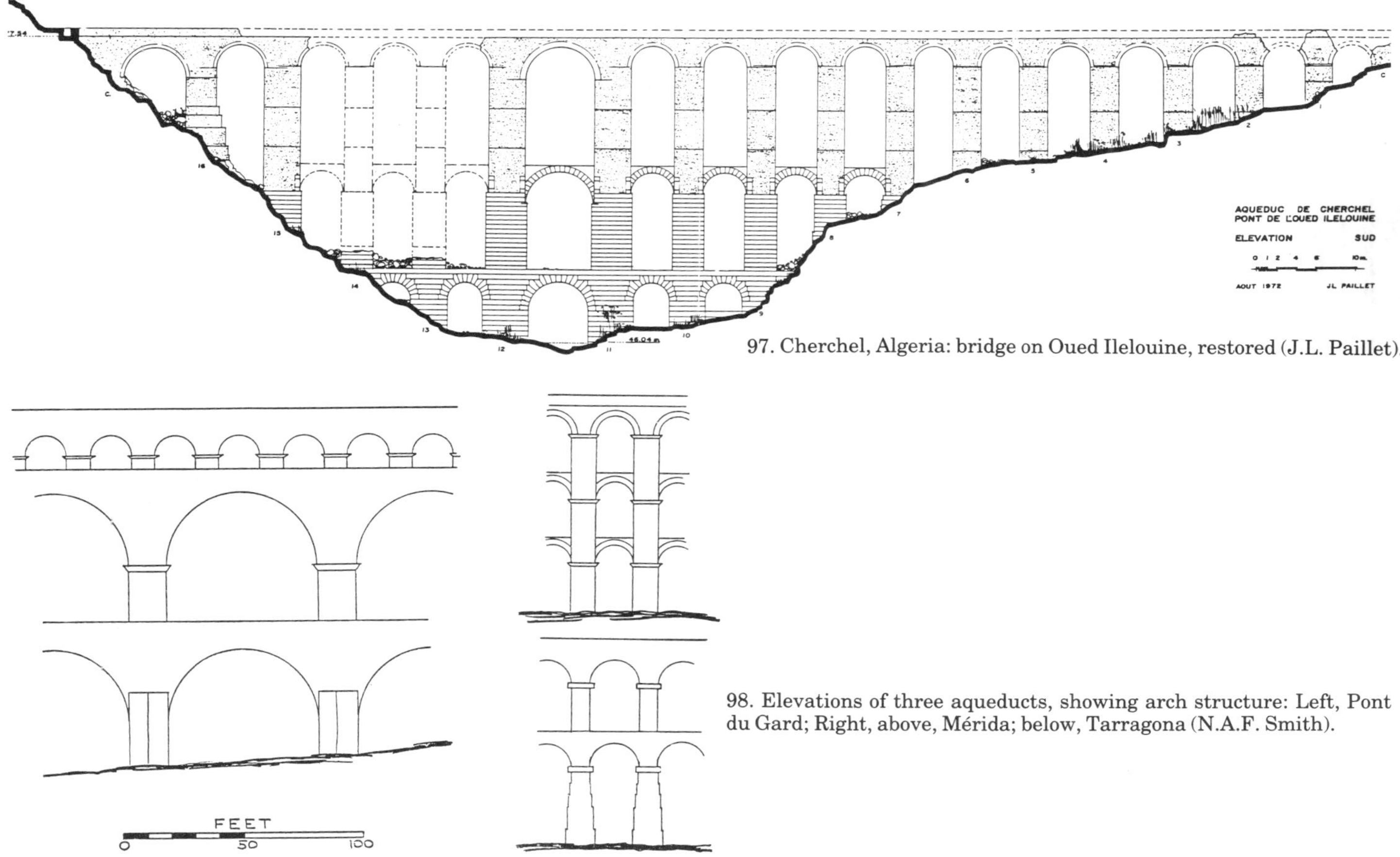

97. Cherchel, Algeria: bridge on Oued Ilelouine, restored (J.L. Paillet).

98. Elevations of three aqueducts, showing arch structure: Left, Pont du Gard; Right, above, Mérida; below, Tarragona (N.A.F. Smith).

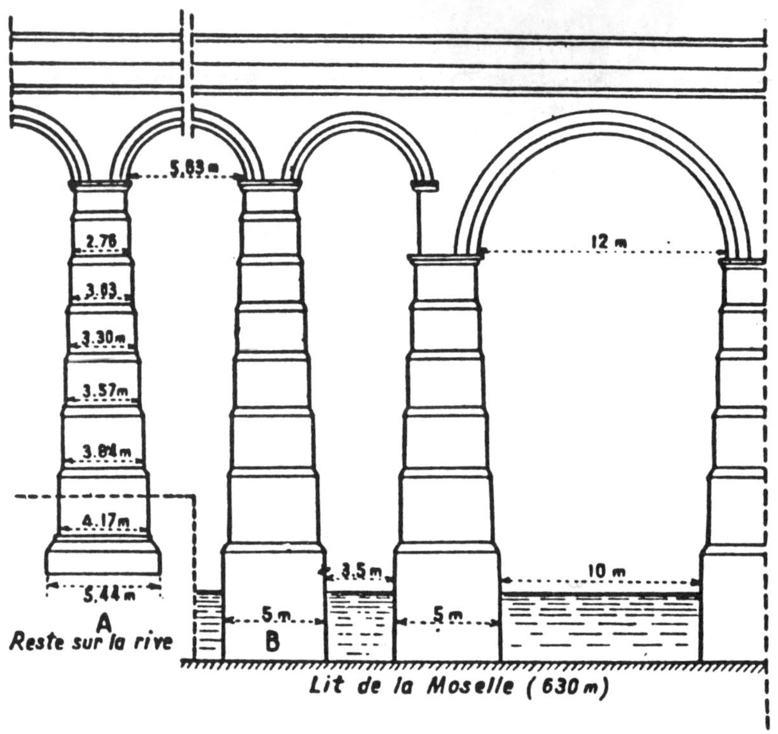

99. Metz: aqueduct bridge over the river Moselle (Doell, 1904).

Gard the independence of the three tiers is stressed by the arches of the top one being much smaller. The other solution (Fig. 98) was to build the bridge of one single set of tall piers running through right up to the top, and then to provide the necessary stability by cross-bracing each one about half way up to its neighbours on either side. The cross-bracing took the form of a brick arch, so that this type at first sight much resembles the multi-tier system, in that one sees two rows of arches. On closer examination, however, the structural difference is plain: the piers form each one continuous element, running from top to bottom, so that this is basically a single bridge divided vertically into a series of high modules, while the other is divided horizontally into two or three long low modules, each forming a separate bridge superimposed on the others. The best examples of the cross-braced type are Mérida (which has cross-braces at two levels), Fréjus and Cherchel (Pont de l'Oued Ilelouine, Pont de l'Oued Bellah).[18]

Indeed, the Pont de l'Oued Ilelouine (Fig. 97) is a remarkable and unique hybrid, for it combines both systems: the lowest level (6.91 m high) is an independent bridge of five arches (Pont du Gard style), on which is superimposed a superstructure of tall piers some 24.68 m high, cross-braced a little under half-way up.[19]

100. Lyon, Beaunant: piers of the *venter* bridge carrying the Gier aqueduct, with transverse archway at the bottom of each pier (here used as garage).

The second problem arising from height was that of lateral stability: even if the piers were stayed by cross-bracing to each other, the whole bridge, reacting like a wall that had been built too thin, might bodily fall over sideways in its entirety, particularly when subjected to the pressure of strong crosswinds.[20] Two remedies for this were involved. In a multi-tier aqueduct one could broaden the base by making the tiers progressively thicker towards the bottom (see Figs 88, 97, 99). Thus in the Pont du Gard the bottom tier of arches is 6.3 m thick, from side to side, the middle one 4.5, and the top one 3.06.[21] A further noteworthy exception may be noted in the *venter* bridge of some siphons. Because they often carried a whole series of lead pipes side by side, *venter* bridges were usually very broad (this is one way of distinguishing an isolated section of *venter* from an ordinary bridge or arcade) and, given their low height, no problems of lateral stability arose. At Beaunant, however, on the Gier aqueduct, the *venter* bridge was so wide (7.35 m) that the architect, no doubt to save material, thought he could set into each pier an archway, crossways to the line of the bridge (Fig. 100), leaving the bridge in effect supported by a row of stilts, one along each side. Later someone evidently had doubts about lateral stability, for seven of the highest piers subsequently had their archways blocked up, 'pour garantir la stabilité de l'ouvrage'.[22]

Alternatively, the piers could be propped and reinforced by buttresses. The most striking example of this is the very prominent sloping

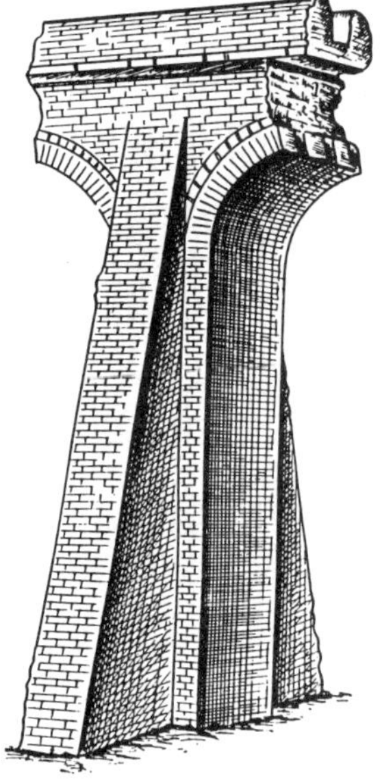

101. Fréjus: 'les Arcs Bouteillière', showing sloping buttresses (A. Blanchet).

buttresses of the aqueduct of Fréjus (Fig. 101), particularly at 'les Arcs Bouteillière', but also observable elsewhere.[23] Sometimes the actual piers could themselves be tapered to provide a broader base; rather than giving the pier a sloping face, it was normally done by a series of set-backs, the pier retaining a vertical face in between.[24] The effectiveness of this has been calculated by Leveau and Paillet for the Pont de l'Oued Bellah (288 m long and 26 m high) at Cherchel. This bridge was particularly exposed to high winds, since it was 'disposé perpendiculairement au vent marin dans un goulet d'étranglement et à quelques centaines de mètres de la côte',[25] and the figures showed that a wind of 234.6 kph could blow it down. A wind of this strength is conceivable in a cyclone or violent gale. However, it is also possible that a wind of some 150–180 kph, blowing in repeated gusts, could set up in the bridge a potentially dangerous oscillation,[26] and since this is not too uncommon over a long period, the authors not unreasonably consider that in this case wind pressure was probably at least partly responsible for its eventual collapse.[27]

It must be pointed out that this particular bridge had no buttresses, and was somewhat flimsy, not to say badly engineered. In the south of France, the northerly blasts of the Mistral are of considerable ferocity and present a constant hazard to the bridge engineer,[28] particularly affecting any bridge or arcade in an exposed spot and on a roughly SW-NE orientation (the Mistral blows with very great regularity from the N-NW). This explains the particularly prominent buttresses at Fréjus, both at La Bouteillière and in the approaches to the city, for Roman engineers were fully conscious of wind effects.[29]

It also must be seen as a particularly cogent testimonial to the excellence of the engineering of the Pont du Gard, which not only is exposed broadside on to violent gusts of the Mistral coming down the Gard valley, but in the sixteenth century had all its piers in the second tier cut back on one side (upstream) to only half their thickness, so as to provide a carriage road across at this level (Fig. 83). This incredible mutilation lasted for at least two centuries, possibly three or even four[30] with no apparent ill effects. Faced with a bridge 48 m high built of unbonded cut stone, with no mortar, concrete, or even clamps, that is then cut back to half its width like a tree ready for felling, exposed for perhaps 400 years to the pressures of turbulent floods and violent winds – and at the end of it all was still standing: faced with this, the commentator can only state as a fact that it was so; he cannot attempt to explain it. Of course the whole bridge ought to have fallen down, many times over. If the failure of the Pont de l'Oued Bellah reminds us that Roman engineering was not always perfect, it is only fair that the Pont du Gard should in turn remind us of the superb heights of excellence it could achieve at its best.[31]

The channel across the top of the bridge maintained something rather more than the usually downhill slope of the aqueduct. This has seldom been expressly studied, but for some of the longer bridges may be worked out where different levels are published for the channel at each end of the same bridge. Thus the Ponte S. Pietro (Marcia) falls 26 cm in a length of 88 m, giving a gradient of 2.95 m per km. At Cherchel the Pont de l'Oued Bellah loses 1.53 m over 288 m, or 5.3 m per km, and the Pont de l'Oued Ilelouine drops 36 cm in 136.6 m (=2.6 m per km).[32] That a bridge, particularly a long one, should not have a level top but should continue the downward slope of the aqueduct is, of course, not surprising, but there was a further factor that sometimes accentuated this tendency. This was the fact that a bridge, especially a large bridge, is often not part of the original aqueduct but a later modification. In its most characteristic form, one sees this phenomenon where the original line ran up one side of a valley and back down the other, and the bridge was then introduced to cut off this loop by going straight across. This resulted in a great saving in maintenance on the abandoned loop, but created a problem of level for the bridge. At each end, it had to join up satisfactorily with the old channel, but as the upstream and downstream halves had originally been part of an aqueduct running for perhaps two or three kilometres in between, round the loop, and losing height steadily all the way, one now lay far below the other. Had the bridge been an integral part of the original design there would have been no difficulty, but, as it was, it had to lose in the hundred metres or so of its length all the height previously lost in several kilometres, so as to effect a junction at its downstream end. One is therefore liable to find a sharp downhill drop associated with bridges of this kind (as well as a right angle bend at each end, where it leaves and picks up the old channel.) At the Pont de l'Oued Ilelouine (Cherchel) this was achieved by having the bridge immediately preceded by a four-part cascade, tumbling down which the water lost no less than 12.28 m. At the Ponte S. Gregorio, where the Anio Vetus was abbreviated by a cut-off of Hadrianic date, the result is even more striking. The bridge is 155.5 m long and 24.5 m high, and for eighteen of its twenty-two arches maintains an orthodox, gentle fall of 1.08 m in 141.6 m (or about 7.66 m per km). At this point the top of the bridge itself bends sharply down, losing 4.09 m in 25 m (or 163 m per km, the steepest gradient on the whole of the Rome aqueducts) as the channel leaves the bridge and plunges underground. This was again a device for losing height rapidly, the onrush of the water down the slope being immediately retarded by running into a tunnel with a right-angle bend in it.[33]

Normally, the water channel (or channels) running along the top of it was the only traffic that an aqueduct bridge carried. One might feel that, granted that the bridge was there anyway, it was a pity not to use it for road traffic too, but this seldom happened. Built and sited to conform to the line of the aqueduct, the bridges seldom corresponded to the natural routes of road communications, which tend to follow valleys rather than cross them; nor did the aqueducts usually cross sizable rivers, where their arches could also have served road travellers. Indeed, only three seem to have carried road traffic, and all can be admitted only with qualification. The first is the Pont du Gard. The case for this rests on pure probability. The fact that in subsequent ages it always *has* been used as a river crossing, even to the extent of cutting away the piers to facilitate traffic, demonstrates that it does fill a real need in natural local communications.[34] To cross was always possible, either by walking across the very top, on the flat slabs roofing the water channel, or at the level of the modern bridge, where the piers of the middle tier of arches did not quite come out to the edge of lower, thicker ones, leaving free a passage just over a metre wide on either side. We may therefore reasonably surmise that it was much used in antiquity too, even though it carried no formally dedicated roadway. The second is the Ponte Lupo. Here there is no doubt about the roadway, for a reconstruction of Severan date sought to buttress its eastern or upstream side with a vast mass of brick-faced concrete, the flat top of which automatically provided across the valley a continuous roadway from 6 m to over 10 m wide.[35]

The only question is whether this highway was ever used by anyone. Its resemblance to the Pont du Gard in its present state is striking, but its provision was purely fortuitous. The purpose was to reinforce the aqueduct, and the magnificent promenade that the flat top offered was not apparently connected to any road network, nor did it lead from anywhere to anywhere worth going. Isolated in this still remote and little visited corner of the Campagna, it was presumably then, as now, used as a crossing only by occasional local herdsmen. The third is Aspendos. Here we see a long, low *venter* bridge crossing a valley some 850 m wide. The bridge is again very broad, 5.5 m, but here the water is carried in a single stone pipe only 86 cm thick.[36] Why the extra width? The presumption is that it also served as a roadway, but apparently the width is all that this conjecture rests on, though its use as a road is stated by Lanckoronski as fact.[37]

Siphons

It is scarcely possible to look at the great aqueduct bridges such as the Pont du Gard without sooner or later asking why the Romans did it. Why did they not use siphons instead?[38] Did they not know that water would rise to its own level? Did they never think of putting the water into metal pipes that ran down one side of the valley, across and back up the other as an inverted siphon? Or did they think of it, but couldn't do it because they could not make pipes strong enough to contain the pressures generated? All of these explanations have been seriously offered, the last being the commonest.[39] All are wrong. They do not even need refutation. It is almost enough to point out that, by common consent, the Hellenistic engineers made wide use of siphons, including the giant one at Pergamon, and ask how could it conceivably happen that the Greeks could make pipes strong enough but the Romans could not?

But the real answer is that the question has been wrongly framed. The Romans did in fact use inverted siphons. They were both numerous and successful, on large and on small scale, and from the earliest days formed a standard recourse in the repertoire both of the aqueduct engineer and the city plumber.[40] It follows that the real question to put is not, Why did the Romans not use siphons?, but Why do modern scholars so often write as if they did not? There are two explanations for this lacuna. The first is ignorance of the evidence, arising from the circumstance that siphons are very rare on the Rome metropolitan network, and this is where study has been concentrated. The second is misapprehension of the hydraulics involved, and in particular what Vitruvius has to say about them: Vitruvius says that siphons generate pressure and steps have to be taken to deal with it; this is then garbled first into statements that the Romans tried to avoid pressure systems, and second, by extension, into flat assertions that they did avoid them and that such things did not exist.[41] We may thus rephrase our initial query why they did not use siphons, and ask why they did not use them more often. This is a fair question. The answer to it is to be found not in engineering or hydraulics, but in economics: bridges were cheaper.[42]

Siphons existed throughout the aqueducts of the Roman empire, but they were specially numerous in Gaul, and relatively rare at Rome. About two dozen have been identified,[43] far more than we know of in Greek

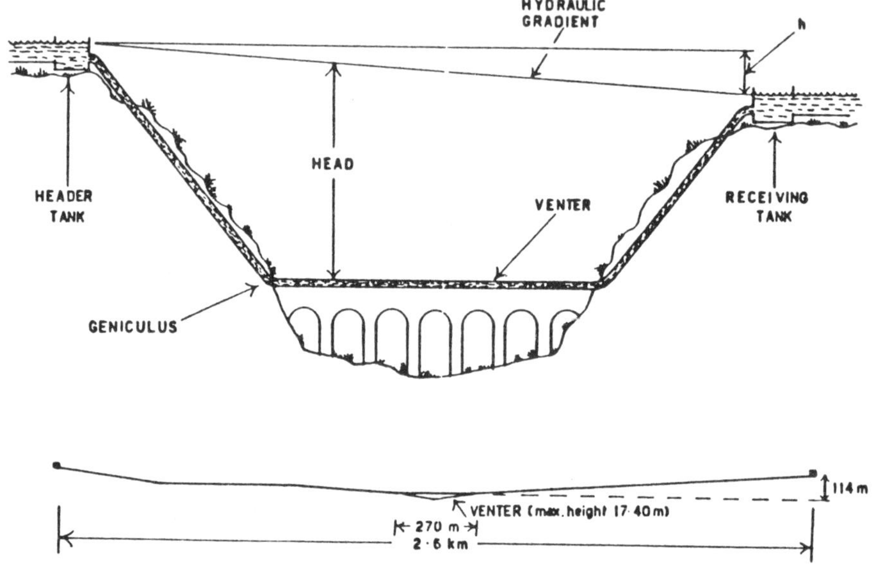

102. A typical siphon: diagram of layout (h = loss of head). Below, profile of Beaunant siphon, Gier aqueduct, Lyon; partly to scale, and showing actual gradient profiles.

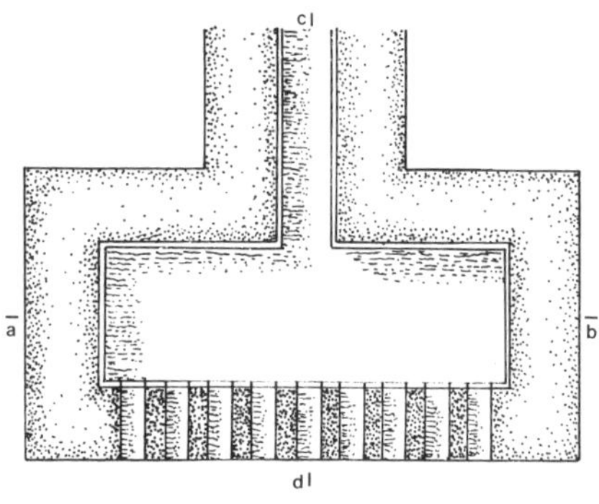

103. Lyon: Soucieu, Gier aqueduct, header tank of siphon, in plan and section (de Montauzan).

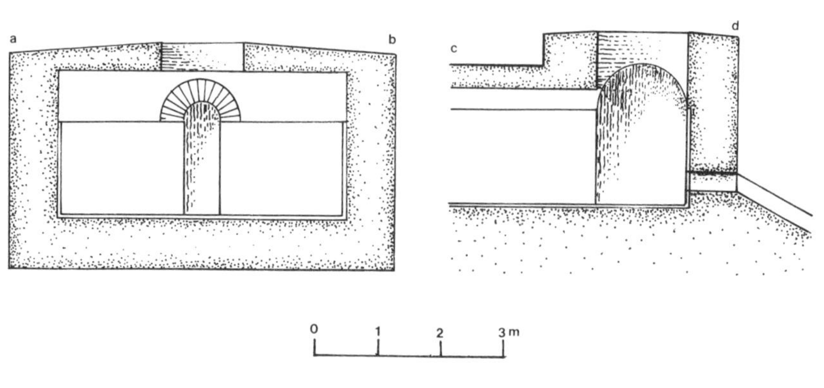

104. Lyon, Soucieu: header tank at the beginning of the Soucieu siphon, on the Gier aqueduct.

105. Lyon, Soucieu: close-up of holes in header tank to carry lead pipes of siphon.

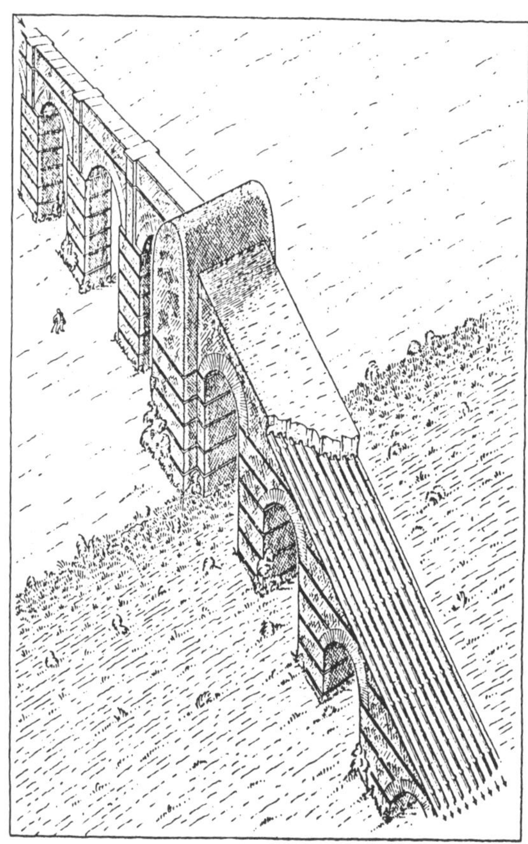

106. Lyon: Beaunant, Gier aqueduct: header tank of siphon, restored by W. Haberey.

107. Lyon, Beaunant: header tank of Beaunant siphon, with ramp carrying siphon pipes down to ground level (compare Fig. 106).

work. They were particularly numerous at Lyon, where are also to be found both the largest and best preserved examples. A typical example is illustrated in Fig. 102, which is based jointly on the two principal siphons, at Soucieu and Beaunant on the Gier aqueduct (Lyon).[44] It begins with the water arriving by conventional aqueduct into a header tank, which is an oblong, vaulted distribution chamber set crossways to the main aqueduct line (Fig. 103). From the far side of the header tank (Fig. 104) the water leaves by a series of lead pipes, of 25 cm external diameter and perhaps 3 cm thick, arranged in a row, side by side; on most of the Lyon siphons there are nine of them (see Figs 103, 105). From the header tank, the pipes run down a concrete ramp to ground level (Figs 106, 107) and thence, running perhaps a metre or so underground, over the edge and down the side of the valley. Being buried they avoided problems of heat expansion in hot weather (a benefit perhaps unforeseen and possibly even unappreciated), and were protected from interference; presumably they had to be dug up for maintenance. Possibly they may have been anchored at intervals by being embedded in concrete blocks, to stop them sliding downhill,[45] but it is unlikely that the slope was often steep enough to require this, as it only becomes a danger when the slope is over 25%.[46]

Arriving at the bottom of the valley, the pipes might often simply continue on underground, following the profile of the terrain, and up the other side. They were, however, at this point subjected to considerable pressure. The pressure inside the pipes depended on how deep the siphon was below natural water level (i.e. the head, or the vertical column of water supported), so it could be reduced by reducing this depth. This was done by, in effect, putting in a bridge (*venter*) to cut off the lowest part of the valley. It should be emphasised that, as a matter of hydraulic physics, this was the only way the pressure *could* be reduced; it could not be done by some arrangement of valves, as has often been suggested. Although this procedure did reduce pressure, it had disadvantages. There was a sharp bend (*geniculus*) as the pipes came off one side of the valley and on to the bridge, and another at the far end of it. Inertial thrust could cause stress here on the outside of the curve as the running water hit the corner. At the *geniculus* the siphon pipes came out from underground and ran, nine abreast (at Lyon), across the *venter* bridge (Figs 108, 109), which is built appropriately wide to accommodate them (the great width of *venter* bridges, as compared with conventional bridges which support only a single channel, is striking: Beaunant *venter*, width 7.35 m; regular width of arcade on same aqueduct, 2.25 m). Indeed, the bridge often seems to be too wide even for that, the most likely explanation for the excess width being that it was also to provide a catwalk for maintenance. The *venter* bridge also, like conventional bridges, is not completely level, but the slope runs the other way, being higher at the downstream end, though it is very slight: this may be a device for getting rid of air pockets.[47]

108. Lyon, Beaunant: *venter* bridge on the Beaunant siphon, Gier aqueduct, from upstream end.

109. Lyon, Beaunant: *venter* bridge, north-east end. The great width of the bridge (7.35 m), as compared with any ordinary aqueduct arcade, is due to it carrying a bank of siphon pipes, nine abreast, as well as catwalks for maintenance. The nearest pier has had its transverse archway blocked up to increase stability (compare Fig. 98).

110. Lyon, Soucieu: general view of the Soucieu siphon. The siphon begins 1.2 km away at the header tank (Fig. 104; here, left-hand arrow), and, after crossing the intervening valley (92 m deep), terminates in receiving tank (remains in foreground, right-hand arrow).

On leaving the *venter* the pipes again go underground and climb the side of the valley until they reach the receiving tank where the siphon ends (Fig. 110). This was similar in design to the header tank and in it the water emerging from the pipes resumes open flow along the conventional aqueduct conduit.

It will be noted in Fig. 102 that the receiving tank is markedly lower than the header. Water, of course, will rise to its own level, and if the two tanks were at the same height, the water would in fact come through. It would not, however, come through very satisfactorily, because of the retarding effect of friction in the pipes. One of the unwelcome by-products of channelling the water from an open conduit into nine closed pipes is a great increase in the contact area and hence friction. Should the siphon be built with both tanks level the flow through it would be so retarded that the water would back up and overflow at the header tank. To keep it flowing with satisfactory velocity and volume, therefore, the receiving tank must be sited low, so that the siphon in effect runs downhill, the imaginary line connecting the two tanks being known as the hydraulic gradient. Because of this, when a siphon is used instead of a bridge to cross a valley, it may mean losing much more height (h. in the diagram), depending on what gradient the rest of the aqueduct observes. On the Gier aqueduct, for example, the average gradient (it is very uniform) is 1.47 m per km, while on this same aqueduct the siphons at Beaunant (2.6 km long) and Soucieu (1.2 km) each lose over this distance a height of 9 m, giving a siphon gradient of 3.4 m per km and 7.5 per km – more than two to five times as steep[48] as the rest.

A vital question remains the pressure generated in the pipes. The solution here seems to have been simply to make them strong enough to contain it, nor was there any difficulty in doing this. In the 1870s the French engineer Belgrand produced and tested to destruction some replicas of Roman lead pipes: they failed only at 18 atmospheres, demonstrating that they could handle the pressure in a siphon over 200 m deep, far in excess of anything normally required.[49]

However, there was evidently a limit to the size that the pipes could be made, purely as a matter of engineering practicality. Most siphon pipes are of around 27 cm external diameter, and anything over 30 cm is rarc,[50] which doubtless explains the battery of multiple pipes. The pipes usually came in ten-foot lengths, and were soldered together at the joints, the solder being a mixture of 84 parts lead to 60 tin. Vitruvius also mentions the use of terracotta pipes for siphons, their joints strengthened against the static pressure by being sealed with a mixture of oil and quicklime, but it is doubtful if in practice they were often, or even ever, used. At Angitia the conduit was of masonry, but this was unique.[51] At corners, or *geniculi*, the situation was a little different, for the force to be countered was not static pressure (which operated whether the siphon was running or not, so long as it was full of water), but the inertial thrust of running water hitting the bend.[52]

This was contained by anchoring the bend in some solid mass – Vitruvius, somewhat enigmatically, recommends a 'red rock'[53] – and was not too much of a problem in normal operation. Draining the siphon for maintenance, and refilling from dry, was a different matter. Potentially this is quite a dangerous operation, to be undertaken with great caution, for both the shutting off and the admission of water have to be done very gradually. If the water is admitted in a sudden rush, one may imagine the damage done when, after an unchecked downhill run in an empty pipe, it hits the first *geniculus*. If the sluices are suddenly closed, the moving column of water is abruptly arrested and, being incompressible and so incapable of taking up the slack, transmits a shockwave in micro-seconds reverberating back and forward along the pipe: this is known as water hammer, and can also wreck the piping.[54]

Along the course of the siphon were installed a series of puzzling devices called by Vitruvius *colliviaria*. It has been debated at length what these can possibly be, the commonest explanation being that they are some sort of escape valve either to reduce water pressure or to blow off accumulated air.[55]

The truth of it is, however, that as we have seen, water pressure could not be reduced in this way, and there was no air in a Roman siphon, for the pipes were full of water under pressure and the hydraulic conditions were unfavourable to the formation of air pockets.[56]

The most likely(?) explanation of the *colliviaria* is that they were drain cocks for emptying the siphon, and perhaps used also in maintenance and cleaning. We have already noted the heavy calcium carbonate deposits that formed in the open channels and the ceaseless labour required to keep them clear of it. A channel is cleaned by chipping, but how do you clean inside a closed pipe? One way may have been to use some sort of pull-through device – a ball of rags on a long cord or something of the kind – and this would require access holes at frequent intervals for its operation.[57] These would be the *colliviaria*, which were probably no more than round holes in the pipe normally closed with a wooden bung. They may have leaked a bit, but the average Roman siphon was possibly quite leaky anyway, with the leaks accepted as inevitable. The ordinary conventional aqueduct on arcades seems to have leaked quite a lot, with everybody considering this as relatively normal. Another solution to the problem, of course, would have been simply to wait until the pipes were completely clogged and then replace them with new ones.[58]

Another interpretation, particularly favoured by the German authorities, sees in the *colliviaria* a series of open-air tanks installed at intervals along the siphon to permit air pockets or entrained air bubbles to escape, and on this basis the 'pressure towers' at Aspendos and Les Tourillons on the Craponne, at Lyon, rank as *colliviaria*. *Colliviaria* like these would have the advantage of doing exactly what Vitruvius specifies for them (and nobody other than Vitruvius ever mentions them at all) and also being in indisputable existence, whereas 'drain-cock *colliviaria*' are entirely conjectural. The disadvantage, it seems to me, is that it makes of the *colliviaria* an impossibly large and costly structure (for of course the siphon must be brought up to natural water level if the open tank is not to overflow; hence the towers, as at Aspendos), while Vitruvius speaks of it as if it were some small device to be installed as a matter of routine at regular intervals (such as that in Fig. 171 below). Moreover, such major structures could not disappear easily without trace, and since, Aspendos and Les Tourillons apart, there are no such traces, we would have to conclude that, in spite of Vitruvius, nearly all siphons did without them; and this includes not only all the Lyon siphons, but specifically the largest Roman siphon preserved, Beaunant, which would need them if anything ever did. I therefore feel that the true explanation remains uncertain (but see also pp. 241-5 below).[59]

In the west, at least, siphons were usually large, as shown by the following list giving their approximate height (or depth):[60]

Beaunant	123 m	Grange Blanche	97 m
Alatri	101 m	Soucieu	92 m
Rodez	100 m	St Genis	90 m
Les Tourillons	100 m	Ecully	70 m

These figures may be compared with those given above for conventional bridges (n. 9 above), and two points will emerge. First, the great size of the siphons. Beaunant, the most remarkable, is more than twice as high as the Pont du Gard, and almost ten times as long (2.6 km/275 m), but there are plenty of others scarcely less impressive. Second, there is no overlap between siphons and bridges. The dividing point comes about a height of 50 m. All the bridges are below that and all the siphons above it, so that one can almost enunciate a rule: the Romans preferred bridges, but felt confident about building them only up to a height of 50 m, after which they perforce fell back on siphons. The ability to make pipes strong enough to handle the pressure does not seem to come into the question. Indeed, it is plain that siphon pipes were used only where pressures *were* going to be high; where they were low, they were replaced with bridges. Instead, the governing factor seems to have been expense. The quantities of lead required for a large siphon were enormous. The Beaunant siphon alone, 2.6 km long with nine parallel pipes, accounted for over 23 km of piping, just to cross one valley. The total length of the nine siphons in the Lyon area comes to 16.6 km. At nine pipes each, that gives a total of 150 km of piping, all of it under pressure, long enough to stretch from Rome to Capua; and this is for the siphons alone, not the whole length of the aqueducts. The Beaunant siphon alone used over 2,000 tons of lead, and the total of the lead for all the siphons in the Lyon area must have come to around 10,000-15,000 tons.[61]

The cost of the lead itself was not perhaps prohibitive. Lead was a regular and perhaps unwanted by-product from the production of silver, and the ancient world seems to have had a positive glut of it. But transport was a different matter, given the weight of lead, and for amounts of this magnitude could be ruinously expensive. Stone for a masonry bridge, on the other hand, could be quarried locally – the stone for the Pont du Gard comes from a quarry only 600 m away.[62] Nor can it have been easy to make and install the large pipes of a siphon. There is some evidence for local manufacture, and at Alatri the pipes may have been not only made on the spot, but cast, instead of being bent and soldered in the usual way.[63] This might facilitate transport a little, for though the lead would still be the same weight it would at least not be in awkward, ten-foot lengths. But the chief difficulty must have been joining up the pipes when they were being installed. Without modern portable blow torches, how was the metal heated up enough to melt the solder at the joints, particularly on the underside of a pipe already laid at the bottom of a trench? Portable furnaces could bring the heat source close at hand but actual application to the joint in a pipe already in position must have been difficult. And the Beaunant siphon alone had 11,000–12,000 such joints.[64] In short, building a multi-pipe siphon was an expensive and difficult job, calling for not only plenty of money but a large pool of skilled labour that might not always be available.[65] By comparison, stone for a bridge could be cut locally with low transport costs, and brick and concrete construction could be entrusted to slaves and semi-skilled workers.

There is therefore every reason to believe that normally a bridge would be preferred, as being both cheaper and easier, to say nothing of being more spectacular. In modern practice the position is reversed. Siphons are cheaper, because of cast-iron pipes, and one sometimes finds locations (particularly in North Africa) where the line of an ancient aqueduct is followed by a modern conduit, but with siphons substituting for the ancient bridges.[66] The ancient world, unable to get a furnace temperature high enough to melt iron, and hence being unable to cast it, was denied this resource.

The siphons which we have been considering represent the siphon in its most typical form. There are, however, exceptions and variations. Underwater siphons were known, where the aqueduct was carried across a river simply by running it through a siphon with the pipes laid on the river bed. At Arles the west bank suburb of Trinquetaille was supplied in this fashion across the Rhône, and so was the villa of Manlius Vopiscus at Tibur (Tivoli), by a siphon running from the Marcia across the river Anio; Statius admired it as a feat of engineering audacity.[67] Again, once the basic principle is accepted that water rises again to its own level, it will be appreciated that there is no necessity for it to fall very deep in between. The typical siphon is in the form of a V or U, where there is a valley to cross, but the pipes could instead have crossed a broad and shallow depression, or undulating terrain, and still have worked, provided both ends were more or less level. We have already seen one 'flat siphon' of this sort at Caeserea (p. 117 above), and there were probably more not yet identified.

There is also the question of the double or repeated siphon. What this entails is two siphons joined end to end, with no more than a shared receiving-header tank between them. The best known example of this is the double siphon at Les Tourillons on the Craponne, Lyon. At Aspendos the sequence is multiplied to form a triple siphon, and in the later Turkish waterworks of Constantinople and elsewhere the system was expanded to form an entire aqueduct made of consecutive siphons joined end to end, with open tanks between them. At Les Tourillons (restoration, Fig. 111) all that is preserved is the remains of two brick piers about 16 m high which supported on top an open tank. The pipes at the downstream end of the first siphon climbed a sloping ramp to the level of the tank, into which they then discharged. At the other end of the same tank was the offtake for the second siphon, which ran down a similar ramp to ground level. This second siphon must have run all the way to the Roman city of Lugdunum, on the heights of Fourvière in Lyon, because there is nowhere in between where the ground is high enough to bring the water back up to the natural surface level. This gives a siphon no less than 6 km long; the first 1.5 km ran across a plateau as a flat siphon, and then dipped to a

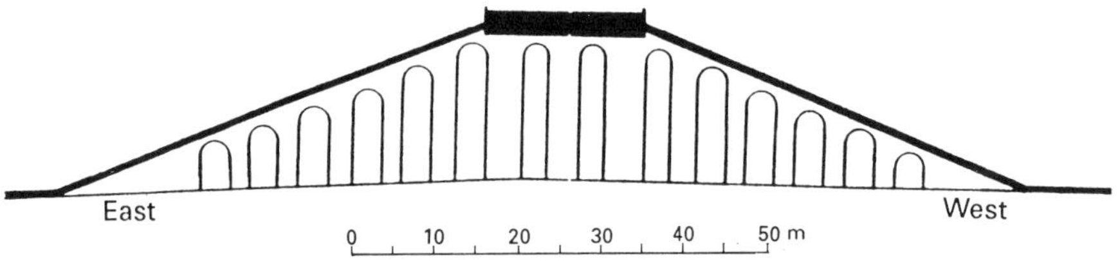

111. Lyon: Craponne aqueduct (Yzeron), restoration of raised tank at Les Tourillons, forming link between two siphons.

depth of 100 m to cross a valley.[68] The Les Tourillons tank presents problems. Why bring the water up to surface level only to plunge it immediately back into a siphon again? The most likely explanation is that it was a settling tank to give a last chance to get sediment out of the water before it got carried into the long 6 km closed pipe.[69]

At Aspendos in Pamphylia (Figs 112, 113, 114) we again meet the phenomenon of consecutive siphons. This time there are three of them, separated by two open tanks, with the added complication that the entire arrangement is located on top of a continuous masonry bridge. There is no true valley, but rather a wide, shallow depression, and the water arrives

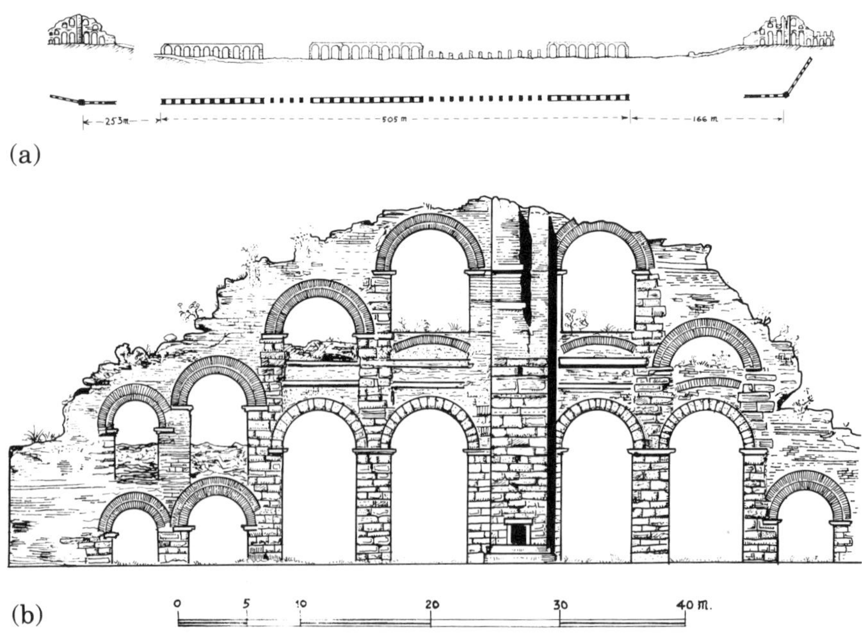

112. Aspendos: (a) elevation and plan of aqueduct; (b) elevation of north 'pressure tower': at the top of the tower was an open tank, with inspection access by a spiral staircase inside the central pier, entered through the small door visible at the bottom (A.T.H., after Lanckoronski).

113. Aspendos: general view from acropolis, showing, in foreground, south tower, and in distance, north tower.

114. Aspendos: north tower.

at its edge already running in a siphon and under pressure (the beginning of this siphon has not been traced). As at Les Tourillons, the pipe climbs a sloping masonry ramp up to natural water level and there pours into an open tank from the other end of which it then immediately re-enters a siphon and descends a second sloping ramp. At a level 29.75 m below the tank it enters upon what may be termed a very long *venter* bridge, its top some 16 m above the ground and 850 m long, though perhaps colonnade would be a better term. On this it crosses the depression, rising at the far end to spill into a second tank like the first, again descending a ramp on the far side, to continue on for 130 m at *venter* level until it meets the rising hillside of Acropolis of Aspendos.

Of particular significance, though it is hard to say exactly how, is the fact that, in plan, the line of the aqueduct has two bends, each at the mid-point of one of the tanks and its supporting structures; both these are themselves bent to accommodate the change in direction, the tank itself being not rectangular but with an obtuse angle at the middle of it.[70]

The purpose of the breaking up of the siphon into three can only be conjectured. The two structures supporting the tanks and their attendant ramps are generally referred to as 'pressure towers' (German publications use the more neutral term *hydraulische Thurm*) but this is misleading as there is no way the towers could affect the pressure in the siphons, either by their presence or absence. The most probable explanation is either that they were a clumsy device to eliminate sideways inertial thrust at the bend in the aqueduct (by ensuring that the water was at surface level and not in a pipe at all when it actually went round the corner, a great deal of the trouble would be avoided); or that it acted as a surge tank (Fr. *reniflard*), a sort of safety valve that could accommodate any abrupt variation in volume and velocity of flow through the siphon pipe (i.e. a surge), resulting from, say, a sluicegate being suddenly closed or a flood coming down the aqueduct channel. Whatever the explanation, one feels that this was an awkward expedient and there must have been an easier and better way of achieving the desired end.[71]

Cascades

When we consider elevation, the principal problem of the engineer, we instinctively feel, was maintaining height: hence all those arcades across the Campagna to ensure that when the water reached Rome it was still high enough to perform some useful service. If the source was originally so high that the problem was one of losing height, of getting the water down to the city, usually an easier answer would be to find another source that was closer and lower. Sometimes, however, the terrain might be such that this was not possible and the aqueduct had to lose height rapidly. The situation was not common, but it does occur. The normal solution was to use cascades. These were of two different types.

One was an open cascade, in which the water ran in an open channel, often on the surface of the ground; the channel was perhaps inclined at a steep slope, interrupted from time to time by a short vertical fall to stop the current building up too much downhill momentum. The best known example is on the immediate upstream approach to the great Ilelouine bridge at Cherchel.[72]

In the second type, the aqueduct ran underground (though perhaps only a metre deep or so) and in through the side of a vertical shaft, which often continued straight on up to an inspection manhole above. The bottom of the shaft terminated in a pit to accommodate the turbulence of the fall, and the conduit proper continued on from an exit a short distance up on one side. This type differed from the first, it will be observed, by being totally underground and enclosed, the falls being inside vertical shafts rather than in open cascades. The best attested example of this type is at Recret/Grézieu on the Craponne aqueduct at Lyon (Fig. 115), where there seems to have been a whole series of such falls – there may have been as many as fifty of them, arranged in an 'escalier' – each one around 2.5 m high, and from 30 to 100 m apart. A similar installation existed on the Montjeu aqueduct serving Autun, which has however suffered from being little studied or published. It is known nevertheless that though it was only 6 km long in its entirety, it boasted no less than 24 cascades, to achieve a total loss of altitude of 160 m; the shafts were remarkably large, 2.80 by 2.20 m in section, while the channel was only 0.8 by 1.57 m, and the best preserved shaft was 4.4 m deep. Other batteries of cascades have been found at Beaulieu (Aix-en-Provence) and Rusicade (Algeria).[73]

It must none the less be confessed that we do not know too much about cascades. Not too many of them have been published, and frequently their very existence is only to be inferred because the channel loses so much height between two recorded datum points that the only reasonable explanation is that there were cascades somewhere in between.

Arcades

As the aqueduct left the hills and approached the city it frequently left ground level and was carried aloft on a continuous arcade, such as those endless lines of arches that dot the Roman Campagna and have always been one of the most familiar and spectacular monuments to Rome's greatness (Figs 116, 117, 118, 119). The arcade was necessary to maintain the water level high enough to provide reasonable service within the city, particularly as the approach to it was often across a plain. The need was accentuated if, as often happened, the city had some kind of acropolis within it (such as the Capitol and Palatine at Rome; Aspendos; Lyon (Fourvière)), so that the arcading became, in effect, a sort of long bridge between the top of the acropolis and the distant hills. It is, of

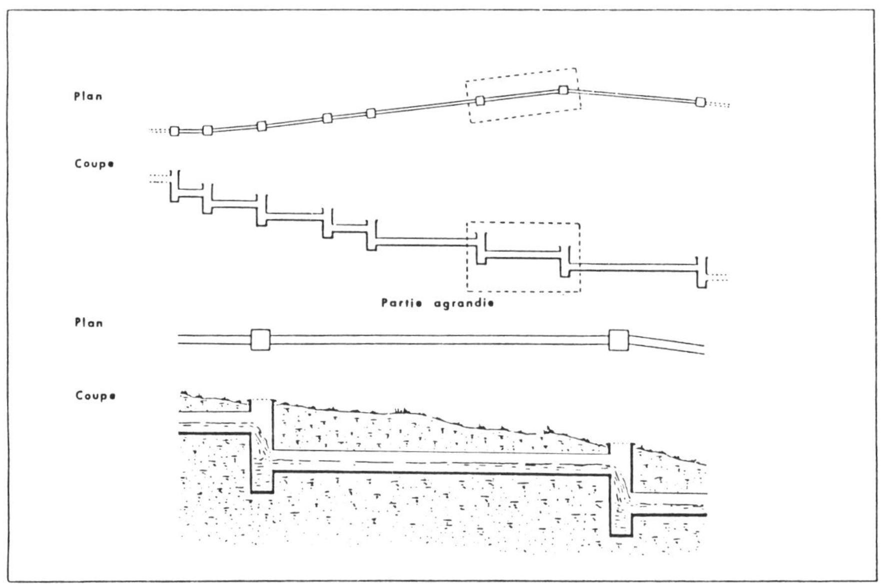

115. Lyon: Craponne (Yzeron) aqueduct, cascades at Recret/Grézieu. There may have been up to fifty such steps in this 'hydraulic stairway' (A. Hernoud).

116. Caesarea, Israel: arcades of Herod's and Vespasian's aqueducts (the two are built side by side) approaching the city.

117. Mérida, Spain: arcades of 'Los Milagros' aqueduct (30 m high), in the outskirts of the town. This is the same aqueduct that originated from the Proserpina reservoir (Fig. 41).

118. Carthage: arcades crossing the Oued Miliane (photo C.M. Wells).

119. Rome: arcade of Aqua Claudia crossing the Roman Campagna, near Roma Vecchia. On top it carries also the Anio Novus, 'piggy-back'.

course, in the engineer's interest to keep the conduit on the ground as long as possible, so that the arcades do not usually extend across the entire plain (which is seldom dead level), but begin at whatever is the latest point consistent with the channel attaining a reasonable height on arrival at the city walls. This beginning is marked, especially at Rome, by the channel seeming to surface from underground and climb slowly on to an arcade of ever-increasing height; in fact, of course, it is not the channel that is rising but the land that is falling away.

The arcading is so impressive a monument that one is sometimes tempted to ask why the engineers went to such an expense. Would not an earthen embankment, carrying the channel on top, have been equally effective? Various explanations suggest themselves – the arcading offered free access across the line to farmers with land on both sides, and did not interfere with the natural drainage patterns of surface water, but I do not find either of these arguments too cogent, particularly in view of the relative ease with which, in a later age, railway embankments confronted and dealt with the same problems. But it is probably easy to underestimate the simple labour costs involved in building a big embankment,[74] and perhaps also to overestimate them for building the brick-and-concrete arches of an arcade. A great deal of Roman building construction was based on the principle of mass production by semi-skilled labour, a procedure to which the repeated construction of identical arches would have lent itself.

One must also note, however, that raising the channel aloft on an arcade did also entail some disadvantages. In particular, it made it nearly impossible to install any kind of tanks or regulatory devices. Vitruvius recommends that tanks (*castella*) be established at regular intervals, presumably, as Callebat surmises, 'grâce auxquels la conduite pouvait être interrompue en cas de fuite, de rupture, ou d'autre avarie', but specifies that they must not be on *venter* bridges;[75] if one were really isolating a section of channel for repairs one would have to do it at a point where there was somewhere for the diverted water to go. This almost rules out doing it on top of a bridge, and hence, by extension, anywhere along an arcade.

Did each arcade operate as one continuous unit from beginning to end with no possibility of intermediate diversion or interruption? Perhaps. Frontinus, 124, for example, mentions one expedient, whereby a section requiring repairs could be temporarily by-passed by lead troughing, ensuring continuity of the supply. Certainly the fact that the conduit was elevated 15 m or so above the ground must have made the provision of sluices, settling tanks, inspection windows and the like more complicated than when it was running on the surface.

Another possibility would, of course, have been simply to run the aqueduct at ground level, in pipes as a flat siphon. That the elevated course of an arcade was always preferred can have nothing to do with the pressure, for, as we have seen, the pressure in the pipes would be only that generated by their depth below the natural water level, i.e. the height of the existing arcade. Usually the highest arcades existing are nowhere higher than 20 m or so,[76] and as we have seen from our study of siphons this would offer no problems at all to Roman piping. But again, the manufacture of the pipes and their jointing would be a different matter, and no doubt the same economic factor that prevented siphons replacing bridges also prevented siphons replacing arcades.

The point is worth mentioning, because in a later age exactly the opposite solution was adopted by the Ottoman Turks. At Constantinople and elsewhere they often built aqueducts on the *suterazi* principle instead of on arcades. These consisted of a long series of consecutive flat siphons, the water running in a lead pipe at ground level (see Fig. 167). At intervals of around 200 m were a series of water towers – brick piers each with an open tank on top – and each siphon ended by running up one side of the pier and discharging into the tank, the offtake for the next one being set into the opposite end of the tank. It was thus very much the same system that we have already seen at Aspendos, except that the pipes and piers were smaller and the siphons repeated indefinitely over a long distance. Sometimes these raised tanks were used as distribution reservoirs, but their chief function was probably to act as surge tanks to relieve sudden variations of velocity and pressure in the pipes. Between each pair of piers the siphon lost 18 cm head, or an overall gradient of

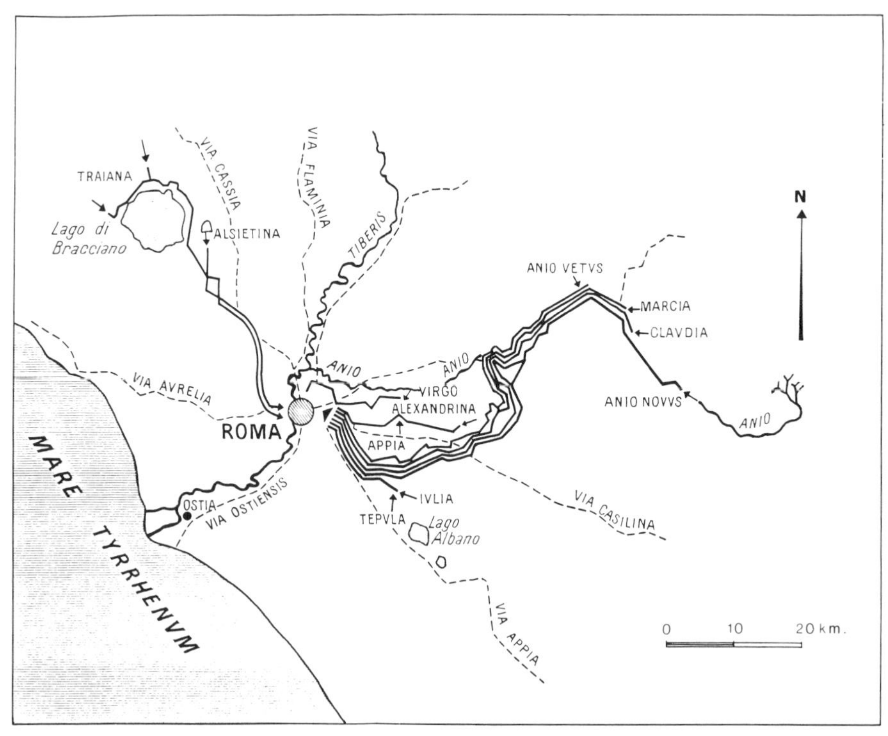

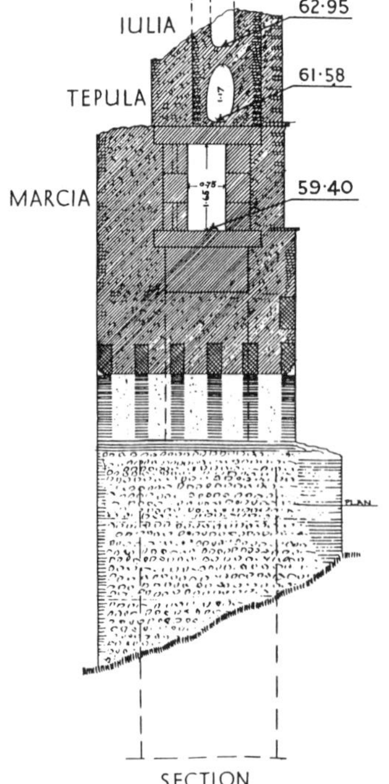

120. Rome: map of aqueducts (after M. Pediconi, *L'approvvigionomento idrico di Roma*, in *Studi Romani* XV, 1967).

121. Rome: cross-section of arcading in the Campagna, carrying three superimposed conduits (note that the two upper channels are not centred over the lower, which caused the Marcia's cover slab sometimes to crack) (Ashby).

122. Rome: close-up of channels in Fig. 119 showing different style of construction. The Aqua Claudia, below, is built of cut stone masonry, while the Anio Novus, later added on top, is of brick-faced concrete.

123. Metz: the water channel over the Moselle bridge was divided by a wall running down the middle, but this was not a side-by-side version of Fig. 122, for the division existed only on the 1 km long bridge; before entering the bridge, the water ran in a single conduit, and reverted to one on leaving it. The conduit is here seen entering a circular *castellum* at the end of the bridge, from which it continues on to the left out of the picture. The reason for the divided channel is not known; see n. 78.

0.9 m per km. The Turks used this system because it cost only one-fifth of a conventional aqueduct to build, but no doubt could do so only because it was on a small scale, employing a single lead pipe; to try to replace the channel of a conventional Roman aqueduct with enough continuous lead pipes to carry the same volume over some ten or twelve kilometres would be a different matter. (See also p. 237 below.)[77]

At Rome one sees a further development of the arcade system, without parallel elsewhere. When increasing needs required more water, the extra requirement might be met by tapping new springs near the source and adding their discharge to the water already coming down an existing aqueduct. More often, however, a fresh source had to be found. When it was, its water was not fed into the existing system but a completely new aqueduct was built to carry it. In the hills this would mean no more than digging an extra channel, but if it was in the plain and following the same route as an existing arcade (Fig. 120), instead of building a second one alongside, the engineers usually simply built the new channel on top of the old one, which it rode, piggy-back fashion, all the way to Rome.[78] Thus many of the arcades did double duty, a distinction often reflected in the use of the differing construction techniques on the same set of arches. The Claudia, for example, ran on an arcade of cut stone; on top of it was later added the Anio Novus, of brick-faced concrete.[79] The Marcia, also of cut stone, later acquired as passengers both the Tepula and the Julia and ran as a three-decker (Fig. 121, compare Fig. 122). The Tepula channel (and also the Julia, on top of that), however, was a good deal narrower than that of the Marcia, but was not even centered over it: one side-wall was directly above the side-wall of the Marcia below, while the other rested upon the Marcia's cover slabs and with its weight, unanticipated when the Marcia was built, often cracked them.[80] This was not an unusual problem. The arches in particular, called upon to carry this unexpected load, often settled or cracked, causing leaks in the conduits above, and Juvenal's famous comment on the Porta Capena as dripping wet (from leaks in the aqueduct above it)[81] could have been repeated almost indefinitely. Various repairs and reinforcements sought to correct the trouble, and Hadrian, not otherwise conspicuous as an aqueduct builder at Rome, was prominent in this work.[82] Arches were strengthened by cross-bracing (more or less such as we have seen in some aqueduct bridges), and supported where necessary by inserting in the archway a second brick and concrete arch as a liner, upon which the original cut stone voussoirs could rest, much in the way that scaffolding is used in an arch under construction (Fig. 124).[83] This was not always wholly effective, for it has sometimes now been found that the brick liner has settled more than the arch it is supposed to be supporting, leaving a slight gap between the two, and the arch still standing.

The reason for this complexity of channels was, of course, the vastly differing quality of the various waters. The Marcia was universally

124. Rome: arches of the Aqua Claudia (Fig. 119) reinforced with brick liners, to help support the extra load of the Anio Novus, added later.

lauded as the best, clear, cool and pure. At the other extreme was the universally despised Alsietina, so unwholesome that it was used only for the *naumachia* and watering gardens, though occasionally in emergencies, when the main supplies from the city were cut off for repairs to the Tiber bridges, the unfortunate residents of Trastevere had to drink it.[85]

There was therefore good reason to keep the channels separate and this was usually done. Connections and junctions were, however, regularly provided so that in case of need water from one channel could be diverted into another, either to drain a section of aqueduct for repairs or to supplement supplies as needed. The level of water in the channels varied with rainfall, and since their sources were in different areas, a further complication was introduced: not only was there a seasonal peak, but the different aqueducts might well peak at different times, even regularly so. Nothing has been published on the Rome aqueducts, but at Lyon this has been noted.[86] The diversion procedure was not always intelligently applied, so that it led to good water being wasted – in a famous aside, Frontinus, with obvious wrath, complains that he found even the Marcia being used for 'purposes too foul to mention' – or to being contaminated by its admixture with bad.[87]

As the actual channels were so close together, junctions were easily arranged. All that was needed was a sluice-controlled down-shaft from one conduit to the other at some convenient location, such as a water-tower inserted in the line of the arcade.[88]

Naturally, such connections worked in one direction only; for example, water diverted from the Claudia into the Marcia channel so as to clear part of it for repairs, could not be switched back into the Claudia once the affected section was passed. The aqueducts were all at different levels, and obviously one could divert only from a higher to a lower one. The Tepula must often have been switched into the Marcia, for though its water was relatively unpalatable, it ran right on top of the Marcia all the way and the transfer would have been very easily made – we cannot say more, for with so much of the arcades now gone, there must have been junctions and *castella* for which there is now no evidence. But the real threat was the Anio Novus. It was highest in level, largest in discharge, and, the Alsietina apart, worst in quality. A large amount of readily available bad water constituted a standing temptation to those willing to correct deficiencies in the supply by adulteration. We have already seen how, at Grotte Sconce, arrangements were made to turn water from this channel into the Marcia; no doubt similar facilities, now lost, existed also in the arcades crossing the Campagna.[89]

Arcades, naturally, are not confined to the plains surrounding cities. They can be found anywhere that the aqueduct needs to maintain a certain height while crossing a flat terrain, such as the kilometre or so of arcading at Chaponost on the Gier (Lyon), immediately before the Beaunant siphon. But it is in the city approaches that they are most commonly encountered: Rome, Caesarea Maritima, Carthage, Minturnae, Mérida, Aspendos – the list could go on. And there they achieve a new psychological significance. No doubt the engineers avoided arcades when they could, but once they were there anyway they told their own story of greatness, grandeur and sheer spectacle. As the traveller came into Rome along the Via Appia or the Via Latina and found himself gradually escorted, then almost encompassed about, by these mighty arcades, striding endlessly, mile after mile after mile, like a great army, across the fields on their way to Rome,[90] how could he not feel that the city he was coming to was in truth the centre of the whole civilised world? The bridges and siphons might be impressive engineering feats, but they were usually out in the country or the hills, unseen by most people. It was the arcades that were an outward public statement of Rome's greatness and (to a Roman's ears it would surely be unrealistic not to add it) the greatness of Rome's emperor.[91]

7
Planning and Surveying

The surveyor, a good man and true.

Frontinus, *Agrimensores*

When we come to the whole question of how the aqueduct was laid out and built, the reader may feel that the hitherto natural sequence of our narrative has been invaded by anomaly. He may well be right. The first steps in building any aqueduct must have been to survey the land and determine the line to be followed. Should we then not have dealt with this first, before proceeding, as we have done, to a description of the various features of the finished aqueduct? This may be so, but there is another way of looking at it. When an engineer or surveyor set about planning an aqueduct, he already knew all the characteristics of the standard design, its possibilities and limitations, and these formed the pre-existing basis on which his survey was conducted. There is therefore some logic in us doing the same thing, and mastering what an aqueduct is before confronting the problems of how to plan one. Only thus can we reasonably attempt to understand the surveyor's aims and methods. In this there is no immediately obvious plan to follow, but we may conveniently break our study into three sections. First we will consider what we can reconstruct of the theory of aqueduct surveying and the issues involved. This will take us some time, not because the surveying is inherently a more complex or important task than building the great bridges and arcades, but because hitherto it has not been much studied and we have to start from first principles, while other aspects of aqueduct building are more familiar and the reader can, if desired, inform himself further on the details by referring elsewhere. Then, abandoning theory, we will turn to practice. How was the plan, once surveyed and laid out, executed? This will lead us to look at the surveying instruments and tools available, which naturally affected and governed the execution of the task. We will end by noting briefly some of the problems in underground surveying and the effect such things as gradient profiles had on tunnelling techniques.

Planning the line

The terrain

The first step for the engineer was to familiarise himself with the terrain through which the aqueduct would run. For this, both in the preliminary survey and in later more detailed studies of the actual line selected, there was no substitute for walking over the ground and observing first-hand. Measured figures and computations had their essential role to play, especially in such matters as determining the slope, but for the general overview the most important thing must have been for the engineer to have in his mind's eye a clear mental picture of the lie of the land. In a later age, archaeologists, especially those working in topography, have also learned the crucial value of tramping over the country on foot. However, it is also plain that such trips would be supplemented by written notes and drawings.

Maps were known in the ancient world,[1] but they do not seem to have been in general use and were always, so to speak, a minority taste. Small-scale maps of the world were sometimes used by geographers and philosophers, but were evidently considered by most people something of a curiosity and *tour de force*.[2] Large-scale city plans (including architectural plans of individual buildings, very much in the modern idiom), and surveyors' maps of rural landholdings, particularly in centuriated areas, were also known; indeed, 'it was part of the duty of a land surveyor to make a map (*forma*) of any land he had divided up'.[3] There are even known plans of rivers and irrigation schemes. One well-known example survives in a stone-cut inscription, in the form of a diagrammatic representation of part of a river, carved immediately below the accompanying text.[4] Moreover, Frontinus himself tells us that he had made plans of the aqueduct system of Rome, on which were shown bridges across valleys and rivers, and other engineering features, so that he could, as he puts it, constantly have the whole network before his eyes and take decisions as if actually there on the spot.[5]

However, although all this does attest to a certain degree of surveying and mapping skill in the ancient world, there do appear to be two major limitations that largely nullify its value to an engineer laying out the line of an aqueduct. First, some of the 'plans' may well have been schematic diagrams of a known and existing network rather than true-to-scale and topographically accurate maps proper. The river on the inscription may be no more than a stylised representation, and Frontinus' 'plans' may well have looked something like the artificially schematic route-maps of underground railway lines one sees often exhibited in large modern cities. Indeed, the only Roman road map that we have, the Peutinger Table, is exactly that: compressing the entire Roman empire into a long, narrow scroll, it throws directions, proportions, scale and geographical truth of every kind to the four winds, and becomes a pure diagram, not a map.[6] For its purpose it works well enough, but such an approach is useless for the precise comprehension of topographical relief.

The second limitation concerns the nature, purpose, and hence the capacity, of ancient surveying. The name of its practitioners, the *agrimensores*, or 'field-measurers', itself indicates the essential and basic thrust of the profession. Surveyors were primarily concerned with measuring and dividing up land, with establishing and recording boundaries. The work was thus largely two-dimensional and was rarely concerned with heights. In the works of the *gromatici*, the ancient surveyors' manuals,[7] the problems given and illustrated are almost all concerned with the measurement of irregular and awkwardly shaped areas, polygonal fields and the like. There is a section on how to measure the width of a river. Mountains are indeed sometimes mentioned, but chiefly as an obstacle to be circled around, on the flat. Roman surveying is basically a product of plane geometry, and the Roman surveying instrument *par excellence*, the *groma*, could not measure differences in level. We must not exaggerate. There were, of course, other instruments that we shall shortly consider, the *chorobates* and the *dioptra*, that *could* measure levels, and even vertical angles. And relative heights were, of course, vital to laying out roads and aqueducts, and they were in fact surveyed. But they did not come nearly as naturally to Roman surveying as laying out lines and angles on the flat, and I doubt if at this stage in his work maps or cartography much assisted the aqueduct engineer. In particular, we never hear of anything like contours, or even topographical relief maps. At sea, we may note, the mariner likewise had no charts, relying only on *periploi*, written coastwise itineraries listing all the ports and the distances between them.[8] No doubt the aqueduct engineer, in familiarising himself with the terrain, also relied chiefly on written notes and recorded figures, with perhaps just the occasional explanatory diagram, to refresh his visual memory.

The source and the castellum

The first and most important thing for him to decide was the points for the beginning and end of his aqueduct, for these governed all else. It might be thought that the end was self-evident. Surely, indeed automatically, that was the city. But it is not quite as easy as that. Cities were often founded and built long before anyone thought of service by aqueduct, and hence in locations very unsuitable in hydraulic terms. To take a purely imaginary case, one could well foresee the difficulties in getting water to the Acropolis in Athens by a gravity-flow conduit. Even if it was on a hill, of course, a city might yet be served by a siphon if there was water available on another hill close enough and high enough to provide the necessary head: this happened at Aspendos. If the city was itself built on an uneven site, it was also possible that some of it would not be served, being uphill from the terminal *castellum* that formed the end of the aqueduct. A great deal thus depended on the location chosen for the *castellum*. Normally it would, of course, be put at the highest point in the town, so as to serve all of it conveniently (as, for example, at Pompeii), but this might not be possible. Two or three metres extra height for the *castellum* might cause no problem at all, if the line of the aqueduct was approaching from a reasonable altitude and on a reasonable gradient. If, however, the gradient was already so slight as to be only just feasible, an extra metre at the receiving end could be quite enough to make the whole scheme impossible.

In selecting his *castellum* site, the engineer therefore might have to balance conflicting needs: service to the full city (to say nothing of upper storeys) might have serious consequences back along the whole line of the aqueduct, even to the extent of changing to a different source at a higher level. A higher source would quite probably also be further away. How much higher, then, and how much further? What effect would it have on the overall gradient? An extra metre's height at the delivery end could be very costly indeed, given the wrong circumstances. It might be what was needed to serve the houses a hundred metres or so up the street, but to get the water that hundred metres might entail an extra five or ten kilometres in the length of the aqueduct, or expensive additional engineering works. A good example is Nîmes, where the gradient was so gentle that the water only just reached the town. It must have been very tempting to the engineer to locate the Nîmes *castellum* somewhere else, at a slightly lower level, and so ease the situation – but it would have been at the cost of leaving some of the city unsupplied. Lyon (Figs 125, 126) also was a difficult city to supply. Most of the industrial area was on the lower slopes and on the island (where the amphitheatre is now) between the two rivers,[9] and presented no problem. But the centre of the Roman city was on the heights of Fourvière, and the summit was at 300 m above sea level. The disposition within the city of the four aqueducts serving it (the Mont d'Or, Craponne, Brevenne and Gier) is very uncertain, but given their respective levels of arrival (260, 280, 280 and 300 m) only the last, the Gier, could serve the whole town, and until it was built some parts must have remained without aqueduct water.[10]

The selection of the source to be used called for a balancing of rather more factors than did the *castellum*. That it had to be tested for the production of good quality water (see pp. 73-4 above) goes without saying. It also had to produce enough of it. How, then, when he had located a suitable spring, did the engineer assess its outflow? Local peasants, whose directions in finding the spring would be invaluable, would be unlikely to provide any evaluation of its discharge more precise than a cheerful 'Oh, yes, sir, plenty of water, plenty!' To be sure that the spring could supply his aqueduct, the engineer would have to measure it. How

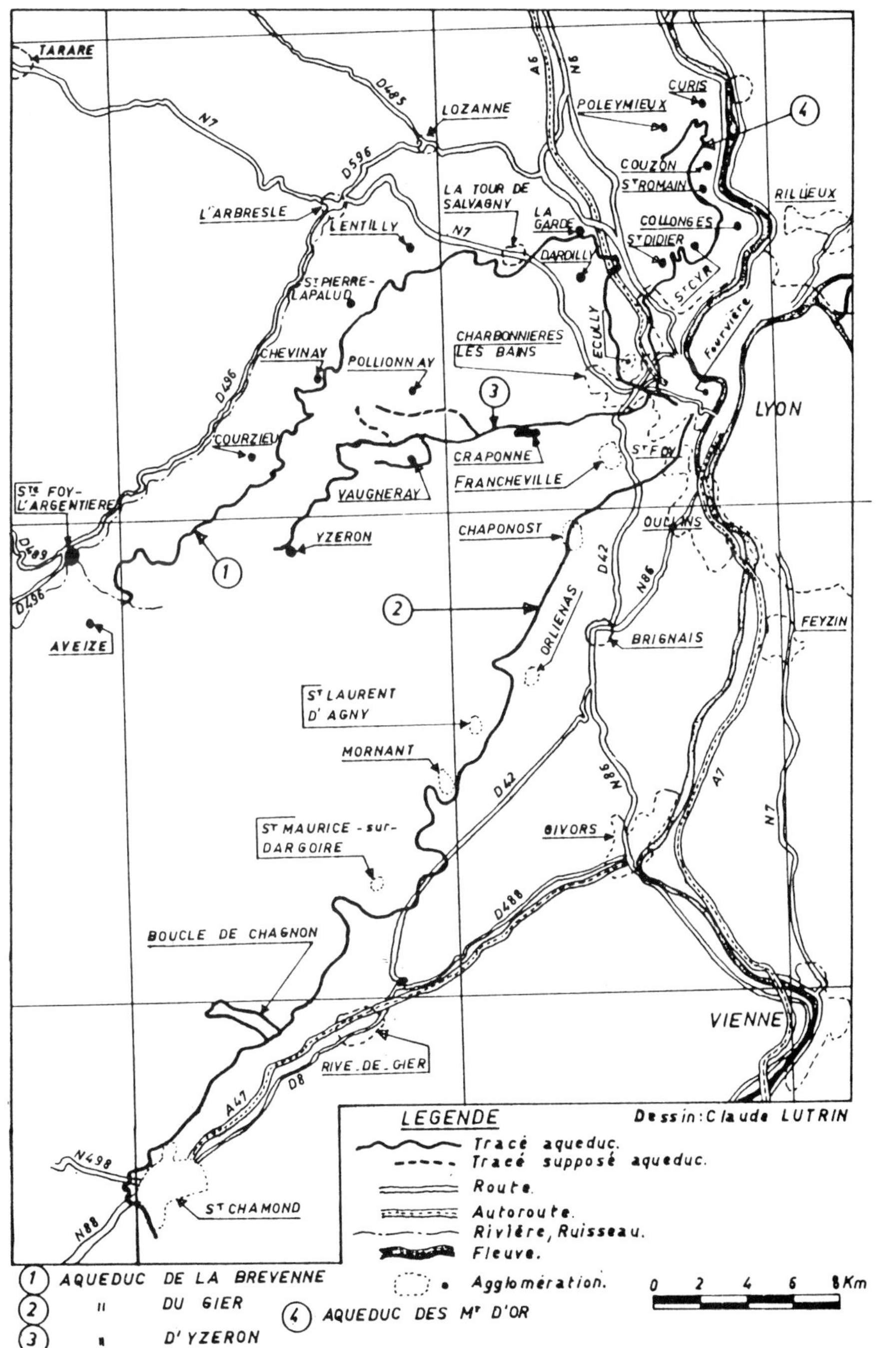

125. Lyon: map of aqueducts (Cl. Lutrin).

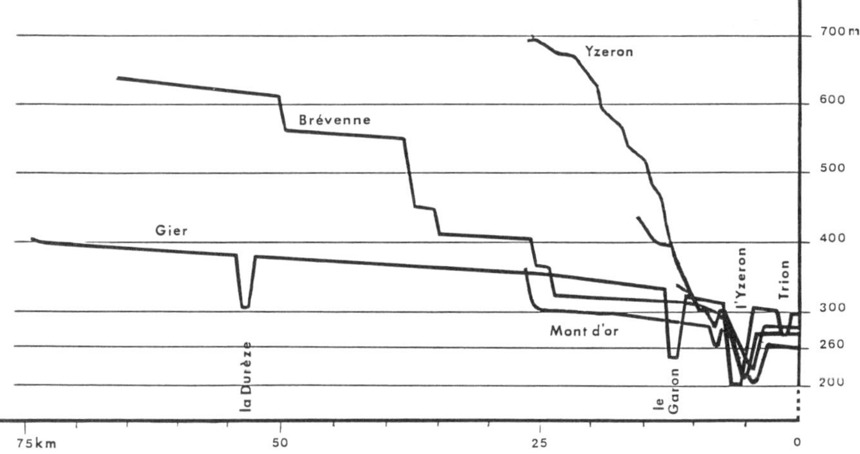

126. Lyon: gradient profiles of the four Lyon aqueducts, Mont d'Or, La Brevenne, Craponne (Yzeron) and Gier. The precipitous slope of the Craponne is partly accounted for by the use of cascades (Fig. 115), not indicated here; the V-shaped notches mark siphons.

one measured volume of discharge is another question that remains obscure,[11] but it is plain that the ancients in general and Frontinus in particular thought they could do it, and had specialised equipment of some sort that they installed and used for this purpose. Whatever it was, one presumes that the engineer installed it at the selected spring and, again presumably, kept it there for at least a year; he would have to know whether the volume of water produced in all seasons was adequate for his needs. In the process of planning the aqueduct, therefore, simply verifying the discharge of the spring may in itself have accounted for quite a large amount of time (though, of course, other work could proceed simultaneously).

The elevation of the source, and the feasibility of the terrain between it and the city naturally also must have bulked large in the engineer's calculations. No real rule on priorities can be laid down. There were a large number of variables, all interrelated (as we saw with the level of the *castellum*) and affecting each other. What really happened in the engineer's mind must have been a kind of mental juggling act in which all the factors were balanced against each other, trade-offs and compromises decided on, until finally the best all-round solution emerged. The closer the source to the city the better, since that meant a shorter and cheaper aqueduct. Yet elevation was paramount, since a slope was required sufficient to deliver the water. A high-sited *castellum* might rule out some local spring that would otherwise be quite suitable, compelling the engineer to look further afield. Yet the further he went, the higher a source he had to find, for if the height did not increase at least as quickly as the distance, then the overall gradient would become flatter. He might thus have to rule out several springs that would have been quite acceptable if they were only closer to the town, where their height would be sufficient to give a proper slope. This is why aqueducts often originate in the mountains. It is not, or at least not chiefly, a question of the purity of mountain water. It is because if the aqueduct is long (e.g. serving a city in a plain, such as Roman Campagna), then it has to start at an appropriately high level to give the necessary slope. This means finding a hill, and the further one goes in search of it, the bigger the hill has to be.

We may here pause to make a remarkable observation. If an otherwise suitable source is only slightly too low – two metres or so – it may sometimes be possible to raise it, so as to serve an aqueduct at a higher level than the spring. If this sounds like making water run uphill, this is effectively just what was done. It was achieved by building a dam immediately downstream from the spring. The water built up behind the dam, forming a spring-fed lake with its surface (if there was enough pressure in the spring) well above spring level. Thus the aqueduct was fed from the top of the lake and the lake was fed from the bottom, so raising the water from one to the other. This expedient, ingenious but questionably efficient (because it might inhibit the natural flow of the spring), was adopted at Caesarea (Israel) and Leptis Magna. It was never used widely, however, and remains an isolated curiosity.[12]

Let us assume, therefore, that the mental juggling process in the engineer's mind has come to fruition, and the spring to be tapped has been selected. So has the position of the *castellum*. With both ends of the aqueduct fixed we now come to the main problem, the choice and laying out of a route from one to the other. Let us, then, consider the factors relevant to it. First, there is the beginning and the end. These, as noted, are fixed points. Nothing can be done about them, and they govern what is to follow. They are fixed geographically, in that they determine the general course of the aqueduct – at least, in so far as they determine where it has to begin and end. They are also fixed in level, and to some extent determine the aqueduct's gradient. This statement is soon to be modified, but at present we may note that they do set height limits that the aqueduct must not exceed: to oversimplify, it must never rise higher than its source or sink lower than its city delivery point (except by siphon). And if the two are in fact fairly close together in level, then a shallow-gradient aqueduct is inevitable. We may note that the converse is not true: if the source is much higher than the delivery point then the aqueduct is not necessarily steep, because its route can be circuitous, and the extra length will ease the gradient.

As well as the fixed points for beginning and end, there is also a third variable factor governing the calculations, though this time it is one to be fixed by the engineer himself. This is the absolute minimum gradient which he deems acceptable. It must be made clear that I am not here talking about overall aqueduct gradients, or averages. I mean the gradient which is so slight that, if it occurs at any single point, it will in effect wreck the aqueduct, because it is not enough to keep the water properly moving, and it will instead back up and overflow. There is no set figure that hydraulics will enable us to assign to this limit. It depends on the individual aqueduct and, even more, on the judgment and decision of the individual engineer. What this decision was depended on the man himself, his experience, local tradition and practice, judgment, rule of thumb and heaven knows what else, but there had to come a point where he would draw the line and say 'No flatter than *this* – ever!' Again, the converse is not true. There was no danger of the gradient being too steep, because it could always be reduced by such measures as cascades or a more circuitous course; but it was *not* always possible to steepen a gradient that was too shallow, if the terrain itself was flat.

The theory of aqueduct gradients

Let us now consider how these three fixed factors – source, delivery point and minimum gradient – affected the line of the aqueduct. Let us say that on a modern map both source and delivery point are marked. Their levels are indicated, say 150 m apart. The two spots are, say 50 km apart in a direct straight line. This then gives a theoretical gradient of 3 m per km, or 0.3%, or, to put it another way, if the water ran absolutely straight downhill, from one end to the other, that would be the slope, all the way along. This therefore sets a theoretical upward limit for the aqueduct average gradient. The figure of a 150 m height difference is fixed, no matter what the engineer does, and let him lay out the route how he will, there is no way he can make it shorter than the straight line. 3 m per km is thus the absolute limit for the overall average gradient of the aqueduct. There is no conceivable way that an aqueduct can be built with a higher figure than that. In actual fact, the course of the aqueduct will be a good deal longer than 50 km, allowing for detours around valleys and contouring around hills, and the increased length will decrease the slope to something well under the 3 m per km figure. This upper limit is therefore something the engineer can calculate before he even starts to consider the route to be followed.

One may also work the other way round. In Fig. 127 is drawn a graph, with kilometres on one axis, and on the other metres above the delivery point (it should be noted that these are *not* heights above sea level). On it are plotted for convenience two of the commonest and best attested aqueduct gradients, 3 m per km, which is around the average gradient of the Marcia and the Anio Vetus at Rome, and 1.5 m, the much shallower gradient of the Gier aqueduct at Lyon, which is noted for the uniformity of its slope. Given then a desired slope of (for example) 3 m per km, it is plain that at a distance of 30 km from the city, the source must be 90 m or more above the *castellum*, at 40 km, 120 m, and at 75 km, 225 m. One may also see that a source, say, 75 m above it will work only if it is 25 km

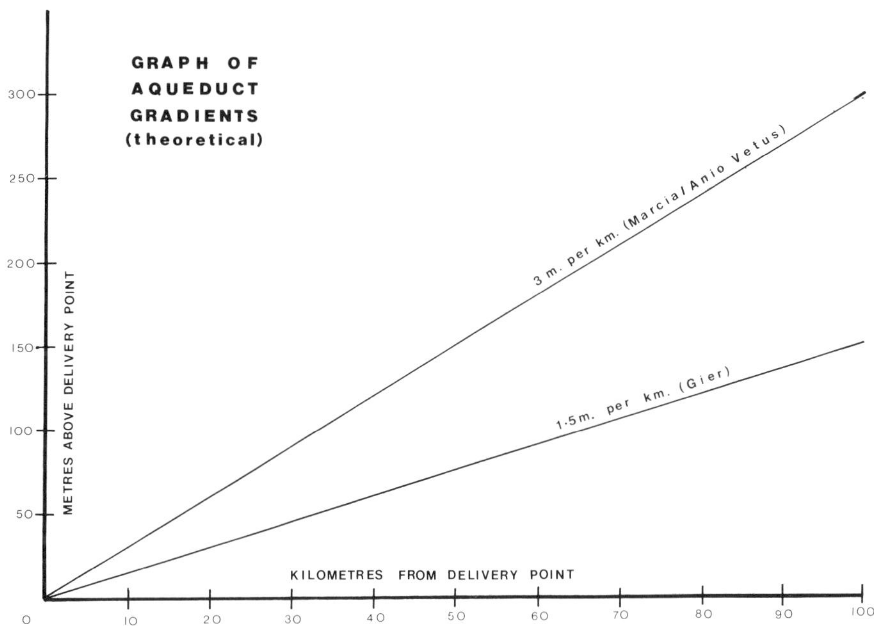

127. Graph indicating interrelation of altitude of source and its distance from city, for any given aqueduct gradient.

(or less) from the city; a source 150 m higher has to be within 50 km, and one 350 m higher within 83 km. Otherwise, an aqueduct, at that gradient, is not even *theoretically* possible. If one accepts a lower gradient, such as the 1.5 m per km of the Gier, then one can also read off the relevant figures from the lower line plotted. The reader is also free, of course, to draw in on this graph a line for any particular gradient that is relevant to his own studies, and then read off from it either how close to the *castellum* (the 'O' on the graph) a source of given height must be to be usable; or, alternatively, at varying distances from it, what height is then required for the source.

It will be noted that on this graph the gradient illustrated is calculated simply in terms of kilometres, and it is not specified whether these are direct, straight-line distance from source to *castellum* (a purely artificial concept), or the much longer distance along the actual route of the aqueduct. The graph will therefore work equally well for both. A 150 m fall over a 50 km distance produces an average gradient of 3 m per km no matter where it is measured, whether along an imaginary straight line through the air, along the winding course of an aqueduct, or even round in a circle.

Hitherto we have dealt largely with the overall average gradient as a kind of ideal figure, based on a straight line on the map. In fact, some publications seem to print this as the 'slope of the aqueduct'; where the

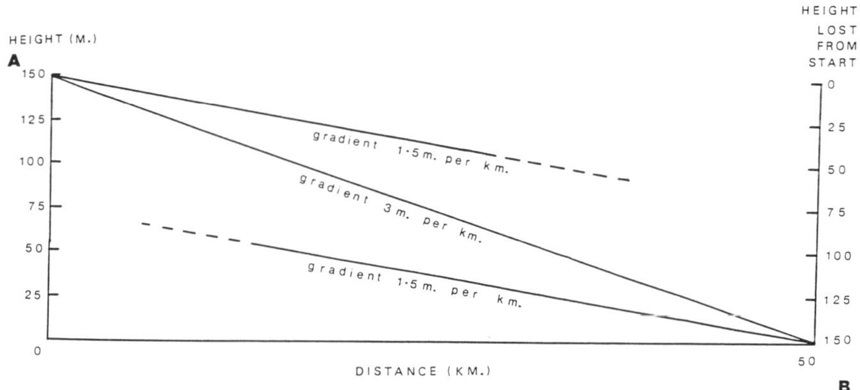

128. Diagram of theoretical direct gradient, and upper and lower parameters within which it must remain (see text).

real route of the aqueduct is largely unknown, and in any case was probably not too devious, it is not very unreasonable simply to divide the total difference in level by the rough distance between the two ends, and give the answer as a percentage gradient. Where the route of the aqueduct is known well enough for a fair estimate of its real length to be made (particularly when the course is a winding one, and hence very different from the direct distance), then a calculation based on relative height divided by this figure will give an average gradient for the aqueduct itself, a much more valuable figure than the theoretical upper limit derived from the direct distance. This also is sometimes done, and is, perhaps, the usual basis on which overall gradients are published. But now a further complication appears.

The theoretical overall average gives us a *terminus sub quo*, an outside maximum. The real overall average, based on real aqueduct distance, is a much more valuable figure, but, like all averages, it too can be very deceptive. A steep drop can be compensated by easy gradients. A medium-grade aqueduct can even include sections, in cascades, where the gradient is 100% vertical. How, then, can the engineer, as a practical matter, so plan his line that these variations can be accommodated, and still be sure of bringing it into the *castellum* at the right level? There are two ways. One is to note the distance, the difference in level and hence the average gradient, and to build to the average gradient throughout. This means (starting from the source) losing a set number of metres height every kilometre and arriving at the *castellum* just as one had lost the last of them, thus bringing the line in on level. This results in a completely uniform slope throughout the aqueduct. It was evidently used in the Siga aqueduct (see p. 192 below), on most of the Gier (exclusive of siphons), and was probably welcomed by the engineer as the easiest and

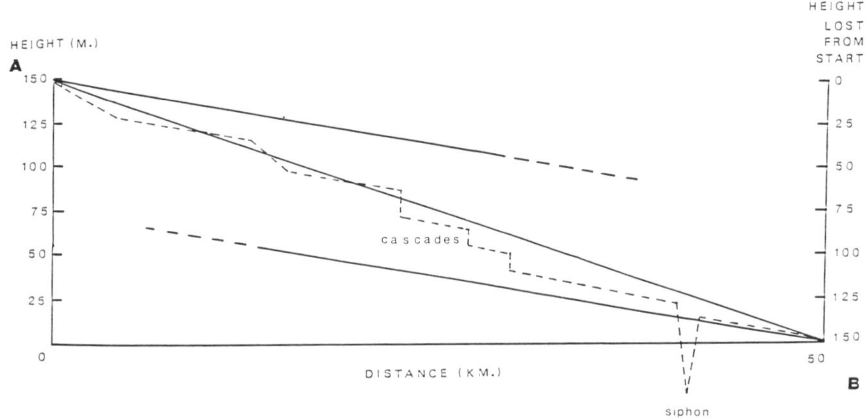

129. Graphs of (above) possible aqueduct profile; (below) two impossible profiles (x, too shallow, y, too steep).

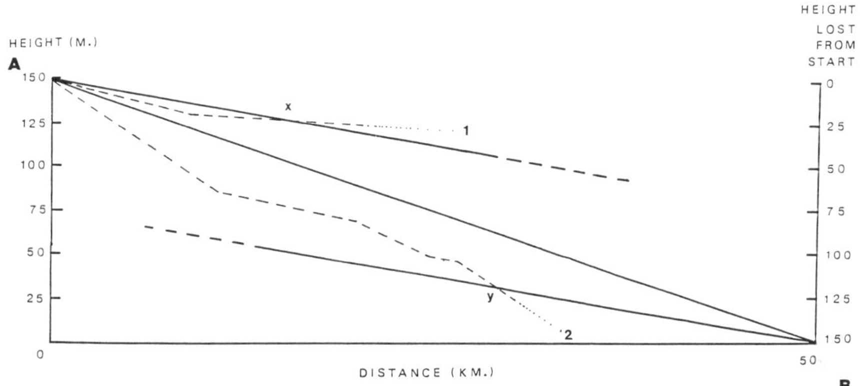

simplest method, when possible. Usually, however, it was not. Valleys, hills and plains combined to produce an irregular aqueduct profile, and the more normal situation is represented in Fig. 128.

Let us suppose a source, A, is to be connected by aqueduct to a *castellum*, or delivery point, B. Naturally, the same arguments and principles will also apply if this is not a complete aqueduct but only a section within a longer one, which continues on either side. A is 150 m above B, and they are 50 km apart, by real aqueduct route-distance. The average gradient will thus work out at 3 m per km, the straight line joining A and B. A and B are fixed. We now recall our third fixed point, the minimum gradient accepted by the engineer. If, as it happens, 3 m is also his minimum figure, then there is no more to be said. The aqueduct *has* to be built to a uniform 3 m gradient all through and no deviation from the straight profile, however slight, is possible. Usually, however, the engineer prefers to have height in hand with which to accommodate inequalities in terrain, and in that case he might well have picked a source higher than A to start with. Let us, however, assume he has picked a lesser figure (say 1.5 m per km, the actual gradient of the Gier) as his minimum gradient acceptable. Fig. 128 then illustrates what happens. Effectively, there are two parameters – the upper and lower lines on the diagram – and he has to keep his aqueduct between them while also getting it down from A to B. It will be noted that these two parameters (or limits) are parallel, and the vertical distance between them remains constant as the line of the aqueduct falls. They are set by the relative levels of A and B, and are a profile at minimum gradient leaving A, and one at minimum gradient arriving at B. We will continue to consider this question first graphically, then in practical terms – how could the engineer actually do it?

Fig. 129(a) shows dotted in the line of a hypothetical, but wholly possible, aqueduct. As long as the dotted line does not cross either of the two parameters, nor ever become shallower than either of them in angle (in which case the aqueduct would have an unacceptably flat section), the result will work. As will be seen, the profile in fact fluctuates on either side of the direct, or ideal, gradient (i.e. the overall average), and near the end in fact does dip below the lower parameter: this is all right as it is in a siphon, and not gravity flow. Finally, it arrives at B running along the lower parameter, that is, by a stretch at minimum (1.5 m per km) gradient. By comparison, Fig. 129(b) shows what can go wrong. In it two aqueduct profiles, again hypothetical, are marked in and shown by the dotted lines 1 and 2. Both of these profiles are impossible, and cannot be completed. Aqueduct 1 has crossed the upper parameter at the point marked x, and is now a lost cause. Of course, it could lose height rapidly and get back within the two parameters, but the damage has already been done. It is impossible for it ever to get above the solid line (as at x) unless somewhere, at some point, its gradient is *less* than 1.5 m per km, and there is therefore now no point in continuing it on towards B. Aqueduct 2 is quite satisfactory as far as point y, but here it dips below the lower parameter. Once this happens, there is no way to complete it and bring the water in to delivery point B without the gradient again becoming shallower than 1.5. Aqueducts 1 and 2 can each be saved by only one of two things. Either the engineer goes back and re-surveys the profile from the start,[13] so that it now stays within the two parameters (i.e. 1.5 in Fig. 128), or, yielding to necessity, he decides that the parameters are not, after all, to be rigidly enforced (i.e. 'Well, maybe we *could* get away with a stretch at less than 1.5, provided it's not too long'). That is, he either corrects the situation, or lets it stand and hopes that his original judgment had enough of a safety margin in it for the aqueduct to work anyway.

At this point, however, the reader may be tempted to put a pointed

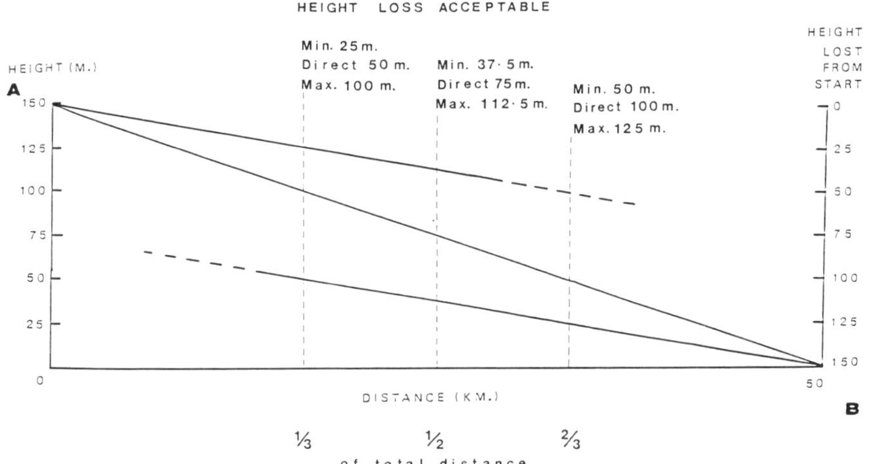

130. System of checking on aqueduct profile by keeping track of height lost as layout progresses (see text).

question: 'This may all be very logical, but is there any evidence that ancient engineers ever worked or thought in this way? Is there, for example, even the slightest archaeological evidence for anything resembling these diagrams and graphs?' The answer is a plain and definite 'No', but the question is itself misconceived. All the preceding material has been presented the way it has not as an exposition of ancient surveying methods, but because it seemed to me that this was the best and clearest way of presenting the realities of the situation. And realities they are, whether the ancient engineer thought of it this way or not. If he ever did get an aqueduct profile into either of the shapes shown in Fig. 129(b) then his aqueduct really was wrecked. Whether he understood the position or not, and what techniques he followed in doing the job, are irrelevant. The aqueduct was irrevocably shallower than what he had deemed to be the minimum gradient possible, and this is a matter of established fact. It must be noted that this position will hold good *whatever* figure the engineer fixes as his minimum. In my example, I used 1.5 m per km for ease of calculation and exposition (in fact, aqueducts *will* flow at a much lesser slope), but the principle is universal. And the engineer does have to draw the line somewhere, to set in his mind a minimum: he cannot have the whole aqueduct dead level.

The final diagram, Fig. 130, does not have any claim to historical reliability, but does outline one method that could perhaps have been utilised in actual practice. Set in the context of the limitations described above, it depends on keeping a careful tally of the height lost by the aqueduct from the start, as a rough guide to whether its profile is still within safe limits. Thus, if there is altogether a height difference of 150 m between A and B, an aqueduct following the direct, straight line (3 m per km) will have lost 50 m height at one-third of the distance, 75 m at half-way, and 100 m at two-thirds. At the one-third mark, it *must* have lost at least 25 m to have the minimum slope, and if it has lost over 100 m it is already too far down. At two-thirds of the way, the figures are minimum permissible 50 m, and maximum 125 m. It is thus possible to form some idea of whether the profile is acceptable on a basis of calculation, without recourse to scale diagrams, though perhaps the concept is more readily communicated visually. The calculations, based on height difference, could of course be conducted from either end: one could either calculate on the basis of what height one was actually at, or how much more height one still had to lose (as shown on the two scales on my diagram).

A practical illustration: Nîmes

All of this has dealt with the question of the gradient in a purely abstract or idealised setting, without considering the effect of real, existing, relief features. It may therefore now be instructive to consider one particular case, where we can see some of these factors in play. We turn to the aqueduct at Nîmes.[14] It was built around 19 BC, possibly under the supervision of Agrippa, the minister of Augustus. The *castellum* was sited in the highest part of the city, and the chosen spring was the Source d'Eure, near Uzès (ancient Ucetia). It was far from ideal, but then so was the location of Nîmes, hydraulically speaking: to the south and east lay plains, where any water found would be at too low an elevation to flow to the city, while to the west the terrain was too hilly and would involve too much engineering. A source in the general direction of Uzès (Fig. 131) was almost the only possibility. The source was at elevation 76 m, the *castellum* at 59 m, giving a 17 m difference in height; in a straight line, they are 20 km apart, giving, under the circumstances noted above, an ideal overall average gradient of 0.85 m per km. Immediately we know where our trouble lies. The aqueduct is going to be on an almost impossibly shallow slope, for 0.85 is the absolute limit and there is no way to make the gradient steeper; yet the Gier channel at Lyon and the Marcia at Rome have averages of around 1.5 and 3.0 – and these are real averages, measured along the line of the aqueduct. Real averages are always less than the ideal figure, so the real figure for Nîmes will be even less than 0.85, and the disparity with ordinary, typical channels even more marked.

Obviously every effort had to be made to keep the gradient as steep, and hence the conduit length as short, as possible. But a direct route was out of the question. From Uzès to St Nicholas would have been no trouble, but then the way south was blocked by the massif of the Garrigues de Nîmes, which would have necessitated a tunnel of impossible length, over 10 km. Following the valley of the Alzon to Collias and turning south

131. The aqueduct of Nîmes.

from there would again mean tunnelling, this time for 8 km. The only practical answer was to go round by Remoulins, but this meant an L- or V-shaped route that brought the total length up from 20 km, the direct distance, to around 50. As the height difference remained constant, this made the gradient yet shallower, giving a real average of around 0.34 (*one-tenth* the average gradient of some of the Rome aqueducts). This was bad enough, in all conscience, but worse was to follow. Not only had the 0.85 figure of the direct Uzès-Nîmes route (which was already very low, even to start with) shrunk to the miserable 0.34 of the only route that

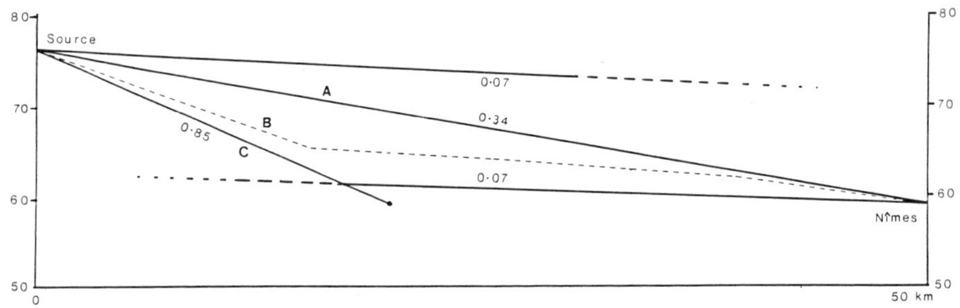

132. Possible profiles and gradients for the Nîmes aqueduct.

could actually be used, but even this 0.34 is an overall average. It can be attained only if the gradient profile is a straight line, losing height with absolute regularity. If there is any variation in it at all, to cope with local problems, then while some parts of the profile may be steeper (and hence no trouble), others will inevitably be shallower still, less even than 0.34. The engineer must therefore determine on a ruling gradient (i.e. the minimum he deems acceptable, *at any point*, if the water is to be kept flowing),[15] and this will inevitably be less even than 0.34. In fact, the absolute minimum gradient on the Nîmes aqueduct is something like 0.07 (7 cm per km!), and from the fact that it is there we must assume that the engineer intended and accepted it.

The situation may thus be graphically represented as in Fig. 132, following the same format we have already used. The upper and lower parameters are now set at 0.07, this being the real minimum gradient on the aqueduct, between the Pont du Gard and St Bonnet. The straight line, A, gives the 'ideal' slope of 0.34 m per km, that is, the average overall. Line B gives the profile of the existing aqueduct. It will be noticed that it has almost hit the lower parameter as early as the Pont du Gard, and from there on, all the way, is almost at the limit. There is next to no margin of height in hand. By comparison, the straight line C, drawn over a shorter distance, and hence steeper, shows the ideal 0.85 gradient that, theoretically, should have been attainable had one been able to come from Uzès to Nîmes direct, via St Nicholas (20 km, for the same height loss).

It was therefore decided to accept a shallower gradient and run the route via Remoulins. Now entered a further complication. The mountains, the Garrigues de Nîmes, were not the only barrier blocking the way. There was also the river Gardon. The mountains might be avoided but the river could not, and somewhere or other the aqueduct would have to cross it. The Gardon is not too big a river, as compared with, say, the Moselle, crossed without trouble by the aqueduct of Metz. The problem was not one of span, nor of length of the bridge – those endless arcades across the Roman Campagna will remind us of just how little Roman engineers were worried about long bridges – but one of height. The Gardon presented two difficulties. First, the river, that is to say, the actual surface level of the water, was very much lower than either the aqueduct source, at Uzès, or its delivery point at Nîmes. These levels are marked in the map (Fig. 131): at St Nicholas, the Gardon is 37 m above sea level, 24 at Collias, and 18 at the Pont du Gard; Uzès and Nîmes are at 76 and 59. It therefore follows that the bridge (assuming there is one, and not a siphon) will have to be a very high one, irrespective of local terrain. Second, as it flows downstream the river loses height much more quickly than the aqueduct does (or can, given its required level of arrival in Nîmes). Therefore, since the difference between the two is continuously increasing, the longer the aqueduct waits before swinging south to cross the river, the higher is the bridge required to do it.

Fig. 133 illustrates this situation. Three possible, but hypothetical, bridges are shown, as they would be at St Nicholas, Collias and the existing Pont du Gard. The river Gardon is shown at its real level at each of these locations. At each, the aqueduct level is conjectural, based on a calculated uniform slope from Uzès to Nîmes by this route. The details are thus entirely imaginary, as no aqueduct was ever built at St Nicholas or Collias – I can only reconstruct roughly what level the channel *would* have been had it followed that route – but the main point is clearly shown: as one moves downstream, the bridges become progressively higher, and if one delays the crossing as long as the Pont du Gard one ends up with an absolute monster, 52 m high.

One may also at this point appositely reply to a query that may trouble the reader: why is the hypothetical aqueduct shown as running at the same height (70 m) at Collias and the Pont du Gard? A glance at the map will show that the second is much further from the source at Uzès; would, then, the channel not have lost more height by the time it got to it, and be lower than at Collias? This is, in fact, an excellent object-lesson in the inter-relation of route, distance and gradient. The Collias level of 70 m is postulated on the basis of an aqueduct running via Collias and thence directly by tunnel to St Gervasy, for a total route-length to Nîmes of around 40 km. The length being less than the 50 km route via Pont du Gard/Remoulins, the gradient will be steeper, and it will already by Collias be down to the same level that the other, shallower route only reaches by the more distant Pont du Gard. The secret is that the altitude of the aqueduct here depends not on the distance it has come from the source, but on the proportion of actual route that has been covered. The respective levels at start and end remaining fixed, one-third of the height difference must have been lost by the time one third of the journey has been accomplished, whatever the route or total distance – assuming, that is, a uniform slope throughout. It follows that at the one-third-of-the-way mark (or any other given proportion) between these fixed points, the channel will be at the same level on *all* possible routes, irrespective of their length or gradient. And Collias is about one-third of the way to

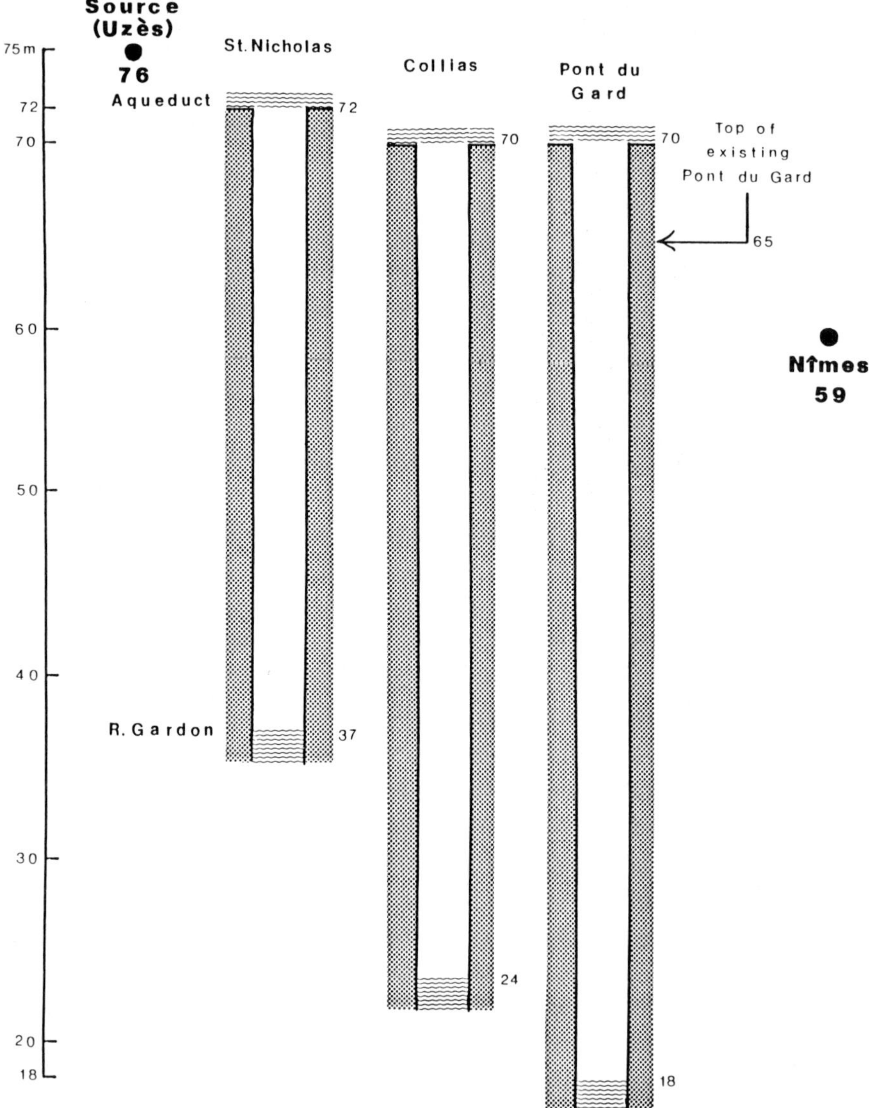

133. Comparative heights that would be required for the Pont du Gard if built at various locations.

Nîmes on the shorter Collias/St Gervasy route just as the Pont du Gard is on the longer one via Remoulins. This is why in both the channel is restored at the same level, 70 m. Indeed, there would be a good case for doing the same thing at St Nicholas. In fact, of course, the channel crosses the real Pont du Gard at a rather lower level, 65 m. It is now time to explain why, and in so doing to leave the theoretical realms of the ideal

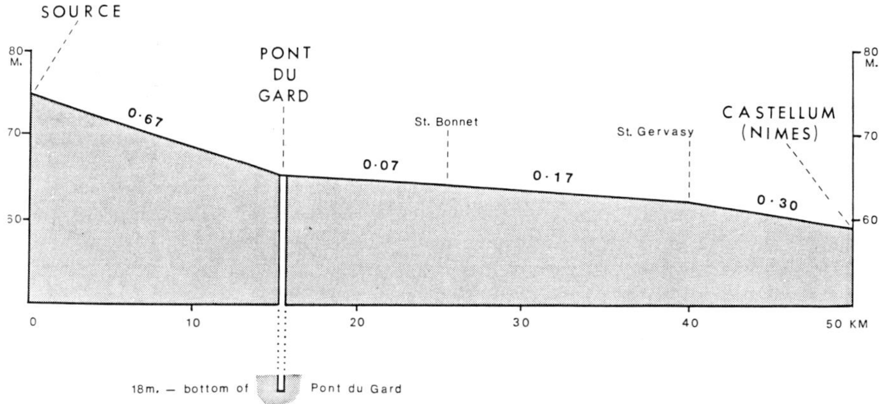

134. Gradient profile of Nîmes aqueduct.

gradients and entirely uniform slopes, based solely on calculated averages.

Hitherto we have been dealing largely in terms of gradients of uniform slope, or, to put it another way, a gradient figure determined by dividing the distance into the height difference. No doubt the engineer would prefer such a route if he had the choice. It was easier to calculate and easier to lay out on the ground. Usually, however, he had to use a profile of varying gradient because of the exigencies of the terrain. The exigencies were usually simple ones that the reader will readily comprehend, such as mountains and valleys, but sometimes the situation could be more complex. Nîmes again offers a good object lesson. The problem was the Pont du Gard. As we have seen, this was the only practical route, and the aqueduct, given its gradient problems, would have to be built as close as possible to a uniform slope. But a uniform slope would bring the aqueduct across the Gardon at a level of 70 m (Fig. 133) on a bridge 52 m high. For a Roman bridge, this would be a record-breaking height, and no doubt many engineers would have considered it unsafe or even impossible. It was vital to reduce the height, even by a small amount. One might feel that on so large a work a small reduction would make no great difference, and might even be a case of spoiling the ship for a ha'porth of tar. This would be true with a smaller work that was already within the limits of engineering capacity. There, all a reduction in size would achieve would be to reduce costs and trouble, and it might well not be worth it. But with a structure as large – in particular, as high – as the Pont du Gard, already on the very edge of what was possible and what was not, even a small reduction might make all the difference, for what was at stake was not expense but whether the bridge could be built at all. At any rate, what happened was plain. The height of the bridge was reduced by about 6 m and built only to a height of 48.77 m above the river.

The result is shown in Fig. 134. The aqueduct now crossed the bridge in

135. Pont du Gard: the levels of the beginning and end of the aqueduct (at Uzès and Nîmes), as marked in the margins, are scaled to the relative levels and proportions of the bridge as shown in the photograph. Note: to arrive at the proper level in Nîmes, the conduit can descend no more than from on top of the bridge to around the top of the middle tier of arches, and that spread over a distance of 34 km. To the source at Uzès, on the other hand, the distance is much less and the height difference greater, resulting in a steeper gradient.

the channel existing today, at a level of 65.3 m above sea level instead of the 70 m of the uniform slope. The gradient profile then did exactly what the engineer had no doubt been hoping to avoid, and developed a sag one-third of the way along, where it dips down to cross the Pont du Gard. The height limit on the Pont du Gard is thus the key factor that governs the profile and gradients of the entire aqueduct, from one end to the other. The first section, at a gradient of 0.67 m per km, is relatively steep and affords no problem to water flow. It is from the Pont du Gard on to Nîmes that the price is paid, for that leaves a fall of only 6 m to be spread over a distance of 25 km. The point is perhaps more clearly made if one envisages the Pont du Gard itself (Fig. 135). The height of the topmost tier of arches is 7.4 m (including the height of the channel itself). Thus the channel, over the next 25 km, has to come down only as far as about the top of the middle row of arches: that is the actual level of the terminal *castellum*, far away in Nîmes. The gradient over this section is itself not wholly uniform, but its shallowness is at once apparent, particularly on the long run from the Pont du Gard to St Gervasy. The worst section is from the Pont du Gard to St Bonnet, where the gradient is a mere 7 cm per km, the figure used for the two parameters (0.07) in Fig. 132 (or 0.007%). Part of the reason the gradient here is so slight is that the route is very sinuous, winding in and out of a series of steep valleys that inevitably lengthened the distance. This must also have made sighting, by *chorobates* or any other means, almost impossibly difficult as the channel was constantly twisting. Indeed, so tiny was the gradient to be established, and hence so vital the need for precision, that in these circumstances one would be tempted to change the phrasing of my last sentence and to describe the feat as not almost but wholly impossible, except for the fact that the Romans did it. Not much more than 100 m of the conduit is ever visible at once, and the amount of fall to be measured and established over this distance is *seven millimetres*, or about $\frac{5}{16}$ of an inch. The degree of fine accuracy required for surveying in these conditions and to so close a tolerance is almost unbelievable, but it was achieved and the aqueduct did flow. It is natural that bigness should always attract our admiration, but smallness can also be a function of greatness, and the Roman achievement in surveying and building this stretch of aqueduct within such very narrow limits must rank as an achievement almost on a level with the design and construction of the Pont du Gard itself.

Executing the plan

Contracts

Once the route was chosen, the gradients calculated and agreed upon, the location of any special engineering works pinpointed, and the whole route surveyed, it was presumably marked out on the ground with stakes. The next step was to start construction work.[16] A certain amount of study has been devoted to the organisation of the labour force and the contract system as it applied to Greek temple-building, but little to Roman, and none at all to aqueduct construction.[17] It does nevertheless seem plain that while some of the shorter aqueducts were built under one contract, with a single construction gang starting at one end and working straight through to the other, longer ones were divided up into two, three or more sections, contracted for separately and constructed simultaneously. 'Contract' is perhaps not the *mot juste* for, unlike most Greek buildings, the Roman aqueducts seem to have been built largely by military rather than commercial labour. The world of the building contract, so common in Greek architecture, is therefore of questionable relevance to Rome. We really do not know how the labour force building a Roman aqueduct was organised, how it was paid, if at all, or whether there was a civil contractor as opposed to an army engineer. The reader therefore must not here interpret the word 'contract' too closely. All I mean by it is the responsibility for building a given stretch of aqueduct, and it does seem that on a long aqueduct this responsibility was shared out or divided up. The proof comes from the Eifel aqueduct at Cologne. There, points in the line can be identified where two different sections met and, because of an error in the surveying or construction, did not meet properly: one section was higher than the other, necessitating an abrupt and unintended drop of 40 cm or so to make the junction[18] (Fig. 136(B)). By comparison, the very short (8.2 km) aqueduct at Siga (near Oran in Algeria: Fig. 136(C), (D)) was apparently built under one contract and by one gang. The inference comes from the fact that after they had correctly built one 1.45 km long section of it, an error then crept into the next section, which became progressively too high. The mistake, when realised, was then compensated for by steepening the gradient on the third and last section, which brought the channel back down to the level required for arrival at the *castellum*.[19] This seems to prove that the aqueduct was not only built as a single operation but that work started at one end and progressed to the other, for the procedure of altering one section to compensate for errors in another is only possible if the incorrect section has been built first.

Conversely, when the workmen on the Eifel aqueduct made their mistake and finished with a channel that reached the junction point too high, it could not be compensated for in the following section because that was already built. The presumption thus is not only that on this aqueduct the construction was divided up into separate contracts (which, on so large a project, would in any case be no more than a reasonable assumption), but that in each section the workers attacked the task as they did at Siga, beginning at the upstream end and working down to the lower one, rather than dispersing in groups and tackling the job simultaneously all along its length. There is no way that the observable 40 cm discrepancy at the meeting point could have happened if the two gangs responsible for the sections on either side of it had been working on the site together and at the same time. The reader will no doubt have a question to ask. Given the existence of a junction point, at which the upstream section of the conduit is higher than the downstream one, how do we know which is at the wrong level? Could it not be the upstream section that was at the correct level, and the downstream one too low? This rapidly opens out into a wider question. Suppose the work in each section began at the downhill end and progressed uphill (the opposite to what is postulated above), would this not mean that when there was a discrepancy at the meeting point, it was the upstream section that was right? The junction point was where, on this basis, work started on the upstream section, and finished on the lower one; and work on any section is going to at least *begin* at the right level, though not necessarily end at it. And, in any case, that only brings up two further points. First, it assumes that the gradient was established and marked on some repetitive principle, so that cumulative errors were possible; if the various stakes and markers used by the gangs were levelled individually, errors would be localised, and the end of the section no more liable to mistakes than the beginning, or anywhere else. Second, what reason do we have to believe that work *did* progress in the downhill direction, from higher to lower?

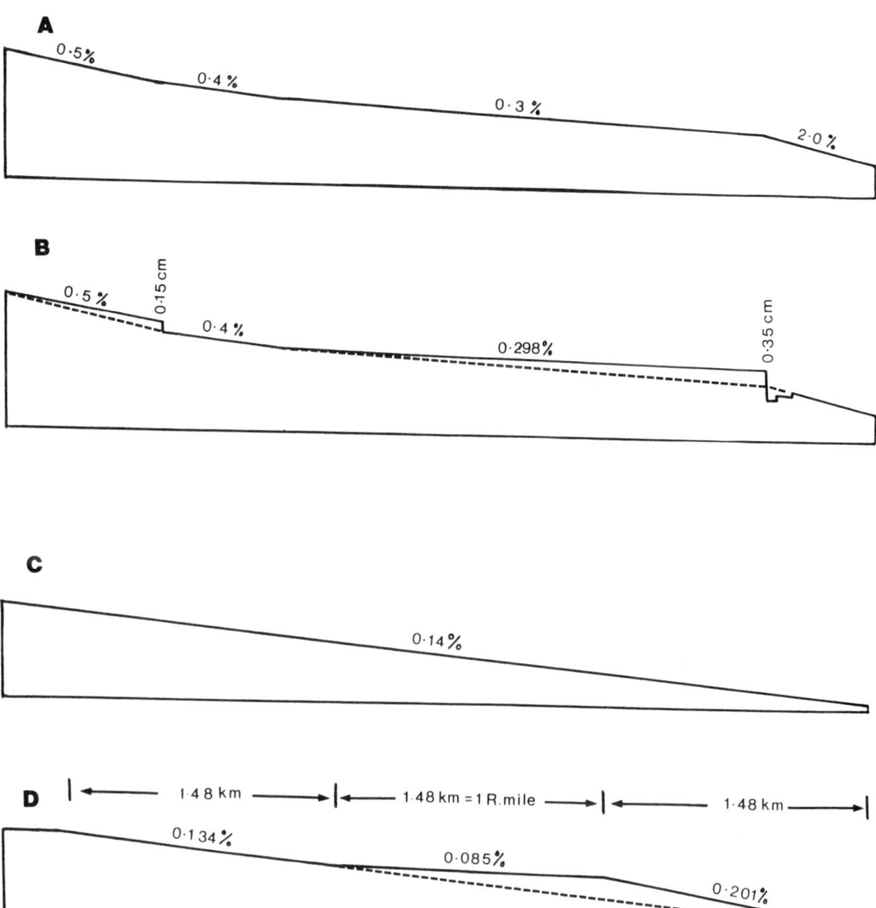

136. Cologne: the Eifel aqueduct was built in separate sections, and at Mechernich the intended profile is shown in (A); (B) shows what actually happened, the steps at beginning and end of the section (0.15 and 0.35 m high) reflecting mistakes in the levelling. The aqueduct at Siga, Algeria, was built straight through from one end to the other, but deviated from the intended profile (C) as shown in (D), the actual profile. Note that in D the changes in gradient occur at points 5,000 Roman Feet (= 1 Roman mile) apart (K. Grewe).

The answers to both of these questions come from recent work on the Eifel aqueduct. Engineering works, such as settling basins and bridges, were constructed first, and the aqueduct channel afterwards. Several times we find a height correction in the conduit to enable it to match the level of the bridge; and such corrections are always on the upstream, never the downstream side, just before the bridge and never just after it. The clear conclusion is that on the downstream side the gang was working away from the bridge, and on the other side approaching it.[20]

This seems convincing, and though one would hesitate to apply the evidence of one single aqueduct to Roman practice in general, the Eifel (and Siga?) seems to be the only place where the direction and sequence of work has been studied at all. Other engineers on other aqueducts may well have done things differently, but until we know about it we have to accept the Eifel as representative, *faute de mieux*, and to assume that aqueducts were generally built in sections, with the gangs proceeding downhill in each. On the Eifel, the sections seem to have been each between 4.44 and 5.33 km long (4.44 km = 3 Roman miles), and there were thirteen to sixteen of them.[21] This brings us to our second question, how was the gradient actually marked out for the workmen?

Marking the gradient

Here we embark largely on conjecture. There are two possible ways of setting and marking out the gradient (within the general theoretical limitations already discussed). One is to follow the lie of the terrain, steepening and reducing the slope of the aqueduct (and bending its course, in plan) to accommodate rises and depressions in the land as it encounters them, but without trying to maintain any given, uniform slope. Presumably a lot depended on on-the-site decisions and judgment by eye, with a constant tally kept of the amount of height lost so far. The result would be a wholly irregular gradient profile of infinite variation, reflecting the variations in the natural terrain. The second way is to build the aqueduct to a fixed gradient, uniform and calculated, evening out the terrain where necessary by engineering work. This need not be extensive. It may entail nothing more than in places digging the trench for the conduit a little deeper, or shallower, or perhaps, as it crosses a hillside, choosing a course that will keep the gradient right. A fixed-gradient aqueduct of this kind can be built in two ways. It can be based on a uniform gradient throughout, giving a gradient profile of the kind we called 'ideal'. This is usually practical only for short aqueducts, such as Siga. Longer ones were divided up, as noted, into sections, and each of these would be built to a uniform gradient, not necessarily the same from one section to another. Where required by topographical exigencies, a section could be subdivided, with different, but internally uniform, gradients for each of the subsections.

This procedure results in an overall gradient profile that is variable, but the variations come at set, often regular, intervals. The magic figure you have to look for is 1.48 km or any multiple of it, in horizontal distance along the line. 1.48 km is one Roman mile (or 5,000 Roman feet) and it was natural that the section lengths should be defined in this unit. At Siga the correspondence is clear. Three consecutive sections of 1.449, 1.473, and 1.504 km are identifiable, each break or junction being marked by a significant change in gradient.[22] Moreover, since the gradients were artificially determined by calculation, rather than by the natural configuration of the ground, they usually resulted in a slope easily expressible in round figures. This holds true both of the gradient throughout a section, and of the gradients of individual subsections where the section is subdivided. Ancient standards of measurement are different from modern, but this raises fewer difficulties than one would expect. The basic way of measuring a gradient is by taking a given horizontal distance, and saying how much height is gained or lost in it. Provided the units in each are the same, their relationship to each other remains constant no matter what they are. In other words, a gradient of 3 in 1,000 = 0.3%, = 3 m in 1,000 m, = 3 m in 1 km, = 3 Roman feet in 1,000 Roman feet, or 3 of anything in 1,000 of the same thing. If, therefore, the Roman engineer calculated his gradient in terms of so many feet per thousand, this will show up in our modern figures as an even percentage (e.g. 0.2%, 0.3%, 0.4%, = 2, 3, and 4 R. ft per thousand). Of course, the basic Roman measurement was the mile (5,000 ft), so that 0.3% = 15 R. ft per R. mile, but the two systems are essentially compatible and an even figure in one will produce an even figure in the other. Thus, the planned gradients in the sections of the Eifel aqueduct already mentioned, around Mechernich-Lessenich (Fig. 136(B)) work out, according to modern surveying, at 0.5%, 0.4%, and 0.3%, giving a gradient in Roman terms of 5, 4, and 3 R. feet per 1,000 ft, or 15, 12, and 9 R. feet per R. mile.[23] As these work out so even, we can therefore assume that we are here dealing with an aqueduct built to a planned and calculated gradient. It is at this point tempting to try to extrapolate to other aqueducts, indeed to try to classify them into two categories: those with gradients dictated by the natural terrain, and those built to a calculated slope. Surely those in the second category will stand out, with the slope regularly changing at multiples of 1.48 km, and with gradients expressible in round figures? Unfortunately, it is not so easy.[24] Most aqueducts have minor variations in course and gradient, the result of slight error, that may yet be enough to confuse the issue. More to the point, publications of most aqueducts are incomplete, often necessarily so since large stretches of the line are either inaccessible or destroyed, and therefore a large part of both route and profile, as published, represent conjectural reconstruction. It is thus often impossible to locate the points of gradient change in sufficient numbers or with sufficient precision for us to say that they came at 1.48 km intervals.[25]

Granted, however, that the aqueduct was being laid out in this way, how was it done? Both the Eifel and the Siga aqueducts suggest what is, in any case, no more than a reasonable and obvious technique. The operation took place in two stages. First, the general line and gradient were measured out and marked at a series of carefully surveyed checkpoints. These, like benchmarks or triangulation points in modern surveying, served as points of reference. They came at intervals of one

Roman mile, though of course the intervals might be shorter if circumstances dictated it. Thus a contract section, as considered above, might include several checkpoints within it, or might simply run from one to the next.

The section between two checkpoints then could be surveyed in two ways. It could be continuously surveyed, perhaps from both ends, marking the levels throughout, and at each spot the proper level calculated independently of the others; this procedure ensures that any error will be localised and isolated. Alternatively, one can start at one end, carefully set the desired slope, and simply carry on digging through to the far end, guided by some simple apparatus that projects and repeats the gradient of the first stretch. Thus, if a gradient of 0.3% (3 in 1,000) is desirable, one gets a stake, marks it off with 3/10 of a foot indicated, and sets it up vertically 100 feet down the line from the start of construction. Arriving at the stake and ensuring from it that this end of the trench is indeed 3/10 of a foot lower than the start, one moves the stake a further 100 ft on, and continues. Theoretically, losing 3/10 ft every 100 ft, one is building on a gradient of 0.3% and will therefore arrive at the far end on the right level. One hopes that errors will cancel out. In fact, they are more likely to be cumulative. This procedure has the advantage of being quick, repetitive and simple. It requires no trained engineer or surveyor on the site, for a working foreman would be quite competent to manage it. Its relative inaccuracy (resulting, in the Eifel line, in sections that do not wholly match) is not important if the gradient is a medium one, where slight imperfections in the uniformity of the slope, or bumps or drops where one moves from one section to the next, are of no great consequence. But where one was dealing with an aqueduct like Nîmes where the slope was so shallow that even the slightest error would have been disastrous, this method cannot have been employed. Other, more careful, measurements must have been taken, with other, more precise, instruments.

One may offer a further observation. Plainly the first method, that of maintaining a constant gradient, is the more inflexible. In theory, a constant gradient – even a dead level – can be maintained in any terrain, no matter how irregular: that, after all, is no more than what is a contour line on a map. But, like a contour, it can usually come in only one place, which in practice may involve some very difficult construction. The second method, always keeping a margin of height and losing bits of it from time to time as required, clearly gives much more flexibility in choosing the alignment, and should therefore be more suitable for an aqueduct passing through broken country. It ought to be possible, therefore, to divide most aqueducts into two categories, constant gradient or variable gradient, and this ought to be discernible from a detailed study of their gradient profiles. This, however, is again something that would require a more detailed and comprehensive range of published data than is generally available to us, at least as yet.

There is also a third possibility. I am told by a modern hydraulics engineer that in fact the chief problem in laying out a fixed gradient from source to delivery point is that it is very hard to get truly reliable figures for what their respective altitudes actually are.[26] Even the contours and heights on modern topographical maps, being partly filled in by guesswork between known points, are not wholly to be trusted, and the ancients, particularly over a long distance, must have been hard put to it to get reliable figures. This being so, he suggested, the most sensible way of going about it would be to begin the aqueduct at minimum gradient (the upper, 1.5 m parameter in Fig. 128), and simultaneously to begin construction from the other end, also at minimum gradient (the lower, 1.5 m parameter, working uphill from B). When both halves had got to the middle of the aqueduct and it had become apparent just how great was the vertical interval separating them, they would then be joined by a steep downhill section. This system, which essentially means starting from both ends while keeping a margin in hand to be expended (or what is left of it) where they meet at the middle, ought to result in a characteristic three-part profile: shallow, steep, shallow. I do not know of any existing aqueduct that readily matches this specification, but, as noted, our available data are very imperfect.

Surveying instruments

The T-board

The T-board is a simple measuring device for which no ancient evidence, archaeological, literary or other, exists, and which is postulated because something like it is needed for establishing the repeated or projected gradient just outlined. Its other justification is that something of the sort is used today in similar modern work.[27] It consists of nothing more than an upright wooden stake with a horizontal cross-piece at the top (Fig. 137), and in its simplest form is used in groups of at least three (though more would probably increase precision). Its purpose is simply to reproduce and project downhill the sloping line of an already calculated gradient. Unlike more sophisticated surveying methods, it measures nothing. Plumb lines, water levels, true horizontals and such are disregarded, and the only concern is to keep projecting the slope already established. The procedure is shown in Fig. 140. A properly measured and surveyed checkpoint, A, is established, and at, say, 100-ft intervals downhill from it are carefully set three T-boards (1, 2 and 3), aligned so that their cross-pieces are exactly on the gradient desired. As excavation work for the trench to take the aqueduct conduit proceeds, T-board 1 is removed and, leap-frogging over numbers 2 and 3, replaced 100 ft further on below number 3, being hammered into the ground until its cross-piece

137. Surveying by use of the T-board (K. Grewe).

exactly corresponds with the established slope, as verified by a careful sighting over the tops of numbers 2 and 3. Number 2 is now in turn removed, and again leap-frogs forward to a new position downhill of the other two, and the procedure is repeated. In this way the gradient already set is extended indefinitely, until eventually it reaches the next checkpoint, B. It is only here that the workmen find out whether everything has worked as it should, and the level is still correct. If it is not, it has at this point to be compensated or corrected, resulting in the situation reflected in Fig. 136(B).

The chorobates

The use of the T-board in antiquity is entirely conjectural. Three other instruments, however, are well attested. We know their ancient names, and have descriptions (notably in Vitruvius) of their operation. They are the *chorobates*, the *dioptra* and the *groma*, and we will consider them in that order.[28]

The *chorobates* corresponds, in modern surveying, to the surveyor's level. It is principally used in establishing a true horizontal and is accordingly of cardinal importance in levelling and establishing gradients. Essentially it is nothing more than a long trough of some kind

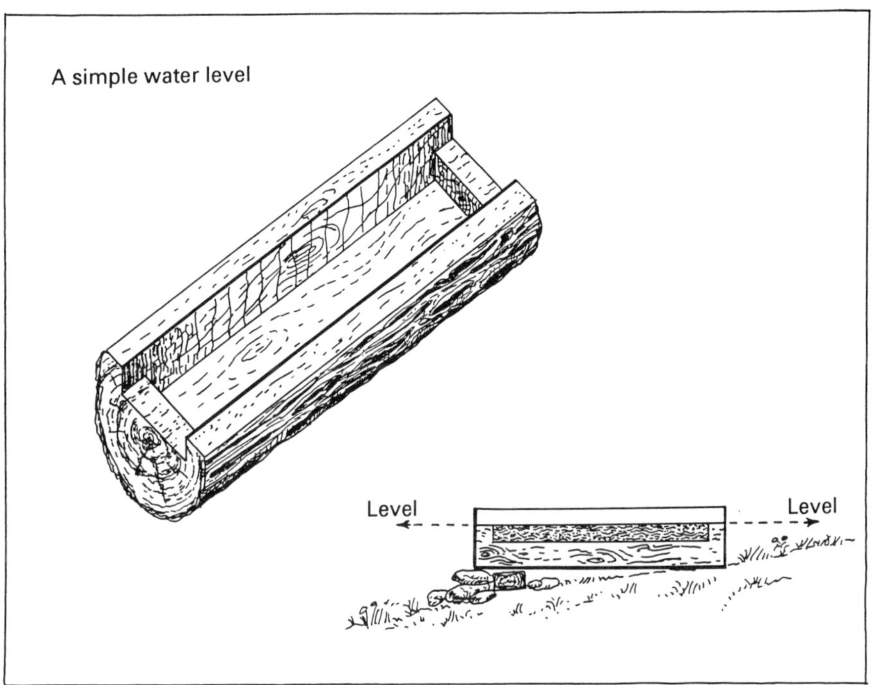

138. Water level, for sighting.

that can be filled with water, thereby giving a level surface across which distant objects can be aligned. In its simplest form it need be no more than a primitive water level. A section of tree-trunk, hollowed and scooped out to contain the water, will do (Fig. 138). Propped up by stones at one end, if the terrain is sloping, no great accuracy in the positioning is needed. The water surface itself will, automatically, be wholly true, and as long as the sighting is taken over that it will be entirely accurate. One suspects that, particularly in the provinces, there may have been many stretches of aqueduct built with such simple devices, rather than the more elaborate instruments generally described.

In its more sophisticated form, the water level became the *chorobates* (Fig. 139).[29] This was basically a narrow table some twenty or thirty feet long. Sunk in the middle of it was a long trough to be filled with water. When the water was flush with the rim of the trough all round it followed that the table-top now formed a completely level surface, and sightings could be taken across it, either of distant objects or of measured and calibrated rods held upright at set distances by assistants. To facilitate observation, a metal sight was usually mounted in the middle of each end of the table. To confirm that it was set level and true, the legs were often fitted with plumb-lines, though they must have suffered from being blown about by the wind; I suspect that it was the water in the trough that

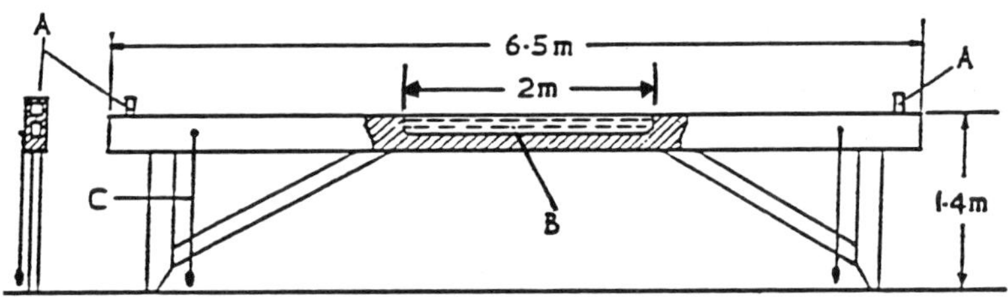

139. The *chorobates*. A = sights; B = water trough; C = plumb line

provided the most reliable and accurate indication. The great advantage of the *chorobates* was its length, which increased the accuracy of observations taken by means of it.[30] For both long-range sightings, such as establishing the relative levels of the two sides of a valley, and short-range observations where fine and extreme precision was required, it would serve well. Its disadvantage was that it was cumbersome. A twenty-foot-long table is not easily transported or set up, especially in confined areas. One thinks of that section of channel between the Pont du Gard and St Bonnet. The tiny gradient means that great precision in the surveying and construction is required. For that, the *chorobates* would be the ideal tool, but what of the very twisty course the aqueduct followed? Could the *chorobates*, for example, be set up actually in the conduit itself and could it be used effectively to check the gradient of a channel that was always curving – given that it could sight only in a straight line? We do not know.

The technique of using the *chorobates* is illustrated in Fig. 140. Again, previously measured checkpoints are established at A and B. A sighting is taken with the *chorobates* on a calibrated sighting rod set up by an assistant, and the difference in level can be calculated by simple subtraction. The sighting rod is then moved and the process is repeated, in as many stages as is necessary to get to B. It will be noted that this procedure is a good deal more versatile and more scientific than that with T-boards. It can be used to measure and establish a uniform slope in a planned aqueduct, but also to record and map the uneven profile of the natural terrain, and is throughout a matter of calculating and recording the level rather than simply carrying on projecting blindly an existing line and hoping for the best. The measuring rod is shown in Fig. 141. In the absence of optical glass and lenses, quite reasonable standards of accuracy in sighting over long distances – vouched for by surveyors who have conducted experiments with modern replicas – can be achieved by sighting on a large, clearly seen movable target that can be raised and lowered on the sighting rod, an attached pointer indicating the level on the calibrated side face of the sighting rod. Again, no archaeological remains of such a rod have ever been found, but it is clearly described by Hero of Alexandria (see below).

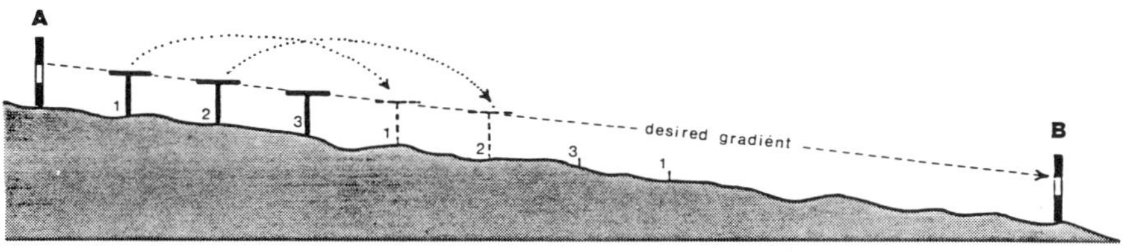

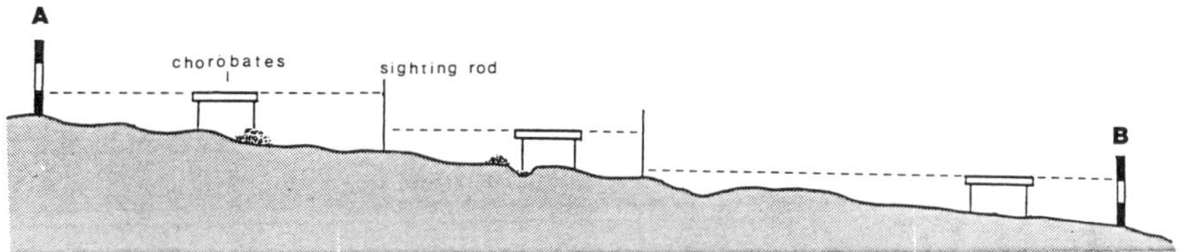

140. Use of (above) the T-board; (below) the *chorobates*.

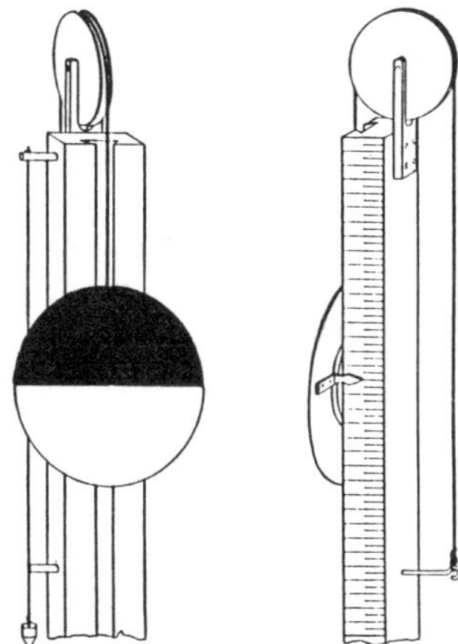

141. Sighting rod with movable target to record height.

The dioptra

Hero of Alexandria was neither an engineer nor a surveyor. Rather he was a professor, one of those Alexandrian scholars whose ingenuity usually exceeded their practical sense, and it is to his written work *On the Dioptra* that we are indebted for detailed descriptions of both the sighting rod (which, though mentioned by Hero for use with the *dioptra*, could also be employed with the *chorobates*) and the *dioptra* itself.[31] Again, the description is so full that we have no problem reconstructing the instrument. Given the inventive ingenuity of the Alexandrians, it may be

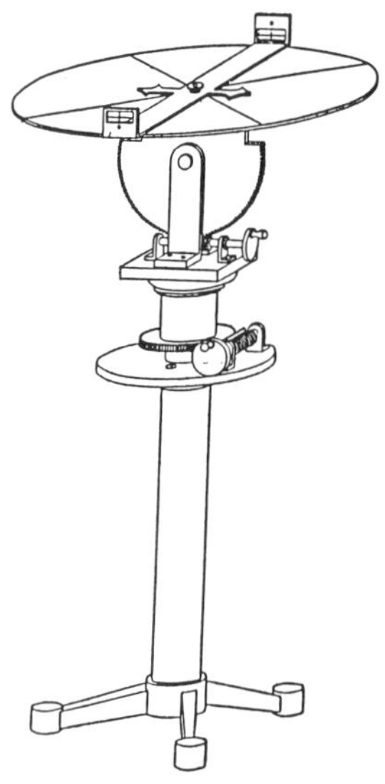

142. The *dioptra*.

that Hero is describing an invention of the Museum rather than an instrument currently in general use. Certainly, his specification that the central support of the instrument be embellished by being shaped like a Doric column seems to have about it the 'banausic' ring that one would associate with an academic mind,[32] but it is also true that the *dioptra* is mentioned by Vitruvius. There can thus be no doubt that it was known and in use. Given, however, the sophisticated and complex design of the *dioptra*, we may surely suspect that it was not standard equipment for the ordinary foreman or even working engineer, and many aqueducts may have been built without it. Even Vitruvius, after all, is something of a perfectionist, and his text reflects not so much real current practice as what he would ideally recommend. I can only echo the verdict of the leading modern authority on surveying, that the *dioptra* 'was presumably regarded as too elaborate, expensive and unwieldy for regular use'.[33] On the other hand, it certainly did exist, was known, and was used, presumably by the better equipped and better trained engineers.

The illustration of the *dioptra* (Fig. 142) is based upon Hero's account.[34] It is essentially a flat disc mounted on top of a pedestal or tripod and can be tilted in the vertical plane as well as rotated horizontally. By means of two sights mounted at diametrically opposite points on the rim, the alignment and relative angles of distant points can be read off; the tilt and rotation are imparted by a pair of worm-and-mesh screws, which are calibrated so as to record the number of turns given. The sights themselves are either some such cross-wire arrangement as shown in the illustration, or are replaced by a sighting tube. In effect this was a telescope without lenses, a narrow, empty tube that one looked

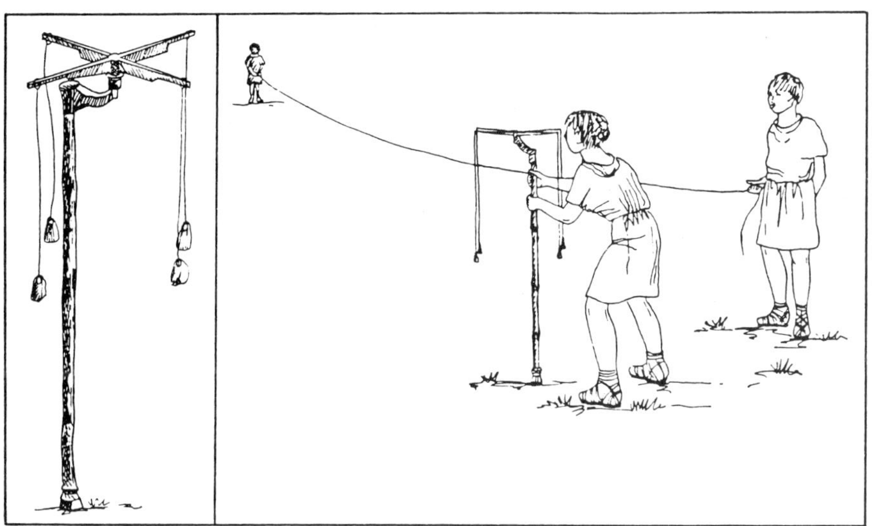

143. The *groma* (Stilwell).

through and aligned upon the distant target. Because it could be tilted in any direction, the *dioptra* was a much more versatile instrument than either the *chorobates* or the *groma* (though less precise than the first), and was roughly the equivalent of the modern theodolite. It could measure differences in height, and also, in the horizontal plane, different bearings. It was by far the best tool an ancient surveyor had at his disposal – if he possessed one, and knew how to use it. It was also extensively used by astronomers for celestial observations.

The groma

The *groma* was the instrument *par excellence* of the Roman surveyor, and was practically the emblem of his trade.[35] This time archaeological evidence does exist, for the *groma* is shown on several grave stelae, and fragments of an actual *groma* were found at Pompeii.[36] The instrument is simple, nothing more than two short spars, intersecting at right angles and with a plumbline attached to each extremity, the whole being mounted on a pole (Fig. 143). Its purpose was to assist in the marking out of boundaries and alignments across the land, and essentially it was nothing more than a square to mark off right angles. Its commonest use was in the surveyor's commonest jobs, centuriation and the verification of property lines, but it could be used for other purposes too – one of the most interesting training exercises quoted in the literature involves its use, in conjunction with a little simple geometry, to measure the breadth of a river without the surveyor having to cross it.[37] But the *groma* could only be used to establish straight lines and mark out squares or

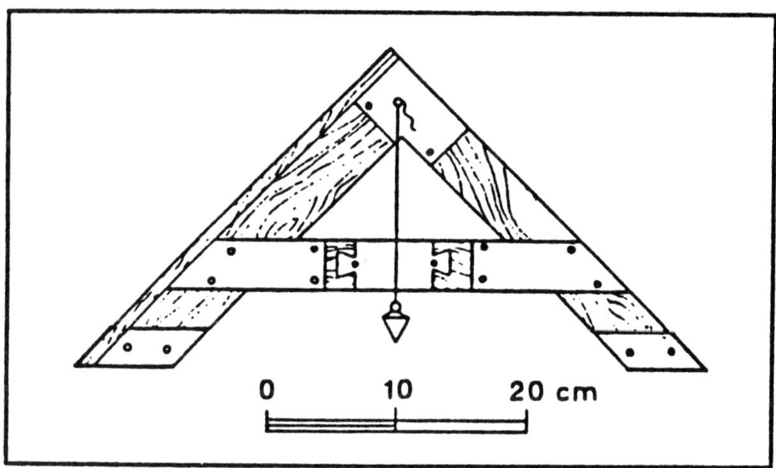

144. The miner's level (J-P. Adam).

rectangles. It could not measure angles, establish a true horizontal or record heights. Though standard equipment in field surveying, therefore, its use in surveying aqueducts must have been very limited.[38]

Other instruments

The *chorobates, dioptra* and *groma* were the big three in the aqueduct surveyor's armoury, but naturally there were other tools as well. Hero describes a hodometer, something like a wheelbarrow that, as it was pushed along, registered on a dial the distance traversed, and a more advanced model is independently described by Vitruvius, but there is no reason to believe that it was ever in common use, at least for aqueducts.[39] Other equipment that actually was in use included such simple things as portable sundials, graduated rods (10 ft long or so), and measuring cords, pre-stretched and smeared with wax.[40] For the drawing of plans, there were set-squares, dividers and compasses. The plumb-line was a simple way of establishing a true vertical, and was used with various modifications. In open-air surveying it might get blown about by the wind, requiring some kind of portable wind-break to shelter it. Moreover, the plumb-line can also be used to establish a true level, for if it is attached to the upper part of some wooden frame like a set square, then when the plumb-line shows the side to be truly vertical, the bottom will be truly horizontal. A simple device of this kind, so often used in underground surveying (e.g. in aqueduct tunnels) that it is sometimes called the miner's level, takes the form of an A-frame, with the plumb-line suspended from the apex; when the feet of the A are set on a level surface,

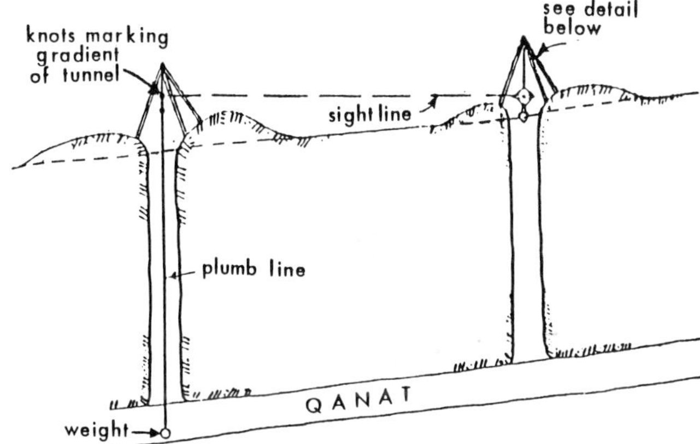

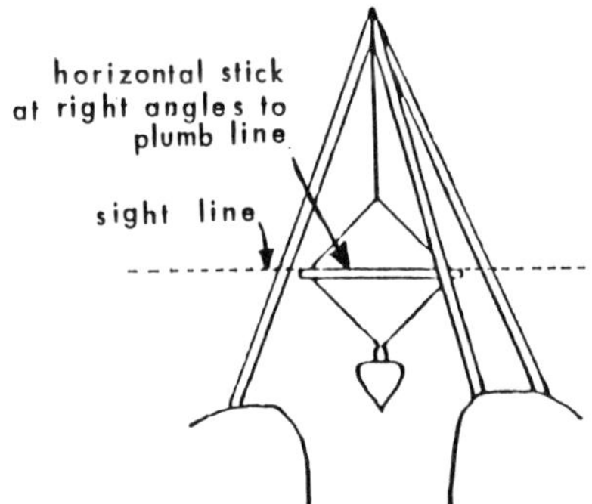

145. Method of sighting level and establishing a gradient in a qanat (Goblot).

the line intersects the A's cross-piece at its mid-point. This instrument was standard mason's equipment and is often shown on tombstones (Fig. 144).

Several other interesting devices are described in the AD 1017 treatise by Mohammad al Karagi on the construction of qanats.[41] This is, of course, well out of our period, but, leaving aside a number of highly sophisticated instruments of great technical interest, it also mentions a number of devices that depend on nothing more than common sense, and hence could have been used in Roman, as well as later Arabic, aqueducts. Most are concerned with establishing a level (or fixing a gradient) over uneven terrain and, though lacking the precision of the *chorobates* or the versatility of the *dioptra*, combine accuracy adequate for most needs with a robust simplicity.

Fig. 145 shows a simple method used to establish the gradient in a qanat. A horizontal sight line is established on the surface between two adjacent shafts, a plumb-line let down, and on it marked by knots the

146. Simple level, as used by the qanat builders.

desired gradient of the tunnel. At the next shaft a plumb-line is again let down, to the same distance below the horizontal minus the gradient allowance (if one is working uphill; plus it, if downhill), and this sets the gradient in the tunnel. It should be noted that all of this can be done without any figured calculations, or even measurements, whatever. The method of establishing the horizontal is shown in Fig. 145 (below), and in Fig. 146. A short double plumb-line is dropped, with a weight at the bottom, and a stick inserted to hold apart the two sides of the string in such a way as to form a square. The measurements do not matter, so long as the sides are the same. The stick, forming the diagonal of the square, is thus at right angles to its other diagonal, which is hanging at a true vertical. One may thus take a horizontal sighting along the stick, facilitated if at each end of it there is also a projecting nail, to be lined up like sights on a gun. My own experience is that this device is both simple and, over shortish distances, accurate. The chief risk is that the string will not form a true square, being displaced by, e.g., a slight misplacing of the bottom weight. Because of this, an alternative method was to use a square wooden frame, hung up by one corner and weighted at the bottom to make it hang straight. Unlike string, the rigid frame could not become deformed, and a horizontal sighting could still be taken across the other two corners. Being less cumbersome than the 20 ft long *chorobates*, this

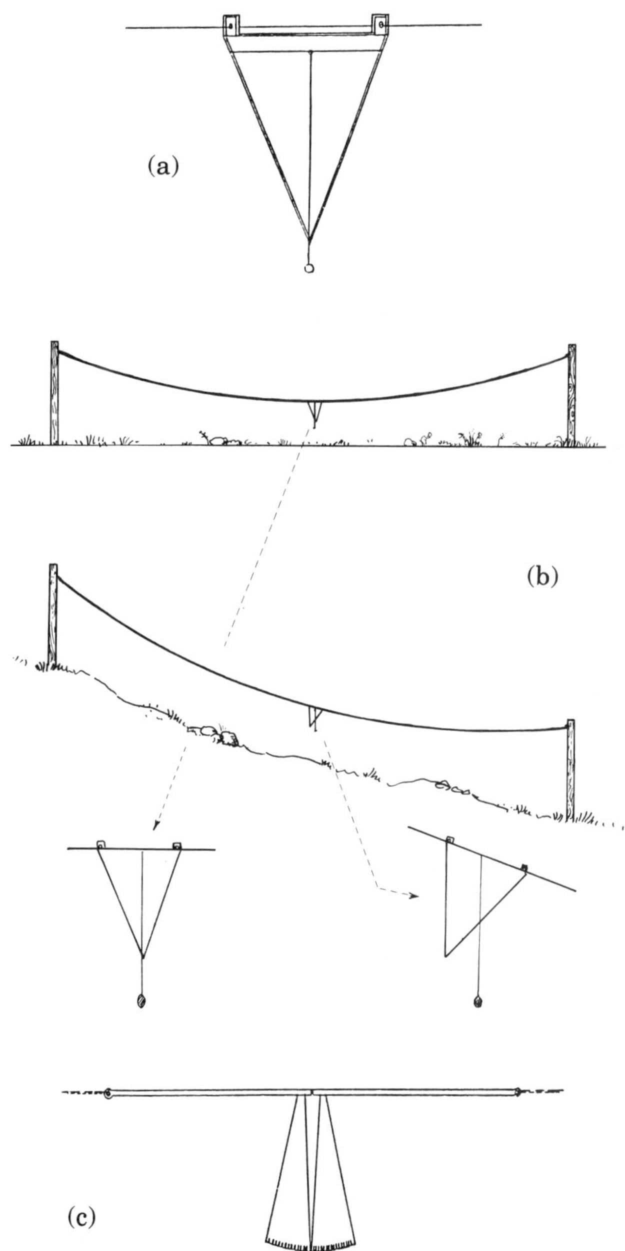

147. (a) Method of establishing level in qanat surveying by attaching metal triangle to a horizontal (but not necessarily taut) cord. Depending on whether ground surface is level or sloping (b), plumb line will pass through apex of triangle or fall clear to one side (see text); (c) an alternative technique for indicating deflection, using a weighted needle suspended from a metal bar inset at the midpoint of the cord.

device would be particularly useful in surveying tunnels.

Another common method of establishing a horizontal is a long cord stretched (but not pulled really taut) between two measuring rods held upright on the ground. Not being taut, the cord dips in the middle, and the system depends on the fact that if both ends of the cord are level, then the centre, or lowest point, of the dip will be at the midpoint, and the cord will here be horizontal. A device of some kind is therefore attached to the mid-point of the cord to indicate whether it is level. If it is, then that proves that the two ends are also on the same level, and the function of a *chorobates* is fulfilled. If it is not, then one end of the cord is moved up or down its measuring rod until a level is achieved.

The simplest way to see whether the mid-point of the cord is horizontal is simply to soak it in water. If the cord is not at this point level, the drops of water collecting on its underside will tend to trickle downhill; if they remain stationary, then the cord is level.[42] Another possibility is to attach to the cord's midpoint a plate or frame in the shape of an inverted triangle (Fig. 147(a)) made of metal or wood; a plumb-line suspended from the centre of the top will pass through the apex below when the cord, and hence the top of the triangle, is horizontal.[43] Another possibility is (Fig. 147(c)) a rigid bar from which hangs by gravity a weighted needle: as the bar tilts, the needle, remaining vertical, registers the deflection on a calibrated scale fixed behind it. A third possibility is to fix horizontally to the cord a long metal tube with a hole bored in the top at its midpoint; water is poured slowly into the hole, and if the tube is not level will run towards one end.

We have no evidence that any of the above methods were in use by Roman engineers, or by anyone before qanat engineers of the eleventh century, but there is nothing difficult about them either in concept or execution, and that something similar was used sometime somewhere in the Roman empire must remain a strong probability.

Underground surveying

The surveying of tunnels had its own special difficulties. Given that the line of the tunnel had been decided and, probably, marked out on the surface, there were two main problems as the main gallery was cut. One was determining, down in the tunnel, what the correct alignment was, and the second was, once that alignment was known, maintaining a straight line as the tunnel advanced. Our only evidence comes from an archaeological study of the surviving tunnels and from the known techniques of the later qanat-builders.

The easiest tunnel to survey was, of course, the multi-shaft tunnel (as opposed to that started from each side of the mountain and simply carried forward till the two halves met, with no vertical shafts). Alignment was not really a problem, as the mouths of the shafts were laid out in a visible line on the surface above. Once each shaft was finished, the tunnel alignment was then calculated from the bottom of it. There were sometimes slight variations, leading to a tunnel plan with minor irregularities corresponding to the spacing of the shafts. The most clearly seen example is the Halberg tunnel near Saarbrücken (Fig. 78), but in the Etruscan *cuniculi* of southern Etruria it has also been noted that changes of direction always come at the location of a vertical shaft.[44]

The easiest way to transfer the alignment, as laid out on top, down to the tunnel below, was to lay across the shaft mouth a plank or beam itself positioned on the same alignment. From near each end of the beam a cord ran down the shaft, and at the bottom a second beam was attached to and suspended from the two cords. It follows that the second beam would then point in the same direction, giving the alignment for the tunnellers.[45] In the open air, of course, the system would not work because of wind disturbance, but in the depths of the shaft the air would be very still indeed. Provided, then, one took time to let the lower beam come fully to rest after the apparatus was set up, accuracy ought to have been at least adequate, particularly since the next shaft would not be too far away[46] and so one could not go too far wrong.

To establish a straight line within the tunnel, a light can be used. All that is needed is to prop up one of the workers' lamps in the middle of the gallery and hang from the centre of the roof a few feet in front of it an iron spike or some other marker: the shadow will be projected in a straight line down the tunnel.[47] Such devices are unique to underground surveying, but it goes without saying that more orthodox equipment, such as water levels, may have been used too. Even the *chorobates* may sometimes have found itself down in the tunnel. It must also be remembered that many of the tunnels are anything but straight. False alignments (Fig. 148), changes of direction and detours abound, and, to be honest, not all of these are to be attributed to faulty surveying: a certain amount may be due to following a rock stratum that was easier to cut.

When it came to establishing the gradient of the tunnel, there were probably two approaches. The gradient could have been calculated and marked out somehow on the surface and then transmitted below, the engineers at the bottom of the shaft being given some kind of template or pattern and told, in effect, 'Dig uphill following *this* angle'. This was how the tunnel gradient was established in digging qanats (Figs 145, 146).[48]

The other way was to calculate the slope down the tunnel itself, by means of a *chorobates*, portable levels or other devices. Inside the tunnel various figures and reference marks, still sometimes visible, were incised or painted on the tunnel walls. They have been found in the Greek tunnel of Eupalinos on Samos; in Roman work they are best attested in the 20 km long Augustan aqueduct – all in tunnel – of Bologna.

Here, three sets of figures have been found. Every ten Roman feet (2.925 m) the distance is marked by an incised or painted X. These are

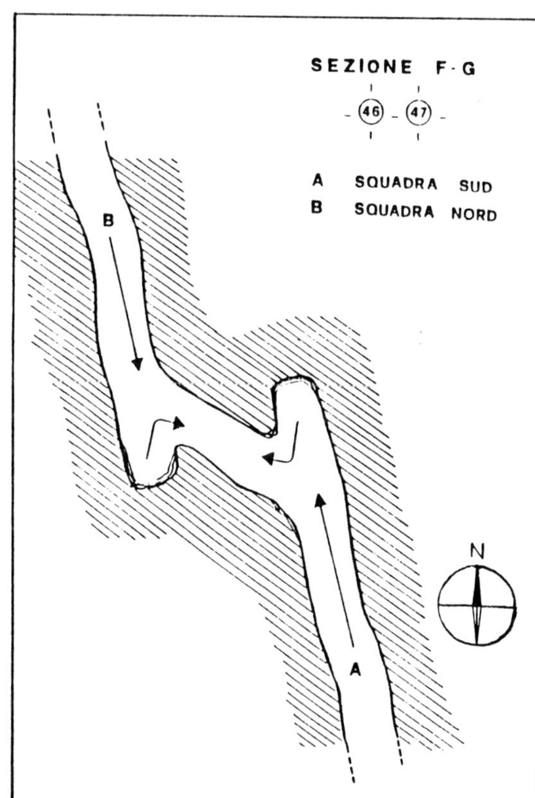

148. Bologna: meeting point in aqueduct tunnel, showing imperfect alignment.

evidently markers for the maintenance gangs, and, like milestones or the *cippi* on the surface, give a reference point in terms of distance. They are sometimes supplemented by higher figures – 260 and the like – giving the cumulative distance from some convenient datum point. Second, at distances of 17–18 Roman ft (5.10–5.35 m), there is a series of incised arrow-heads pointing upwards, exactly like our modern benchmarks. They are usually four (sometimes three) feet above the floor, and marked by a carefully painted figure IIII, in red paint, above the benchmark, indicating a level. The level is also sometimes painted in a continuous red stripe along the tunnel wall. It is thought that this represents the gradient to be maintained (or, less convincingly, the normal water level). Finally, there are figures that seem to be a tally of the number of days spent cutting that section of the tunnel. They are found every 25 Roman ft (7.40 m) and run from 17 days to 38, the time varying with the hardness of the stone to be cut through at that spot. On a very rough average, we may interpret it as a tunnelling rate of about one foot per day.[49]

149. Fucine Lake, Italy: relief showing the draining of the Fucine Lake by a tunnel (*emissarium*). Below are the waters of the lake, with a ship on it; above, three trees, growing along the bank; top right, workmen in the tunnel (photo: Museo della Civiltà Romana, Rome).

150. Detail of Fig. 149, showing in close-up the workmen. They are operating a vertically-mounted capstan, the cables of which apparently disappear up one of the ventilation shafts to the surface; Presumably they were used for lifting and removing excavated spoil.

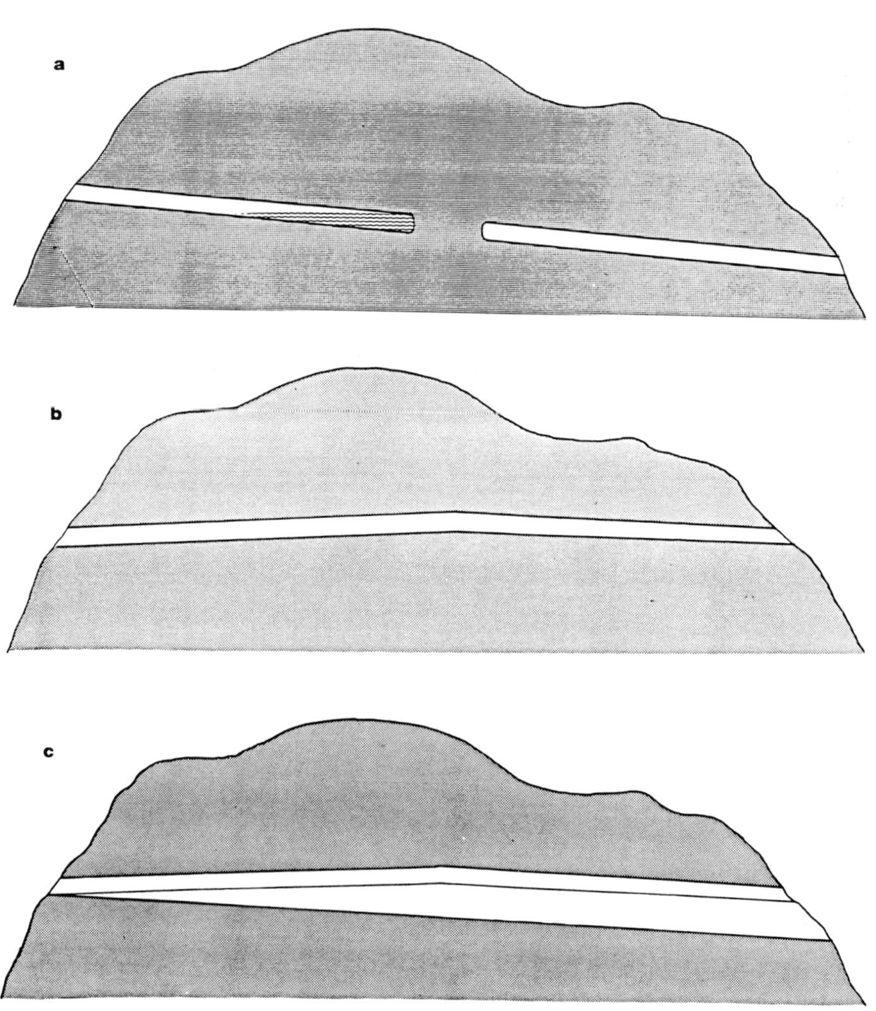

151. Tunnelling: (a) double-ended tunnel cut on continuous slope permits water to collect and flood the workface; (b) the preferred method, resulting in a hump-backed tunnel; (c) a sloping trench is then cut in floor to assure gradient for water conduit.

Tunnelling techniques

We have elsewhere considered tunnels in general and the Eupalinos tunnel on Samos in particular (pp. 27-9). Here it is appropriate to turn to the question of how surveying and planning affected the actual techniques of construction (Figs. 149, 150). The problems of alignment and orientation have already been mentioned. We turn, once again, to maintaining the gradient. If it is to carry water, the tunnel (or at least the conduit) must not be level, but run at a slope. This is self-evident, but given that most tunnels, particularly deep ones, are liable to have water

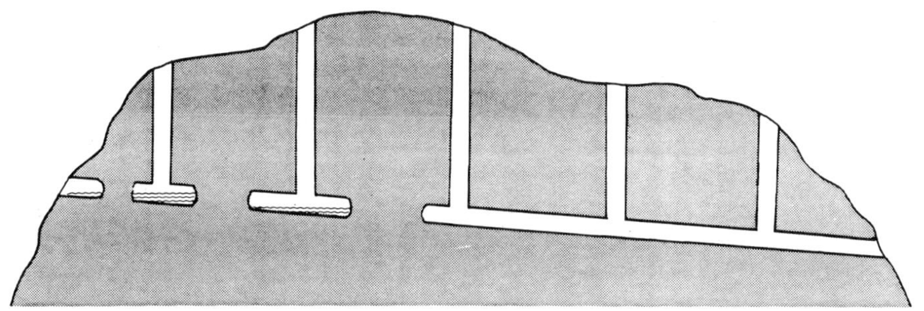

152. Multi-shaft tunnelling: flooding is restricted, and hump-back profile avoided.

seep into them, a clear conclusion follows: in tunnelling, one must avoid cutting a long, uninterrupted section working in a downhill direction. Otherwise one is liable to have water running downhill in it to accumulate at the bottom end, flooding the workface (Fig. 151(a)). Once the tunnel is completed, of course, there is no problem for any seepage can run right through and out the lower end; it is during construction, when it simply runs into a dead-end, that trouble arises. It therefore follows that if the tunnel is being dug without vertical shafts and from both ends simultaneously,[50] the gallery at the upstream end may have to be cut slanted slightly uphill, even though the finished conduit gradient has to slope the other way. This can result in a hump-backed tunnel (Fig. 151(b)) in which surface water or seepage can get away by running out each end, in the way rain water is shed into the gutters by the camber on a road surface. The tunnel then has to be given the necessary downhill profile throughout, by digging the floor deeper as a second and quite separate operation, once the necessary break-through has been made (Fig. 151(c)). If the tunnel is not itself to be flooded but only to carry a piped or channelled conduit, this can be done simply by digging in the floor a trench to take it, ever increasing in depth as it reaches the lower level. This is what Eupalinos did, and the trench inside his tunnel is to be seen as an intended part of the original plan, not a later attempt to correct an incorrectly engineered gradient.

If, on the other hand, the tunnel is shallow enough to be provided with vertical shafts, this problem does not arise. The tunnel is constructed in a number of sections, working outwards from the bottom of each shaft, and all eventually linking up. Because, during construction, each section is isolated and compartmentalised, it cannot acquire enough water to impede work – the trouble with a long gallery is that water from *all* of it runs down to the bottom end. Therefore the tunnel can be cut to the intended final slope right from the start (Fig. 152), and any water accumulating at the downhill end of any section will be small enough in amount to be removed up the shaft in buckets. This gives yet another advantage of the shaft-type tunnel over the two-ended type. To sum up in terms of work needed, the two-ended type required an extra excavation of the floor all along its length, while the shaft type required the digging of the shafts: and the shaft type, irrespective of the amount of work to be done, could get it done more quickly by the multiplicity of the work-faces. When one adds to this the difficulties of alignment and ensuring a meeting point, one can only feel that an engineer must often have preferred a longer and more circuitous tunnel that, by following the contour lines, remained shallow enough to be built on the shaft principle, to a shorter and more direct one that had to be dug straight through from either end.

8
Hydraulics

> What is fluid mechanics all about? Why do I have to study it? Why should I want to study it?
>
> R.W. Fox and A.T. McDonald, *Introduction to Fluid Mechanics* (New York, 1985), 1.

A study of hydraulic engineering is probably our most promising single source for further enlightenment on Roman aqueducts and water supply.[1] This is not because it is inherently more relevant than the other two chief sources, archaeological evidence and literary references, but because it has hitherto been less studied than they have, and accordingly still contains more information yet to be gleaned. Nor yet must it be considered a magic 'open sesame' to all our problems. Often an engineering study will only result in a mathematically precise proof of what had been obvious all along. Another difficulty is that the end result of the calculations, for all its seeming precision, is often of questionable value.[2] Yet, bearing all possible caveats in mind, it is a field that will richly repay our attention, enough certainly to make us go ahead in spite of the Scylla and Charybdis threatening us – that archaeologists will shrink from what follows as too technical, and engineers scorn it as too elementary.

It is hard to know how to organise an exposition of aqueduct hydraulics, but we can hardly go wrong by beginning at the beginning. Water flows downhill. As far as we are concerned, this means two things. First, it can be used as a distribution technique, especially in rural irrigation. Indeed, in the absence of pumps, it is the only technique available, and Roman irrigation in general failed to make any great use of it. Much more could have been done to retain rainfall in the uplands and water them as it made its way downhill through carefully distributed channels, instead of running straight down into the valleys. This was essentially what the Bedouin peasants of the Maghreb did, where local but sophisticated schemes of distribution reflected the lie of the land; or, as it has been put, 'differential altitude creates the natural dynamic which makes possible the use of run-off waters in widely geographical separated zones'.[3] The second use of downhill flow is as a technique not for distributing the water, but for transporting it in bulk, i.e. the aqueduct.

The key position of gravity in moving the water and running the aqueduct, one is tempted to say, cannot be overstated. And yet it has been. Of course aqueducts normally ran by gravity flow – one is particularly struck by Haberey's comment on the Eifel aqueduct at Cologne, that you could roll a bowling ball along it from one end to the other without interruption[4] – but the Romans also clearly understood and applied another basic principle of hydraulics: water always rises to its own level. The principle is explicitly enunciated by Pliny, and is in any case self-evident.[5] As applied to aqueducts, it means that the water could be fed into closed pipes which, unlike the regular open conduit, did not have to remain level: they could rise and fall as convenient (provided they did not rise *above* the natural water level; water will rise to its own level but not above it, which would mean flowing uphill), and the water would rise to and issue from the far end of the pipe at the same level at which it entered it. In practice, this strictly theoretical position had to be modified by a recourse to gravity. Just as an open conduit could not be built level but had to have a slope to keep the water moving, so the far end of the pipe had to be lower than the intake if it was to deliver water in adequate volume and velocity. Again, just as in a conduit, the steeper the slope the faster the water will flow through the pipe. This means that it is quite possible to have, on an otherwise open-conduit aqueduct, a piped section (such as a siphon) where the pipe is much smaller than the conduit supplying it. Looking at the anomaly, one fancies one sees a bottleneck that would lead to overflowing, but in fact the carrying capacities of conduit and pipe are the same as the water in the pipe flows faster.[6]

Gradient

Pipes were always something of an exception on the aqueduct proper, at least, as opposed to the urban distribution system. Most of the aqueducts were open conduits (open in the sense that although they were closed by some kind of roof or cover, and were perhaps even below ground, they were not intended to run full and the cover might not even be watertight), and being gravity-flow, the speed of the current in them was determined by the gradient. What, then, was the gradient of an aqueduct?

This is a natural enough question, but answering it is next to impossible, for it is rather like asking 'What is the speed of an automobile?' In the ancient sources, two quite different figures are given for the minimum acceptable slope. Vitruvius opts for 'one half-foot per hundred feet', or 5 m per km (0.5%), while Pliny specifies 'a quarter-inch per hundred feet', or about 20 cm per km (0.02%).[7] As for the surviving remains of actual aqueducts, we may perhaps risk an oversimplification verging on the reckless and say that for all practical purposes the figure was in the range of 3.0 to 1.5 m per km (0.3%–0.15%). 1.5 m per km is round about the average fall on the Gier aqueduct (Lyon), one of best surveyed aqueducts and of most uniform profile, and 3.0 a rough, but common, figure for the principal aqueducts of Rome.

The difference in slope may reflect factors other than topographical necessity. The steeper the slope, the greater the wear and tear on the conduit from the increased speed of the water; from this point of view there is therefore much to be said for a shallow gradient. On the other hand, a shallow gradient, and a slower current, will let sediment in suspension in the water settle to the bottom, clogging the conduit. In a word, the slower the current, the less need there was for repairs, and the faster, the less need for cleaning: and it was up to the engineer to strike a balance between the two. A lot may thus depend on the quality of the water to be handled, a relatively sandy or silt-laden source requiring a faster stream.[8] For the same reson, the gradient might be made significantly steeper while passing through a tunnel, not because deposit in the channel was more liable to form in a tunnel but because it was harder to clean it out when it did; this, obviously, is particularly true of long tunnels without airshafts, where all deposit removed has to be carried out through the two ends.[9] Another reason for an abnormally steep gradient is a later rebuild or realignment. Often this involved cutting off a loop in the original scheme by a new, more direct line – the commonest case was where a bridge was put in to cross a valley that the original line had contoured around. The new line, being much shorter, also had to be steeper if it was to match up again at its lower end.[10]

Yet another technical reason for introducing a section of steep gradient could be, paradoxically, to slow the water down. The rate of flow on a steep slope will, of course, be increased, but, provided the stretch is short, the forward momentum will not be developed correspondingly. The situation is illustrated and perhaps clarified by the extreme case where the slope becomes so steep it is perpendicular, resulting in a waterfall. The velocity of the falling water is great, but does not cause any great acceleration in the current of the river downstream as the energy is absorbed in the turbulence at the base of the falls, leaving the current to build its forward speed up again from scratch. It will thus be seen that a few steep stretches can be used, like cascades, to retard the momentum the water might develop on a long unbroken downhill run. Steps or settling basins introduced along the line of the conduit could be used for the same function.[11]

A further difficulty in identifying a 'typical' gradient lies in the great variation in slope within the profile even of individual aqueducts.[12] A look at the varying profiles of the four Lyon aqueducts (Fig. 126) well illustrates not only how internal variation can run riot: in these four examples from the same city even the *kind* of variation varies, ranging from the uniform slope of the Gier, to the steep fall of the Craponne and Mont d'Or, and the alternation of plateaus and cascades exhibited by the

Brevenne. A further possible variation is found where the source is high in the hills but on leaving them has to approach the city across a wide and level plain. This often results in an aqueduct profile that breaks into two halves with quite different characteristics, a steep run-down out of the mountains followed by a long flat section over the plains. The best known examples of this are the aqueducts of the Carthage and Segovia.[13] We may also note the most remarkable example where this did not happen – Rome itself. The major aqueducts of Rome all began in the Sabine hills and followed the valley of the Anio downstream to Tivoli. There, they emerged from the hills and confronted an escarpment dropping abruptly 180 m to the level of the Campagna below. The river Anio solved the problem by tumbling straight down, in the celebrated Tivoli *cascate*. The aqueducts, instead of following it in artificial cascades or other means, all swing south-west in a wide detour, sloping down across the face of the escarpment and foothills and effectively maintaining the same uniform gradient as they had in the Anio valley. One might have expected all the Roman aqueducts to show a steep drop at Tivoli; by this means they avoided it.

A further trap, already hinted at, often lies hidden within the aqueduct publications. Given the frequent variation in real gradients, an overall average figure can be, as we have seen, misleading. The only reliable information can come from an aqueduct carefully and fully surveyed, with the altitude (and hence the gradient) established at a large number of separate points along the line. Often, however, a figure for the aqueduct gradient will simply be quoted without the reader being told of the basis on which it is calculated: it may be an actual observed gradient, or it may be a simple computation of total height loss divided by total distance. Nor is there any guarantee that the channel followed a uniform gradient even between observed locations. It was always possible that some irregularity, even a cascade, intervened to vitiate the calculations; indeed, cascades have sometimes been postulated just because the gradient indicated by pure calculation seemed to the excavator unduly high. Conversely, it is equally possible that some of the sections to which is attributed a very high gradient in fact contained an as yet undetected cascade, so that the ordinary running gradient of the conduit was in reality much less than that printed in the publications.[14] And, of course, there is always the problem of the way the gradient is expressed, for which there are many different conventions.[15]

Yet, in spite of all these qualifications, the question 'What was the slope of an aqueduct?' does remain a reasonable one, and one that, in general terms, may be answered in the spirit in which it is asked. And the answer is what we said at the start. The usual gradient is somewhere around 1.5 m to 3.0 m in the km, or 0.15 to 0.3% (that is, between 1 in 666 and 1 in 333). When one looks at the profile of the Eifel, say, or Carthage aqueduct, it is plain that 80% of the line is built to something like this figure,[16] and the exceptions, usually concentrated in one locality, stand out like a sore thumb. We may also note that the Rome aqueducts were characterised as being rather steeper than the rest, and, further, that chronology seems to have nothing to do with it; no correlation between date and gradient is observable.

Velocity

One thinks of the speed of flow in the channel as being determined by the slope, and this is essentially true. It is, however, far from being the whole truth. The velocity of the current in the channel – be it noted that we are here considering water running under gravity in an open conduit (even if it is roofed over, provided the water does not fill the channel), and not water under pressure in a closed pipe, as in a siphon or in the street distribution network – is also substantially affected by surface friction with the conduit walls. This friction itself depends on two factors – degree and magnitude. By degree I mean how smooth and friction-free, or the reverse, the channel walls are. By magnitude I mean how large, or small, an area of them is in contact with the water, and retarding its passage. What this last factor comes down to is the cross-section of the conduit, and its proportions relative to the volume of water flowing down it. This cross-section in turn is governed by three things. First, there is the intended cross-section and proportions of the channel as designed and built. Second, there is the variation temporarily introduced by the water running high or low in the channel, depending on seasonal rainfall. The channel itself, of course, remains the same, but hydraulically the only part that counts is the part that the water touches, and this configuration changes, and its proportions relative to the volume of water, as the water level in the conduit changes. Third, the configuration of the channel may be permanently changed by accretions of *sinter*, calcium carbonate deposit accruing from the hardness of the water. All of these varying conditions significantly affect the friction, and hence the velocity of flow, quite irrespective of gradient. We will consider them all, and those perhaps still obscure from this necessarily brief summary we will try to elucidate by further explanation and illustration.

First, however, we must emphasise one key point. Even within the same aqueduct, speed of flow was not uniform. On steeply graded sections the water ran fast, while on slight gradients the current was much slower. Since at all points on an aqueduct the discharge, the throughput, must be the same irrespective of velocity or depth of water in the channel – were it not so, were one part of the aqueduct transmitting more, or less, water than the rest of it, then it is obvious that this would eventually result in sections of the channel either overflowing or running dry – variation in velocity is automatically compensated for by variation in depth: when the water runs fast, downhill, it also runs shallow, and

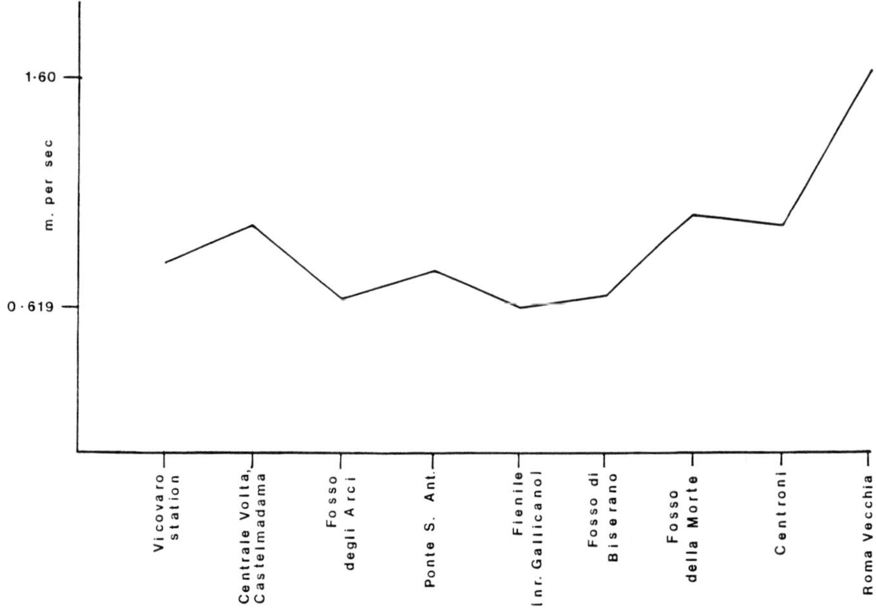

153. Speed of current at various points on Anio Novus, Rome, as estimated by A. Pace.

conversely (to state a law of hydraulic physics enunciated with a technical accuracy rare in folk proverbs), 'still waters run deep'.[17] As is shown therefore in Fig. 153, even for the same aqueduct no single figure, other than an average, can be given. Granted this caveat, however, can nothing be done? Can no estimate at all be formed?

To the basic question 'How fast did the water flow in an aqueduct?' it would be tempting therefore either to give no figure at all, with the excuse that 'it all depends', or to calculate an actual figure on the data from one given aqueduct. Both are unhelpful, the first because the reader deserves an answer, the second because it is liable to result in a figure in several places of decimals and give a wholly misleading impression of precision where no such precision exists. Let us say, therefore, simply and clearly, that the usual speed of the water in the channel was around 1.0–1.5 m per second.[18] This gives a speed of about 3.5–5.5 km per hour. With an aqueduct 70 or 80 km long, this means that the water would take something around 14–20 hours to run from one end to the other. If one takes the slowest possible speed and the longest possible aqueduct (say

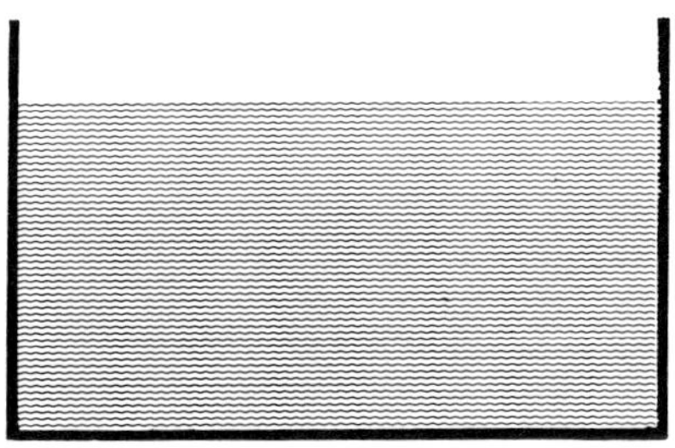

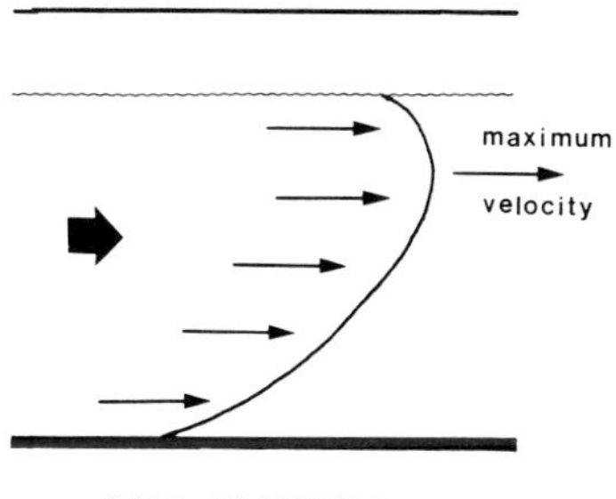

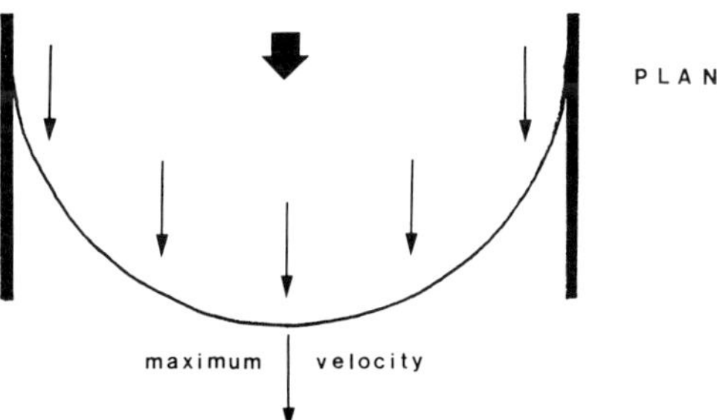

154. Distribution of velocity of flow in an open channel. Greatest velocity achieved is in centre of channel and slightly below the surface (A.T.H., after L.C. Urquhart).

the Marcia, the longest Rome aqueduct at 91 km, though even here the water speed is likely to have been higher, in view of the average 0.27% gradient), one can just imagine it delivering to the city water that has already been in the system for twenty-four hours, but normally the time must have been much less, only two or three hours for cities with short aqueducts.[19] We may also note that speeds in this range are just about within the limits rated by modern engineers as acceptable without causing excessive wear and damage to the channel walls.[20] The actual speeds in modern aqueducts do not normally go above 1.2–1.3 m per second.[21]

Friction

The retarding effect of friction can be considerable. It is also not locally uniform. What do I mean by that? Let us assume that an iron rod is being pushed down on a close-fitting tube, the frictional contact with which acts

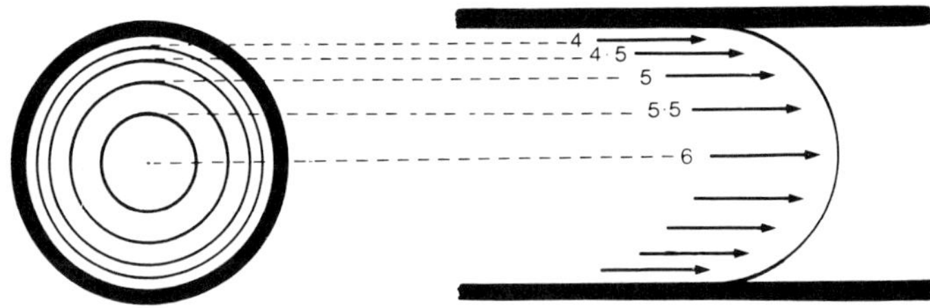

155. Distribution of velocity of flow in a closed pipe: (above) straight; (below) curved. Fastest flow is at the centre in a straight pipe, near the outer edge in a curved one (A.T.H., after L.C. Urquhart).

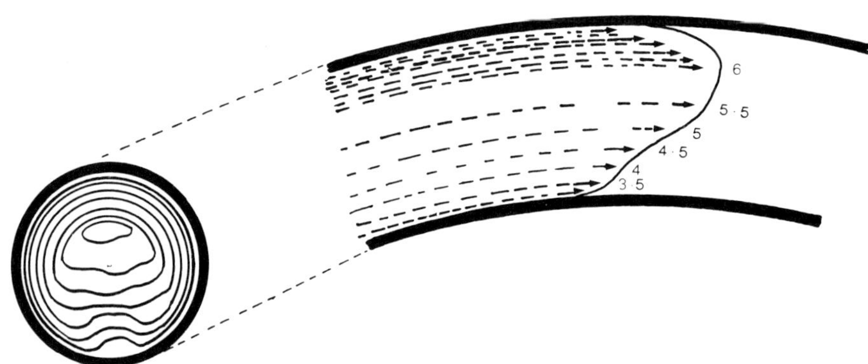

as a brake and slows it up. The entire rod will be slowed up and to a uniform degree. But this is not true of water, for it is fluid. Anyone looking at a river can see that though the water along the banks, retarded by frictional contact with weeds and the banks themselves, may flow but slowly, yet out in midstream the current runs strongly. The same is true in straight aqueduct channels. The difference may be less obvious, but (Fig. 154) the current moves progressively faster towards the middle of the conduit. But it is not only the sides that retard it with their frictional contact. So does the bottom, so that, for any given volume of flow through the aqueduct, the deeper you go, the slower it gets. One would therefore logically expect the water flowing faster to be in the centre of the channel and on the surface. But this will not work either, though it nearly does. A new consideration now enters the scene. This is surface tension, a phenomenon produced by the cohesive forces between the molecules of a liquid, which results in a skin or film forming on a free water surface.[22] This skin retards velocity of flow, though not very much, so that, as is shown in the figure, in an open conduit maximum velocity is attained in the centre and about one-quarter to one-fifth of the way down. Of course, should the water level rise high enough to fill the conduit completely, then we have a very different situation, for we have added the retarding friction from the conduit roof, with which the water is now in contact. In effect the conduit is now a pipe and, there being no free surface, maximum flow will be at the point most distant from all the containing surfaces. In the case of a round pipe (Fig. 155(a)), this point is at its centre, except when the pipe is curved and centrifugal force displaces the stream of maximum velocity to the outside of the bend (Fig. 155(b)). Depending on the roughness of the pipe (especially from corrosion and decay), the difference in velocity from centre velocity to mean, overall velocity runs from 98% to 80%. In point of fact, in aqueduct calculations it has usually been found both easy and acceptable to disregard the variations in velocity in different parts of the channel and to refer only to the mean velocity of the whole cross-section. Thus, when we speak of how fast the water is flowing, it is this mean velocity (and not the maximum velocity, in the centre) that we are talking about.

Of course, this does not mean that the retarding effect of friction is to be neglected, only that it is both practical and simplest to average the figure out, as it would apply to the whole flow of water coming down the channel. A modern civil engineer, speaking of the Nîmes aqueduct, has calculated that a conduit of rough as opposed to smooth concrete would retard the velocity of the water by as much as 80%. Other calculations, based on the Mont d'Or aqueduct, Lyon, show that the retarding effect of the *sinter*, the calcium carbonate deposit, judged solely on the frictional difference afforded by its rougher surface, and neglecting any way that its build-up altered the configuration of the aqueduct channel, was around 40%.[23] So much for the quality, or degree of friction applied to the moving water, and caused by the roughness or smoothness of the walls and bottom of the conduit. We now come to the quantity of friction, a function of the shape or proportions of the conduit.

Plainly, a given volume of water can be carried in containers of vastly different shapes, and the perimeter, the outline in cross-section, of such containers can again vary enormously while still containing the same cross-sectional area of water. Once the water is in motion, this affects the frictional effect upon it, for the greater the contact area, the greater the amount of friction. An extreme example is seen in the case of the Gallic siphons, where the multiple pipes through which the siphon runs have so increased the surface contact area and hence frictional resistance that it has been necessary to overcome it by siting the receiving tank very low and, by increasing the head, so increasing the velocity in the pipes. The most hydraulically efficient conduit is thus one in which (in terms of cross-section) the perimeter of the enclosing channel walls and bottom is as short as possible relative to the amount of water contained. In terms of strict geometry, the shortest line enclosing a given area is a circle and hence the most efficient conduit is a circular one. This is true in theory (and in terms of pressure supply, where pipes are needed, in fact also), but in practice there is a modification. A gravity supply running in an open conduit, like most Roman aqueducts, does not need to be enclosed at the top, only at the bottom and sides. This means that friction is greatly

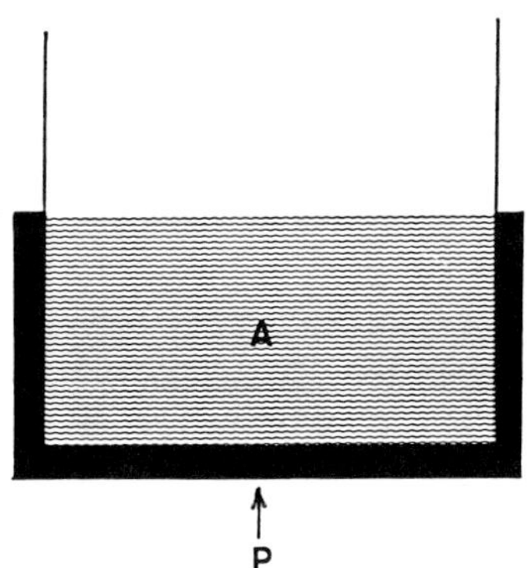

156. Open channel flow: cross-sectional area (A) and wetted perimeter (P).

reduced in an open channel, for on one of its four sides (the top) there is no friction at all since the water is not touching anything. To this extent, an open channel is more efficient than a closed one (such as a pipe), and, all other things being equal, water will flow through it faster.

We now apply this argument to the optimum configuration, in cross-section, of an open conduit. Given that there is now no need to enclose above, what we are left with is the bottom half of our circle. The most efficient form of conduit is in fact semi-circular in section, or, if one prefers it, the bottom half of a U.[24] There is no reason to believe that the Romans ever realised this, even empirically, though the effect may sometimes have been achieved accidentally, as with pipes which, for one reason or another, ran half-full.[25] However, given that Roman conduits were rectangular in cross-section, the same principle applies: the shorter the enclosing perimeter relative to the amount of water it enclosed, the less will be the friction. The situation is illustrated in Fig. 156. In a partially filled aqueduct channel, A is the area, in cross-section, of the water carried (known to hydraulic engineers as the 'wetted area'), and the heavy line around the edge, P, is what I have called the enclosing perimeter ('wetted perimeter'). It is the relationship between these two that matters, and it is normally expressed by dividing A by P; the resultant figure is then called the 'hydraulic radius' of the conduit (or R = A/P).[26] As long as A remains the same (i.e. for a conduit carrying the same given volume of water), the smaller P is, then the less will be the friction slowing down the current. To put it another way, the smaller the figure in the lower line of the A/P fraction, the higher the total figure, and the more efficient the conduit. In actual fact, the optimum proportions for a conduit are for the breadth to be twice the depth.[27] Thus, Fig. 157 shows a typical Roman aqueduct, of breadth 0.5 m, filled to a level of 0.25 m. This, with the breadth twice the depth, represents the conduit operating at maximum efficiency. It will of course carry more water if more is put in – there is plenty of room for it – but at this point a remarkable paradox arises.

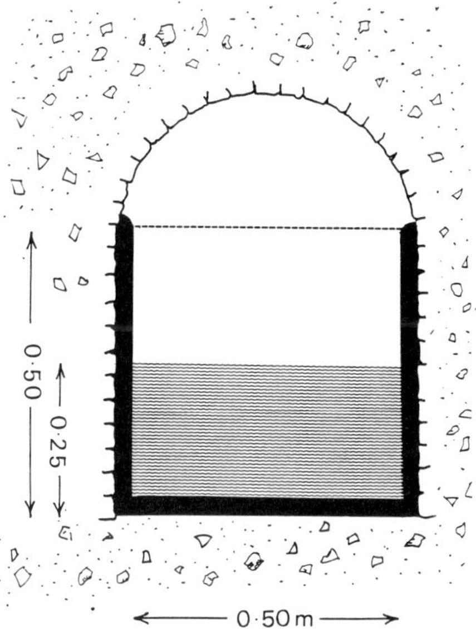

157. Open channel flow: most efficient flow in an aqueduct is achieved when depth of water is half width.

Let us assume that we double the amount of water in the channel, raising the depth to 0.5 m. One might expect this to double the amount delivered to the *castellum*, but in fact the amount delivered, the discharge, will be substantially more than this. It will be observed that, in the table of discharges for the Mont d'Or aqueduct reprinted in n. 23, when the water depth is doubled from 0.3 to 0.6 m, the discharge more than doubles, from 9,400 m^3 to 23,000. This is because, in non-technical terms, when we added the second, upper 0.25 m thick layer of water, it was retarded by friction only from the sides of the conduit, while the lower layer was generating friction from both sides and the bottom of the channel as well. In other words, P, the friction-generating 'wetted perimeter' increased only from 0.25 + 0.5 + 0.25 (=1.0 m) to 0.5 + 0.5 + 0.5 (=1.5 m), while the cross-section of water in the channel doubled. Twice the water with only one-and-a-half times the amount of friction spells an overall increase in speed of flow, resulting in the disproportionately larger discharge already noted. In mathematical terms, for the first situation the formula to find the hydraulic radius, R=A/P, works out as 0.125/1.0 and the answer is R=0.125. For the second, with doubled area and increased perimeter, it is 0.25/1.5, or R=0.166. And as we have noted, the higher the hydraulic radius figure, the faster the flow, which is just what we have found.

But this is where the paradox comes in. For was it not clearly stated that the optimum proportions for a conduit were for the breadth to be twice the depth? How then are we to reconcile this with the fact that by doubling the depth of water, thus making the cross-section of the stream a square instead of the 2:1 arrangement, we more than double the discharge? Does this not mean that the square configuration is more efficient than the 2:1? And if so, are we not here contradicting our dictum that 2:1 represents optimum efficiency? The contradiction is only apparent, and the key to it is that the 2:1 configuration is the most efficient for any given volume (i.e. cross-sectional area) of water. In our example the efficiency was indeed increased, but only by changing, doubling, the volume of water. True, it is carried more efficiently, but if it too were carried in a 2:1 shaped conduit then it would be carried yet more efficiently still, and the discharge would be even greater again. In absolute terms, the 2:1 is always the most efficient, and once we have any given volume of water, if we can arrange for it to be carried in a conduit of such a width that the water will fill it to a depth half the breadth (whatever the actual size), then that is what will carry the water with minimum friction, maximum efficiency, and as quickly as possible.

In fact, no Roman aqueduct channel is ever built to these proportions. They are all much higher than they are wide, but this is not really relevant, because what counts is not the proportions of the conduit itself but that part of it occupied by the water. What shape or form the conduit walls took once they rose clear of the water surface is, of course, of no hydraulic importance at all, and was instead generally determined by the need for human access for maintenance. Hydraulically, therefore, the frictional efficiency of the conduit was determined by where the water level came. Sometimes this can be determined by the incrustation on the side walls, sometimes estimated by calculation, sometimes it remains guesswork, or unknown. If, however, for any reason the water level rose, or fell, it automatically and effectively altered the shape and proportions, and hence frictional qualities of the channel containing it. If therefore, hypothetically, we have an aqueduct channel 1 m wide and the water running in it 50 cm deep (optimum proportions), and for some reason the water depth and hence the amount of water in the aqueduct is halved, then the discharge will be reduced not by 50% but by more, because, proportionally speaking, the influence of friction will be increased, retarding the velocity and cutting the discharge. This can happen either all along the aqueduct, caused by seasonal drought or heavy rainfall, or locally, where narrowing of the channel or a change in gradient causes the current to run faster (and hence shallower), or vice versa. This happened all the time, and Fig. 158[28] graphically shows how, on the Anio Novus, the depth of water in the channel could vary not only widely (from around 30 cm to over 2 m) but repeatedly. It will be noted how the sections of the graph where the water depth is lowest (e.g. around numbers 26-7, 75-92, 151-4) are usually also marked by R (for 'rapid flow') and followed by a steep rise (especially number 48): this indicates where the water, after a fast shallow run, returned to deep, slow, flow – perhaps because it had reached the bottom of a downhill run, or because it was

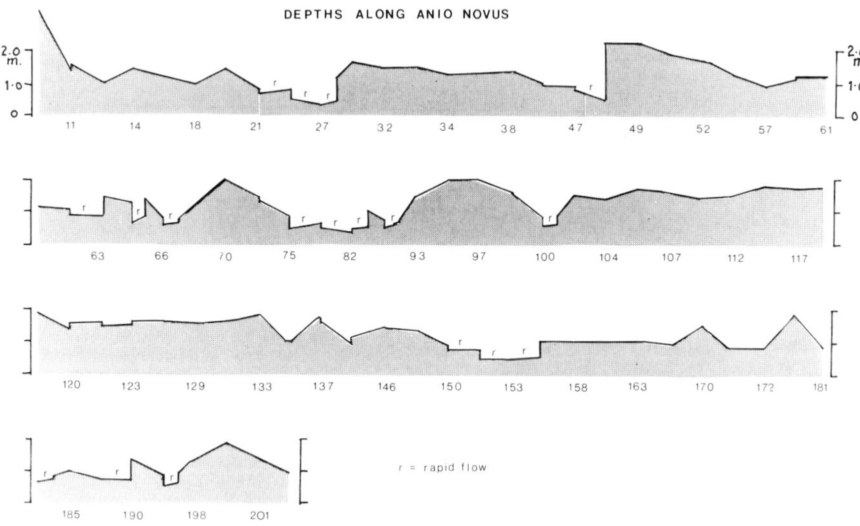

158. Depths of water along the Anio Novus (Rome), as calculated by D. Blackman.

already blocking back, piling up before some bottleneck or obstruction (such as a narrow channel or a very shallow gradient) further downstream. The whole picture is very complicated, even, dare one say, fluid: one gets the impression that the current is always, so to speak, in the process of either recovering from the last obstruction or getting ready for the next one.[29] Even modern hydraulic engineering has not yet fully elucidated all the mysteries of fluid flow in open channels, and it has been pointed out that in aqueducts 'in general, to predict by intuition which mode the flow will adopt is almost impossible. Satisfactory analytical methods are now available for the solution of this kind of problem, but it is certain from their nature that no Roman (or Greek) engineer had available to him any rudimentary form of this method for predicting the depth in any part of his aqueducts.'[30] The reader who has faithfully followed through the labyrinthine multiplicity of variables governing water flow in aqueducts will doubtless agree with these sentiments, and with a depth of feeling that might have surprised their author, but we are not out of the wood yet.

Incrustation (*sinter*)

The final factor affecting channel flow was the calcium carbonate incrustation (more simply referred to by the Germans as *Sinter*) that actually formed inside the conduit. *Sinter* is an interesting topic that is only now beginning to receive the attention it deserves; here it will suffice to confine ourselves to those aspects of it that directly affected the hydraulics of the channel. The incrustation is caused by the hardness in the water which deposits a coating of calcium carbonate (or lime carbonate, $CaCO_3$) on the conduits and pipes it runs through.[31] Some aqueducts suffer from it, some do not, depending on the water, but its occurrence is very much commoner than its absence. The process by which the channel becomes progressively blocked by *sinter* has however been studied in detail in only two aqueducts, the Eifel (Cologne), and Nîmes.[32] The actual causes affecting the build-up of the *sinter* are still debatable, the chief factors under discussion being speed of the water, temperature, and chemical composition of the various substances involved.[33] The rate of build-up on the Eifel, at least, has been measured at rather less than 1 mm per year, or, in figures more readily comprehensible, 1 cm in 13-14 years, and 10 cm in 140.[34] A really heavy layer of *sinter*, a foot or so thick, thus signifies up to two centuries' neglect. The story from Nîmes is essentially the same, where an annual accumulation rate of 1.15 mm is generally accepted.[35] Moreover, though the rate of accumulation might be uniform, the area of the conduit walls so affected was not. The higher the water level in the conduit, the greater was the surface area in contact and therefore the greater amount of *sinter*. The parts of an aqueduct line where the water ran deep were those where it flowed slowly. Other things being equal, one would therefore expect the greatest *sinter* build-up on the sections where the gradient was least; these were already the most vulnerable to clogging from sediment, so the *sinter* was an extra hazard.[36]

The *sinter* build-up, as well as increasing friction by offering to the water a rougher surface, affected the carrying capacity of the channel in two other ways. First, if not removed by the maintenance crews, it simply made the channel ever smaller and smaller. The amount of water being fed into it by the springs remained the same, so the level in the channel rose until eventually it began to overflow at various points along the line. When this happened, the water continuously cascading down the outside of the aqueduct in turn began to form concretions of *sinter* adhering to it, and these, freed from any artificial containment, in time became (Fig. 51) great, shapeless masses almost indistinguishable from the natural rock.[37] This extremity, however, would be reached only after a prolonged period of neglect and accumulation. For the Nîmes channel it is calculated that in 200 years the channel was narrowed by 46 cm, more than a third of its width, but since the channel was originally much larger than the volume of water it carried, even with this reduction it could still accommodate it.[38]

Second, as well as reducing the cross-sectional area of the channel, the *sinter* build-up changed its shape, for though the annual thickness of the deposit formed on any contact area remained the same, the areas affected did not. The floor of the conduit was always in contact with the water, and so acquired every year a new layer of *sinter* 1 mm thick, progressively accumulating. The same was true of the lower parts of the side walls, so,

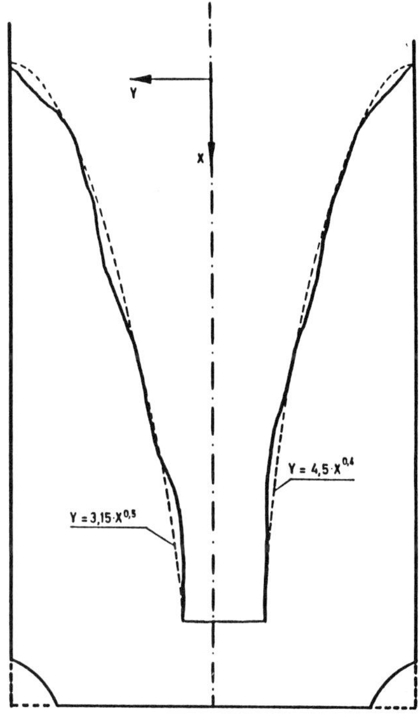

159. Cologne: profile of *sinter* (incrustation) accumulation in conduit of Eifel aqueduct (W. Brinker).

as the floor began to rise, so did the side walls begin to close in. This constriction raised the level of the water, so that it now began depositing *sinter* on the upper levels of the side walls, as yet untouched. The build-up here, however, was thinner than that in the lower parts of the channel, which, so to speak, had already got a head start. As the level rose further the progress continued, the accretion always being thinner (because it had been in progress for a shorter time) at the top. The result was that the original U-shaped channel, as it diminished in size, did not simply become a smaller U, but instead became V-shaped, and the accretion on the side walls took the shape of wedges. This may be clearly seen in the incrusted cross-sections from both Nîmes and the Eifel (Figs 159, 160), but as stated is something of a simplification. As may be observed from the Eifel reconstruction, the sides of the V are not straight, but form a convex curve, which on averaging out the irregularities, is seen to be a parabola.[39] The changed, and constantly changing, profile thus introduces yet another factor into any calculations for rate of flow. To go into detail is probably not worth the trouble, but the general effect may be noted: by having the water run in a deep cleft (Fig. 161) the hydraulic radius is greatly increased and accordingly the friction-generating contact area. The accumulation of *sinter* thus retards the flow of the water both by giving a rougher surface for it to run over, and by increasing the contact area for a given amount of water. It also greatly

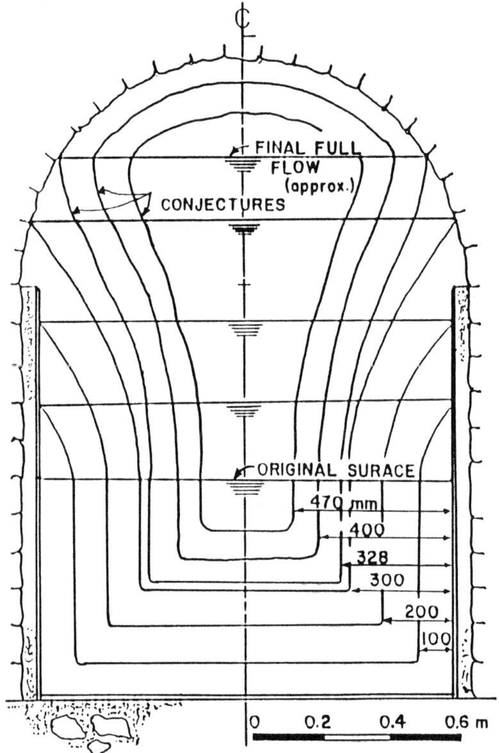

160. Nîmes: successive profiles of *sinter* incrustation in conduit (Hauck).

reduces the capacity of the aqueduct by reducing the size of the channel.

The first two factors will certainly reduce discharge, no matter what the volume of water carried, for the water will be coming through more slowly. The third will become effective only if the aqueduct, when new, was already running full, and had no capacity to spare. Almost all of them did have, and until this spare capacity was filled by the constantly growing *sinter*-accumulation, discharge would remain unaffected; at this point, further growth would reduce discharge, the excess water being forced over the side as an overflow. As it is often impossible to determine the original depth of water in the channel, only rarely can we estimate how much of the original capacity was spare. It is therefore difficult to say how much the discharge was reduced by the existing *sinter*, though we can always form theoretical estimates of the capacity of a given section of conduit with and without *sinter*, on the basis of the water filling the conduit (or running at some other assumed level). The real problem is that discharge and capacity are not the same thing. They were not even determined by the same factors, since the size of the conduit (hence capacity) is usually governed by the needs of maintenance access, while the amount of water carried in it (discharge) is often set by how much the springs can produce. Nevertheless, a rough figure is again desirable and

161. Cologne: a heavy accumulation of *sinter* in the conduit of the Eifel aqueduct at Kreuzweingarten near Euskirchen.

perhaps permissible. We may therefore close this section by recalling the conclusion of Bailhache, based on a study of some of the leading French aqueducts, that in their closing years, because of *sinter*-accumulation, they were delivering on an average only 50% of the water they did when new.[40]

Before we leave this topic, *sinter* and its later developments deserve a final paragraph. For one thing, when it builds up over a long period of neglect, a layered structure (Fig. 50) is discernible, reflecting chronological growth in the same way that do tree rings, and making possible correlation on the general principles of dendrochronology.[41] Even more interesting is the subsequent use of *sinter* from the Roman aqueducts as a building material in the Middle Ages.[42] This phenomenon was limited to Northern Europe, for after the age of Charlemagne marble from the quarries of Carrara was no longer conveniently accessible, and *sinter* from the aqueducts (of all things!) was often pressed into service to replace it in architectural ornamentation. The idea of aqueduct incrustration being used for such a purpose sounds preposterous, but in fact it can be cut from the conduits in large plates which, when dressed smooth and polished, assume a light brown or creamy finish, often with dark brown veining. The whole appearance is very like travertine, which *sinter* also closely resembles chemically. In this form it was often used in churches as a decorative veneer and even for columns, though since the

162. Kreuzweingarten, Germany: the altar facing (carrying the monogram IHS) is a monolithic slab of *sinter* from the nearby Eifel aqueduct; in Kreuzweingarten church.

thickness of the plate was limited by its method of formation, the columns often have one side incomplete, which is turned to the wall, out of sight. In Fig. 162, taken in the church of Kreuzweingarten (some 30 km south-west of Cologne, near Euskirchen), the facing of the altar, carrying the letters IHS, is made of *sinter*, and so is the headstone from the nearby cemetery (Fig. 163), where the date shows that *sinter* was still being used for this purpose as late as 1964; it will also be noted that the headstone retains the slightly curved shape of the *sinter* as naturally formed inside the aqueduct conduit.[43]

Pipes and pressure

Hitherto we have been considering the hydraulics of gravity flow in open channels. When we turn to closed pipes, the situation and the laws governing them are completely different. Pipes, in Roman water supply, will be found in two contexts. There are the big pipes, usually lead or stone, found in inverted siphons; and there are the small-gauge lead pipes of the urban distribution network. Terracotta pipes, though numerous, can to a considerable extent be disregarded, for (though some were employed in siphons and pressure systems) they were in general little used where there would be any significant internal pressure, and the frequent provision of inspection holes in the top of the pipe (p. 114 above), through which the water would overflow, shows that they often

163. Kreuzweingarten, Germany: headstone in a cemetery made of a plaque of *sinter*; the slight curve of the headstone reflects the shape of the *sinter* layers in the aqueduct channel. The date, 1964, shows that *sinter* was still being used for such purposes as late as this.

did not run full and hence were subject rather to the laws of hydraulics in open channels, as previously discussed.

Pipes, whether in siphons or in urban use, are in some ways simpler to deal with than open channels. Since they are, in principle, permanently full, *sinter* builds up uniformly around the circumference (Figs 62, 164), in contrast to the asymmetrical accretions in open conduits; and we do not have to worry about the depth of water in them, as we do in open channels, when estimating discharge (though velocity is still variable). And, from the viewpoint of the engineer, a tap is much better than a sluice for regulating or cutting off the supply; a sluice will only work if there is available to the still oncoming water some diversion or overflow, otherwise it will simply rise higher and flow over the top of the sluicegate; while a tap, on a closed pipe, will simply stop the water flowing, with no further complications. And hydraulics in closed pipes are also relatively simple, for in the absence of pumps, the two most important features, pressure and velocity, are governed by the same thing, head.

Static pressure

Head is, in layman's terms, the distance a pipe (or other aperture carrying water) is below the water's natural surface level. It is easily illustrated. One has only to imagine a container full of water (Fig. 165 (a))

164. Antalya, Turkey: section of large-gauge modern concrete piping, with heavy *sinter* deposit inside; on display outside the offices of the Turkish Hydro-Electric Authority.

with an escape hole at the bottom;[44] familiarity will suggest the resemblance to a tea or coffee urn. The head, h, is the distance the escape orifice is below the water level, and this will determine both the static pressure pressing against the sides and the bottom of the container, and also the velocity of the jet of water escaping from the hole: the greater the head, the greater the pressure, and the greater the speed of the jet.[45] By contrast, when the level is low (b), and the head small, then the static pressure will be less, and the jet of escaping water will move more slowly and with less force. This is no more than self-evident. What needs to be emphasised is that if (c) a pipe is then attached to this hole (as the siphon pipes are in the header tank), then the static pressure in the pipe will be governed by the head, h, or, to give a more technically correct definition, the pressure is that generated by the vertical column of water supported (the velocity of the water flow will be considered later).

Even more to be emphasised, because it has often been misunderstood and misstated, this head, and hence the static pressure within the pipe, remains the same as long as the pipe remains the same depth below the natural water level. The horizontal distance along the pipe has nothing to do with it, and the head being the same at the two points marked h in Fig. 165(c), the pressure in the pipe will be the same. It does not build up, or diminish, as the pipe gets longer. On the other hand, it does increase if the pipe dips deeper (d), thereby increasing the head. But what, one may ask, if we have a pipe connecting two containers, with a different level of

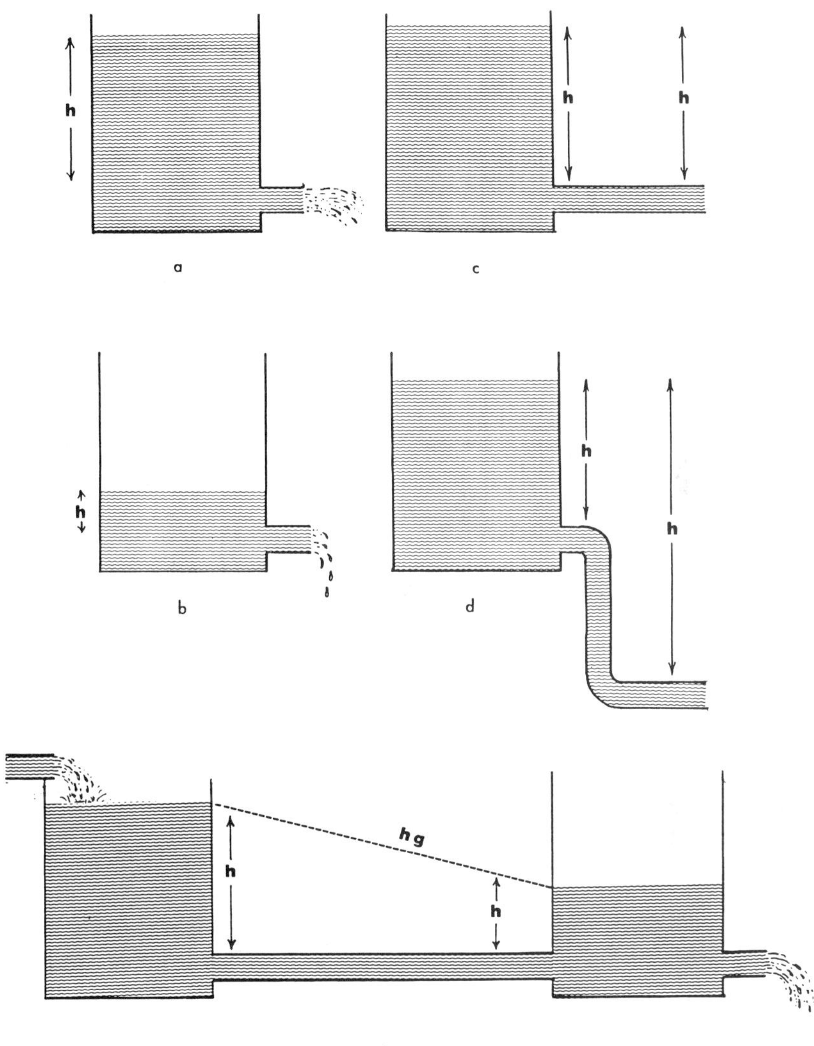

165. (a), (b) Relationship between head (h) and static pressure. So long as head remains the same (c), so does pressure in the pipe; but pressure increases (d) if pipe dips lower and head (h) becomes greater. In (e), hydraulic gradient (hg) governs head, and hence static pressure, at any point in pipe below.

water in each (e)? With still water, this of course would never happen, for water always rises to its own level, and water would run from one container to the other until the level was the same in each. With flowing water, however, as in the diagram, such a situation is quite possible and indeed often encountered in aqueducts. What then governs the static pressure in the pipe, since it is attached to two quite different water levels? The answer is that there is an imaginary line connecting the two

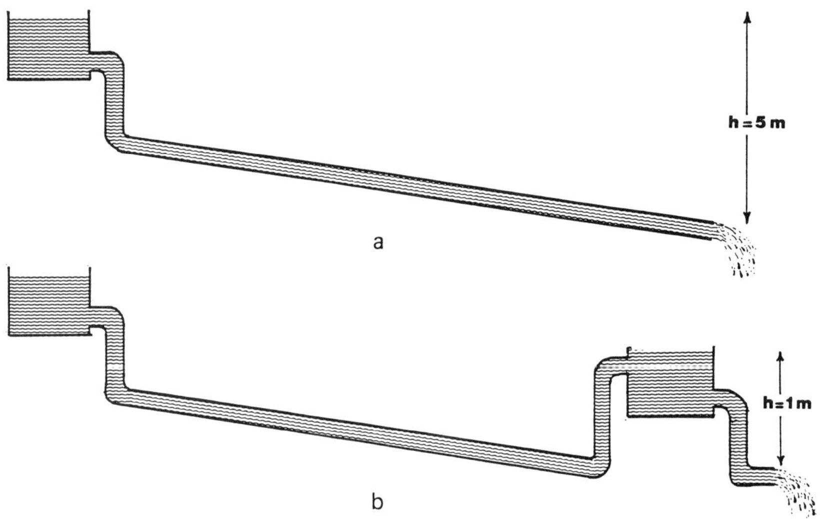

166. Reduction of natural head of 5 m (a) to 1 m by inserting into system before delivery point a second tank (b). See also Fig. 167.

levels, known as the 'hydraulic gradient' (hg), and the pressure at any point along the pipe is governed by the vertical distance that point is below the hydraulic gradient. Thus, in the diagram, the static pressure in the pipe becomes progressively less from left to right. Likewise, to go back to siphons, Fig. 102 makes it clear why the greatest point of stress in a siphon is the first (rather than the second) *geniculus*: there the distance below the hydraulic gradient is greatest, and hence the head, and hence the static pressure.

It therefore follows that as hydrostatic pressure is generated by the head, reducing the head is the only way to reduce the pressure. This is relevant to siphons, for suggestions have been made from time to time that perhaps the Romans reduced the pressure in the bottom of the siphon by introducing water cushions, pressure reduction valves, and other strange and half-understood contrivances.[46] All of these are impossible. The only way to reduce static pressure is by installing a *venter* bridge, which, by in effect cutting off the bottom of the 'U', reduces the depth and hence the head.[47] However, although this is true for siphons, there is one qualification that must be borne in mind.

Suppose (Fig. 166 (a)) a pipe leaves a header tank and runs downhill for a considerable distance before ending in an open spout. The head operating at this spout will again be h (say, in this example, 5 m), the vertical column of water supported, or the sum of the distances the header tank is raised above the ground and the height lost by the pipe in its downhill course. Now suppose (Fig. 166(b)) that at this point the pipe discharges instead into a tank raised 1 m above the ground (compare the situation at the secondary water towers in the urban distribution

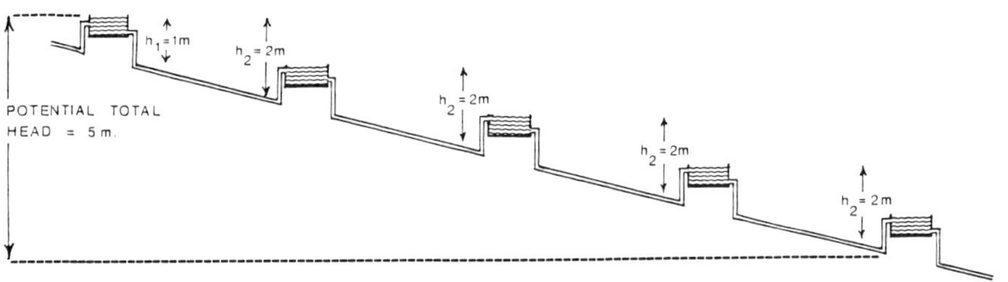

167. Repeated use of raised tanks to limit head; this system, much used by the Ottoman Turks, was known as the *suterazi* (= water balance).

network, Fig. 203). The head immediately before this tank will be 15 m, but on the downstream side of it, 1 m. Thus the head has been reduced from 5 m to 1 m, and the static pressure in the pipe is accordingly less. Now let us suppose that this arrangement is repeated indefinitely (Fig. 167). The head on each section of pipe is measured from the level in the tank immediately before it. If each tank is one metre above the ground, the head at the start of each section will be 1 m (h_1). Assuming a downhill height loss of 1 m before the next tank, that gives a total head (h_2) of 2 m. This will be the greatest head encountered anywhere in the system, whereas if the intermediate tanks did not exist the accumulated head would be 5 m by the time the water got to the last one. It will be seen that in this way not only can the static pressure in the pipe be reduced, but this reduction can be maintained over a pipeline indefinitely long, providing that one keeps putting in intermediate tanks. This principle was often used in the waterworks of the Ottoman Turks (see Fig. 168), where it was called a *suterazi*.[48] The tanks were raised on brick piers, and as they continued in indefinite sequence, effectively governed the pressure in the pipeline and kept it within manageable limits. It is quite possible that the same thing was done with the secondary water towers at Pompeii (and elsewhere), and that instead of all drawing their supplies independently and directly from the *castellum divisorium* at the Porta Vesuvii, they emptied in sequence one into the other, gradually working downhill and so reducing the pressure in the pipes. The routes of the various pipelines in the city have as yet been insufficiently surveyed for us to be sure.

However, although this system will effectively reduce the static pressure in the pipes, we must note that it is subject to two limitations. First, it will not work for siphons, for it is a one-way-only operation, and will operate only on a continuous downhill, since the height lost can never be regained. True, the configuration between each pair of tanks does constitute an inverse siphon, and the arrangement overall a whole

168. Akko (Acre), Israel: *suterazi* tower on nineteenth-century Turkish aqueduct.

succession of siphons, but it cannot be used to replace one single large siphon, such as that at Beaunant. Second, it depends on continuous flow, for the water must be drawn off from one side of each tank as fast as it pours in the other, otherwise it will overflow. Within these limits, it is a simple and acceptable way of governing static pressure in the pipes.

Where the static pressure cannot be (or, at least, has not been) reduced further by reducing the head, whether by *suterazis* or by the interposition of a *venter* bridge, then it can only be contained. Effectively, this means that the pipes must have walls sufficiently thick to withstand it, and on an archaeological site one sometimes finds fragments of terracotta pipes evidently belonging to two series, both of the same diameter but one with much thicker walls; in which case the second set belong to some pressure system, and the first to gravity flow, probably running only partly full. For the same reason one often finds stone pipelines used in an urban setting, far from Beaunant-type siphons out in the country. These may come from pressure lines joining secondary water-towers or *castella* linked in series on the *suterazi* principle or from siphons used to serve the hilly parts of town from an appropriately elevated reservoir; it may be significant that a large number of such pipeline stone blocks has been found at Ephesos (Fig. 11), a particularly hilly site. Normally, however, in the West at least, urban pipes were of lead, and there does not seem to have been any difficulty in making their walls thick enough.

Velocity of flow

Velocity of flow is (or can be) governed by two factors, head and gravity. Head has been explained above. The rate of flow in the pipe will be increased by the distance it lies below the natural surface of the water, though, of course, if the pipe should later bend upwards to something approaching that, thereby greatly reducing the head, the velocity throughout the pipe will fall to a speed appropriate to the lesser head. This is why siphons, even if they go deep, still have to have the receiving tank sited well below the header, because it is the difference in level between the two that determines the rate of flow. This also illustrates a further point. Water is incompressible, therefore throughout the whole length of a pipe, whatever its profile, up, down, or sideways, the water will at all points be moving at the same speed, as long as the pipe itself is always the same size. In effect, the water behaves as a solid column from one end to the other; since, to put it simply, it can neither be compressed nor stretched, it has to move at a uniform speed. In this it differs fundamentally from water flowing in an open conduit which, in response to circumstances, is free to vary its speed since it can compensate by running shallow or deep, the cross-section varying inversely with the velocity; in a pipe running full the cross-section remains constant, so the velocity must remain constant as well.

Gravity influences the speed of flow in that if the pipe, on leaving the tank, runs steeply downhill and then discharges, either into the air or into another tank, the speed of flow will be accelerated beyond that which would obtain in a level pipe. In such a case, the gravity is added to the influence of head, so that the water in the pipe is, to oversimplify, pushed from above and pulled from below. An example of this is to be seen in the ten large pipes leading out of the *castellum* at Nîmes; they run steeply downhill so that when the effect of gravity is computed into the calculations they result – at least partly from this factor – in a velocity of flow so high that the pipes must have run only about one-third full.[49]

However, there is yet another physical phenomenon largely related to velocity of flow which goes far to create pressure and stress in the pipes. This is inertia. Although in popular parlance inertia usually means just standing still, it more properly means that a moving object will, unless deflected, tend to keep moving in a straight line, and if it is deflected some of its energy will be transmitted in the form of stress or pressure to whatever deflected it. Under this definition water counts as an object, and as applied to both pipes and open conduits it means that when the water comes to a bend (whether in the horizontal or vertical plane is immaterial), it exerts force on the outer curve of it; the sharper the bend, and the greater the velocity, the greater the pressure.[50] It must be noted that while static pressure is omnidirectional (that is, its effect is as if a balloon were being blown up inside the pipe: it presses in all directions equally), inertial stress is highly directional, pressing only on the part of the pipe on the outside of the bend. Moreover, it is also highly localised, being found only where the bends are. Since the pipe at these points is already subject to static pressure (if it is a pressure line at all), inertia thus becomes an extra source of stress added to static pressure. In a low-pressure or gravity-flow pipeline, where there is little static pressure, inertial stress at the bends is not by itself usually enough to call for any reinforcement or special countermeasures. In high-pressure lines, however, such as big siphons, the addition of static pressure makes all the difference, so that the bends become recognised danger points. Thus at Segovia (Fig. 73) the aqueduct, running as a surface channel (and hence under no static pressure) negotiates a 55° bend without any buttresses to the arcade carrying it or other supports to counter the sideways thrust;[51] while at the bottom of siphons, the *geniculus*, the bend leading on to the *venter* bridge, usually much less than the bend at Segovia, was always a recognised danger point, because it was also subject to static pressure.[52]

Vitruvius, in a celebrated passage,[53] specifies that at such bends the pipeline should be reinforced by being embedded in a block of 'red stone' (some kind of porphyry?). He is here speaking of a terracotta pipeline rather than the commoner and more robust lead or stone variety, but the principle is the same, and it is significant that in a similar position modern cast iron or steel pipelines are often anchored by being fixed to a large block of concrete; though, so far as I know, no such 'red stone' anchors have yet been found on an actual ancient aqueduct siphon. Before leaving the topic, we must also note that although this inertial thrust is present at the *geniculi* (or any other bend) in a siphon all the time that the water is running (static pressure remains constant whether the water is moving or not), a very special case develops when the pipe is being filled up from dry. If the sluice at the beginning of the siphon is simply thrown wide open, the water rushing unchecked downhill through an empty pipe can hit the *geniculus* at the bottom with an inertial force so explosively violent that it can burst and wreck the entire siphon. In both filling and emptying a siphon, or other closed pipe system, the process must be taken slowly, for the situation is inherently and radically different from normal operation.

Sharp bends and elbows are also to be found in the smaller-gauge pipes of the urban distribution network, including drains. Where the pipes are of lead, they could simply be bent round the corner, but terracotta pipes might employ a specially fashioned right-angled elbow-piece. Sometimes, however, they adopted a solution reminiscent of the siphons' 'red rock', and the right-angle was contained entirely within a cubical stone block to which the terracotta pipes were joined on either side (Fig. 72). This may have been done to help resist the inertial sideways thrust at the bend through greater solidity, though it may also have been inspired by no motive more arcane than a desire to get a good firm joint at an awkward place: there is, after all, the parallel of Ephesos, where a lead pipe had each joint encased within a stone block (Ger. *Marmormuffen*), and that on a straight section where there was no question of sideways inertial thrust (see ch. 2 n. 42 p. 395).

Irregular flow and corrective devices

In an open conduit the water is usually free to compensate for any irregularities in flow that may develop. It can run deep or shallow, thus in turn varying its speed (while keeping the discharge uniform). In a closed pipe that is running full, there is no such flexibility. Irregularities are thus liable to become much more serious. Air may be entrained into the closed pipes in the form of bubbles, and there collect to form air locks. Air may be dissolved within the composition of the water and come out of solution to form air pockets when subjected to low pressure. A sudden interruption in the flow, as by the abrupt closure of a sluice, may cause the shock-wave phenomenon known as water-hammer. And there may be surges in pressure of various kinds.[54] In turn, various devices may be employed to correct these anomalies, though the reader must be warned that we are here entering upon the most confusing and least understood aspect of the whole of ancient aqueducts. Evidence is indeed scanty, and analytical deduction often perforce supplanted by theoretical speculation, if indeed the scholar does not simply throw up his hands in despair and retire from the field, confessing himself worsted.

The problem is greatest in siphons, where it has always been recognised that the pressure caused difficulties. Vitruvius specifies that they be alleviated by the provision of *colliviaria*, and a lively controversy has raged over just what these are, for Vitruvius does not say. The two commonest views are that they are some form of open tank at natural water level to which the pipeline ascends, subsequently descending again on the other side to normal siphon depth, as at Aspendos or La Craponne (Lyon); this solution finds its champions chiefly among the Germans. The alternative is to see in the *colliviaria* some form of drain cock, escape valve or other similar mechanism, and quite a few studies have accordingly prescribed 'pressure reduction valves' or the like, the author usually following the precedent of Vitruvius and not explaining further what he means.

There is nothing too difficult about installing in a pipeline some kind of valve to reduce pressure in the pipe following it. The problem, as applied to siphons, is that once the pressure has been lost it cannot subsequently be recovered, and the water will not rise high enough to come out of the top end of the siphon. It would also presumably not have been beyond the theoretical capacity of ancient technology to design some sort of safety valve to release excess pressure, and the use of such valves has already been discussed (n. 47) and the objections noted. A further objection must

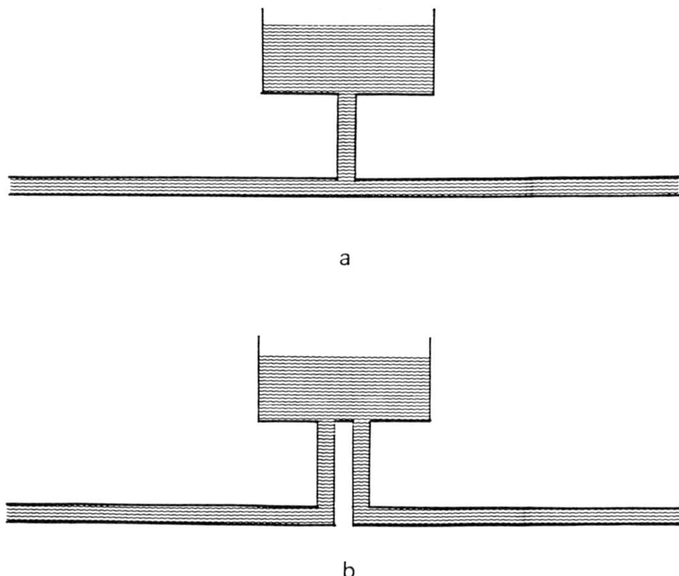

169. (a), (b) Surge tanks (*reniflards, colonnes piézometriques*).

be raised here. Not only have no archaeological remains of such a mechanism ever, to my knowledge, been found, but the ancient world in general seems to have been very reluctant to make and use machines with moving parts. Nothing could have been simpler than the design of ancient water taps, but even so the most remarkable thing about them is how seldom they were used.

The counterpart in ancient technology to this mechanical reluctance was its eager expertise in civil engineering. And so we come to surge tanks. Other names have been used in learned articles, such as standpipes, water cushions, and, in French, *reniflards* and *colonnes/tubes/cheminées piézometriques*. The terminology is not always clearly defined and sometimes differs from that used by engineers. It will therefore be best if I simply describe the structures with which we are here concerned and discuss their purpose. There are two of them, and though very similar in appearance they are essentially different in function. In both (Fig. 169 (a), (b)), somewhere along a pipeline running under great or little pressure, is a vertical pipe communicating with an open tank, which, of course, has to be at natural water level. Normally this tank and the upright pipe leading to it will be housed in a brick or masonry pier, which in itself ensures that this device will be employed only where the pipeline is not too far below natural level, otherwise it would need an impossibly high tower to reach it. The purpose of type (a) is to relieve any momentary surges or local increases in pressure in the pipeline, which are dissipated as the water rises up the pipe like mercury

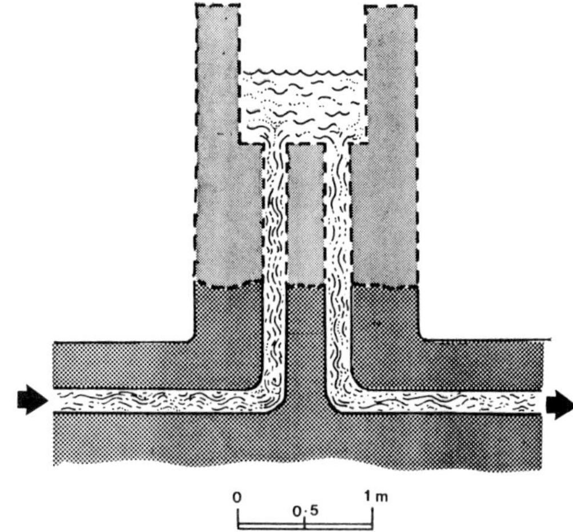

170. Caesarea, Israel: surge tank (?) on piped section of north aqueduct (A.T.H., after Y. Peleg).

in a thermometer, sinking again as the pressure subsides. Thus, although the water is always flowing along the pipeline, that in the vertical pipe remains stationary except when responding to a surge. The tank at the top could in theory be omitted, but is a useful reservoir to accommodate a large volume of temporarily displaced water. This is the surge tank. Although such a device has occasionally been suggested in theoretical restoration of some Greek or Roman aqueduct, no firmly identifiable remains of such a thing have ever been found.[55]

The second type, (b), differs in that the tank is served by two vertical pipes with no direct connection at the bottom, so that all the water going through the aqueduct has to pass through the tank, going up one pipe and down the other, a system radically different from (a). Unlike (a), (b) is well attested, and quite certainly formed a part of the Roman hydraulic engineer's repertoire. The elevated tank at Les Tourillons on the La Craponne aqueduct at Lyon, though served by slanting rather than vertical pipes – it does not affect the principle – conforms to this pattern. So do the tanks on the 'pressure towers' at Aspendos, though there the whole thing is constructed on top of a viaduct. The *suterazis* so common in Turkish aqueducts (see p. 165 and n. 48 above) are regularly built to this design, and for classical Roman work we now have a very fine and completely clear example on the north aqueduct of Caesarea in Israel.

As a result of a rebuild, the aqueduct was here running inside three terracotta pipes laid side by side in the old conduit. This had sagged somewhat, necessitating the piping, but was still not far under the natural water level. On this section, remains of two 'water towers' have been found, about 100 m apart, though only one is now still preserved.[56] At the base of the 'tower' – it was only a metre or so high – all three pipes were fitted with elbow bends, directing the water vertically up into the

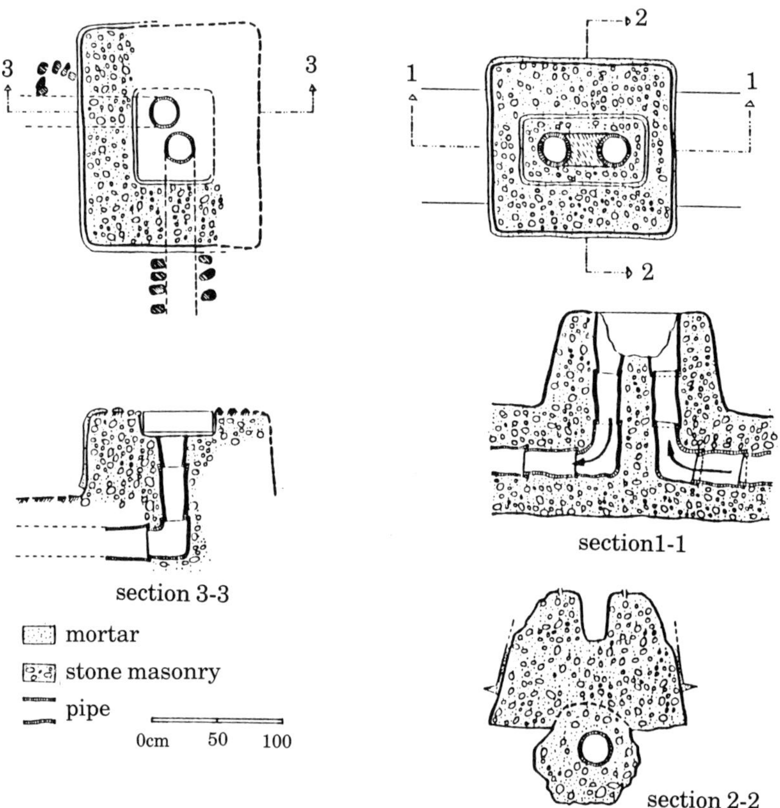

171. Caesarea, Israel: surge tanks (*colliviaria?*) on (left) bend, and (right) straight section of south aqueduct (third-sixth century AD) (Y. Peleg).

tank, while a further three close behind them received it on its downward course and sent it on its horizontal way (Fig. 170).

An even more striking example (because of its more complete state of preservation) is afforded by the south aqueduct, also at Caesarea. The date is late, third-sixth century AD, but its relevance to ancient hydraulics is too great to be overlooked. The pipeline again runs just under natural water level, though this time it is a single pipe. Two small basins are preserved (Fig. 171), one of them on a right-angled bend, the other on a straight section of the pipeline. In each there is no direct connection underground, so that all the water passing through the pipeline has to rise and pass through the tank before continuing on its way. It is therefore not a surge tank, and, from its small dimensions, can have fulfilled no function other than ensuring that the pipeline water was briefly exposed to the open air (Fig. 172).[57]

But why? To relieve the pressure (but how?)? To release entrained air? To aerate the water? Or was it nothing more complex than a visceral reluctance to run a pressure pipeline in a long unbroken section without

172. Caesarea, Israel: 'hydraulic device' (*colliviaria?*) on south aqueduct (see Fig. 171) (photo: Y. Porath)

some sort of intermediate break, so that, say, in the event of blockage it could be more easily located? And are these Vitruvius' *colliviaria*, at last come to light? Certainly they match his description much more closely than anything else yet found, and one can only wait in hope of the identification being confirmed by the discovery of more of these devices on an aqueduct of classical date. It is surely particularly ironic that, as we look back at the two devices shown in Fig. 169, we have a clear explanation for the purpose of type (a) but no evidence for its existence, while for type (b) we have clear evidence of its existence but the explanation for it remains uncertain.

Finally, to have led the reader thus far into the labyrinth and there to abandon him without any Ariadne's thread to even the most likely direction for the exit, may seem ungracious if not irresponsible; but I have to confess that reliable guidance beyond this point is also beyond my capacity.

9
Special Uses

WASHING IS GOOD FOR YOU

Mosaic floor inscription in the Baths at Sabratha, North Africa

Most people, when they think of water supply, instinctively conceive of it on the basis of their own personal experience, and envisage a sequence running: catchment, aqueduct, urban distribution, domestic use, drains. This familiar sequence has thus been adopted for the framework of the present book. And yet, after a moment's reflection, the reader will recall that there were also demands other than domestic on the water supply, demands that might require even more water, and themselves be even more important, or at least be adjudged so by contemporary society, then as now. These uses I am here treating as special cases. I do so as a matter of organisational convenience. The reader must not assume that the topics dealt with are exceptions, incidentals, or of minor importance. Depending on where we are talking about, one of these special cases could provide an aqueduct's whole *raison d'être*. We may classify these uses as agricultural, industrial and recreational.

Irrigation

Irrigation must have consumed vast amounts of water, and had an importance out of all proportion to what we actually know about it. The only study specifically devoted to it is a very minor four-page article dating to 1919 in which the author justly complains that he has no established corpus of data to work with, even the great German encylopaedia, Pauly-Wissowa, being wholly deficient.[1] True, much of ancient farming, especially Roman, seems to have been 'dry farming', a series of techniques designed to grow crops whenever possible *without* watering, thus making a virtue out of the arid necessities of the Mediterranean climate.[2] But two factors argue persuasively for our consideration of agricultural irrigation as a major feature in water supply. One is that very dryness of the climate just mentioned. Where it was at all possible, irrigation, even locally, would make a very great difference. Dry farming was not a substitute for irrigation. It was merely the best you could do if you had no water anyway, and if it was a standard technique in Roman farming, the consequence is clearly seen in the judgment that the Roman farm only just worked, and was never far from total failure.[3] It is plain, therefore, that anywhere irrigation could be provided, it would have been. One has only to look out of the train window practically anywhere in the Mediterranean today and note the ubiquity of irrigation channels, of one sort or another, in our more technically advanced age, for the point to be clearly made.

The second factor is that irrigation, in comparison with other forms of water use, requires water in quite enormous quantities, so that quite a modest local scheme would yet use as much as the whole city of Rome. A few modern figures will give an idea of the order of magnitude, though they must be used only subject to an important proviso. Modern irrigation schemes, unlike tap-regulated urban water supply, run continuously, twenty-four hours a day. This creates a very great disparity in the quantities of water used, making the irrigation figures seem to modern eyes correspondingly large. But ancient urban supplies also operated continuously, so that there the difference between urban and rural consumption must have been less marked. Likewise, if ancient urban and modern rural consumption both strike us as high, it may be because of the same reason, continuous running. With this in mind, it must yet be admitted that modern irrigation figures are high. North America is notorious for its rate of water consumption, and this is generally attributed to industry, or sometimes to a wasteful domestic lifestyle. In fact, 'in the prairies 72% of the water consumed is for irrigation, and that goes to 1% of the land under cultivation'.[4] A 1976 UNESCO study puts it in perspective:

> Vast quantities of water are needed merely to supply, say, 20 mm to a field of one hectare: 20 mm on one hectare is 200 m^3 of water, and a tanker can carry only a few cubic metres. A quantity of 200 m^3 of water would last a family of five several years. Each individual needs an intake of no more than about 3 litres per day in addition to water for dishwashing, hygiene, and laundry. In the developing countries, 25 litres per person per day, i.e. 125 litres per day for a family of five, is usually enough. Thus, one cubic metre is enough for eight days, and 200 m^3, the amount we need to irrigate one hectare of land, would last the family of five more than four years! Irrigation uses very large amounts of water.[5]

And a field of one hectare is not very big. It is 10,000 m^2, or 2.47 acres. One hundred hectares is one square kilometre, not in itself a particularly large area as irrigated tracts of farmlands go. The amount of water required to provide such an area with the 20 mm of the UNESCO study would thus be 20,000 m^3, or the entire daily output of an average, medium-sized Roman city aqueduct.[6]

In hydraulic terms, therefore, irrigation was big business. Yet very little archaeological evidence has been found, or at least published, as compared with the volumes on urban aqueducts. Indeed, the aspect of rural irrigation that has been most studied is the various water-lifting machines asociated with it,[7] rather than the actual conduits and functioning of the irrigation networks themselves. It is partly because the machines – the Archimedean screw, the noria, the compartmented wheel, the shaduf, and so on – are of greater interest, offer more evidence, and are perhaps more important, as playing an important role in the study of ancient technology; and partly because the simple ditches of farm irrigation usually leave no archaeological trace.[8] But the volume of water such machines could handle was limited, and I doubt if they could ever have played more than a minor part in the irrigation picture as a whole, though no doubt often locally indispensable. The major moving power in Roman irrigation as a whole must surely have been natural gravity flow. In Egypt one sees many references to the shaduf and the Archimedean screw, but what really kept the country going was the Nile flooding, and the ancient agronomists – Varro, Columella, Cato – do not mention mechanical irrigation at all.

Mechanical irrigation suffered from three drawbacks. First, there was expense – both the capital expense of building and installing the machinery and the running expense of operating it: some water-lifting wheels might themselves be turned by the current in a river, if a suitable one was available nearby, but most devices such as the screw, the compartmented wheel and the sakia, had to be worked by slaves or animals, which accordingly had to be bought and maintained. This made them uneconomic where the need for water was not truly pressing, or the average size of the farms too small to carry such costs. For this reason there was no mechanical irrigation in Italy.[9] Second, the height to which the water could be raised was very limited. Lifting wheels could raise water only to something rather less than their own diameter, after allowing for part of the wheel to be immersed in the river or other water source. Moreover, this assumes a wheel with the lifting receptacles, such as the pots on a noria, mounted on the rim; with a tympanum or compartmented wheel, with its discharge apertures near the axle, the total lift was no more than the wheel's radius. The length of an Archimedean screw, and hence the height to which it could lift the water, was likewise limited both by the increasing friction and the increasing weight of water inside the device, which had to be moved at each revolution. Where the need was sufficient this could be overcome by installing the devices in successive relays, such as the sequence of eight pairs of wheels in the Rio Tinto[10] gold mines in Roman Spain, which lifted the water 30 m, or the multiple shadufs shown in Fig. 24(a). But this again only multiplied expense, as is shown by the arrangements to irrigate a fertile ridge in Egypt by a battery of wheels and screws that took a gang of 150 prisoners to operate it.[11] Pumps, which would have been the real answer, existed, but were apparently too unreliable or hard to construct to be seriously used in irrigation. Instead, we find them employed in situations where only intermittent use was required, such as fire-fighting or pumping drinking water.[12] Third, the amount of water such devices could raise was also limited. A single shaduf could provide 2.7 m^3 daily, enough to irrigate 0.4-1.5 hectares, an Archimedean screw, it has been calculated, can produce around 288 m^3 daily (twenty-four hours), and a tympanum 1,152 m^3;[13] it must be noted that these estimates come from different sources and are therefore not necessarily even internally consistent, let alone accurate in any absolute sense, but they do give some idea of the order of magnitude we are dealing with. For a long time there was also thought to be a further and conclusive testimony, now however challenged and apparently discredited. This was the famous inscription from Lamasba, near Lambaesis in Algeria, listing the regulations for rationing out the irrigation water among the local farmers. Rationing is on a time basis, each beneficiary having the right to open his sluices during set hours. It appeared that the text differentiated between 'lower' and 'upper' farmers, and the latter, whose lands lay uphill from the supply channel and had to use lifting machines, were granted longer hours, in manifest recognition of the inefficiency of the machines. It now seems that this is quite wrong, though it is still hard to say what the correct interpretation is.[14]

For the small, rural conduits running under gravity, however, that must have formed the great bulk of Roman agricultural irrigation, there is little direct evidence. On the Mediterranean coast of Spain some of the original Roman conduits may still be in use. Such, at least, has been stated[15] and may well be true, though one would hesitate to place too much faith in a dating and identification that must be largely speculative. But if they are not original, the Roman ones must have been very much the same thing, and so we can take their modern counterparts as reflecting ancient practice. To get their water, they may have tapped a local spring, drawn it from a convenient stream, or even from an established city aqueduct passing nearby. This last was no doubt unusual, but recent flow studies on the Gier aqueduct at Lyon show that at various places, notably just before the more important siphons located along its course, it might often be expected to overflow, the siphon being incapable of accommodating the amount of water the aqueduct could deliver when running full, and it is reasonable to assume that the surplus was diverted to agricultural use rather than simply wasted. At Rome, the Anio Vetus, because of the poor quality of its water, was extensively and officially used to irrigate gardens and vegetable plots.[16]

A more usual situation was perhaps that of the Aqua Crabra, near Tusculum in the Alban Hills. Though it is something of a *cause célèbre*, our knowledge of it seems to be entirely literary, not archaeological, and it is not clear exactly what it was. It was certainly a local water channel used for irrigation, but whether a natural stream or, as White boldly states, 'an aqueduct which ran from Tusculum to Rome, a distance of fifteen miles, supplying irrigation water', I would prefer not to say.[17] Its waters were shared out among the local farmers, and a surviving inscription gives us the details of the rationing scheme, which operated on a basis both of set hours allowed to each landowner, and supply channels of specified (but varying) size, rather like the *calices* used to regulate domestic supply.[18] Unlike the otherwise similar Lamasba inscription, there is no time differentiation between 'upper' and 'lower' landowners. Other inscriptions likewise record local rationing schemes for irrigation water, though, like the Crabra, one often cannot tell whether the '*aqua*' that provides the supply refers to a natural stream or an artificial irrigation conduit.[19] But the main principle seems to be clear: in ancient irrigation, as in modern, all the outlets were never open at once, and fields were watered in rotation.[20] Where the water supply was continuous, rationing was normally by time. How the relevant hours were determined, given the general imprecision of ancient time-measuring equipment, we can only guess, but the obstacles cannot have been insuperable. A water-clock of some sort could have served, and we may get some idea of the possibilities by observing the system employed in antiquity for water rationing at the Gadames oasis in North Africa, and still in use without modification. The time is determined by filling a pot with a small hole in the bottom to allow the contents to escape. Each owner is permitted to divert the entire flow of the nearest irrigation ditch on to his land for a time-span measured by a given number of 'potfuls', on the principle of the hour-glass or the ancient clepsydra. When the time has expired, into the channel is dropped a small bundle of straw, which, floating down with the current, arrives as a signal to the farmer that his turn is over; his sluice is closed, and that of the next participant opened. The complete cycle of all participants takes about twelve days to complete.[21] There must have been many such primitive (but simple and efficient) systems of water-rationing in use in antiquity, particularly in the remoter reaches of the empire.

However, if in general we know all too little of Roman irrigation, we also must note three exceptions, fields in which our knowledge is fuller than elsewhere. One is in the arid lands of the Middle East and North Africa, especially the latter. Here, we benefit from extensive French studies of the colonial era, dedicated to remains of Roman hydraulic works of every kind and motivated by the optimistic expectation that many of them could be put back into service. Upon this foundation of assembled factual evidence modern studies have now built a more accurate comprehension of the principles on which they functioned and how they served the local community.[22] The argument may be simply expressed. In North Africa, the picture has been clouded by the monumental Roman works devoted to urban supply – aqueducts, bridges, arcades and the like – which has obscured the importance to the regional economy of the more humdrum rural irrigation schemes. These, moreover, were built and operated on traditional lines not by the Romans but by the native Berber tribesmen, and the Romans, accustomed to a different set of principles formed in a land where water was more plentiful, often did not understand them. Finally, modern scholars have been so convinced of the superior efficiency and organisation of the Romans over all who preceded or followed them in the occupation of these lands, that any structure or arrangement reflecting a modest level of competence is automatically described as Roman, on the grounds that there was never anybody else in the area who had enough brains, or at least organisation, to build such a thing; thereby clinching the argument, even if it is only a circular one. A few fragments of conduit, or a row of rough foundation blocks, are not inherently easy to date, and it is quite possible that a number of works identified as Roman are not Roman at all, and some other people or society should get the credit. The same may also hold true of areas of the empire other than North Africa.[23]

The actual principles of irrigation in the Maghreb region of North Africa were also substantially different from what obtained elsewhere in the empire. Indeed, the distinction between drainage and irrigation almost breaks down in these arid lands. Instead, one finds an integrated attempt to capture and redirect what rainfall there is to where it will do most good, a procedure that one could perhaps better refer to by some term such as 'water management'. Even this does not tell the whole story, for it leaves out the vital factor of erosion. This is greatly misunderstood. To the general reader, all erosion is bad and should be prevented. The hydraulic engineers of the ancient Maghreb and the Negev knew better.[24] Cloudbursts were going to wash away topsoil anyway, so the important thing was to choose and control where it was eventually deposited. Viewed thus constructively, the run-off became a convenient means of transporting soil, even of gathering it all together where it naturally lay too thin on the ground to be of any use, and assembling it in land-parcels of viable size located in suitable areas which could be adequately irrigated. It was not a question of stopping erosion, but of harnessing it. This was normally achieved by building a series of dams across the wadis, or gullies, that carried the run-off. Their purpose is often described as flood control, or even irrigation, but one might more appositely say it was 'to trap soil and water in compartments where it can be successfully cultivated'.[25] Silting up is always a problem in reservoirs behind dams, an average figure being 4.5 cm yearly. In this case it was the whole object of the exercise; a successful conclusion was reached when the reservoir was completely full of silt and so ceased to exist, turning instead into a terrace of new arable land. The wadis of the ancient Maghreb and the Negev were full of whole series of these terraces, all formed in this manner.[26] It was small wonder that the Romans, arriving from a land where water was more plentiful, did not grasp such subtleties. On a Roman farm, ditches and conduits were usually for drainage, not irrigation. Given the dry farming techniques, the problem was usually to get rid of water (i.e. at times of temporary flooding) rather than the reverse; Cato lists it as one of the emergencies liable to face a farmer, calling for prompt and vigorous action.[27] Moreover, even dam construction has its own rules in the Maghreb, which the Romans seem not always to have understood. Used to building dams on the fixed watercourses of perennial streams, they were often confounded by the wadis, which not only delivered their water in seasonal torrents but were also liable to frequent changes of course and often, once a dam was built, outwitted it by a simple outflanking movement. This protean response usually proved more than a match for the Roman builders, and the best dams were simple earth dykes that, when the pressure got too great, were washed away and then cheaply and easily replaced by local farmers.[28]

The end result of all these techniques seems to have been a level of agricultural production that has caused frequent astonishment in ancient times and modern alike.[29] It should, however, be noted that we are here speaking of rural areas, on the evidence of known crop yields. The common belief in the prosperity of North Africa generally in Roman days, on the evidence of urban archaeological remains (i.e. cities such as Timgad or Volubilis) is quite a different matter. The assumption that the land must have been prosperous since it maintained so many fine towns where none have existed since, has now often been contested on the grounds that the towns were for, in modern terminology, a foreign colonialist élite, while the Berber rural peasantry constituted the real economic backbone of the country. We do not here want to get into political theory, but I cannot help wondering what the Berbers thought and said when they saw the precious water that they depended on for their crops and livelihood being put into aqueducts and carried off to a new town for Roman army veterans to use for endless bathing. In this context, aqueducts and water use raise a serious social issue.[30]

The second exception, an area where we know something of irrigation, is represented by the law. A study of the law hardly falls within the scope of this book, but we may offer one or two comments. First, the provisions of Roman law abundantly confirm our generalisation about the inversion of priorities on North African and other Roman farms. All are concerned about *aquae pluviae arcendae*, keeping rain water under control and legal liability for flooding: that is, they are concerned, at time of cloudbursts, with getting rid of the water rather than gaining access to it. Conversely, the body of traditional North African customary law is concerned with disputes over a farmer *preventing* surface run-off on to a neighbour's property, and Roman *agrimensores* in North Africa found themselves in a topsy-turvy world with which their law was out of step.[31] For irrigation from a river or stream, rather than run-off or surface rainfall, Roman law held firm and clear to one simple principle. Once a right to the water was established, it was vested in the property and was inalienable. The Digest specifically lays down that *hauriendi ius non hominis sed praedii est*, 'the water rights belong to the property, not the proprietor', and in case of sale automatically went with it. Where water rationing was in force, the legal allocation was based on the area of the property to be watered and might be measured either in volume units or in hours of flow.[32] It should be noted that this is the complete opposite of Roman urban practice where, as recorded by Frontinus, a grant permitting a resident to tap into the water supply was personal and individual, and did not pass on to anyone who bought the property, or even to the owner's heirs after his death; at which point an application for renewal or transfer of rights had to be made.[33]

The third exception is references in literary sources. From these a certain amount of fragmentary information is forthcoming, but a disproportionate amount of it – at least for our purposes – relates to the early oriental empires, whose feats in irrigation and canalisation of waters made a great impression on classical writers acquainted with them, especially Greek ones. For the rest, references in Latin writers to Roman irrigation form something of a rag-bag, contributing little to any coherent, overall picture. They have mostly been assembled in the article by Knapp.[34]

Industry

The reader will look askance at this category, for the ancient world is not famous for its heavy industry. Yet there was a significant volume of water consumed for purposes to which, if we did not mentally affix the label 'industrial', we would be hard put to it to think of anything else. In towns, the chief commercial users were probably the fullers, the ancient equivalent of laundrymen. Like most tradesmen, however, they operated out of small workshops which no doubt drew their water from the ordinary city network with no great need for special arrangements.[35] In the country, some mining operations required an abundant water supply. We have already, when considering aqueduct conduits, noted the aqueduct serving the Dolaucothi gold mines in Wales, where a lot of water was needed for 'hushing', washing away the overburden to expose the lower auriferous deposits. In the Las Medulas gold mines of Spain, private aqueducts (which included tunnelling through a mountain ridge) were likewise constructed to serve the miners' needs.[36] In the Greece of classical and later days, water was also needed in large quantities for silver refining in the area around the Laurion mines, but this came from rainfall locally stored in cisterns.

Water-mills

However, the main industrial use to which water was put, apart from mining, must have been the provision of power – water-mills, for grinding grain. How extensive, then, was the use of such mills? Until recently, this was a question that could have been – and generally was – confidently answered: 'Very little.' But this picture, even if it has not yet changed, is now being submitted to probing that calls much established doctrine into question. Established doctrine rests on two points. The first concerns the type of mill. Two basic types are known, the vertical wheel (subdivided into undershot and overshot), and the horizontal, also known as the Norse or Greek mill.[37]

The horizontal mill, which has a set of paddles or blades arranged horizontally and turned by a jet of water playing on to them on one side, is the most simple, primitive and inefficient type. Its simplicity and inefficiency both derive from the fact that the blades were mounted on the bottom end of a vertical shaft, the top end of which carried and rotated the millstone. There was no gearing, which simplified construction, but meant that the millstone turned at the same rate as the waterwheel below it. This was much too slow for efficient milling, though an inefficient mill can of course produce as much as an efficient one if its working hours are extended. Because of the superior efficiency of the vertical wheel, scholars have seen in the two a natural sequence of development from one to the other. This doctrine is clearly enunciated by, for example, R.J. Forbes, to whom it is clear that the primitive Greek mill 'inspired' Roman engineers to invent the vertical type, which then largely replaced it. However, mechanical or even economic efficiency was seldom a major goal in the ancient world, and this argument neglects one of the great advantages of the horizontal mill – its simplicity and ease of construction. It now appears more likely that the two types, instead of being successive, co-existed contemporaneously throughout the Roman world, with perhaps a slight predominance of the horizontal mill in Northern Europe. As for its efficiency, we may note that the horizontal wheel has always been the standard form of water-mill preferred by the Chinese, not a people historically noted for their technical backwardness; it is still in use down to the present day.[38]

The essential difference with the vertical wheel, the form of waterwheel now generally familiar, is that this is mounted on a horizontal axle that then employs gearing to transmit the drive through a right-angle up to the mill-stone above. The gearing is also a convenient means of increasing the speed of rotation, usually, in Roman mills, by a factor of five. The actual wheel can be built with its lower edge dipping into the river and rotated by the natural current pushing past it. This 'undershot' type (Fig. 173 (left)) is the easiest to build, since it involves no

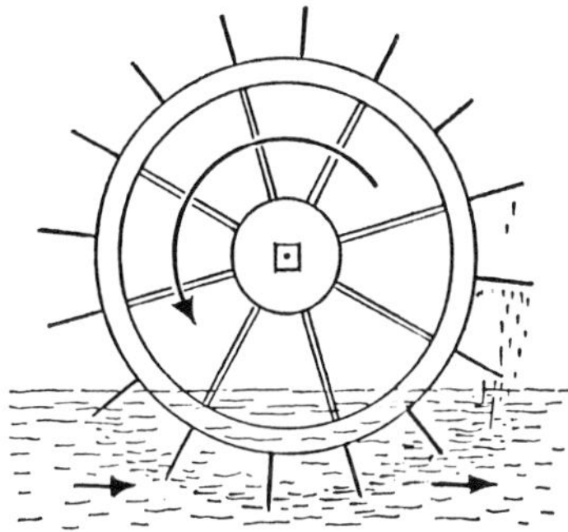

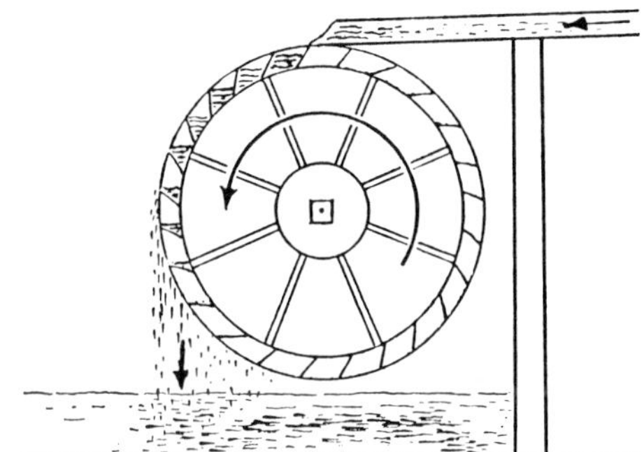

173. Water wheels: (left) undershot; (right) overshot (Landels).

special hydraulic arrangements, but it does need a strong and rapid stream, and is in any case of low efficiency, about 22%. The 'overshot' wheel (Fig 173 (right)) is turned by water delivered from a chute to its upper rim and is turned not merely by the horizontal impulse of the onrushing water, as is the undershot, but also by gravity as the water cascades down to a lower level – that is, harnessing both the water's momentum and its weight; it operates at higher efficiency, around 65-70%. The actual power output depends on the rate and volume of flow of the river or water channel, but in average circumstances it still calculates out to a depressingly low figure – around 1/20 hp for an undershot, and 2-2½ hp for an overshot, wheel. For comparison, we may note that the power output of one man, on a treadmill, is around 1/10 hp.[39]

The second pillar of conventional doctrine is that though the ancients, especially the Romans, understood waterwheels and knew how to build them, yet they hardly ever did, and a good deal of thought has gone into explaining why not. Of the fact of their existence there can be no doubt. The earliest reference in literature seems to be a Greek poem in the Palatine Anthology (9,418), attributed to one Antipater of Thessalonika and dating to the late first century BC.[40] Vitruvius clearly describes the vertical wheel (so clearly that it is sometimes referred to as the Vitruvian mill), and, from a much later viewpoint, in the fourth century AD, Ausonius equally clearly (if poetically) describes a waterwheel-powered marble-sawing works on the river Moselle, which can only be called an impressive achievement.[41] Archaeological remains of several water-mills have also been found, notably in the Athenian agora (but of late Roman date, in the fifth century AD) and at Venafrum, near Cassino, between Rome and Naples.[42] They are also attested in the multi-wheel installation at Barbegal, near Arles (a special case, to be considered below), and in installations of various kinds, known to us by varying evidence, notably at Rome on the Janiculum and in the basement of the Baths of Caracalla. A remarkably high number of small mills also seems to be attested in Britain.[43] Yet, goes the argument, for a total throughout the Roman empire, this is a remarkably low figure, and waterwheels must have been correspondingly scarce. The scarcity is most usually explained away by the hypothesis that rivers suitable to the undershot wheel were few in the Mediterranean area, and that the overshot wheel required hydraulic arrangements that were expensive and technically difficult, namely a mill-pond and mill-race to ensure a regular, controlled supply of fast-running water.[44]

This is very unconvincing, though still widely believed. The requirement of a steady continuous stream of water artificially delivered, for operating an overshot mill, is one that Roman engineering could readily fill. Indeed, it sounds like a dictionary definition of an aqueduct, a branch of hydraulics in which the Romans have always been thought, rightly, to excel; certainly there is something very incongruous about reading, as one does in so many handbooks, of how the Romans had no mills because of lack of exactly the hydraulic facilities that the same book then goes on to explain existed (for other purposes) in vast numbers throughout the empire. Aqueducts ought to have been ideal for water-mills, and it was aqueducts, bringing the water from afar rather than a nearby mill-pond or reservoir, that supplied most of the Roman mills we know about – Janiculum, Barbegal, Venafrum, the Agora. It might be an individual aqueduct expressly used to drive the mill and for no other purpose (as at Barbegal),[45] or, as at the Janiculum, the supply could come from an urban aqueduct. By the fifth century AD this was evidently a common enough solution to require legislation to stop millers doing it.[46] With the technical difficulties thus disposed of, the next point to be faced is, why then were there so few mills? This argument is met head-on by questioning its truth.[47] The only reason that we think water-mills were rare is that we have not found many, and this may reflect simply a failure either of archaeological evidence, or of our interpretation of it. A parallel from mediaeval England, indeed, argues compellingly that it is so. The Domesday Book lists the existence in England of 5,624 mills (though of all types, not only watermills); in Norfolk and Suffolk alone there were 500 of them; 'Yet', it has pointedly been asked, 'how many of these are attested archaeologically?'[48] The answer is, scarcely a dozen. The implication is plain, and the conclusion at present justified. In antiquity, water-mills were far more numerous than is now generally believed. We cannot tell what was their distribution among the various types, but some at least must have been overshot and supplied by aqueducts. These may have been individual aqueducts to supply the mills, or water may have been taken from existing urban aqueducts. Since water is not really 'used' by a water mill, in the sense that it is not consumed or polluted, but passes on and is available for further use downstream, one would like to think that sometimes a mill may have been set into the middle of an aqueduct's course, where it would not interfere with the flow or total discharge. It would be particularly appropriate in the hills, where aqueducts were in any case often obliged to lose height and often had recourse to cascades in achieving it. What would have been simpler, or more economical, than to drive a mill while doing so? However, it must be made clear that this is pure speculation, and no evidence for such an installation has to my knowledge yet been found.[49] The general, overall picture, of course, remains both conjectural and far from plain. The great majority of references to mills are of late date, and one might well suspect that mills were relatively rare in, say, the first century BC, becoming increasingly common in the first four centuries AD to an extent not hitherto appreciated. If we grant this, we must then also assume in the Roman empire a whole network of small aqueducts specifically serving mills, or other analogous arrangements. It postulates a whole field of water-supply that we have not yet covered, and can only guess at.

Multiple mills

A special case must be made for what we may call multiple mills, where a number of wheels are arranged to turn either in series or in parallel. Considering that it seems such a simple and obvious development, very few examples are known. One is on the Crocodile river, at Nahal Tanninim (near Caesarea) north of Tel Aviv, in Israel.[50] The installation consists of a dam built to create a reservoir to supply the 'low-level' aqueduct to Caesarea, in which were housed a number of horizontal (Norse) mills, set in parallel, side by side. Some of these were Turkish, but two have been identified as Roman. Each was in a penstock (aruba), a round, vertical shaft; the horizontal wheel, with blades slanted like a turbine, is mounted at the bottom, and the water admitted at the top so that it strikes the wheel with its momentum increased by the fall. The second site of a multiple mill to concern us is at Chemtou (ancient Simitthus), on the river Medjerda, in western Tunisa. Here it was three horizontal mills set side by side in the abutment of the Trajanic bridge/dam crossing the river (Fig. 174); though they have not been properly published, they were again apparently housed in penstocks, and have consequently been described as 'turbines'.[51]

But the outstanding example of a multiple mill – indeed, the only other one known – is Barbegal.[52] Barbegal is 2 km south of Fontvieille, north-east of Arles. Probably of fourth-century date, the mill complex is served by its own 9 km aqueduct, running parallel to a second one that continues on to supply the city of Arles. The actual mills are laid out in two parallel rows (Figs 175, 176, 177) on a hillside, with a service road running up between them. In each row there are eight millhouses, to a

174. Chemtou, Tunisia: the abutment of the Trajanic bridge from above, with, cutting across it, slots housing the three water wheels (photo: J. Gallagher).

175. Barbegal: reconstruction of multiple mill (Benoît).

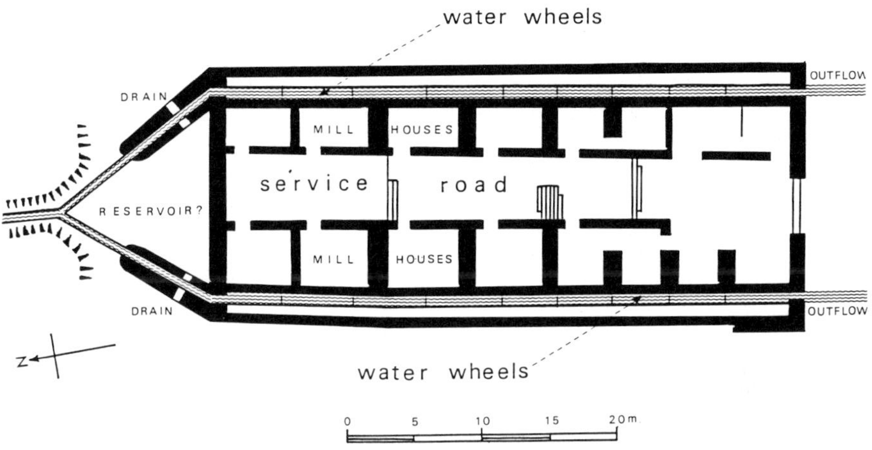

176. Barbegal: reconstructed plan, partly conjectural.

177. Barbegal (Arles): general view of milling complex. The notch in the skyline is the cutting through which the aqueduct entered to serve the mills.

total of sixteen, each fitted with a single vertical wheel, probably overshot, though undershot has also been suggested. The water, arriving by a cutting some 3 m deep through the crest of the ridge, divides into two streams, each then descending the slope along one row of millhouses and turning all eight wheels on the way. There has been a good deal of discussion about the power output of the wheels, how much grain they could grind, and where was the population to eat it all.[53] The water discharged by the aqueduct has been calculated at around 7,000 m^3 daily, and is believed to have driven wheels of 2.1 m diameter and 0.7 m width, at 10 rpm. The chief points of controversy are whether the wheels were undershot or, as the excavator believed and probably rightly, overshot; and whether at the top of the complex the aqueduct fell into a large triangular reservoir that in turn fed the two rows of wheels, or whether, as is more likely, it simply split, in a Y-junction, to serve the wheels directly, with no intervening reservoir.[54]

By a happy coincidence, an inscription is preserved in the Alyscamps cemetery at Arles on the gravestone of a certain Quintus Candidius Benignus, a famous local engineer who was 'clever like none other, and none surpassed him in the construction of machines and the building of water-conduits'. The excavator of Barbegal, F. Benoît, maintained that this was more than mere coincidence, that Benignus was actually the engineer who built Barbegal, and this was the fact that gave him the reputation for hydraulic skills celebrated in his epitaph.[55] There is no evidence at all to support the identification, but it is so attractive an idea one would like to think it was true just the same. If it were so, Benignus would deserve to be celebrated for more than an outstanding feat of engineering. The true significance of Barbegal is much greater. It is the only example known from the ancient world of a power-driven mass-production factory and is hence of great importance to the ancient economy and our understanding of it. The whole economy of Greece and Rome was based on small-scale industrial production, the back-street workshop being the norm. This was due largely to difficulties of distribution and the possible inefficiency of the harnessing of draft animals, which increased transport costs to a level that outweighed any economies of scale of mass-production at a central location. But as for the feasibility of building and operating a big central factory, Barbegal shows that it could be done. It is a feat both of technology and psychology. In a world not noted for innovative thinking, this revolutionary new concept was here accepted and embraced. It was two-fold in scope, seeking productivity both from centralisation of the work force and their efforts, and from economy in the use and provision of power – one aqueduct driving sixteen mills. Above all, if this was a technical triumph, it was a practical, not a theoretical one. One cannot but compare the ingenious but impractical devices created in the equally gifted brains of the scholars of the Museum of Alexandria. Half the time we are not even sure if these devices actually worked, let along doing any good to anyone. There can be no doubt about Barbegal, on either score.[56]

But this in turn brings up another question. As an example of a power-driven factory, Barbegal is unique (if one excepts the small installations at Chemtou and Crocodile river). It is located in a densely populated area of a modern, developed country. It was not buried, and did not need to be excavated, other than a minor clean-up It is a large, imposing site that cannot easily be overlooked. Yet, with all these advantages, it was only with Benoît's investigation and publication in 1940 that anyone recognised what it was. The lesson is obvious. If as prominent and unmistakable a site, in the heart of France, can go unrecognised for so long, how many more Barbegals are there still awaiting discovery in the more remote and less studied regions of the Roman empire? Barbegal may be not nearly as unique as it currently appears, and this may be a form of industrial water use much more prevalent than we at present think. Future study will tell, but in the meantime we may note that, with all its implications for mass production, technological history and ancient economics, this may well be the largest and most important question raised by this book.

Recreation

For all practical purposes, recreational water use in the Roman world means baths. As sober a work as the *Princeton Encylopedia of Classical Sites* can speak of 'the Roman passion for bathing establishments', and it is no exaggeration. Roman baths can be divided into two categories. The first is those establishments which were patronised because of the medicinal qualities of the local water, baths served by hot springs or waters with some mineral content. The vital difference is that here we are dealing with a spa, where one came, as later ages put it, 'to take the waters', or even 'to take the cure', as opposed to the second category, an ordinary bath where one bathed in ordinary water, purely as a matter of comfort, pleasure and relaxation. The differentiation was never absolute, in that in a spa, baths of traditional Roman type were often built and the water was not necessarily drunk, but a differentiation there nevertheless was. For one thing, such a spa was built wherever the spring was, and there was therefore normally no aqueduct or other means of supply from a distance.[57] With hot springs, indeed, it had to be. If the heat of the water, which constituted its major attraction, was to be enjoyed, it had to be consumed locally, for running through a long conduit it would have been lost. As in subsequent ages, therefore, the existence of such springs was often enough to cause a town to grow up, or even be founded, at that spot. When the first Roman town in Gaul was founded in 122 BC, at Aquae Sextiae (Aix-en-Provence), it owed its foundation and no doubt much of its location to political and military motives. But though these might account for locating it somewhere down in the plain below Entremont, the *oppidum* of the Gallic tribes that it replaced, the precise position chosen must have owed something to the existence of the hot springs, since the town was named after them – Aquae. Indeed, the very name tells us something of the Romans and their priorities. One could well imagine a fortified guardpost, opening up the border territory, being given some

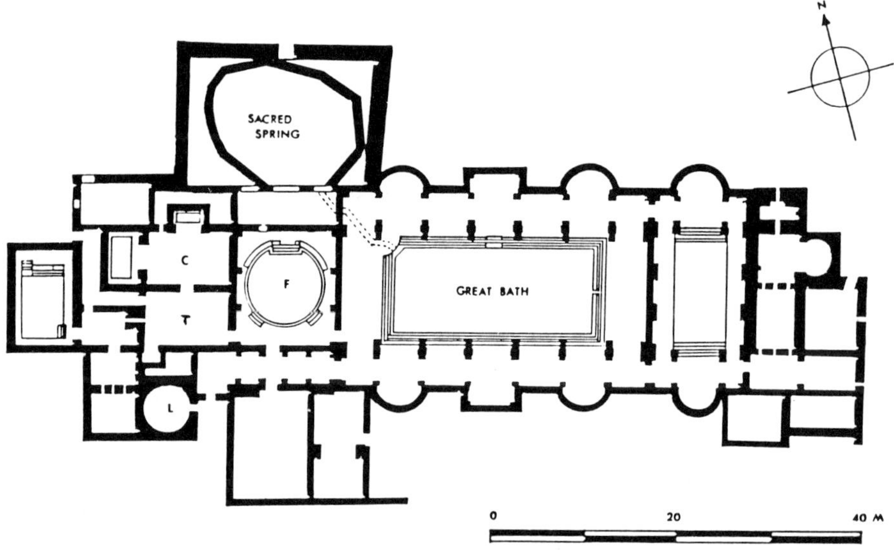

178. Bath, England: plan of Roman bath; C = *caldarium* (hot room); T = *tepidarium* (warm room); F = *frigidarium* (cold room, with round plunge pool); L = *laconicum* (sauna).

such name as Fort Julius, or even New Rome. But 'Sextius' Spa'? There is a lesson in the name itself, yet even a cursory search will show that this case was far from unique. In the *Princeton Encylopedia* already mentioned, there are listed some seventeen towns or cities in the Roman empire all named Aquae this-or-that. Hammond's *Atlas of the Greek and Roman World in Antiquity*, in its index, lists twenty-nine of them. These, of course, are only such cities as laid claim to repute as spas by actually including the word in their official name; the count does not cover the many others, no doubt much more numerous, where mineral springs of similar beneficence were yet held to exist.[58] Many of these cities, named by the Romans after their waters, subsequently enjoyed a prosperous history based on their spa. Many are still well known and operational. They include Aix-les-Bains, Dax and Aix-en-Provence, in France (Aix = Aquae); Aachen, or Aix-la-Chapelle, and Wiesbaden, in Germany; Baden, in Switzerland; and, in England, Buxton and Bath. Grenier lists some sixteen such *villes d'eau* in Roman Gaul alone.[59]

The actual installations at such an establishment[60] usually consist of pools and basins built either on top of the spring or communicating directly with it, in which the bathers could profit from the waters by varying degrees of immersion. The local waters, usually hot and often with a pronounced mineral or (as at Aachen) sulphur content, though they were of prime importance to the spa, might prove insufficient in volume, and have to be supplemented by ordinary water (for ordinary

179. Bath, England: the Great Bath (photo: Bath Tourism).

bathing; one presumes that the precious mineral waters were not adulterated by mixing) brought in by aqueduct; this happened at Aquae Segetae (Sceaux du Gatinais, Loire, France), Aquae Statiellae (Liguria), and, above all, at Aquae Sextiae, where the city was eventually served by four conventional aqueducts as well as the local hot springs.

Bath requires special notice, for two reasons. It is probably the best preserved installation anywhere, and certainly the most familiar; this means that many visitors and readers, not fully recognising that it is of the spa type and hence illustrative only of that, take it as representing Roman baths in general. This is incorrect, and even of baths of the medicinal type it is not wholly representative, being in fact mainly a military spa used by the troops, like Aquae Mattiacae (Wiesbaden).[61] At Bath (Fig. 178) there were two separate sets of facilities: a *frigidarium, tepidarium* and *caldarium* of the conventional sort, grouped closely together in a tightly-knit complex at the west end of the site, and a much larger and imposing set of structures devoted to immersion in the healing waters. The hot spring poured forth daily some 1,150 m^3 of water at 48°C and, to assist control, was surrounded by a wall forming a roughly oval-shaped reservoir. Thence, by lead-lined ducting of rectangular cross-section, the water was led off to fill the Great Bath, whence in turn it left by an outflow, controlled by a bronze sluice-gate, to escape into the drains and so into the river Avon. The Great Bath (Fig. 179) was something like a swimming pool, or *natatio*, about 24 × 12 m and 1.65 m

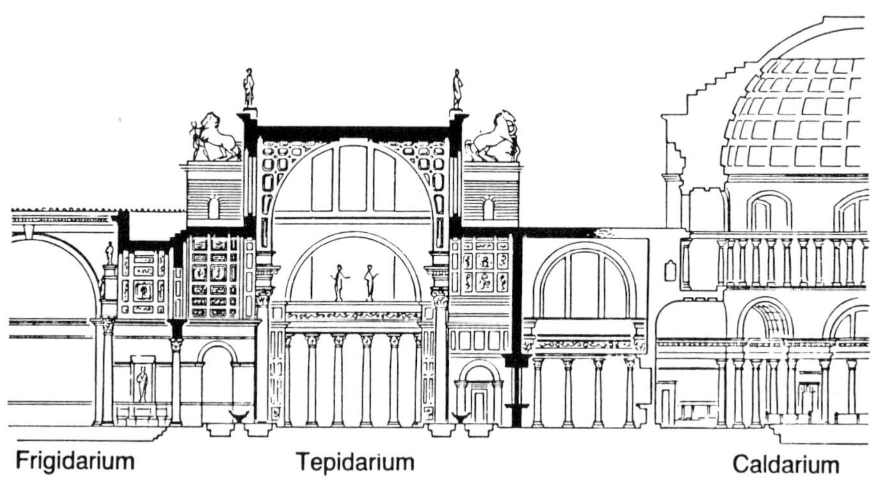

180. Rome: Baths of Caracalla, restored cross-section.

deep, and was lined with an enormous quantity of lead sheeting which, at 2.5 cm thick, came to a total weight of no less than 86,000 kg. Originally the bath was either roofed in wood or simply left uncovered, but was later covered in by a magnificent concrete barrel vault of 12 m clear span, the ends of which seem to have been left open to let out the steam. In addition, there were two smaller, subsidiary pools supplied through lead pipes with hot water from the spring, which was also fitted with an overflow to accommodate the excess whenever the natural discharge, which fluctuates considerably, rose too high. The medicinal and conventional bathing facilities could of course be utilised in any combination of ways by visitors to suit their individual needs, pleasures or infirmities. The bathing complex, about 100 × 50 m in its latest phase, is of monumental magnificence, and one often sees the Great Bath, with its steaming swimming pool, illustrated in tourist literature under such titles as 'The Roman Bath', giving the impression that that is what conventional Roman *thermae* were like. They were not, and it is to them that we must now turn.

The conventional Roman baths, as exemplified by the Baths of Trajan, Caracalla (Figs 180, 181) and Diocletian in Rome and the great provincial complexes such as the Imperial Baths of Constantine at Trier (Fig. 182), Leptis Magna, Djemila, Timgad and elsewhere, were centred upon a series of large halls heated by the underfloor hypocaust system. Plunge baths and swimming pools (*natationes*) were indeed provided, but were always subsidiary to the main purpose. The chief difference between a spa bath and a conventional bath was thus that in the second the main

181. Rome: Baths of Caracalla, exterior. The cars and human figures give an idea of the scale.

rooms were, so to speak, dry, with an ordinary floor upon which the visitors walked, instead of a pool for immersion. That said, we must hastily move on to the paradox that, nevertheless, the conventional bath consumed enormous amounts of water. This was because although the plunge baths, fountains, and the like, were of secondary importance, the water flow through them was continuous and consumption correspondingly high. In any case, we must not overstate: given the nature of the baths, there was no way they could operate totally without water, any more than can their modern descendants, the Turkish baths. However, in the absence of an aqueduct many smaller baths could and did make do with some water-lifting apparatus, such as a compartment wheel or bucket chain, to keep filled a tank on the roof, from which the head would provide water under sufficient pressure for fountains, showers, hot-water heating tanks, and other needs. Baths supplied by lifting devices in this way have been found at Pompeii, Herculaneum, Ostia, Cyrene and Abu Mena (Egypt).[62] Sometimes, as at Pompeii, the issue seems to have been the simple availability of the water. The water table was 25 m below ground level, and until the aqueduct was built in the Augustan age, there was no other way of getting at the water than by bucket chain. When the aqueduct arrived with a copious supply at ground level, the lifting installations may have gone out of service. Perhaps not, however. At Ostia, where the water table was very close to the surface, water was in

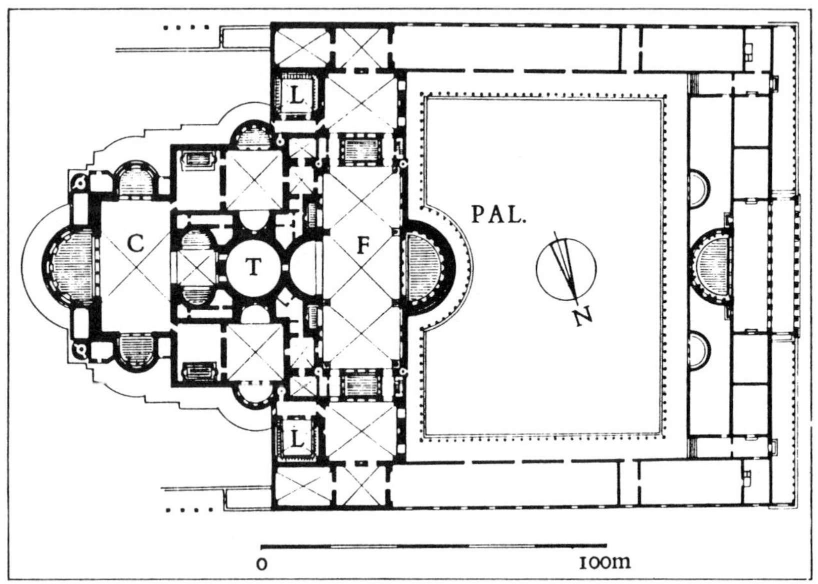

182. Plans of typical large provincial baths: above: the Imperial Baths, Trier, early fourth century AD: C = *caldarium*; T = *tepidarium*; F = *frigidarium*; L = latrine; PAL = *palaestra* (athletics ground); below: the South Baths, Cuicul (Djemila, Algeria), second century AD: *caldarium*, *tepidarium*, *frigidarium*; C = courtyard; latr. = latrine.

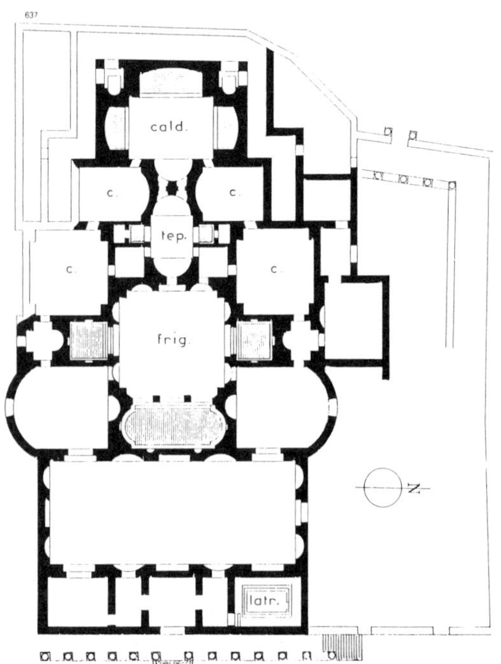

effect available at ground level whether supplied by the aqueduct or not, yet in the baths lifting wheels were provided to raise it to the roof tank. Here not availability but level was the issue. The water had to be delivered in the baths under some head if it was to be of any use and this meant a roof tank, unless the aqueduct itself was at a reasonable height. Estimates for the discharge from such a lifting device, wheel or bucket chain, range from around 120 m^3 to 1,200 m^3 per day, the difference depending on the size of the wheel, and on one-man or two-man operation.[63] These figures are only very approximate, but they do make it plain that a lifting wheel delivered much less than even the smallest of aqueducts, and, since they are based on twenty-four hour operation, at infinitely greater cost. The volume was enough to serve an establishment of, at best, only moderate size, and the absence of an aqueduct was thus a limiting factor in the size of baths.

For really large baths, an aqueduct was a necessity. Indeed, an entire aqueduct might be built specifically to serve a new bath, though the builders usually took advantage of the situation to fulfil other needs as well. All the great baths of Rome were supplied from an aqueduct as a matter of course, and the emperor responsible for building them, if he did not build a new aqueduct and simply drew his supplies from an existing conduit, at least usually supplemented the volume of water in it by tapping new sources. Thus the Aqua Virgo was built, in 19 BC, by Agrippa to supply his baths. Diocletian, when he built the baths bearing his name, renovated and expanded the Aqua Marcia, from which they drew their water. Caracalla did the same thing and for the same reason, adding a new spring, the Fons Antonianus, to the Marcia.[64] The supply was usually achieved by running a branch of the aqueduct from some convenient point on its main line to a reservoir alongside the baths. This is not typical of normal metropolitan Roman practice, which usually ran the aqueduct into a *castellum*, a mere distribution tank, and recalls rather North Africa, where large city reservoirs were common. But it did apparently happen with baths. The Baths of Diocletian were served by the Botte di Termini, a reservoir which, with dimensions of 91 × 16 m, went far beyond anything we could reasonably call a *castellum*, and the Baths of Caracalla near the Via Appia had a reservoir, 'a huge erection with thirty-two chambers in two stories',[65] which is enough to remind one of the reservoir of Carthage.

What happened to the water inside the baths is much more obscure. The main principle is clear enough. It was fed into a large swimming pool (the *natatio*), smaller plunge pools that were provided in profusion, and served an unspecified but doubtless large number of showers, fountains and exedrae, before finding its way into the drains and back outside the building. All of these pools no doubt operated on a continuous flow basis, like the street fountains, and so accounted for the large amount of water used. At Rome, Forbes calculates that up to 17% of the total water supply

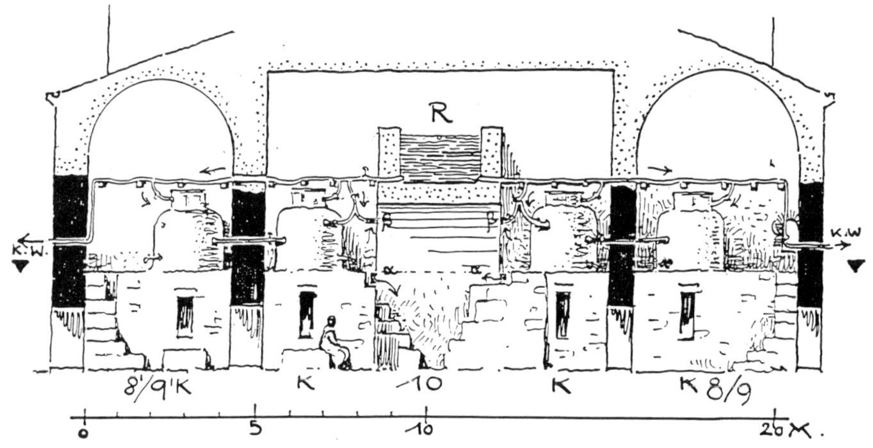

183. Lambaesis: boiler room in baths, showing four boilers (Krencker).

went to the baths. In many cities, where there was an enormous bath-complex to sustain and perhaps only a single aqueduct to draw on for all purposes, the proportion must have been much higher.[66] Studies of the great Roman baths have generally concentrated upon their architecture rather than their technology, and in few can we form a clear picture of the hydraulic layout and trace the actual course followed by the water through the building (Bath being an exception).[67] This is partly because in some locations detailed investigation has for various reasons been impossible,[68] and where the layout is recoverable it usually proves to be highly complex. Distribution was, as one would expect, normally by lead pipes and sometimes open ducting, but the picture was complicated by the addition of boilers to provide hot water. There was usually a whole battery of these arranged side by side in a basement corridor under the central area of the baths, the location usually being shared by the *praefurnium*, the furnace for the hypocausts. What with the rows of furnaces, boilers, fuel bunkers, whole ranges of taps within ready reach for mixing the hot and cold water in various ways, and the general atmosphere of a smoky inferno, the boiler-room (Fig. 183) must have presented an impressive sight when in full operation, something like the stokehold of a large transatlantic liner around the 1920s; no doubt its impressiveness was less obvious to those unfortunate enough to have to work there.[69]

The best-known example of such a boiler is that from the private baths at Boscoreale, now in the Naples Museum (Fig. 184).[70] It consisted of a vertical boiler mounted over a furnace and supplied from a separate cold-water tank. A complicated array of taps mixed the outgoing hot water with cold water drawn directly from the tank to provide warm water of the temperature desired. We may note that although a sort of three-way 'mixer' tap was known (pp. 328-9 below), at Boscoreale each

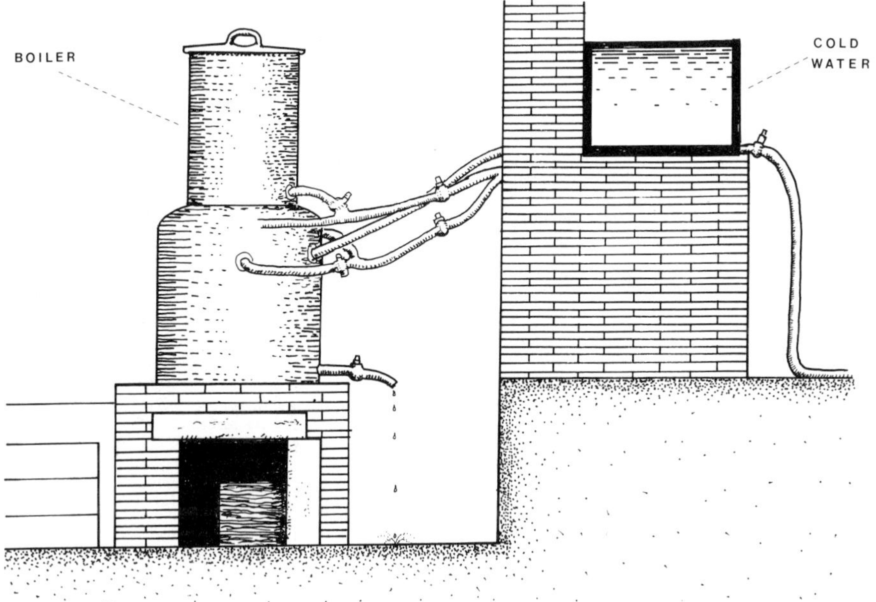

184. Boscoreale (near Pompeii): above: heating boiler for domestic baths; below: close-up of piping: 7 = cold water supply: 2 = boiler; 8 = warm water delivery pipe (Kretzschmer).

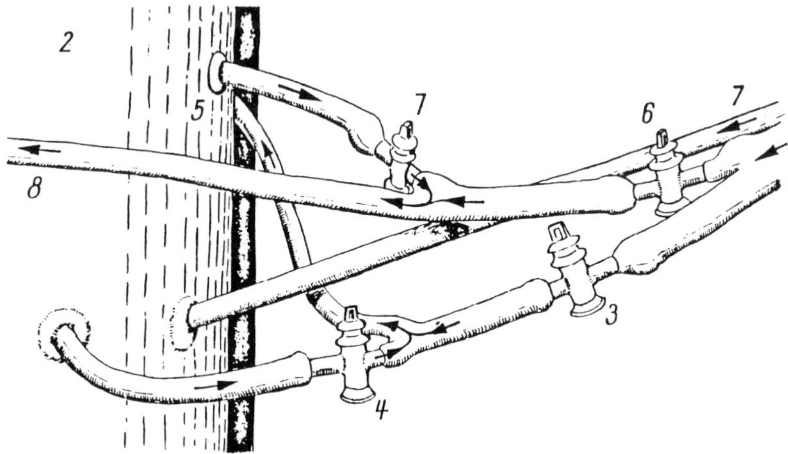

junction in the piping was controlled not by such a device, but by two ordinary taps, one controlling the admission of hot water, the other of cold, into the common pipe. The 'mixer' did not in fact mix hot and cold water, but could only deliver either hot or cold. This may be the reason for the, at sight, more complex and clumsy Boscoreale arrangement: by opening both hot and cold taps, water could in fact be mixed. We may also

185. Ostia: multiple public toilet (demonstration courtesy a group of Carleton University students).

note a variant form of boiler, in which the container was mounted horizontally through a wall; one of the two protruding ends was heated by a fire built under it, the other projected into the waters of a plunge pool and in turn heated them by contact.[71]

The water from the boilers, of whatever sort, entered the baths at a temperature of around 40°C (or a little cooler than the natural spring water of Bath). This was about the same as the air temperature in the hypocaust-heated *caldarium* or hot room, the combination being designed to produce a very humid atmosphere conducive to perspiration. Thence the bather proceeded to the *tepidarium* (warm room), 25°C,[72] and so to a cold plunge in the *frigidarium* (cold room). The water in turn passed into the drains en route to the outside. All of this basement area, through which the drains ran, was honeycombed with various service facilities that often made it a labyrinth defying detailed analysis.[73] However, one principle was clearly understood. Given the vast amount of waste water constantly pouring forth, it was plainly desirable to extract from it any further work possible before letting it go. In the Baths of Caracalla a water-mill was installed in the basement,[74] but this seems to have been a unique experiment that went unimitated. By far the commonest use was for public toilets, provided for those who did not have one at home.

The Roman public toilet has for long aroused an interest in equal proportions shame-faced and puzzled, because of its lack of the privacy

186. Dougga, Tunisia: public toilet (photo: K. Grewe, Bonn).

that modern man deems a civilised essential. The public toilet was a place of communal resort, rather like the barber shops of a later age, and usually consisted of a large open room, often ornately furnished with architectural embellishments, and surrounded on three sides by a row of seats up to ten, twenty or even forty in number, allowing the occupants to consort in happy camaraderie (Figs 185, 186).[75] The seats were usually of wood but often of marble (a highly decorated individual 'throne', in red marble, with armrests and curved back – and, of course, a hole in the seat – is preserved in the Louvre (Fig. 187)[76]) and were mounted directly above a constantly running stream that immediately carried off all sewage and obviated the need for flushing. This highly hygenic procedure was reinforced by arranging for a small gutter or runnel, again carrying a continuous stream of water, to run along the floor just in front of the seats, in which patrons could bend forward and dip their hands;[77] no doubt it also conveniently carried away any spillage, and generally helped in keeping the place clean. Because of the need for a constant (but not necessarily clean) water supply, it was logical to locate such a toilet in some corner of the baths, which was in any case a natural place for it. Usually it was placed just inside one of the doors opening on the street, so as to be equally accessible to patrons of the baths proper, and ordinary passers-by.[78] We must also note that such establishments, though they

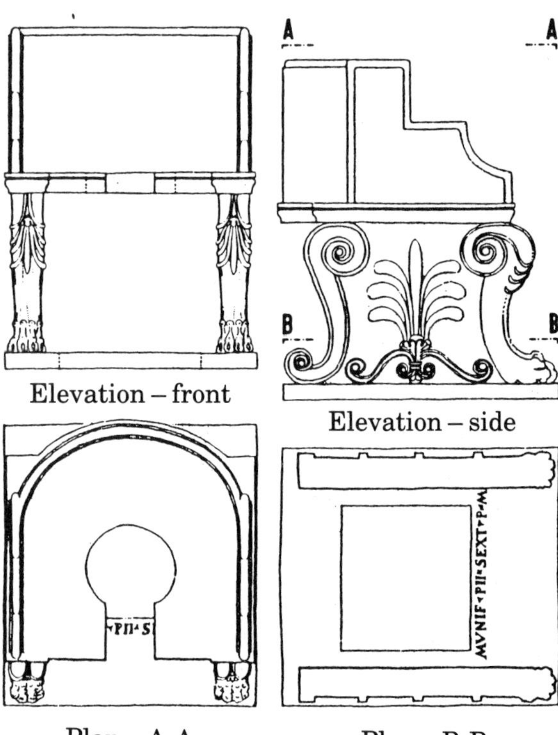

187. Paris: luxurious toilet seat in red marble, now in the Louvre (M. Grassnick).

found a natural place in the baths, also exist independently of them. One often finds public toilets in strategic positions, such as alongside the forum; permanent military camps, such as Housesteads on Hadrian's Wall, had them too.[79]

10
Urban Distribution

> It is very likely that less than half the water that was taken into the aqueducts ever reached the city.
>
> Herschel, p. 195

Filtration

On arrival at the city the water had already been largely filtered, both by small settling chambers cut in the floor of the channel itself and by large settling tanks and reservoirs built, as has been described, at intervals along the line of the aqueduct. By these means a good deal of the sediment had already been removed, though not by any means all, if there was a lot of it to start with. We have already noted the vast heaps of pebbles removed from the Anio Novus by the settling tank at the Villa Bertone, Capanelle, but it is hard to know what conclusion to draw from it. One is tempted to think that the tank was plainly very effective, and how bad the water would have been on arrival at Rome if the tank had not been there. But we have Frontinus' word for it that in fact the Anio Novus

Aqueduct	*Quinariae*	*m^3 per 24 hrs*
Appia	1,825	73,000
Anio Vetus	4,398	175,920
Marcia	4,690	187,600
Tepula	445	17,800
Julia	1,206	48,240
Virgo	2,504	100,160
Alsietina	392	15,680
Claudia	4,607	184,280
Anio Novus	4,738	189,520
TOTAL		992,200

188. Discharge of Roman aqueducts, calculated from the *quinaria* figures of Frontinus (Grimal).

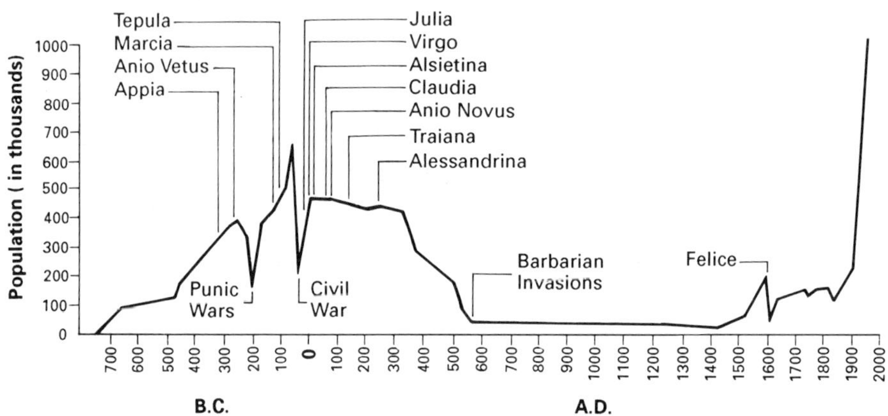

189. Graph relating dates of building Roman aqueducts to growth of population (Pace-Pediconi).

water *was* very bad and contaminated any other with which it was mixed. And all those pebbles that were screened out at the Villa Bertone, since they came from Subiaco, had already successfully run the gauntlet of all the other settling tanks en route, including the big *castellum* at Grotte Sconce. One would have thought that it, at least, would have been able to stop solid impurities like pebbles (as opposed to mud), and yet all those millions of pebbles at the Villa Bertone must have passed through it in order to get that far. One must therefore ask how many more million that we do not know about also got past the Villa Bertone, and were still in the water when it got to Rome? And, of course, at Rome three aqueducts, the Appia, the Virgo and the Alsietina, had no settling tanks at all and the water arrived unfiltered; the Alsietina, the worst of all, was perhaps so bad that it was adjudged useless even to try to clean it up.[1]

However, it was common for such a settling tank to be provided at the end of the aqueduct's run. All that was really needed was a large tank where the speed of the current would be sufficiently slowed in passing through for impurities in suspension to settle to the bottom, but we do find more elaborate arrangements. A common principle was to have two or more chambers so communicating that the water passed through all in turn, becoming progressively purer. With two chambers, it might be possible to close one and drain it for cleaning while the other continued in operation. Where the chambers were arranged at different levels, the water might be made to arrive first at low level and leave from above, to assist the settling process.[2] Otherwise, a typical example is that on the Virgo, where the water circulates through four chambers in turn (Fig. 77).[3] The lower has an opening low in one side running directly into the sewers, to facilitate cleaning. The floor of settling chambers was also

190. Ampurias: filter formed by a circle of amphorae, probably packed with charcoal and sand.

often tilted at a slight angle towards a drain, to facilitate shovelling out the accumulated debris,[4] access for the workmen being provided by a manhole in the roof which, as in most Roman construction, was normally vaulted. Properly speaking, in such installations the water was not filtered, but simply cleared by allowing impurities to settle by gravity. It could likewise be cleared by letting it flow across a bed of sand, or actually filtered by having it penetrate and physically pass through a barrier of some porous cleansing agent, but this was rarely done on aqueducts, where the volume of water to be treated and the speed with which it was passing through ruled out such efficient but leisurely methods.[5]

Nevertheless, it was done. At Ampurias (Fig. 190) we find a filtration plant where, apparently, the water percolated through a circle of amphorae packed with charcoal and sand, though this was in a cistern and treated only rainfall – Ampurias had no aqueduct.[6] More appositely, an actual filter across an aqueduct channel has been preserved at Cirta (Constantine) in Algeria. It was formed by a 60 cm thick wall of sandbags built at an angle across the channel, here 1.5 m wide. On the downstream side of this barrier were constructed a series of eight small channels, each 10 cm wide, separated by planks, and running side by side along the floor of the conduit. They could be followed for a length of 12 m, and bore traces of a fine white sand, which was apparently also what was inside the bags.[7] However, we must remember that this was on an Algerian aqueduct. In North Africa, where water was a rarer commodity than in

191. Baiae: the Piscina Mirabilis reservoir; capacity 12,600 m^3 (photo: K. Grewe, Bonn).

192. Cistern (reservoir) at Bacoli (Baiae), the 'Cento Camerelle' (Daremberg and Saglio).

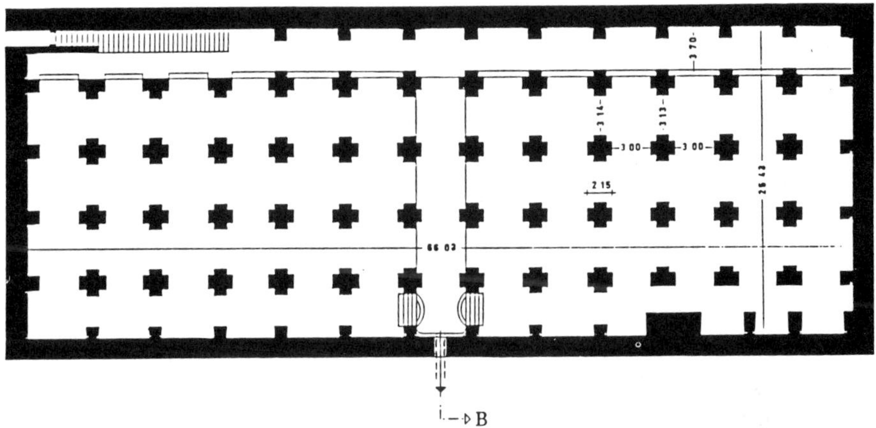

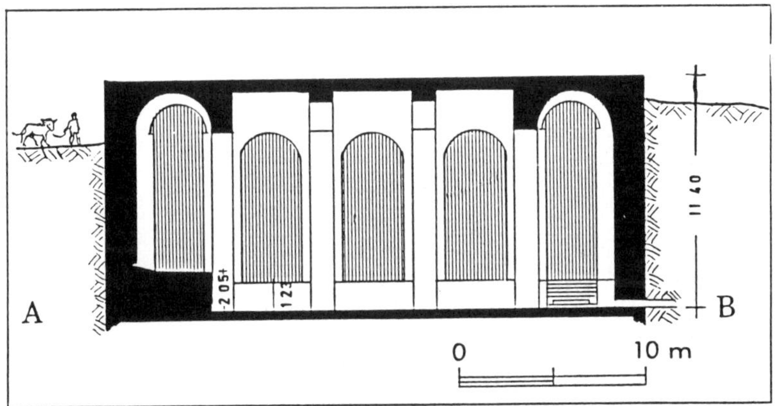

193. Cistern at Bacoli: (above) plan; and (below) section (J-P. Adam).

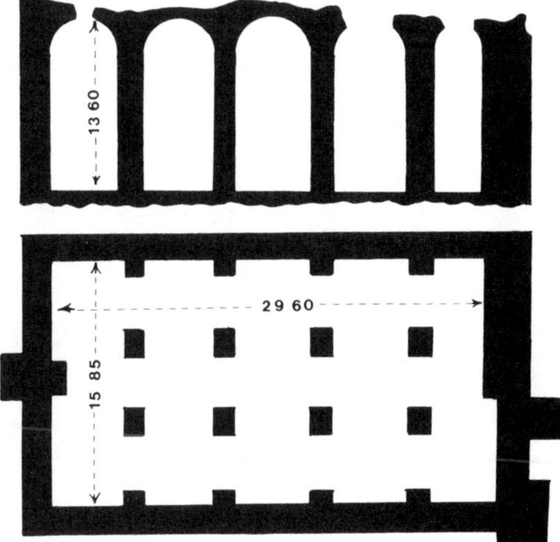

194. Saldae (Bougie, Algeria): cistern, almost 30 x 16 m (Birebent).

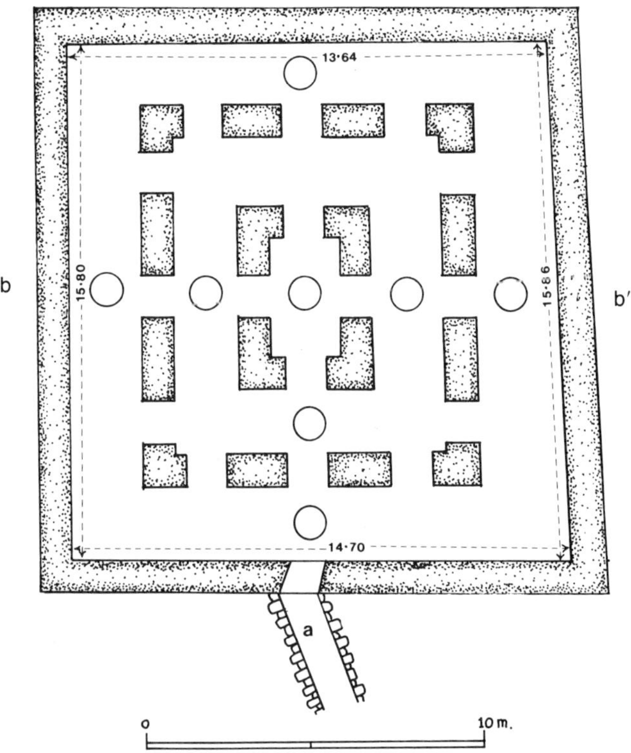

195. Lyon: cistern at the Grand-Séminaire, Fourvière ('La Grotte Berelle'): (above) plan; (below) longitudinal and transverse sections (de Montauzan, revised by J. Burdy and L. Jeancolas).

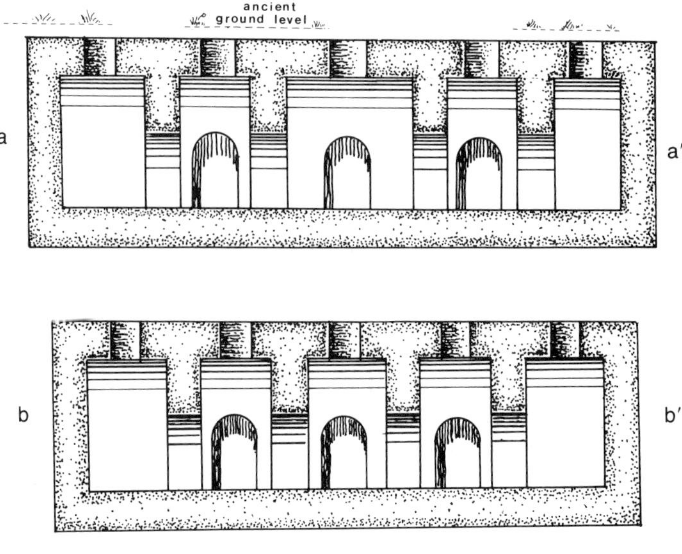

Europe, the Romans do not seem to have been able to afford the luxury of constant offtake as practised in Europe, and on arrival at the city the aqueduct usually discharged into great storage reservoirs, from which supplies were drawn as required. The Cirta filter is on a channel linking two such reservoirs, and presumably the water was fed through only as fast as the filter could deal with it, so that once again we are not dealing with a main aqueduct in full, continuous spate. We must therefore almost rule out filters as part of the normal fittings of an aqueduct. Grilles, on the other hand, were probably common. There is evidence for one on the outflow of the aqueduct into the *castellum divisorium* at Nîmes, and they have also been restored in the *castellum* at Pompeii. No doubt there were plenty of them, and it is perhaps surprising that more have not been actually found.[8]

The *castellum divisorium*

We now reach what is both one of the most complex and least studied aspects of Roman water supply, the urban distribution system. The first step depended on how much water was available. If supplies were limited, the aqueduct might at this point discharge into a large storage reservoir, to be used as needed. As noted, this was particularly common in North Africa, where it ranks almost as standard practice, but these large reservoirs are also found in such well-watered regions as Misenum (the one there, known as the *piscina mirabilis*, is 70 × 25.5 m, 15 m deep, and has a capacity of 12,600 m^3; it was fed by the Serino aqueduct and perhaps supplied the fresh-water needs of the fleet at the Misenum naval base), Baiae (the 'Cento Camerelle'; Figs, 191, 192, 193), Constantinople and Lyon (Fig. 195).[9] Such reservoirs were usually set low in the ground, or actually underground, and roofed over, usually by concrete vaulting. In plan they might look like what one could only call a kind of subterranean hypostyle hall, the roofing vaults supported by rows of columns, piers, or walls pierced with doors to allow the water to circulate. The floor, at Lyon at least, was slightly concave with a drain in the middle, to permit cleaning. In Tunisia an outstanding example is the Bordj Djedid cistern at Carthage, into which the Carthage aqueduct emptied after a run of no less than 90.43 km from its source. This great cistern, reckoned by Rakob the largest in the ancient world,[10] was oblong, 39 × 154.6 m, the size of an entire city block, and subdivided into eighteen transverse compartments. Its capacity was 25,000–30,000 m^3, but to put this into context we should note that this represents about a day and a half's discharge for the aqueduct.[11] For all the size of the cistern, therefore, it only contained something over a day's supply, and we must not consider it as anything like the storage reservoirs in modern work (e.g. at the Marathon dam in Greece) which effectively store enough water for months. On the other hand, it was no mere *castellum divisorium* but a major source of supply.

Unless it was purely for emergencies, I can imagine only two ways of using it. Either it was a reserve for use when the aqueduct ran low, by adding in a little from the cistern every day to supplement supplies until the aqueduct discharge picked up again; or daily consumption did, in fact, exceed what the aqueduct could bring in, at least in the hours of daylight, so that the reservoir was in effect topped up every night to meet the next day's demands.[12]

If this second were true, it would of course imply a whole philosophy of water supply radically different from the continuous offtake principle traditional to Roman aqueducts. It would imply the acceptance of storage against future needs and something akin to the recognition of peak and off-peak hours: an un-Roman philosophy, but they might have been driven to it by actual lack of water. Certainly the widespread use of cisterns and reservoirs, some even aqueduct fed, in minor and rural sites, must have accustomed the North Africans to the idea of drawing upon a store as required, and filling it up as opportunity offered. Is it not then possible that they took the mental leap necessary to extend this principle also to major aqueduct supplies?

Setting aside such storage cisterns, we come to the *castellum divisorium* proper (Fig. 199). As normally understood, this is a small tank (at least compared with the cisterns) located usually at the edge of the city. The aqueduct enters it as one single unit, and leaves it as a number of separate branches, sometimes running direct to their point of use, if it is one, such as baths, using enough water to justify a pipeline of its own, otherwise running to a number of sub-*castella* or water towers, where the branches are further subdivided into individual lines serving fountains and private houses. The *castellum divisorium* thus is in effect a junction box, marking the end of the aqueduct proper and the start of the urban distribution process; it is sometimes referred to as the terminal *castellum*. It is normally to be found on the edge of the city, perhaps on or just inside the walls, and, in order to give service as widely as possible, at the highest point the incoming aqueduct can reach given the level of its arrival. At Pompeii it is at the Porta Vesuvii; at Nîmes on the higher ground at the north end of the city; at Fréjus it abutted on the city walls at the Plateau du Moulin-à-vent – the aqueduct serving it ran along the top of the walls; at Arles it was facing the amphitheatre; at Rome the various terminal *castella* are in different locations, but most fairly high.[13] Modern expositions are sometimes confused, and are based largely on three *castella* – Nîmes, Pompeii, and the ideal *castellum* described by Vitruvius. This last has exerted great influence, for none of the existing *castella* correspond to Vitruvius' prescription, and a great deal of confusion has been caused by efforts to make them do so.[14]

Vitruvius' prescription for a *castellum* is almost better understood by a written description alone, without the illustrations and diagrams that normally accompany it in any standard commentary, for these cannot but create clarity and certainty where Vitruvius may be unclear and ambiguous. To that extent, they may misrepresent the issue. What Vitruvius says is that a triple basin for receiving the discharge from the aqueduct should be attached to the *castellum*; and three equal pipes should bring water from the *castellum* to the three compartments of the basin, so arranged that when the two outside compartments overflow, it goes into the middle one. From the middle one, pipes go off to serve the public drinking fountains: the other two serve the baths and private homes. This is to prevent private users from cutting into the public supply.[15] If the reader finds this account not entirely straightforward or readily understandable, it is because I have tried to reproduce what is actually in Vitruvius. Some points are plain. There is to be a tank divided into three, and each division is to serve one of the three main categories of customer: public fountains, baths and private houses. Priorities are established, and the overflow is the automatic device set up to maintain them.

Other points are not so clear. If the three basins are served by pipes of the same size, will they not get the same amount of water? How will they then ever overflow into each other? The answer is apparently to set the middle one, the recipient of the excess, lower than the other two, so that the overflow goes only in one direction; what happens to its own overflow is not stated. Again, surely the three equal pipes from the *castellum* to the basins must be bigger than the offtake pipelines from the basins to the town, or at least bigger than the two outer ones, otherwise the pipeline could carry everything the *castellum* pipe delivered and there would again be no overflow, except perhaps from the *castellum* itself. Commentators even differ on whether the aim of the arrangement was to protect the private users or the public.[16] It may have been both. The tripartite division of available supplies (by 'three equal pipes' – *tres fistulae aequaliter divisae*) guarantees that both private and public customers have a given, and equal, volume of water reserved for them. It does not, as Callebat implies, recognise the priority of public fountains, in that it makes no effort to ensure that they are served first, and in time of shortage all three – baths, street fountains and private users – will go short alike and in equal proportion. There is no question, as has sometimes been suggested, of ensuring that the public fountains are the last to run dry. Such a course would be logical, but it is not what Vitruvius says. The overflow mechanism comes *after* the division into three, and in time of shortage will not operate because there will be no overflow. What it does ensure is that in times of abundance the surplus, once the commitment to private users and baths has been met, is not divided up but all goes to the fountains, which in this sense do have a priority. The private users have thus not a guaranteed minimum entitlement but an automatically enforced maximum. One may ask why, then, was there not simply a single overflow from the main *castellum*, leading directly into the public service line, without any three basins? This would cut out a lot of complications and produce exactly the same effect. The only answer I can think of is that the plan prescribed by Vitruvius is his own recommendation and purely hypothetical; no *castellum* actually conforming to it has ever been found. Vitruvius, after all, was not writing an encylopaedia of existing Roman building practice (though that is exactly what modern readers often take him for), but rather, based on that practice, a set of guidelines and recommendations: he tells us not so much how things were done as how he thinks they ought to have been done. And sometimes he is wrong.[17]

From Augustan times onwards, Pompeii had a single aqueduct. It drew its water from springs at Serino, near Avellino, whence it ran via Sarno and around the north side of Vesuvius to serve Naples and, eventually, the two big cisterns of Cento Camerelle (Baiae) and the Piscina Mirabilis (Misenum). From Sarno a branch ran to Pompeii, terminating in the *castellum* at Porta Vesuvii.[18] The supply channel was not very big – 30 × 25 cm (Fig. 49(b); ch. 5 n. 7 above), but the distribution arrangements were of remarkable – even needless – complexity. The entire *castellum* was housed inside a large brick building, and consisted basically of a shallow circular basin (Figs 196, 197).[19] After passing through two transverse mesh screens to remove the worst impurities, the water was faced with a row of three parallel exit channels, the entrance to each barred by a low wooden gate; the gates themselves are gone, but their bronze fastenings are extant. The water entered the channel, and so escaped through its exit, at the far end, by spilling over the top of the gate, which thus acted as a weir. From the exit the water ran into an ordinary pipeline of lead pipes, the central one (serving the fountains) being of 30 cm external diameter, and the two side ones 25 cm. The three gates were of different heights, so that the highest gate, which was on the channel serving private houses, cut off their supplies until and unless the water level in the main body of the *castellum* rose high enough to spill over it and start flowing down the channel; conversely, the lowest gate (that in the middle) governed access to the public fountains, which, if the water level sank, were thus the last to dry up. As Eschebach says, the principle plainly reflects the system of Vitruvius, but it is not the same. By introducing the selection mechanism before the water is divided up between the three channels, instead of after, its operation is radically changed. The private users have no minimum entitlement, for until the needs of the other two groups have been fully satisfied, they get none at all. Nor is Vitruvius' scheme simply reversed, for in the event of a large surplus it does not automatically now all go the private users, just as Vitruvius directed it all to the public fountains. When the water is high enough to spill over into the private users' channel, it still continues spilling over even more into the other two, so that the surplus is itself then divided among the three. Thus the Pompeii *castellum* genuinely

196. Pompeii: *castellum divisorium*, at the Porta Vesuvii. Plan of the interior arrangements, Fig. 197. Water mains serving the city came from the three large openings at the base of the wall.

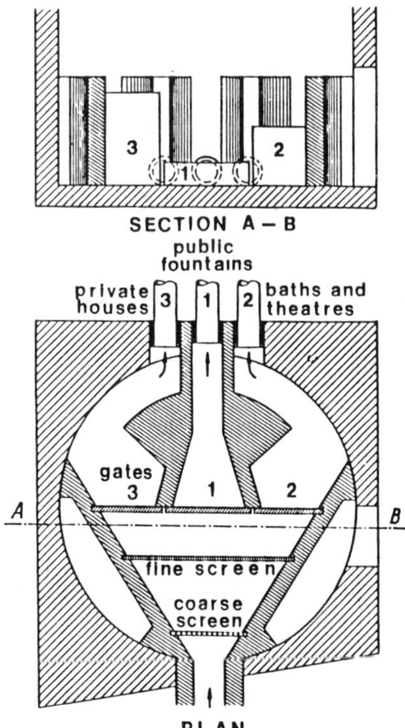

197. Pompeii: plan of distribution arrangements inside *castellum divisorium* at Porta Vesuvii: plan and elevation (Kretzschmer).

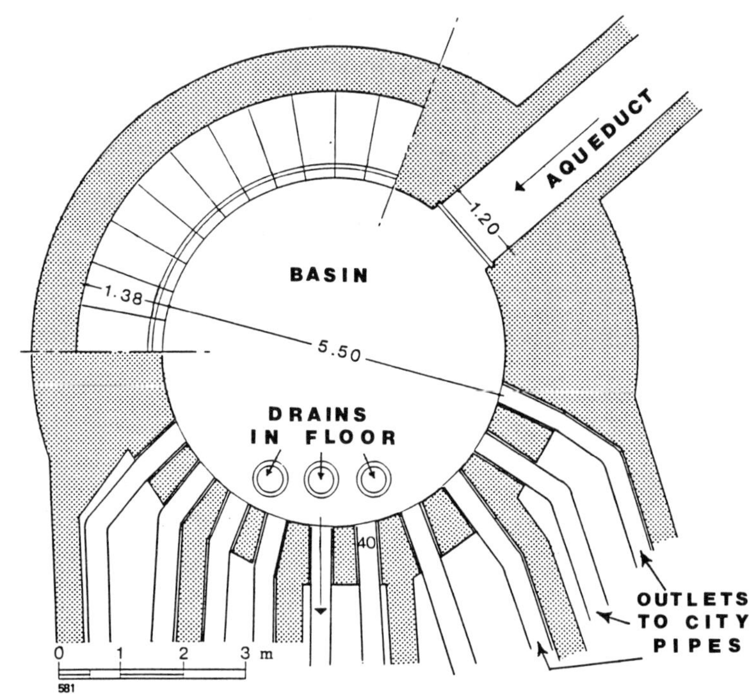

198. Nîmes: *castellum divisorium*, plan (J-P. Adam).

does fulfil the principle enunciated by Callebat, and gives not just a preference but absolute priority to public drinking water: until this need was filled, and in full, all other pipes remained completely empty. Of course, one must not forget that, as in other cities, piped supplies were extensively supplemented by wells and cisterns, so that this restriction was perhaps not as onerous a burden as might be imagined.[20] The real surprise perhaps is twofold. Is not this a very complex arrangement for a very small aqueduct – the conduit is 30 × 25 cm and carries only half the discharge of the industrial leet at Dolaucothi? And could the same effect not have been realised just by building an ordinary tank and setting the three offtake pipes into its side at different levels? Of course, the wooden gates could easily be replaced by others of different heights, thus varying the proportion of water delivered to the three main pipes, and could even be lifted out altogether to permit unrestricted flow, so that the Pompeii system does have more flexibility than three pipes in the side of a tank.

The *castellum divisorium* at Nîmes (Figs 198, 199, 200) is certainly the best-known extant example.[21] Though today somewhat austere, not to say deserted, in aspect, it was in Roman times a place of popular resort. Its heart is an open, shallow, circular basin, diameter 5.5 m and 1 m deep.[22] In antiquity this was surrounded by a metal balustrade, over which visitors could lean and admire the swirling waters. Around it was

199. Nîmes: *castellum divisorium*. The water entered (from the aqueduct) through the large square opening (left), and left by a series of ten large lead pipes through holes in the sidewall; a further three ran from openings in the floor. For plan, see Fig. 198.

built some sort of enclosure with a tiled roof, Corinthian columns and entablature, and a wall decorated with a brightly painted fresco of fish and dolphins, an appropriately aquatic theme.[23] All of these were in a fragmentary state on discovery and have long since disappeared, so one cannot reconstruct the appearance of the edifice, but we would surely not be far wrong in visualising some kind of small but ornate pavilion, rather like that which a later age erected for the eighteenth-century *château d'eau* at the end of the aqueduct in the Promenade de Peyrou at Montpellier (Fig. 1). As at Montpellier, so also at Roman Nîmes was the water supply considered an object of civic pride well worthy of architectural celebration.[24]

The aqueduct entered this basin through an oblong opening to the east, to the left of the entrance doorway of the pavilion. This opening, 1.20 m wide by 1.10 m high,[25] was covered by a (stone?) grille behind which was a sluicegate controlling flow from the aqueduct into the *castellum*, apparently operated by a windlass mounted on the stone paving surrounding the basin, the connecting cables passing through two holes in the stone (Fig. 200). The sluice itself raises great problems. First, why is it there at all? It is natural to see it simply as an isolating sluice to shut off supply to the *castellum*, but when one remembers that it is supplied by a downhill, gravity-fed open aqueduct channel, one realises that the place to cut off the supply is somewhere a lot further back, where the water can

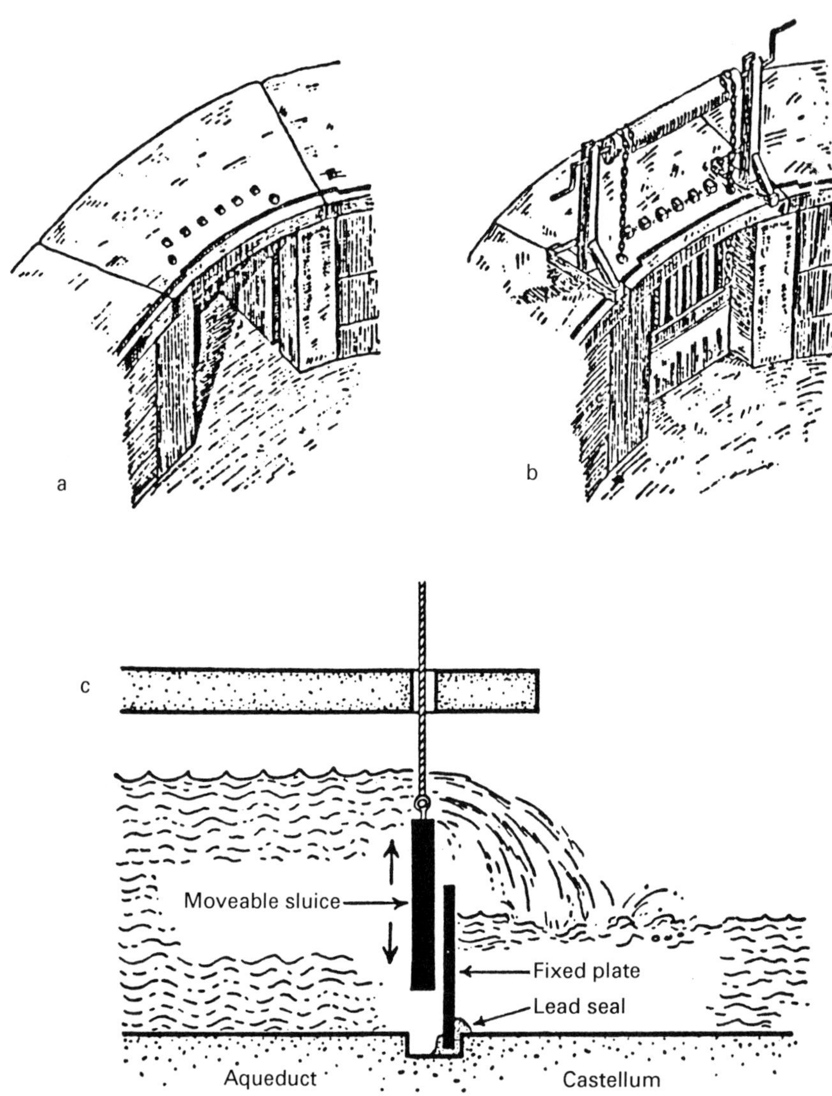

200. Nîmes: *castellum divisorium*, restoration of sluice (a) and (b) at entry point of aqueduct (Stübinger); (c) reconstruction of its possible use in measuring aqueduct discharge (A.T.H.).

be diverted into another channel: cut it off at the bottom of the slope and the result is an overflow and flooding. Second, because the sluice is tucked in underneath the stone cover slab of the paving, it cannot be lifted up to let the water flow, for there is no slot in the stone-work to let it through as it rises. It was perhaps in answer to this difficulty that the sluice was apparently made in two halves (Fig. 200 (c)). The lower half was a fixed lead plate, set permanently in position and sealed round the edges with lead. Behind this slid the sluicegate proper. When it was raised, it and the lead plate between them closed the entire aperture, though it must have been a very leaky arrangement. When it was lowered, to rest behind the lead plate, the upper half of the aperture was opened and the water, provided it was deep enough, could flow into the *castellum*, pouring over the top of the sluice and plate as if it were a weir.[26] The reason for this complicated arrangement remains obscure, and could all have been avoided just by fitting an ordinary sluice in some position where it could be operated, either in the *castellum* itself, across the mouth of the opening, or, if it was desired to keep it tucked away inside the opening for the sake of neatness, then all that was required was a slot cut in the stone cover to let the gate rise clear of the water stream. The effect of the fixed plate would be to maintain a certain minimum depth and slow rate of flow in the aqueduct behind it, reducing wear and tear on the walls, but this would seem to be a minor factor.[27]

In the middle of the *castellum* are still to be seen one or two blocks of stone set into the floor (there are four marked in the *Archives* plan, n. 21 above), themselves carrying sockets for the mounting of some kind of barrier constructed across the *castellum* and through which the water flowed on its way from the entry sluice to the offtake pipes at the other side. This barrier, from the position of the blocks, seems to have been crescent-shaped with the convex side towards the offtake pipes, but apart from that we can only conjecture what it was like or what it did. There are two possibilities. One is that it was simply to create a disturbance in the water and aerate it from the eddies and turbulence thus caused – as the *Archives* publication puts it, 'pour lui donner une agitation qui, dans les idées des Romains, contribuait à sa salubrité ... En arrivant dans le bassin par l'aqueduc, l'eau venait se briser contre cette barrière, tourbillonait, et se mélangeait d'air; elle était pour ainsi dire *battue* au moyen de cet appareil.'[28] The other possibility is that it was an arrangement like the gates in the *castellum* at Pompeii, and controlled the supply of water to the various pipes. Certainly such a provision was needed, for their openings show no signs of being fitted with sluices, and the pipes themselves, at a 40 cm external diameter, were too big for taps.[29] But for it to work, a separate sluice would have to be provided for each pipe, or group of pipes, and if this were installed in the barrier, then there would have to be a channel from sluice to pipe, to stop the water thus admitted also going into the wrong pipes. One thus postulates a metal barrier with a series of sluices, and behind it a row of parallel channels separated by dividers, such as planks set on edge, to connect the sluices with the respective pipe openings. This is not too different from what we have actually seen at Pompeii, though there were only three channels and the whole operation was on a much smaller scale, but, in the absence of any evidence at all, this is probably as far as guesswork can profitably be pushed.

In the wall facing the entry of the aqueduct is pierced a row of ten large holes, which provided the offtake for the main pipelines serving the city. The row formed an arc, and the aqueduct was directly facing one end of it and at a right angle to the other: it was therefore not centred or aligned upon it, for whatever reason.[30] The holes are 56 cm above the floor and of 40 cm diameter. They have no ridge, flange, or recess of any kind for fixing into the hole the lead pipe that it carried, so that this must have been held in place by a ring of cement or plaster. No trace of this survives. The pipes themselves are also gone, though part of the calcium carbonate deposit from inside one has been found, giving the pipe's dimensions.[31] As with many other similar reservoirs, a problem is posed by the apparent absence of any sluices, taps or other control equipment on the pipes leading out of it; the same problem arises with the header and receiving tanks of the Lyon siphons. Setting aside the possibility of individual sluices on the metal barrier as too conjectural (though perhaps correct), how then was water shut off in one of these pipes? The normal answer is that there must have been a tap or stopcock on the pipe itself, now disappeared, but no taps of this size have anywhere been found. The stonework around the holes is completely flat and virgin, with no signs of fittings for sluices or anything else. There is only one other way that the supply could have been cut off without leaving a trace, and this is by lowering into the water a sheet of canvas to cover the hole, against which it would be held firm by water pressure. There is evidence for the use of this system from a later period – in Turkish hydraulics, which also used *castella*, it was one of the standard ways of turning off the supply. I do not, however, offer it here as anything more than an interesting possibility.[32]

Although in the *castellum* the ten holes are equidistant, once the pipes left it they were grouped into five pairs, each pair running in a hollow between two masonry side walls that divided it from its neighbours and also presumably supported the stone slabs of the pavement above. None of these pipes have been traced any further, and we do not know whether they simply represent a geographical division (mains branching off to five different parts of the town) or a division by category of customer, like the threefold division of Vitruvius – private, public and baths. If so, a different political or social principle seems to have been involved, since all ten offtake holes are at the same level and all users, of different kinds, are treated on a basis of equality, as opposed to the efforts of Vitruvius and the Pompeii authorities to establish priorities. Of course, if there were control sluices somewhere they could have been used to regulate distribution; this remains conjectural.

Finally we may note that in the floor of the *castellum* immediately in front of the bank of ten orifices in the wall, is a further set of three. These, unlike the others, have a recessed ring around the edge of the hole to provide a seating for a lid to close off the hole. Moreover, projecting from the edge of the ring into the stonework are three short, recessed slots, evidently to engage with three projecting lugs on the circumference of the lid; presumably they prevented the lid from turning in its seating, though why this should be thought either necessary or advisable I do not know. There is no sign that they could be used in any way to lock the lid down and in place, and presumably it was held there by gravity and the weight of the water above. Presumably that would be enough to hold down even a wooden lid and prevent it floating, but the lid may have been of stone in any case.[33] Presumably also there was some method of unplugging these holes from above when the castellum was full; a rope or chain attached to each lid would be enough. As for purpose, the existence of these holes has sometimes led to misapprehensions, evoking memories of Vitruvius and his two-level overflow system of priorities. On this basis, the three holes in the floor would serve the top priority since, being lowest, they would be the last to run dry: Ashby fell into this trap and asserted that here 'three main pipes catch the maximum flow and ten others draw equally from a higher level in the tank'.[34] But in fact the three holes go straight into a drain. Why, then, are they there? A single orifice would be more than adequate for draining the *castellum*; with the inflow shut off, somewhere, it would be empty in two or three minutes. The use of three implies that either the *castellum* had to be cleared as fast as lightning (which makes no sense) or, if the inflow was *not* closed off, a large volume of water was, on occasion required, and quickly, down this drain. The most likely explanation here is that long ago urged by Auguste Pelet, that it was for a *naumachia*, at the theatre nearby.[35]

These three terminal *castella*, two real and one hypothetical, are the evidence upon which nearly all our discussions of the topic are based. Other examples actually excavated and known are very few. One may cite the small but remarkable *castellum* from Thuburbo Minus (Tebourba) in Tunisia published by de Montauzan[36] (Fig. 201). An oblong chamber, some 1.5 × 2.8 m and 1.5 m deep, was entered by the aqueduct at one end. From the far end and the two sides three lead pipes served, by de Montauzan's account, the baths, the theatre(?) and the cisterns. This is a North African context, where scarcity of water means it has to be used as required from storage cisterns instead of kept running all the time as in the fountains of Rome, but given this substitution what we have here is basically the Vitruvian tripartite division (reflected also at Pompeii). In view of its small size, this is plainly not conceived as a storage reservoir – that is what the cisterns are for. It is a junction box, a convenient way of splitting the main channel into three, with control sluices on each (the vertical grooves for which are preserved in the stonework). It is one of these sluices that is the most noteworthy feature of the tank, for it controls two apertures simultaneously (Fig. 201 (below)) and can be set in four positions, making it a device of remarkably sophisticated flexibility.[37]

From the *castellum*[38] the aqueduct is divided up into branches. These

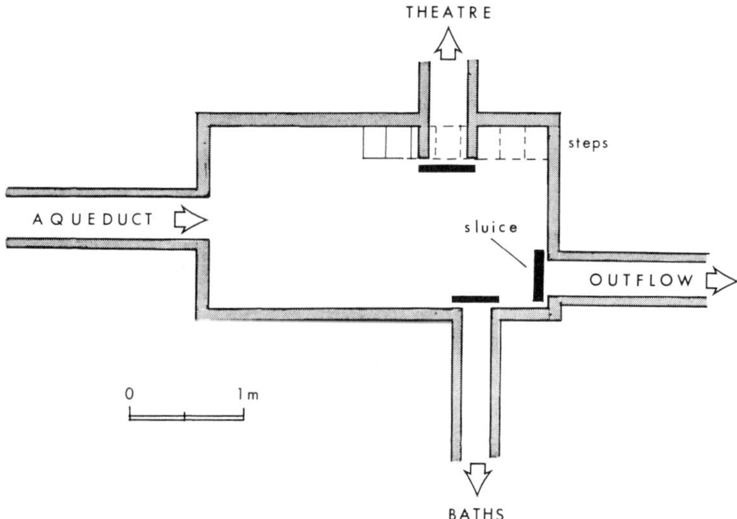

201. Tebourba, Tunisia: (above) *castellum divisorium*, serving theatre and baths. The main outflow to the cisterns which serve the city is controlled by a sluice (as marked) with four possible settings: (below) four-position settings of this sluice.

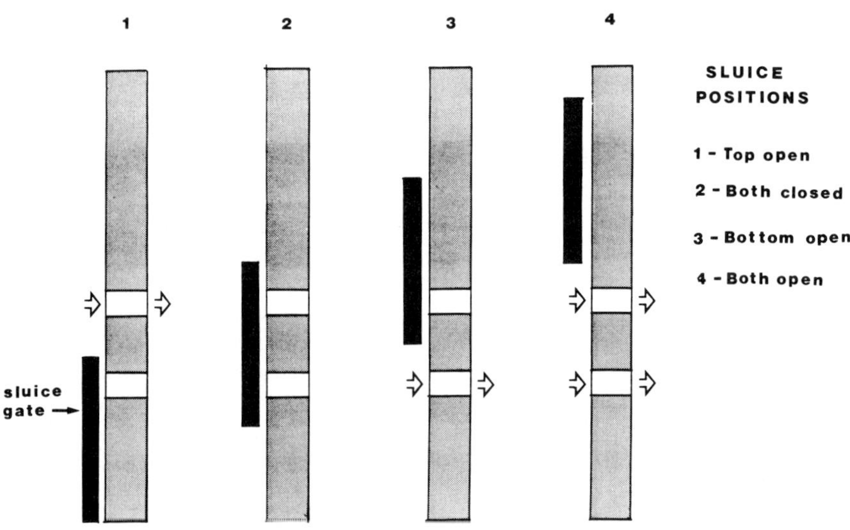

may be either branches to the three different categories of user, as apparently at Pompeii and Thuburbo, or, as at Nîmes and Rome, to different regions of the city, the division by category to be made later, if at all. These branches might be lead pipes, in which case the *castellum* marks the point where the water leaves the open channel of the aqueduct to enter the environment of enclosed low-pressure piping in which it will be carried henceforward, as at Nîmes and in our Fig. 198. Alternatively, the branches, particularly if the aqueduct is so big that to carry the water you would have needed a whole battery of pipes, as in siphons, could be conventional open channels, like the aqueduct itself. Many of the aqueducts of Rome divided up into branches in this way serving different regions of the city (some, like the Rivus Herculaneus of the Marcia, having their own names), their channels running on arcades or subterraneously as terrain and levels might require. Thus ceremonial arches sometimes carry an aqueduct on top as part of the construction, to get it across the street,[39] and the same is true of city gates if the aqueduct is carried on top of the city walls. The best known example of this is the Porta Maggiore (Fig. 202), where the Claudia and Anio Novus are clearly visible on top of the city walls and crossing the Viae Praenestina and Labicana.[40] The city walls did sometimes provide a convenient substitute for an arcade in carrying the aqueduct round from one part of the city to another, but there was a potential snag: built to military, not hydraulic, criteria, and usually in existence before the aqueduct came, they were liable not to be level enough for an open water channel. This might sometimes be corrected by building the walls higher where necessary, but there were limits to how far they could be modified. At the north corner of Fréjus the city wall, here making a right-angle bend, also rises at the apex of the angle to a height hydraulically unacceptable. The water-channel, therefore, which is here running on top of the wall, leaves it to cut across at an angle following the lower ground and pick up the wall again when it has made its return and has descended again to a suitable level.[41]

The secondary *castellum*

As the water thus continued its course, whether by pipe or conventional aqueduct, it next reached a smaller, secondary *castellum*; sometimes this was raised on top of a brick pier to become a water tower (Figs 203, 204). It was from these that pipes then branched off to supply individual customers, and they were quite numerous. Rome had a total of 247 such *castella* (Fig. 205).[42] Despite this, not many of them have been found outside Pompeii: at Rome discoveries are rare, and 'nous ne connaissons en Gaule que bien peu de ces châteaux d'eau: deux ou trois à Lyon et de façon assez incomplète.'[43]

It is also probably again a mistake to seek to reconstruct a uniform

202. Rome: Porta Maggiore, the point at which all eastern aqueducts entered Rome together. The channels of the Aqua Claudia and (above) the Anio Novus are clearly to be seen running on top of the walls.

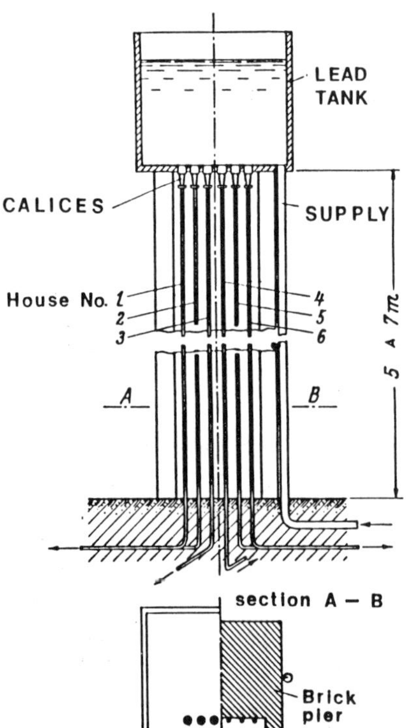

203. Pompeii: water tower, or secondary *castellum* (Kretzschmer).

204. Pompeii: water tower. Clearly visible are the vertical recess carrying the piping up its side to the tank on top, and the public fountain (the square stone basin) at its base.

205. Table showing allocation of the *castella* in Rome among the various aqueducts, in the time of Frontinus (Pace).

Appia	20
Anio Vetus	35
Marcia	51
Tepula	14
Julia	17
Virgo	18
Alsietina	—
Claudia and Anio Novus	92
TOTAL	247

system, consistently employed throughout the Roman empire. They were probably of different shapes, round or rectangular. As well as subdividing the supply among other customers they probably also served each other, for there are far too many of them, at least in Rome, to have all drawn their water directly from the terminal *castella*. A lot of them must have been arranged in series, the water passing through several *castella* in turn with some of it being diverted at each one into the pipes of the users. Again, Frontinus often speaks of water being switched from one aqueduct to another, and throughout the city the public fountains were in general equipped each with two water jets drawing upon different aqueducts, as a stand-by in case service was interrupted. This does seem to indicate a whole network of interconnections and switching facilities that we can only guess at, but which must have been of cardinal importance in day-to-day operation. The role of the secondary *castellum* represents perhaps the greatest lacuna in our understanding of Roman urban distribution.[44]

Frontinus raises yet another point, by quoting a law of 11 BC forbidding private users to take their supply line directly from the aqueduct conduit: it had to be taken from a *castellum*.[45] Two implications are plain. Previously many private users did tap the conduit directly for their supply lines, and the making of the connection was not done by the waterworks staff, and was outside their jurisdiction. Of course, such connections were often made illegally, either by the users themselves or by the waterman for a bribe, but that is not what we are dealing with here. If it were, the illegality of taking the water would itself be enough for a prosecution; and if the waterworks staff had a monopoly of installing the connections, all that would be required would be a directive from the *curator aquarum* telling them not to put the *calix* into a conduit but to go to a *castellum*. Instead, it took a decree of the senate, acting on a report of the two consuls, to put a stop to the practice. We may therefore reasonably assume that at this time private users who had the right to a water service also had the right to make their own arrangements for putting in the connection, without employing the official watermen.[46] We thus have a suggestion that this public service had perhaps been to some extent 'privatised', and the private contractors had adopted the practice of saving money by taking the lead from the nearest stretch of conduit handy instead of carrying the line to a *castellum*, thus cutting out the secondary *castellum* from the system. They evidently had to be stopped, because the main pipes and conduits were suffering ('being ripped apart', Frontinus says) from the great number of small-gauge offtakes being inserted into them, often none too carefully. Henceforth, all leads had to come from a *castellum* where they could be properly supervised both in installation and in operation.[47] On the other hand, when enough private users were involved to justify it, they could club together and build a private secondary *castellum* at their own expense and for their own use, at a location approved by a waterworks inspector; this could be supplied by a mains pipe from the nearest public *castellum*, and the local proprietors could then individually take their supplies from it. On this basis, there was in Rome a further subdivision in secondary *castella*, into private and public. There was presumably no difference in design or function, except in so far as while the public *castella* would, or at least could, supply public fountains as well as private houses, private *castella* were exclusively for private use. Presumably also, being privately built, cleaning and maintenance were a private responsibility and not borne by the water office, but we really have no way of knowing. We do not even know how many of them there were in Rome.[48]

At the point where it left the secondary *castellum*, each one of the lead

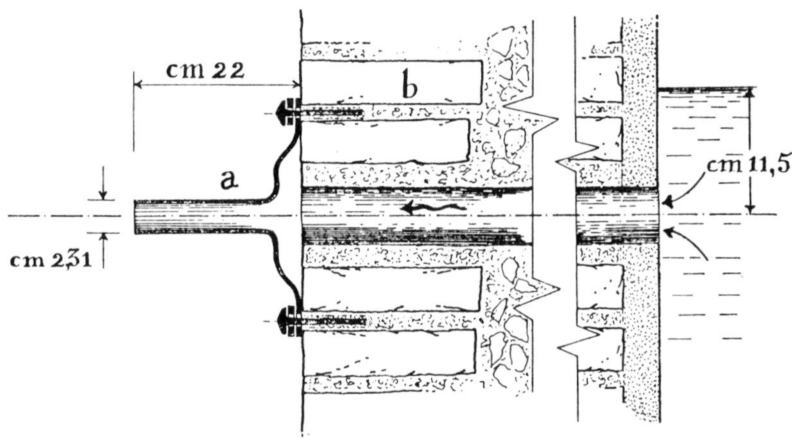

206. *Calix*, *quinaria*-gauge, mounted on a *castellum*; in this reconstruction the *calix* (a) is attached to the outside face of the *castellum* wall (b). The distance below water level (11.5 cm) is in accordance with the calculations of Di Fenizio (Pace).

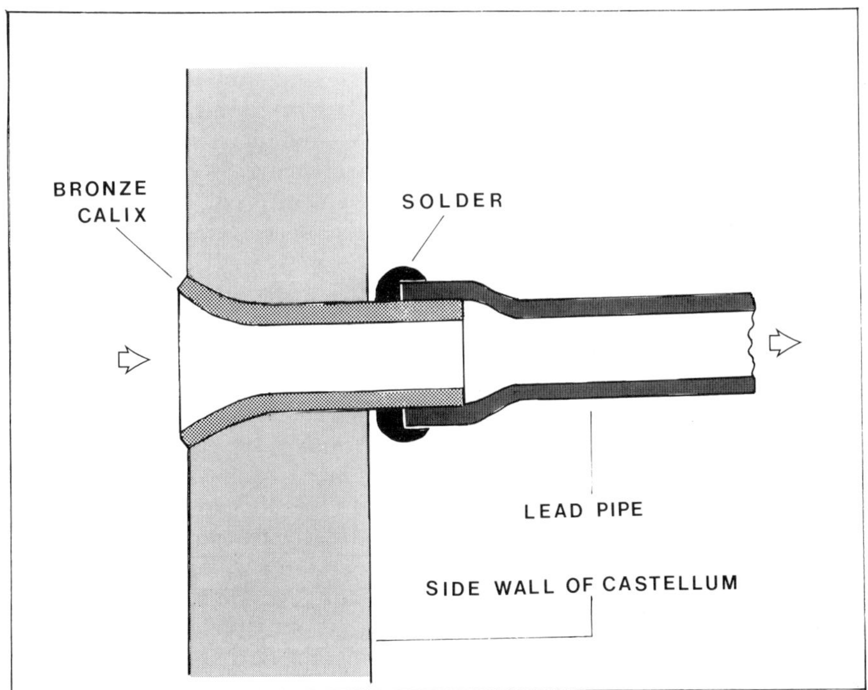

207. *Calix* embedded in the *castellum* wall, with lead pipe attached (right) (Forbes).

pipes into which the water now passed was fitted with a *calix*. The *calix*, sometimes called an adjutage, was a bronze nozzle or tube (Figs 206, 207) at least 12 digits (21.6 cm) long, and of standardised size. It was mounted in the side wall of the *castellum*, so that it received water from the tank, and passed it on to the lead pipe which was fixed to the other end.[49] The *calix* thus governed the amount of water the user could receive, and, in the absence of water meters, was the device whereby his entitlement was officially regulated and enforced. It was based upon, and automatically delivered, the maximum discharge attainable by continuous running, day and night. No one looked askance at that, since the whole Roman aqueduct system operated on the same principle, and no doubt the proprietor, who had paid for his grant of water rights, regarded water flowing continuously night and day not as waste, but as something he paid good money for, and was *entitled* to. At the same time, the unscrupulous were not above wreaking an improvement by taking a crowbar to the *calix* and enlarging its orifice; this is why the calix had to be made of bronze, harder to deform than lead.[50] It was also specified that at least the first fifty feet of lead pipe following the *calix* had to be of the same size as it was. Apparently it was feared that if a large pipe was fitted leaving the *calix* as a short bottleneck only a few inches long, it might act hydraulically as a sort of funnel in reverse and unduly increase flow through the *calix* by the suction so created.[51]

The *calices* came in regular sizes to match the size of the standard pipes in use,[52] and, since they controlled the distribution and consumption of the water, were the key to the aquatic book-keeping on which the whole system was founded, particularly from the viewpoint of a bureaucrat such as Frontinus. The calculations involved were of great (one sometimes feels, needless) complexity, and offer perhaps one of the best examples extant of the cumbersome nature of Roman mathematics.[53] It was not helped by the fact that two quite different standards of measurement, inches and digits, were involved, the digit being 1.8 cm and a measurement commonly used in Campania and a large part of Italy, while the inch was 2.4 cm and of Apulian origin. The waterworks staff had at their disposal a repertoire of no less than twenty-five *calices*, though only fifteen were in use in Frontinus' day, ranging from the *quinaria* (diameter 2.3125 cm) to the *centumvicenum* (diameter 23 cm). To complicate matters further, some four of these, but only four, were by convention actually a different size from what they were supposed to be. Worse yet, of the four, one was smaller and three larger than their nominal dimensions.[54] It is small wonder that Frontinus spent a third of his book sorting out the niceties of this situation, which, however, need not concern us in a general conspectus such as this. The nominal size of the *calix*, whatever it might be, was officially stamped upon it and a pipe of the same size connected to it, itself repeatedly marked with the same stamp for the first fifty feet to certify that it too was the correct size.

Such was the theory. In practice, opportunities for cheating were boundless, and, if we are to believe Frontinus, the Rome Metropolitan Waterworks staff were expert at all of them. The *calix* on a take-off line

Number	Latin name	Diameter digits	Diameter cm	Circumference digits	Circumference cm	Area sq. digits	Area cm²	Capacity quinariae
5	quinaria	$1\frac{1}{4}$	2·31	$3+\frac{267}{288}$	7·26	—	4·191	1
8	octonaria	2	3·696	$6+\frac{82}{288}$	11·611	—	10·728	$2+\frac{161}{288}$
12A	duodenaria	3	5·544	$9+\frac{123}{288}$	17·417	—	24·14	$5+\frac{219}{288}$
12B	duodenaria	$3+\frac{18}{288}$	5·659	—	17·779	—	25·151	6
20A	vicenaria	$5+\frac{13}{288}$*	9·323	$15+\frac{246}{288}$	29·29	(20)	68·265	$16+\frac{7}{24}$
20B	vicenaria	$4\frac{1}{2}$	8·316	—	26·125	—	54·315	13
40	quadragenaria	$7+\frac{39}{288}$	13·186	$22+\frac{5}{12}$	41·425	(40)	136·56	$32+\frac{7}{12}$
60	sexagenaria	$8+\frac{212}{288}$	16·144	$27+\frac{11}{24}$	50·71	(60)	204·69	$48+\frac{231}{288}$
80	octogenaria	$10+\frac{26}{288}$	18·646	$31+\frac{17}{24}$	58·58	(80)	273·06	$65+\frac{1}{6}$
100A	centenaria	$11+\frac{81}{288}$	20·847	$35+\frac{11}{24}$	65·495	(100)	341·33	$81+\frac{130}{288}$
100B	centenaria	12	22·176	—	69·668		386·24	92
120A	centenum-vicenum	$12+\frac{102}{288}$	22·83	$38+\frac{10}{12}$	71·724	(120)	409·35	$97\frac{3}{4}$
120B	centenum-vicenum	16	29·568	—	92·89		686·64	$163+\frac{11}{12}$

*By the system used for the smaller sizes, this should be exactly 5 digits ($20\times\frac{1}{4}$), but Frontinus has corrected it to bring it into line with the other system, so that the cross-section area is 20 square digits.

As stated in the text, Frontinus takes $\pi = \frac{22}{7}$, but in calculating the centimetre equivalents, the value 3.1416 has been used.

Where two sets of measurements are given, the first in each case (12A, 20A, etc.) is the theoretical value calculated by Frontinus, and the second (12B, 20B, etc.) the values used by the *aquarii*.

208. Table of *calix* sizes, as listed by Frontinus (J.G. Landels).

Roman size	Lead required per "10-foot" length (lb)	Lead required per "10-foot" length (kg)	Diameter, allowing for overlap (in)	Diameter, allowing for overlap (cm)
100-digit	864	392·25	22·6	57·4
80-digit	691	313·7	17·9	45·5
50-digit	432	196·1	10·9	27·8
40-digit	346	157	8·6	22
30-digit	259	117·6	6·3	16
20-digit	173	78·5	4	10·2
15-digit	130	59	2·8	7·2
10-digit	86	39	1·7	4·3
8-digit	72	32·7	1·2	·3
5-digit	43	19·5	0·52	1·32

Overlap doubtful, so very approximate in this range

A 20-digit pipe required 1 ton per 125 ft approx.
1 tonne per 37·5 m

209. Table of standard lead pipe sizes (J.G. Landels).

might turn out to be a larger one than that authorised, where the gang installing the connection had been well bribed by the customer (or perhaps the better expression would be 'tipped').[55] The *calix* might carry the wrong stamp, or even no stamp at all. The use of a completely unapproved *calix* Frontinus considered the worst fraud of all since it implied the connivance of everyone, while the men installing an improperly stamped one could plausibly plead ignorance on their part. Again, the *calix* could be all right but an over-large pipe be attached to it causing the *calix*, so to speak, to work overtime (n. 51 above). There even might be no *calix* at all, with the supply pipe introduced directly into the *castellum* to draw off as much or little as the watermen pleased. The positioning of the *calix* was another source of trouble. Properly, it should

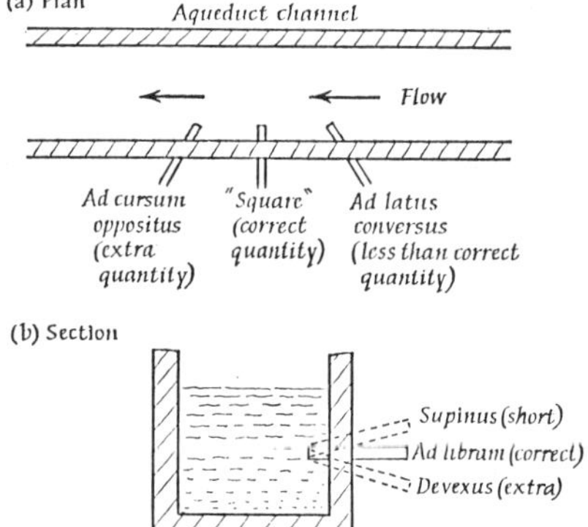

210. Effect of *calices* improperly set at an angle (J.G. Landels).

be set half-way up the wall of the tank into which it was inserted, and at a right-angle to the wall (Fig. 207). Not only will a *calix* set too low take in too much water, because of the head under which it is operating, but Frontinus was quite aware of the fact, and, at least partly, of the physics behind it.[56] Likewise, a *calix* not set straight would affect the volume of water it delivered. If it were angled to face the oncoming current in a conduit it would receive more than its due, and the opposite if it were facing away from it (Fig. 210). The volume would again be affected if the connected lead pipe sloped downhill or uphill.[57] It will be noted that while the point about low or high setting could apply to a *calix* sited anywhere, the angling relative to the current will affect only one set into a conduit with water running along it. Frontinus has just told us that placing a *calix* in such a position was prohibited, so there is a minor contradiction here. Possibly there were still plenty of old conduit-sited *calices* that had never been removed and which coloured his thinking. Strictly speaking, his words should be irrelevant, for they deal with an imperfect manner of executing an operation that is itself totally banned. And Frontinus says that his book is intended to set helpful guidelines for himself and his successors, so the question of the right way of setting a *calix* into a conduit should never arise since the thing was never to be done at all.[58] There were plenty of other abuses as well, such as claiming labour time for work not done, which were purely administrative in scope and need not concern us here. Only one final point need be made before we leave the *calix*.

The *calix*, I have said, controlled the distribution and consumption of the water. It also had a third function. It was the basic unit of measurement for the volume of water handled by the system, the term in which all estimates were expressed and on which all planning was based. More precisely, the smallest *calix*, the *quinaria*, was this unit, and was the only one used. Faced with a large figure one did not ascend to the use of a higher unit, as with cents and dollars, or inches and feet; and

Frontinus (II,64) can list the total capacity of all the Roman aqueducts as being 12,755 *quinariae* according to the official records, and 14,018 *quinariae* by actual measurement (i.e. the records were wrong). This has always caused grave problems. It is not just a question of 'How much water is a *quinaria*?' There are two problems. As a matter of practical engineering, how can a simple pipe be used to measure a volume of water? And, second, how can a linear measurement, the diameter of a pipe, be used to express volume? It is like saying, in modern terms, that a city daily consumes 15 six-inch pipes. The vital element missing, for a figure with any meaning, is time, or rate of flow (which is a function of it). Thus, a six-inch pipe running at five feet per second gives us that concept of volume that, without the time or speed element, we seek in vain. But nowhere in Frontinus, or elsewhere, do we find any reference to speed of flow, without which listing the cross-section or diameter of the pipe is idle and pointless.[59] This has sometimes led to complaints that the ancients did not even realise that the volume of water flowing through a pipe depended quite as much on how fast it flowed as how big the pipe was. The truth almost certainly was that Frontinus fully realised the importance of measuring speed of flow, but simply could not do it.[60] The ancient world, lacking any time-measuring equipment more precise than the water-clock or sundial, could not record speeds. The proof of it is that no set unit of speed, such as kilometres per hour, or feet per second, exists in Latin or Greek. With no way of measuring variation in speed, therefore, the only way a pipe could be used to measure volume was to standardise on speed of flow, so that it was always the same and the size of pipe, the only variable, therefore accurately reflected volume of discharge. On this principle, it will be seen, one did not have to worry about the actual speed of flow; as long as one could rely on it always being the same, it did not matter what it was. There was a simple way of doing this. Speed of flow through a submerged pipe in gravity flow (i.e. with no pumps involved) depends solely on the head. In other words, the speed the water flows out through a *quinaria calix* set into the side of a *castellum* depends on how far it is below the water level inside. This is purely a linear measurement, and can be measured with nothing more complicated than a foot rule. Provided one can therefore standardise the head, the *quinaria calix* will always discharge the same amount of water, and can thus validly be used as a unit of volume. This is almost certainly what was done. Purely as a pipe measurement, we know what a *quinaria* was. It was a pipe of 2.3125 cm diameter. Used as a measurement of volume, it was the amount of water that passed through such a pipe under a fixed, given head. What that head was, neither Frontinus nor anyone else tells us. It has reasonably been deduced by Di Fenizio[61] that it must have been, at an absolute minimum, at least 12 cm. A *quinaria calix*, or pipe, operating under this head, will discharge 40 m^3 per twenty-four hours, and this is accepted by most modern studies as what

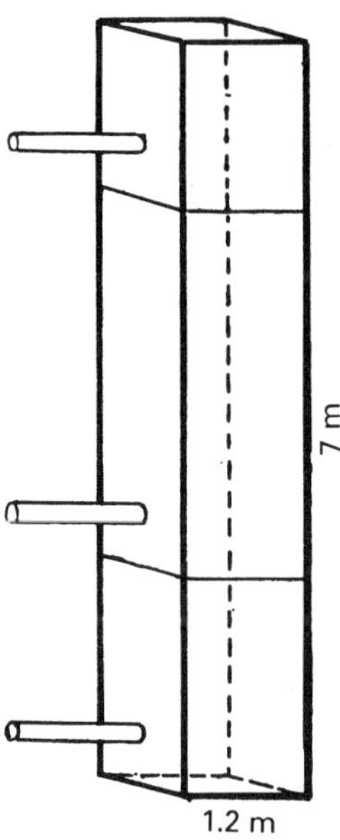

211. Novae (Svistov, Bulgaria): lead distribution tank (Biernaka-Lubanska).

Frontinus and the Romans meant when they used *quinaria* as a unit of volume.[62] It must in closing be emphasised that all of the above material, relative to the *quinaria, calices*, their installation and their sizes, applies only to Rome itself. Other cities may have regulated their water in a similar fashion but we know little or nothing about it. All the evidence comes from Rome, and nearly all from Frontinus. That is, it is literary and not archaeological. Hardly any remains of *calices* have been found, and had not the one single manuscript of Frontinus survived,[63] this whole area of ancient water with all its sophistication and complexity would have remained unknown to us, perhaps even unsuspected. No other city handled anything like the volume of water provided by the eleven aqueducts of Rome; most had three or four as a maximum, and just one was quite common. It is quite possible, therefore, that other, simpler, means of water allocation may have been employed. For actual archaeological remains of the urban distribution system, easily our best source of evidence is Pompeii, just as our best literary source, Frontinus, represents Rome. But it is hard, and sometimes risky, to integrate the two to provide a generally valid overview, for there is no telling how far the watermen of Pompeii operated on the administrative principles of Frontinus, or whether the pipes of Rome were the same as those of Pompeii.

We now leave Rome, and turn to the secondary *castella*. So few have been found it is difficult to generalise about them, but in so far as their function was the distribution of water, presumably they were normally built at ground level. At Pompeii, where most of our known examples are

to be found, they fulfilled another function as well, the control of pressure, and in pursuance of it were built on top of brick piers around 6 m high. These piers are to be found at strategic locations on street corners around the town – at Pompeii twelve of them have been identified – with a lead tank on top, forming the actual reservoir.[64] For such containers, lead was the normal material. Iron could not be used because of rusting, there being no galvanised iron, and in any case the ancient world could not produce the sheet metal needed for the sides of tanks: only lead could be poured out to form a flat sheet (see Fig. 215 below). We have already seen how, apart from pipes, it was also normally used to make the temporary conduits installed as local by-passes in case of repairs, and it was also used for distribution tanks inside houses. Particularly striking is the unique lead tank found at Novae, Bulgaria (Fig. 211).[65] It is used as a collecting tank at the springs serving the aqueduct, and is notable both for the fact that it is upright, much higher than it is wide or deep, and for its extraordinary dimensions – it is 1.2 m square and no less than 7 m high, and was apparently made in three sections, each with a single offtake pipe 18 cm in diameter and fitted with a perforated lead filter. The three sections were made separately and added one on top of the other, extending the tank upwards, apparently during two rebuilds, 'with', says its excavator, 'the simultaneous shutting one after another, of some of its outflows'. The Pompeii tanks were normally served by a large-diameter lead pipe running at surface level from the main *castellum* at Porta Vesuvii, and operating as an inverted siphon. On arriving at a secondary *castellum* (Figs 203, 204) this pipe climbed vertically up one side of the supporting pier, where a recessed slot to take it is often preserved in the brickwork,[66] and discharged into the lead tank. From the bottom of the tank – low down in the side or actually from its floor – smaller offtake pipes carried the water back down the brick pier and along the street to individual houses. The base of the pier was often found a convenient spot to locate a public fountain, directly served by one of the downspouts. A further refinement is urged by Kretzschmer, who carries the Vitruvian tripartite division also into the water towers. Public works (e.g. theatre, circus, nymphaeum), according to this scheme, are served directly from a single main[67] which is one of three coming directly from the primary *castellum*. The other two serve two completely independent sets of water towers; from one set of towers are served the public fountains, from the other the private users. There are thus two completely separate networks co-existing throughout the town.[68] This unnecessary duplication seems also to lack evidence, and I think we may assume that in the normal way each water tower served both private users and public fountains alike, in whatever way was most convenient.

The manner in which the water towers controlled pressure has been often misunderstood, though now clearly explained by Kretzschmer, followed by Eschebach (see Fig. 212).[69] The problem is the static pressure in the pipes of the urban distribution network. Some pressure is necessary, to get the water to rise as high as the taps in the houses or spouts in the street fountains, and there to issue in reasonable velocity and volume. On the other hand, one does not want too much. The reader will remember my insistence that pressure was not a governing factor in siphons and wonder why things should now be different. There are two reasons. One is that while Roman pipes could contain high pressure, Roman taps could not. There were no taps on siphons, so the problem did not arise, but on the city network the taps were the weak link. Static pressure in a pipe (or tap) is exerted equally in all directions. The construction of the tap was such that, though it could well resist pressure at the sides and the bottom, the top was weak. There was nothing really holding down the central rotating cylinder inside the body of the tap (see pp. 322ff. below on taps), and under pressure it would simply blow out through the top. Pressure, therefore, had to be held down to a level that the taps could cope with. Second, in the siphons high pressure was faced because there was no way of avoiding it. Had it been possible to reduce it no doubt the engineers would have done so; it would probably have cut down on leaks and put less strain on the joints; but since it could not be reduced, it was accepted and dealt with. The urban network, however, was different, for here a remedy was at hand.

The basic point is that pressure is governed by head, or the vertical column of water supported. The pressure in the pipes therefore depends on how far the pipes lie below what would be the natural level that the water in them would rise to, given the chance. For our purposes, this natural level is the level the water was actually at, the last time it was running in the open before plunging into the closed pipes, which in turn means the level in the main *castellum divisorium*. If this *castellum* is at ground level, pipe pressure will therefore not be high. The trouble comes if the city itself is not flat, but built on a slope, as so often happens. The aqueduct must, if possible, enter at the highest point of the city, so as to serve it all. The *castellum* will be located there too, and as the pipes branch out along the streets, they get progressively lower below it and the pressure in them increases. To take an example from Pompeii (Fig. 212), the ground level at the intersection of the Via Stabiae and Via dell' Abbondanza is 18 m below the *castellum* at the Porta Vesuvii, producing in the pipes a static pressure of 1.8 kg/cm^2.[70] This is quite enough to burst the taps and, in Kretzschmer's opinion, to harm the pipes as well. However, before it gets to the taps the water passes through one of these water towers. The tank on top of it being open, the water level in the tank now replaces that in the *castellum* as the factor governing pressure at the taps. There still *is* pressure, for the tank is raised some 6 m above ground level, but this produces a pipe pressure of only 0.6 kg/cm^2 instead of 1.8. It will thus be seen that the water tower does, in a sense, both reduce and increase pressure, and so Van Buren and Maiuri are both right;[71] it

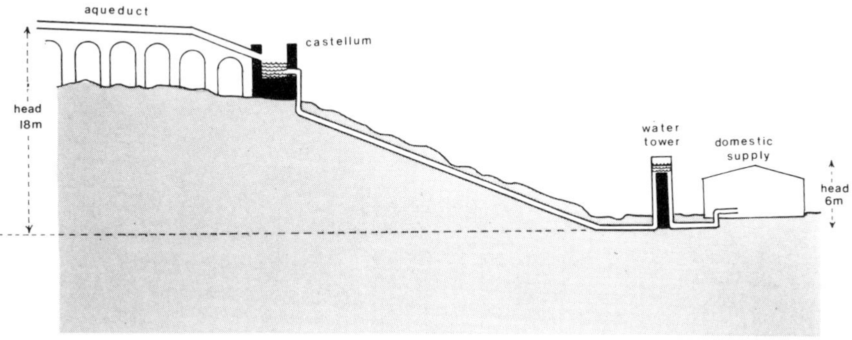

212. Diagram of urban distribution (based on Pompeii), showing how the use of water towers reduces the head governing pressure in domestic pipes from 18 m to 6 m (after Kretzschmer).

reduces pressure in that the pressure on the taps is reduced to that of a 6 m head instead of an 18 m one, and increases it in that by raising the tank on top of a pier one does provide enough pressure to serve the taps properly, whereas if the tank had been built at ground level there would be no pressure at all. Of course, as one looks at Fig. 212, the weakness in this system becomes apparent: if one has an open tank with a pipe running downhill into it, will not the tank overflow? This well exemplifies how the Roman system of constant offtake operates, for the only thing that stops, or can stop, the water tower overflowing, is if the water is constantly drawn off from it as fast as it flows in. A modern engineer thinks in terms of supply keeping up with demand. Here it is a question of demand keeping ahead of supply, otherwise the system will not work. The use of water towers to reduce pressure would not be necessary in a city that was completely flat, but in practice few such existed. The slope did not have to be very great to make its effect felt, and we encounter the Pompeii problem also at Nîmes, Arles, Vaison-la-Romaine, Trier, Metz 'und andere Römerstadt waren daraufhin zu studieren'.[72]

11
The Domestic Supply

> No person shall draw water from the public supply without official permission, i.e. an imperial licence, nor shall he draw more than he has been granted.
>
> Frontinus, *De Aquaeductu* II, 103

Public fountains

From the secondary *castellum* the water entered a further set of pipes, this time of smaller gauge and each equipped with a *calix*, which carried it to its ultimate destination. The location of the offtake *calices* depended on circumstances. At Rome it is plain from Frontinus that they were normally set into the side walls of the *castellum*, or even the conduit, hence his preoccupation with getting them straight and at a uniform distance below water level. At Pompeii where the secondary *castella* were not of masonry, but rather tanks of sheet-lead, usually 0.6 cm thick and set at the top of a tall brick pier, the *calices* were placed vertically in the bottom of the tank, the pipes they served running down inside the pier. At Minturnae the aqueduct, running on top of the walls, crossed a roadway by the arch of the main city gate, at which point the floor of the conduit was evidentally pierced by three vertical pipes (or 'down-spouts') drawing off water to serve fountains below.[1]

The water towers, or secondary *castella*, were normally located at strategic points in the city, such as busy crossroads, to facilitate service to the surrounding area. It was natural that once the water had arrived there for distribution, provision should be made to make it immediately available, and therefore we normally find at the foot of each water tower, and served directly from it, a public fountain. Others, more distant, were served by further piping under the paving of the streets and sidewalks, as were private houses. For this whole secondary urban network, Pompeii offers far and away our best evidence.

In the already excavated part of Pompeii, about two-thirds of the whole, some forty public fountains have been found, served from fourteen[2] water towers; Eschebach estimates a total of around fifty for the entire city, which, with an estimated population of 8,000, works out at about 160

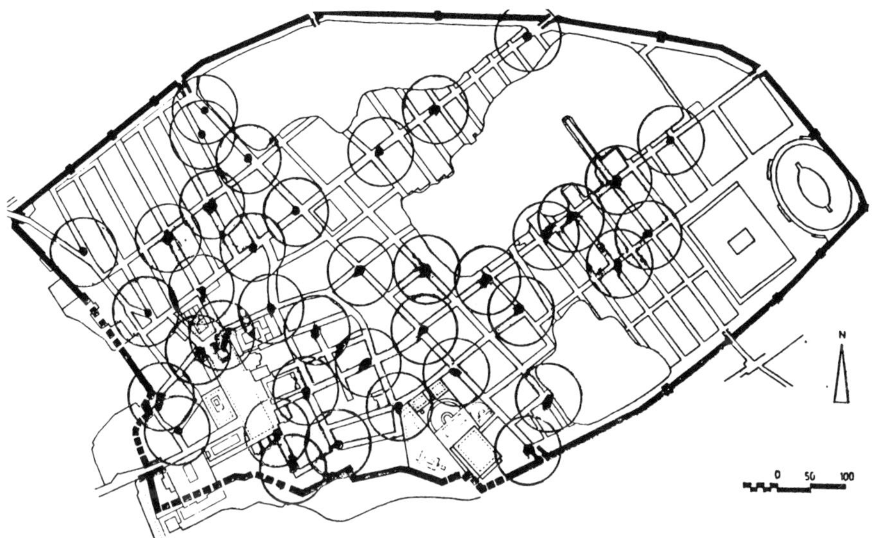

213. Pompeii: plan marking the location of all public water fountains. The radius of each circle is 50 m; few inhabitants had to go further than this to get to a fountain (H. Eschebach).

persons per fountain (not counting supplies to private users). The fountains were located at fairly evenly spaced intervals of about 100 m, and it was rare in Pompeii for anyone to have to carry their water for more than 50 m (see Fig. 213). On the basis of an estimated personal need of 500 litres of water per head of the population per twenty-four hours, the personal use requirement for the whole city works out at 4,000 m^3. The aqueduct, it has been calculated, brought in 6,480 m^3, leaving 2,480 m^3 for public and industrial use.[3] As a daily personal requirement, 500 litres per head is relatively generous, allowing for cooking and washing as well as drinking. As a strictly subsistence allowance, the minimum required for good health, the figure would be closer to 10 litres.[4] There seems to be no doubt, therefore, that the personal needs of the population at Pompeii were conveniently and adequately served by the street fountains. This takes no account of the numerous inhabitants who did not use them because they had private supplies piped into their house, or of the large but unknown amount of water residents drew from wells or cisterns.[5]

On the other hand, it equally neglects shops and streetfront workshops that may have found it convenient (and free) to cover their commercial needs by filling up at the local fountain instead of paying for a private supply, and likewise dwellers on upper floors, who, depending on the terms of their occupancy, may not have had access to wells. This was probably not a major factor at Pompeii, where the *domus* reigned supreme, but the high-rise apartment blocks must have presented a radically different picture in a city such as Ostia, or Rome itself. At Ostia little has been published on the urban distribution network,[6] but one cannot be far wrong in assuming that the street fountains played a much more important role in filling the city's needs than at Pompeii. Even in apartment blocks that did have piped supplies the water can scarcely have risen to the upper floors, resulting in a much greater proportion of the population that had to carry their water, whether from the street fountain or from some point within the building. It would thus appear that residents in large cities (despite their elaborate and costly waterworks), often did not enjoy as ready access to water as those in small towns, purely because of an economic factor: property values enforced high-rise buildings that raised much of the population to a level beyond what the water could reach. Conversely, the coming of an aqueduct did provide water to ground-level residents, whether by street fountain or private supply, in a generous abundance that could affect life in unlikely ways. At Pompeii the style and layout of gardens was completely changed when the aqueduct was built, bringing copious supplies for watering; they can actually be dated into pre- and post-aqueduct periods, the chief difference being that before the coming of the aqueduct they consisted largely of trees, requiring little water, afterwards replaced by flowers and by ornamental pools and fountains. Once available, water became a standard and much prized feature of Roman gardens, both in an urban setting and in country villas. At Ostia, on the other hand, one of biggest water users was the city's taverns, all of which prized the convenience of a 'wet bar' and took care to get hooked on to the piped supply.[7]

The simplest form of street fountain would, of course, have been an ordinary metal spout, set at a convenient level above the ground, as is often done nowadays. In practice (Pompeii is again the model) this was normally accompanied by an oblong stone basin, typically about 1.5 × 1.8 m and 0.8 m high, into which the spout discharged, and which presumably was normally full (Fig. 214). This could expedite matters when several persons arrived for water all at once, for filling an amphora is then simply a matter of dipping it into the tank, instead of having to wait in a queue while it fills up under the running tap; it would no doubt depend on what one wanted the water for, and how important it was for it to be fresh. The delivery spout was normally carried by a squat stone pier overlooking the basin, and bearing around the orifice some decorative bas-relief – a Gorgoneion, shield, rosette, lion's head, silen, cornucopia, or the like. Near the bottom the stone tank was pierced by a drainhole, closed with a plug, for use when the tank was to be cleaned out. In the lip of the tank was cut an overflow, which was, as a rule, permanently running. It flowed into the street, and thence the drains, if any. Eschebach's belief that this overflow was used for street-cleaning no doubt has much to be said for it, for though I know of no regular Roman

214. Herculaneum: typical public fountain.

street-cleaning on the modern model, where sewers were not provided the streets must indeed have been filthy and no doubt benefited from the continuous flushing;[8] and even where sewers did exist, they too would have been kept flushed clean. This indeed was a vital feature of the overflow, and we should not think of it simply as water wasted. It was only the constant flood of water coming down them that kept Roman sewers reasonably hygenic, and the Romans were fully aware of it. The fountains were deliberately designed to overflow, and at Rome there were laws to stop anyone helping himself to the surplus water.[9]

Lead pipes

From the secondary *castellum* to the public fountain, as to private houses, the water travelled through pipes, and these must now be our next concern. Various materials could be used for these, but throughout the empire lead was by far the commonest, so much so that the Latin for lead-workers, *plumbarii*, has entered the language in both English and French (as plumbers, *plombiers*). Exceptions to this rule are few. One of them, surprisingly (in view of the lead production at Laurion, n. 13 below), is Athens, where lead pipes are rare. Another was Britain, equally surprising since British production of lead was the greatest of any province in the empire; however, Britain also produced oak trees in even greater abundance, and municipal pipes were normally wooden, preferring this even cheaper natural resource that did not need to be mined or smelted. The same thing happened in Germany. Copper (or bronze) pipes also exist, but are of such extreme rarity that we may omit them from all practical consideration. In the museum at Metz are to be seen pipes made of concrete (perhaps from a siphon, or some other pressure line?), but these are apparently quite unique.[10]

We may here in passing note one result of the municipal preference for wooden pipes: it made impossible any application of the strict criteria for pipe sizes – *quinaria, vicenaria*, etc. – specified for Rome by Frontinus and Vitruvius. How far these sizes were applicable outside Rome is in any case little known, but when pipes were made by boring a hole down the middle of an oak tree, such niceties were out of the question. The Vitruvian specifications will only work with lead piping, but in regions where the standard piping was wooden, it is unlikely that even such lead pipes as are to be found strictly obey the Vitruvian formulae, and any resemblances in size to an equivalent at Rome are probably fortuitous.

The great disadvantage of lead has always been that it is poisonous. This was fully recognised by the ancients, and Vitruvius specifically warns against its use. Because it was nevertheless used in profusion for carrying drinking water, the conclusion has often been drawn that the Romans must therefore have suffered from lead poisoning; sometimes conclusions are carried even further and it is inferred that this caused infertility and other unwelcome conditions, and that lead piping was largely responsible for the decline and fall of Rome.[11] In fact, two things make this otherwise attractive hypothesis impossible. First, the calcium carbonate deposit that formed so thickly inside the aqueduct channels also formed inside the pipes, effectively insulating the water from the lead, so that the two never touched. Second, because the Romans had so few taps and the water was constantly running, it was never actually inside the pipes for more than a few minutes, and certainly not long enough to become contaminated. The thesis that the Romans contracted lead poisoning from the lead pipes in their water systems must therefore be declared completely unfounded.[12]

Lead, on the other hand, did offer considerable advantages. For one thing, it was cheap and readily available in large quantities, either mined directly as lead, or produced as a by-product in the refining of silver. It was, of course, also very heavy, so that the transport costs might be very high, depending on where it had to come from.[13] Second, it was an almost ideally easy metal to handle. It was malleable enough to form the sheet metal from which ancient pipes were manufactured, flexible enough for the pipes to be easily bent around obstructions, strong enough to contain any water pressure generated within,[14] tough enough not to fracture readily, and with a low enough melting point (327°C) for it to be easily cast or soldered. In practice, the chief difficulty in laying in lead pipes must have been soldering the joints. The work had to be performed *in situ* and without a modern portable blowtorch, raising the question of how

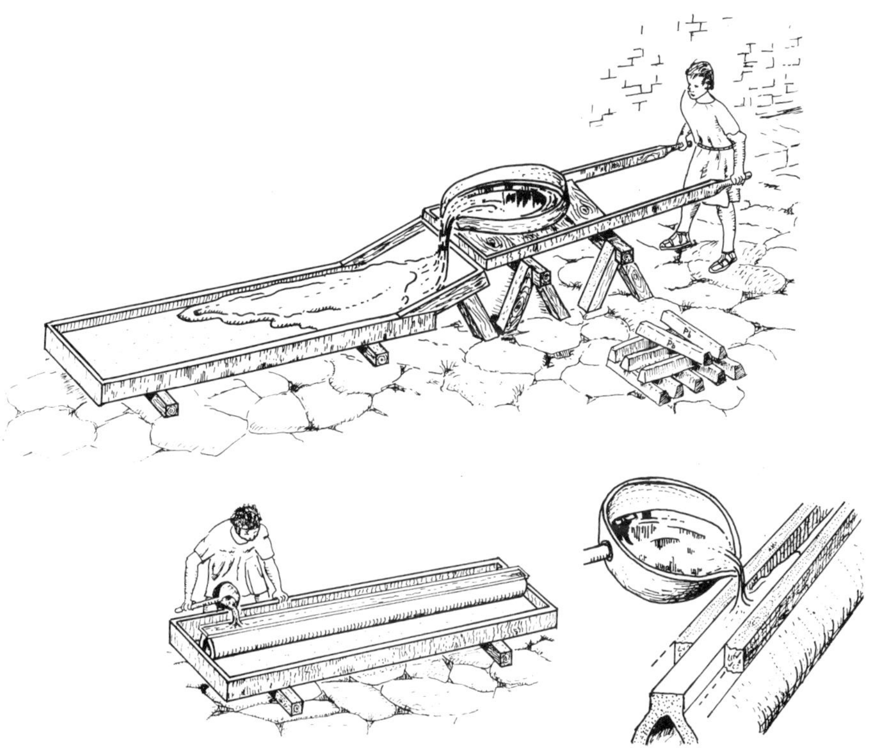

215. Reconstruction of steps in pouring out lead and making a lead pipe (J. Hansen, 1981).

sufficient heat was to be applied to, say, the undersurface of a pipe at the bottom of a trench in the roadway. Presumably the answer was a hot soldering iron, but it must have been a difficult and awkward job, in which clumsy work would be forgivable, particularly if buried out of sight.[15]

The actual process of making lead pipes was described by both Vitruvius and Frontinus, and has often been illustrated since (Fig. 215).[16] The process began by pouring out from the melting pot on to a flat surface enough lead to make a section of pipe of the desired gauge. What the surface usually was is unknown. It may have been stone, more probably sand or clay smoothed flat, with slats of wood of the appropriate thickness to form the sides of the mould and stop the lead spreading.[17] The length was uniform, for all pipes came in lengths of ten Roman feet,[18] so as the pool of lead was run out, the two things that could vary were its width (depending on how far apart the slats were set) and thickness. The thickness of the lead sheet which was thus being formed would vary, but only roughly, according to the size of the pipe being made. Thus the large pipe forming the water main running along the Decumanus at Ostia used lead 2.5 cm thick for a pipe of internal diameter 15 cm, dimensions we see repeated in piping preserved in the Museo Nazionale delle Terme, Rome.

A medium thickness of 1.4–1.5 cm has been found, while the commonest, small-gauge pipes (such as the *quinaria*) used for domestic supplies, had a thickness of around 0.6–0.7 cm.[19] The lead may actually have been measured out by weight (and then spread out on the stone bed to cover a given measured area), for Vitruvius, even though his actual figures are not above suspicion, is very clear that there is a set weight of lead allocated to a ten-foot length of pipe in each of the various accepted sizes.[20] This weight was then poured out and spread to form a lead sheet, say 2.9 m (10 ft) long by 0.6-0.7 cm thick by 9.3 cm wide. This would form a *quinaria* pipe, the smallest in the range. For larger pipes, the sheet would be of the same length and thickness, but wider in proportion, as the pipe is to be formed by bending the sheet over to form a cylinder. The range of pipe sizes has already been listed (ch. 10 n. 52 above), but it is not completely clear to what they owe their names. A *quinaria* is plainly named after '5', and a *vicenaria* after '20', but five or twenty whats? And what is it that they measure? The unit of measurement seems to be the *quadrans*, or quarter, and the unspecified larger unit made up of four of them was the digit (1.85 cm). A *quadrans* was thus about 0.46 cm, and a *quinaria* (1¼ digits) around 2.3 cm. But what was it that was 2.3 cm? The two possibilities are that it was the width of the original lead sheet, or the diameter of the finished pipe. The Romans themselves were not sure which, though Frontinus, quoting both, prefers the second.[21]

It was at this stage in the manufacture that the pipes acquired their inscriptions. Not every length of pipe had one, but they are not uncommon. Normally they are formed by letters raised in relief on the pipe, as opposed to being stamped, engraved, or scratched. The procedure utilised a series of individual movable moulds for the various letters. What happened next depended on whether the lead was being run out on to a sand or a stone surface. If it was sand, the letters were simply impressed into it, perhaps one at a time, or assembled into some sort of stamp. The inscription thus now formed part of the mould. On a stone surface, the letters would have to be assembled into the desired inscription, more or less on the principles of modern typographical composition, and the whole locked into a slot in the stone slab, so that the typeface was flush with its surface. When the liquid lead was run in on to the slab to form the lead sheet, it of course also ran into the type moulds, solidifying in the form of a raised inscription. That the types were movable, and re-usable, is shown by irregularities in their spacing and alignment, recalling those sometimes occurring on a modern typewriter.[22] There is no denying the stature of this technological feat. As with the Phaestos Disc of an earlier age, in which an inscription was imprinted in clay by movable types, it tempts one into speculating how close the ancient world was to making the full-scale breakthrough into printing. Nor can there be any real doubt that that was how it was done. Normal inscriptions, engraved, painted or stamped, do not produce raised lettering. That can only be done by moulding, at the time the pipe (or the lead sheet forming it) is being made, so that as the pipes were mass produced, so were the inscriptions.

At the same time, the phrase 'mass produced' requires qualification. What were these inscriptions? Often they give the name of the emperor or other authority responsible for the installation, often the name of the manufacturer. Sometimes they give the name of the house being served, or its proprietor. It will thus be seen that the inscription changes whenever the job does, and there are a large number of different texts. It has also been pointed out that this underlines a profound economic truth about the Roman world. Even where it was possible, there was no real mass production. Evidently each set of pipes was custom-made for the job actually in hand, individually marked as for it alone. Given the standard sizes and large demand, one might have expected the plumber to manufacture in quantity and carry a stock, if necessary stamping the individual information on the pipes as used. Since the inscription was an inherent part of the manufacturing process, however, we know that this was not done, and the pipes were only made as required.[23]

As the molten lead hardened to form a long, narrow sheet it was no doubt rolled or hammered to make it fill out the lettering and any odd corners.[24] While the lead was barely solidified and still highly flexible, the next step was to bend the sheet up around a wooden (or bronze, to resist the heat?) core laid along the middle of it (Fig. 217), to form a cylinder. The core was then withdrawn, leaving a length of formed piping, as the two edges were now soldered, welded, folded or hammered together. This joint formed a very prominent seam running along the top of the pipe (Figs 216, 218) (pipes were apparently always installed with the seam on top to facilitate repairs, should the joint fail or leaks develop), and the pipe itself in consequence usually had an oval or even pear-shaped cross-section rather than a circular one. The *calix* serving the pipe, on the other hand, being cast in bronze, had a hole which was truly circular. It is a good question, then, whether the complicated series of pipe standards – quinaria, etc. – which were based on pipes of circular cross-section, were not affected when the actual pipes in use were deformed in shape. Since they were, in effect, squashed, was their discharge not less? If they had been squashed enough to be seriously flattened no doubt it would have been, but in actual practice the oval pipes had a cross-section essentially of the same area as the theoretical round ones, reduced width being compensated for by extra height. There was thus no difference.[25] We may also note that when subjected to internal pressure from the water inside, these pipes slowly change their shape, becoming more and more rounded. This was not usually a factor in the urban network, where pressure was not high, but it has been experimentally verified by Belgrand that, under pressure, this is in fact what happens. A pipe retaining its oval or pear-shaped contours is thus one that has probably never been subjected

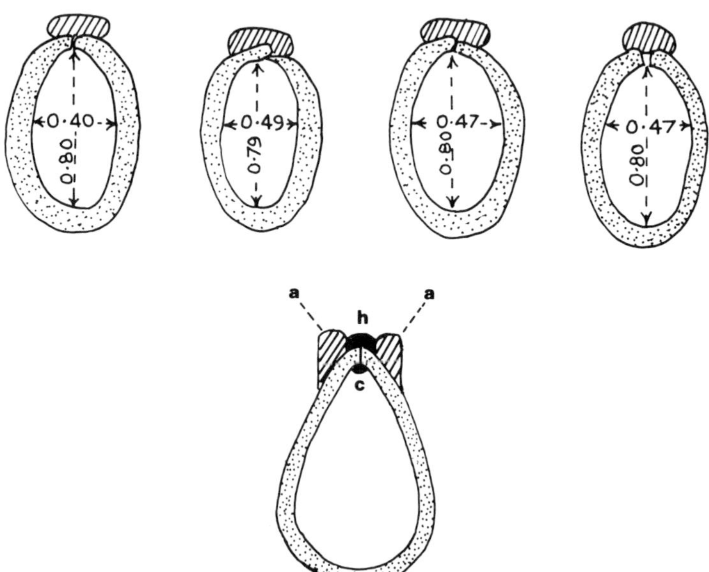

216. (Above) cross-section of Roman lead pipes, showing method of soldering the top seam; (below) two ridges of clay are applied along the top and the solder (h) runs in between them (compare Fig. 215) (de Montauzan). Measurements in decimetres.

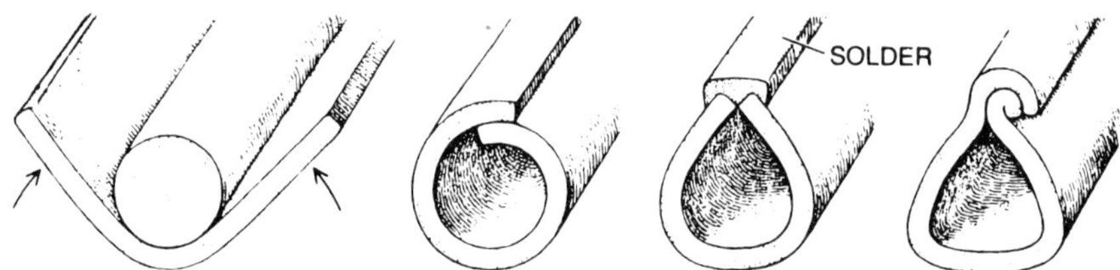

217. Method of forming the pipe (left) and three alternative methods of closing the seam (*Scientific American*). Siphon pipes were made by bending a sheet of lead around a wood core. The core was withdrawn and the joint at the top was hammered or soldered to be watertight, as is shown by the three drawings at the right. The pipes were oval or pear-shaped in cross section.

to high pressure.[26]

Almost the final stage was to fasten closed the open seam along the top. The solder used by Belgrand in his experiment was a mixture of two parts lead and one of tin, the proportions prescribed by Pliny,[27] but there was a good deal of variation. A large block of solder weighing 5 lb 8½ oz, excavated at Silchester,[28] contained 61.83% lead to 38.01% tin (= 1.63 to 1), while solder from a pipe join there gave proportions of 73.6% to 25.44% (= 2.9 to 1). In Britain at least, where the practice has been studied by a modern technical authority, the practice with soldered joints was to use a butt joint, the two edges coming together flush and level, and to solder the

218. Bath, England: lead pipe *in situ*, showing continuous seam along the top.

crack.[29] However, the normal method was to use not solder but molten lead. Presumably it was a matter of expense. Except in Gaul and Britain, where it was native, tin had to be imported while lead did not, being universally available. Since this process involved using lead to join lead, we may call it welding, or, to use Penn's term, 'autogenous soldering'.[30] The molten lead either formed a bond, like solder, between the two existing lead edges, or, if it was hot enough, induced them too to melt and unite with it, so that as the joint cooled it formed a weld, a homogeneous, uniform surface with no joint any longer detectable in it. More than a century later, the experiments conducted by Belgrand still seem to be our best evidence for this technique. The manner in which the two edges of the pipe were joined varied. Those illustrated in Fig. 216 (above) show them brought together in either an overlap or a butt joint. Another common method was that shown in Fig. 216 (below). Instead of overlapping, the two edges were pinched together, side by side, forming a triangular or pear-shaped pipe. Two low ridges of clay (a, a) were then applied to form a long U-shaped mould (as deduced by Penn, n. 30 above) and the solder, or

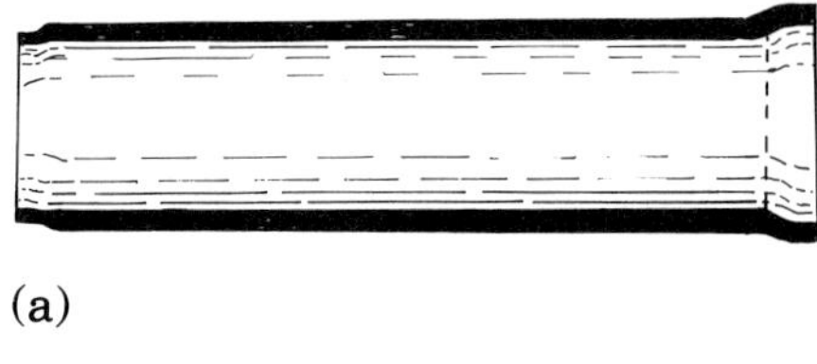
(a)

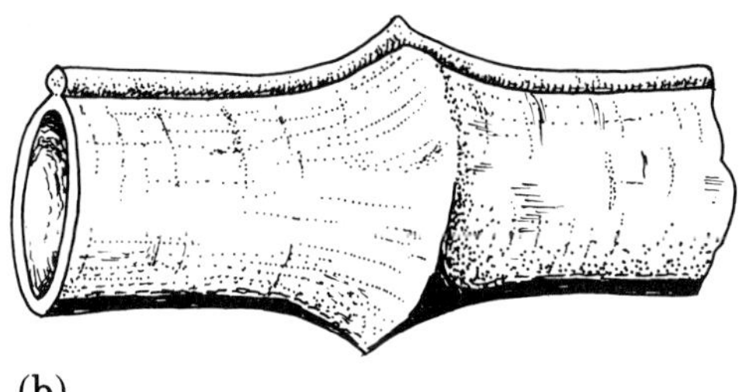
(b)

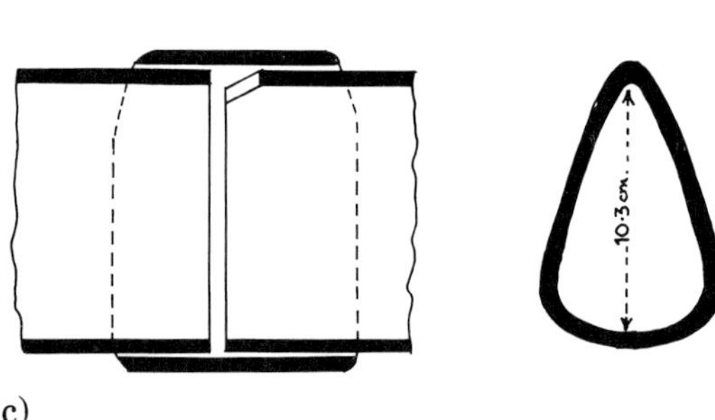

(c)

219. Methods of jointing pipes to each other: (a) pipe with overlapping male-female joint; (b) heavily soldered joint (Lyon); (c) pipe from rue Gay-Lussac, Paris, with joint covered in closely-fitting sleeve (de Montauzan).

molten lead, (h) was run in between them, often leaving a continuous bead (c) along the inside of the pipe where it has run in through the crack. It was primarily this method that left the characteristic heavy seam along the top of the pipe.[31] However, if the lead was hot enough to induce the two sides of the pipe to weld, or melt, together, it was possible for a good plumber to produce a pipe that not only had no ridge along the seam, but no discernible seam at all. Belgrand, though he was able to make such pipes, could make them only of triangular (or pear-shaped) section, not round, though they would later round out under internal pressure. In appearance they duplicated some genuine Roman pipes, presumably made the same way. The necessary heating could be achieved by running in molten lead or solder already at a very high temperature; by the application of a soldering iron; or by placing live coals directly along the top of the pipe.[32] The resultant continuous seam, whatever the form of joint, strikes modern eyes as a ham-fisted way of making a pipe, and surely a gross source of weakness. This is not so. Belgrand points out that a properly soldered joint is as strong as the metals it joins, and when he tested his replicas to destruction, it was the side walls, not the joint, that failed.[33]

The next step was to arrange the joints at each end of the section of piping, by which it was connected to the next line. Here again, procedure varied. The simplest and, probably, commonest, was to use a male/female joint on the analogy of terracotta pipes. The end of one pipe was widened out, probably by hammering a wooden plug into it (it was quite easy to widen lead pipes; it was expressly because people did so that Frontinus required *calices* to be made of bronze). The end of the next was tapered off and fitted inside it, and molten lead run in to fill the crack between the two (Fig. 219(a)). Sometimes the pipes could simply be run end to end in a butt joint and the whole encased in a thick layer of lead (or solder[34]) (Fig. 219(b)). A third possibility was to enclose the joint in a cylindrical metal sleeve slid on lengthways along the pipe (Fig. 219(c)). No doubt this could be, and was, leaded in place too, but from modern practice one would suspect that the sleeve was heated, slid on, and then gripped the pipe firmly by contraction as it cooled.[35] Exceptionally, it is not even unknown for the two pipes, in a male/female joint, to be nailed together, by one long nail that pierces the two thicknesses at the overlap, then spans the interior of the pipe like a diameter, and goes on to pierce them again on the other side.[36] It is these joints between the sections, the *fistularum commissurae*, rather than the continuous seam, that the Romans evidently considered the weak point, in the event of pressure or inertial thrust being generated by the water.[37] In fact, as we know from the mechanics of structures, the weak point in the piping was the sidewalls. In a pipe under pressure, circumferential stress is twice longitudinal, and since fractures develop at right angles to the line of stress, the first cracks to form in the pipe will be running along it lengthways. This was what happened in Belgrand's test.[38] Obviously, the end joints might be more vulnerable to stress of a different kind, such as earthquakes, heavy traffic on the sidewalk above the pipes, heat expansion, and so on, nor are we here taking into account the failure or leaks from bad workmanship (to which joints would obviously be more liable); but, purely in terms of resistance to internal pressure, which is the only thing that can really be calculated, the jointing and continuous seam in Roman pipes appear to have been in general efficient and satisfactory techniques.[39]

Other conduits

Though the vast majority of Roman installations employed lead piping for the urban distribution network, terracotta pipes, wooden pipes, and conduits of other forms are sometimes found. Little need here be said on the piping, which has already been considered in connection with the main aqueducts. For open conduits as part of the distribution network, our best evidence comes from Germany and the North. They were of stone, masonry and terracotta, but perhaps most commonly of wood. In wood they were of semi-circular or U-shaped cross-section, depending

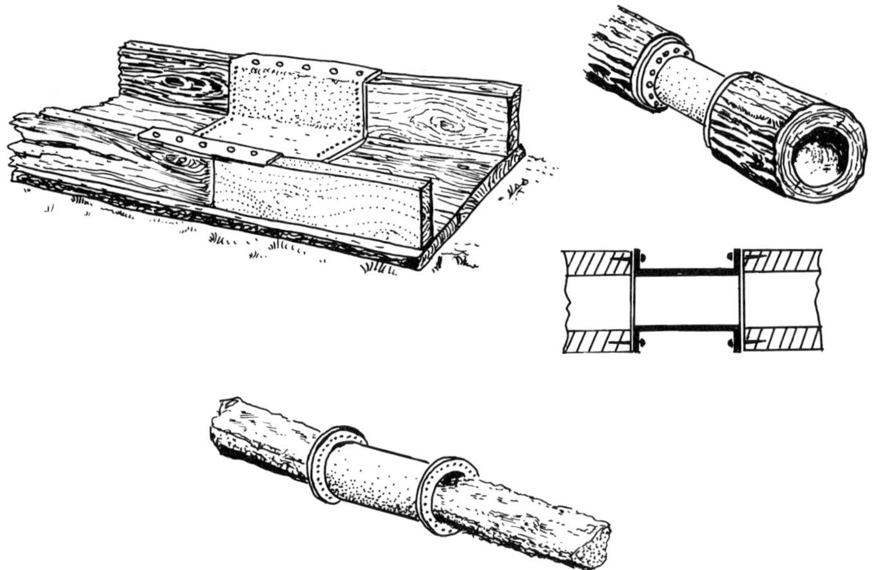

220. Wooden ducting (top left) and piping, showing the metal sections joining the wooden pipes; often (bottom left) the wooden pipes have decayed, leaving *in situ* only the semi-circular bed of the sediment that formed inside them, and the metal junction sleeve between two pipe sections (Haberey).

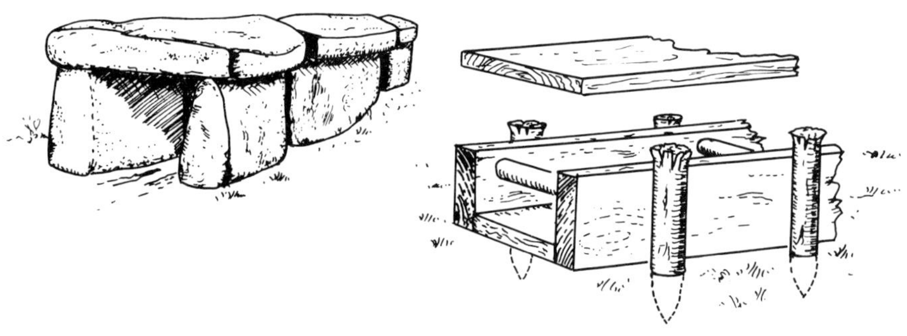

221. Water channels from Roman legionary forts, in stone and wood (Anne Johnson).

whether they were made by gouging out the centre of a split tree-trunk, after the fashion of making a dug-out canoe, or built up of three planks, forming two sides and the bottom. Joints were made waterproof by a sheet of lead applied to the inside of the channel and nailed down to the wood (Fig. 220) (Cologne).[40] Similar U-shaped open conduits come from Aachen (in terracotta) and Biebrich (sandstone), their joints shaped so as to fit together at the overlap in the fashion of classical roof tiles.[41] In Greek work it is of course not uncommon to find small open stone conduits of this kind, or runnels, set flush into the ground and used at large sanctuaries to carry water to all parts of the site; they formed, for example, the chief distribution network at Olympia and Epidauros. Occasionally also lead could be used to form ducting of a size, shape and proportions that one could not really call a pipe. The main bath at Bath was supplied through such a duct, of flattish and roughly oblong cross-section.[42]

Junctions

In laying out the network of piping, junctions were often necessary, where a pipe divided into two, or a branch was taken off to serve some individual house. With lead pipes, the junction could be made as now, cutting a hole in an existing pipe and welding another directly on to it (Fig. 222). This could be done to join a small pipe to a large one, or two of equal size. It could be done at a right-angle, to form a T-joint (as in the example quoted, of a line taken into a house from the main pipe running along the street), or, if the two pipes were running in the same general direction, the branch could be brought in at an angle, for the joint to form a kind of compressed Y.[43] It may be, however, that the Romans had some doubts about their ability to solder (or weld) an angled joint with sufficient strength, for the more normal method was to use a junction box. This was basically the same technique that we saw employed where two aqueduct channels met so that the connection between them was indirect. In the urban network the same effect was achieved by installing a closed oblong box of lead sheeting (Fig. 223). The pipe from the *castellum* emptied directly into it on one side, and from the other an offtake led to its continuation. From the sides led off one or more branched pipes, usually so laid out that they were mounted on the box at a right-angle, facilitating soldering. Sometimes the box was a longer cylinder, laid horizontally, and with the pipes entering and leaving each end, like the muffler or silencer on the exhaust pipe of an automobile (Figs 224, 225).[44] One presumes that this shape was favoured when the pipes to be joined were all, in effect, running in the same direction (as the oblong box would be more suitable to a branch running off at right-angles), particularly if vertical or lateral space was lacking (if it were, for example, to be fitted in under a sidewalk, or along the base of a wall); but the great paucity of the remains makes it unsafe to generalise.

For pipes of terracotta, the arrangements were analogous. Direct junctions, with pipe-sections actually moulded in the form of a T, did exist, but seem to have been (Fig. 227) rare.[45] Much commoner was some form of distribution device (Fig. 226). At Antioch, for example, water lines of terracotta piping joined in a 'distributor', evidently a kind of cylindrical solid block with holes, to which the pipes were attached, all meeting at its centre. More usual was some kind of small distribution tank or box. The

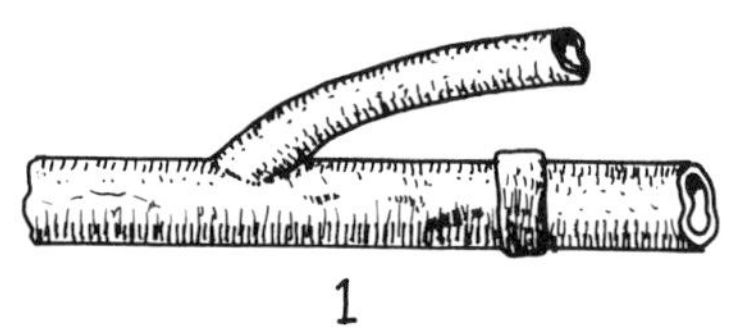

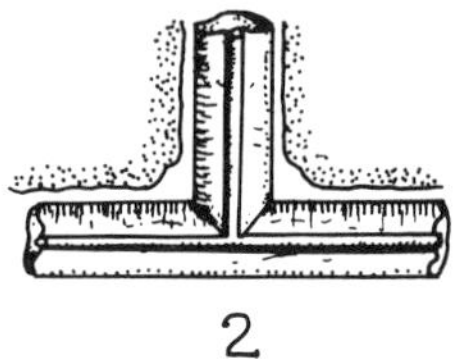

222. Junctions in lead pipes from (1) Pompeii; (2) Cologne.

223. *Right* Junction box as fitted to lead piping: (a) mains supply; (b) small-gauge branch; (c) junction box (Pace).

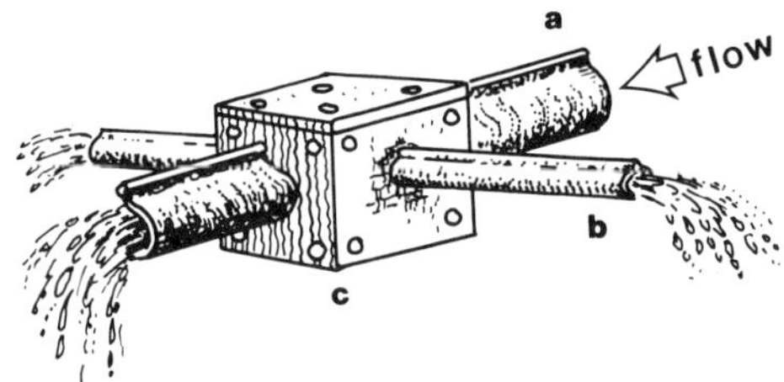

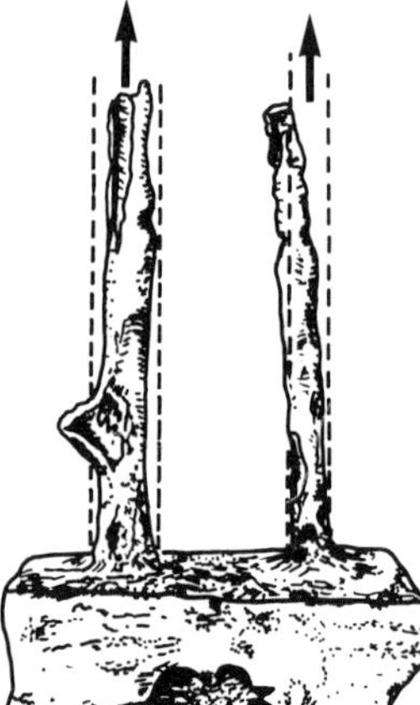

224. *Left* Pompeii: junction box, actual state (Fig. 225) (Pace).

225. Pompeii: lead pipes under street sidewalk, showing (centre left) junction box (see Fig. 224).

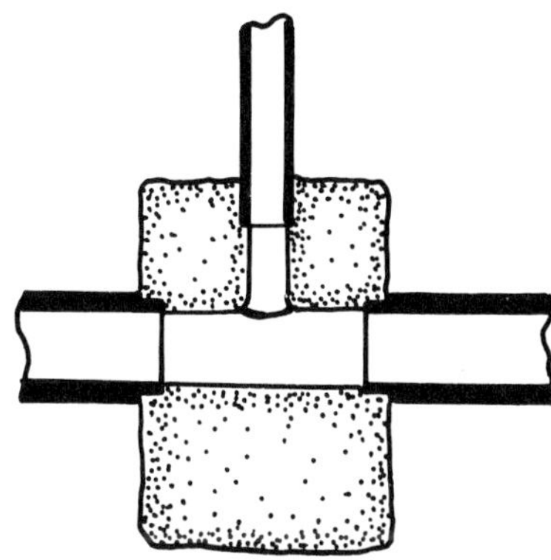

226. Strasbourg: junction on terracotta pipeline, formed by stone block. The small vertical pipe may have led to a surge tank or some other device for accommodating variations in pressure and/or velocity (Samesreuther).

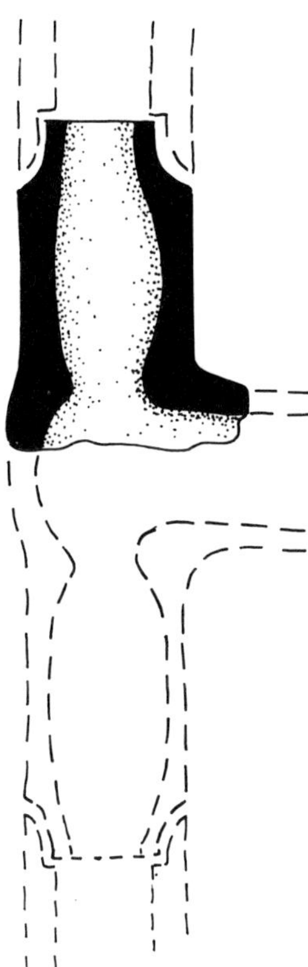

227. Strasbourg: T-junction in terracotta piping (Samesreuther).

best examples of this technique are to be seen at Kourion in Cyprus.[46] There the distribution box was a small oblong receptacle in stone; the top, it seems, was usually covered by a flat stone slab, sealed down by gypsum plaster, but easily removable. The main pipe entered one end and left at the other, with a branch departing from one side, in effect giving a T-junction. Other installations from this same site show a roughly similar layout. Boxes of the same general type but without the branch line leading off evidently served as settling tanks, and the excavator hypothesises that householders living nearby had the duty of lifting the lid and cleaning out the box every week or two, perhaps in return for the right to use the water. A further small box, cut out of one block of stone, has the pipes both enter and leave flush with the floor, so that debris could not settle. This is interpreted as an inspection chamber, accessible from above through a round hole. In principle, it is a more sophisticated version of the rather more familiar terracotta pipe with a removable lid in its top. *In toto*, the Kourion installations seem to reflect what we may enunciate as a general principle of Roman hydraulic design: that when it was desirable to introduce some kind of change or complication into the water flow, many engineers preferred to handle the disturbance by isolating it in a separate tank or box, where movement was slower, stresses less and construction easier, than by strengthening the pipes or channels themselves to resist it.[47]

The layout of the street pipes

As they were laid in position, the lead pipes of the urban network were normally accommodated by being buried in the sidewalk, but not, as a rule, very deep. At Pompeii they were sometimes even left exposed, in the angle formed by the sidewalk paving and the walls of the houses, held in place if necessary with iron hooks driven into the wall. Eschebach optimistically applauds this procedure as facilitating access for repairs. It also increases the need for them, and the pipes exposed in this vulnerable spot were sometimes protected by being covered in semi-circular sections of terracotta piping.[48] Sometimes pipes could be protected by being laid inside the U-channel of a stone conduit. Otherwise, if it was felt that the pipeline needed protection against either external damage or internal pressure, it was provided by embedding the pipes in packed clay. Neither practice was common, and it was rare indeed to have recourse to the desperate measure one finds at York, where a lead city water main was not only placed three feet below the road but 'encased in a mass of concrete four feet thick as extra protection'.[49] The need to bury the pipes was, of course, much more imperative in the northern provinces, where frost was more severe, but I would not like to assert, in the present state of our knowledge, how far the Romans actually did so; in Britain, at least, they usually did, but the whole question of how far Roman hydraulic theory, formed in the sunny Mediterranean, in practice recognised these northern climatic realities, has yet to be fully studied.[50] Wooden pipes, on the other hand, where they were employed, were normally laid in excavated trenches; often, indeed, once the wood has perished, only the trench remains to testify to their existence.

For the actual layout of piping through the town, two approaches were theoretically possible, akin to the phenomena identified by philologists as parataxis and syntaxis. One is to follow the modern practice of running a main pipe down the street, with a small-gauge branch pipe running off into each house served as the main passes in front of it. The second is for there to be no main, and each house to have its own individual service pipe going all the way back to the *castellum*. On the first principle, one will find under the street one large pipe, on the second, a multiplicity of small ones, all parallel. To our thinking there is no question which way is best and most efficient, but it is often stated that the second represents normal Roman practice. This is emphasised by the well-known diagram in Forbes, showing a block of houses, each with their individual supply lines going back to the *castellum*. Kretzschmer offers a variant of this, where a main from the *castellum* serves a series of secondary *castella* (water towers), but each house has a line running all the way to one of these. That something like this was sometimes actually done seems to be proved by occasions where a multiplicity of small pipes, all parallel, has been found under a street by excavation. And in any case the whole system of allocation described by Frontinus at such length depends crucially on each user having a line to the *castellum* or conduit, there controlled by a *calix*.[51]

But it must not be thought that such a layout was the invariable rule, or even perhaps the commonest. At Antioch things went to the opposite extreme and the city was served by a water-pipe grid that apparently matched the street plan. It was based on a main conduit running along the edge of the town and serving a series of parallel secondary conduits aligned on the streets forming the east-west minor axis of the street-grid. When they reached the cross-roads with the colonnaded main avenue, each conduit ran into a distributor-block (see p. 317 above) from which emerged further terracotta pipelines to serve individual houses or groups of them.[52] It will be noted that, as outlined, this system has no water towers, and almost completely follows the pattern of a modern installation. A very similar system is found at Priene. At Ostia too the layout, though not so revolutionary, strays far from tradition. The town was served by a terminal *castellum* where the aqueduct reached the city walls, near the Piazza della Vittoria. From there the water ran by a massive (16-20 cm dimaeter) lead pipe under the length of the Decumanus, from which smaller pipes again branched off directly, without the interposition of water towers, *castella*, or other such devices, to serve public buildings, fountains and private houses. Once again, this can only be called a true water main, in modern terms.[53]

Pompeii operated largely on the system outlined by Kretzschmer, with large gauge pipes running along the line of the main streets from the main *castellum* at Porta Vesuvii to the various water towers, whence radiated small pipes serving the various users. So far this looks like a completely conventional system, but one must quickly enter two caveats. One is that even in Pompeii, our best known example, the system has not been fully studied or published, resulting in various anomalies. Thus, the main *castellum*, at Porta Vesuvii, is normally explained in Vitruvian terms (pp. 280-1 above). This is done because its triple division and three offtake pipes have led scholars to recall his specification and apply its principles, attributing to the three different pipes three different priorities and functions: private houses, public buildings and fountains.[54] Two points must be made clear. For the Pompeian *castellum* to work in this way, the gates controlling access to the three chambers must be of different heights, to impose different conditions of admission for the water. Although they are indeed shown this way in all published reconstructions, this is wholly conjecture, for no fragment of the gates is preserved. Second, there is no evidence at all that the three exit pipes, supposedly serving three different classes of user, went where they were supposed to. One should note a cardinal weakness of the Vitruvian system. Because, presumably, all three categories of user were equally distributed throughout the town, it rquires three independent networks, each covering the entire city, with, presumably, three separate mains running down each principal street.[55] This is not only wasteful, but apparently never happened: main city water pipes have never been found in threes, even if offtake holes in the *castella* sometimes are. The implication is clear. On leaving the *castellum* the pipes went to different places, and their distribution is geographic, serving different areas of the city, not by category of user. In Pompeii, the course through the city of only one of these three pipes has been traced, and it, in fact, seems to serve (through water towers) just about everything in its path irrespective of category.[56] It is not known where the other two went, but I would suggest that probably one served the unexcavated and unexplored area in the northern part of the city, *Regio* III, IV and V, and the other perhaps *Regio* VI at the western corner, the insulae around the Porta Ercolano, in which water is known to have been available.[57]

The second caveat is that though Pompeii did in general draw its supplies through water towers, presumably fitted with *calices*, it did not always do so. One does find there (and also at Herculaneum) houses or shops served by a pipe running directly off either the main pipe or some smaller gauge supply running along the street. In other words, Pompeian users do not all have an individual line going all the way to a *castellum* – *any castellum* – but often tap directly into the system as it passes their front door (see Fig. 228). Conversely, junctions in the Pompeii pipework, as one pipe branches off from another, are relatively common, though by the conventional account such things should not exist.[58]

Taps

As the water arrived at the house or shop, we encounter another problem: taps. Taps were much less used in Roman water supply engineering than they are today, for in principle in the Roman system the water flowed continuously, without ever being shut off. This has led to the role of taps being neglected, to the extent even of one modern standard handbook stating as a fact that even their existence is dubious.[59] We should therefore make it quite clear that it is not. Taps have been found in large quantities, fulfilling a number of different functions and in different situations, and over a wide geographic area. Today, some are preserved in museums, while some are still *in situ* on the pipes in the ground. Given the continuous offtake factor, in the house at Pompeii that reportedly had thirty of them the supply seems to have been quite uncharacteristically excessive,[60] but they are not too uncommon.

Taps serve two quite different functions[61] depending on where they are located on the water network. First, there is the kind of tap we find

228. Herculaneum: water pipes in street; from the left-hand one, a connection is taken sideways, into the neighbouring house, passing under the doorstep.

mounted over a sink or bath. There is a control wheel or handle and a spout, from which the water pours forth when the tap is turned on. Second, there is the kind of tap that one finds set into the middle of a length of piping, or where another pipe branches off, and which, when closed, shuts off the supply, isolating part of the plumbing circuit; every household has a number of these, the best known being the one where 'you can turn the water off at the mains'. In English, the same word, 'tap' is applied to both devices, but their purpose is quite different: the first is to give access to the water, the second to regulate its movement through the pipes.[62] To avoid confusion, I will here refer to the first type as a 'discharge tap', and to the second as a 'stopcock', for not only is their purpose different, but so is their actual operation. Discharge taps, where fitted, are kept normally closed. They are turned on only when someone wants to draw water, and are hence a device to avoid waste and cut down consumption. Stopcocks serve more varied uses, and their operation varies accordingly. They may be purely an emergency device, to isolate part of the system for repairs, or when a customer has not paid his bill. In this case, they will be left permanently open and rarely touched (like the mains tap in a house). They may also, for example, control some cross-link in the piping, opened by the waterworks staff only when the regular main is out of service and an alternative supply has been provided for a street fountain. In such cases, they will be left permanently closed, and opened only when needed. They may be opened and closed in various combinations, to direct the flow of the water in different directions as needed; opened and closed at different times, they can be used in effect to ration water to different establishments, where there is not enough to serve everybody all the time. These different functions both control water use and reflect policies towards it. We must therefore be careful, when we hear of taps on a Roman network, that we know which kind we are dealing with. If they are discharge taps, they definitely do reduce consumption and cut waste; if there are a lot of them, it implies water is not abundant. If they are stopcocks, it *may* mean the same thing, if they are used to share water out in turn; if they are emergency isolating stopcocks they have no effect on consumption at all, and are hence irrelevant to the amount of water available.

In Roman work, for all these different purposes, there was only one design of tap. Invariably made of cast bronze, it was used for them all, the tap itself being large or small depending on the size of pipe to which it was fitted.[63] This was the type of tap today known as 'rotary plug' (Ger. *Zapfhahn*, Fr. *robinet à boisseau*). Essentially (Figs 229, 230), it consists of a circular hollow casing; inside this turns either a solid metal plug pierced by a horizontal hole, or a cylinder with a pair of holes in the sides, facing each other.[64] When the hole (or pair of holes) is aligned with the pipe, the water flows through; when the tap is given a quarter-turn in either direction, so that the hole is now set crossways to the pipe, the flow is cut off. It is still in common use today, usually on small-gauge pipes, such as gas-pipes or garden hoses. Its chief advantage is its simplicity, and its chief disadvantage that, though in theory an intermediate setting, partly open, is possible, in practice it tends to be a two-position, off-on, operation, as opposed to the infinite variation achievable with the screw tap as normally fitted in modern plumbing.[65] From this comes one serious result. We have already seen, in connection with siphons, how, on a pressure circuit where the water is running in a closed pipe, shutting it off abruptly can produce water-hammer sometimes sufficiently violent to be dangerous. We must now note that the type of tap used by the Romans was, in effect, incapable of doing anything else: by comparison, a modern screw tap cannot be closed so quickly, and is therefore safer. Whether the Romans actually experienced any trouble from this source, we do not know, nor whether they fully understood the factors involved. As we have seen, taps were in general used only on lines where the pressure had first been reduced by going through a water tower, making them less susceptible to water-hammer even if the taps were abruptly closed. And in any case, it must be emphasised, if such trouble did arise it was the pipes that suffered, even if it was the taps that caused it; and, as we have seen, Roman pipes were surprisingly strong. Just the same, it is probably lucky that pressure was low where the taps were installed, and one would like to know how the really big taps, such as the 20 cm one on the Decumanus at Ostia, worked out in practice, in positions where a modern

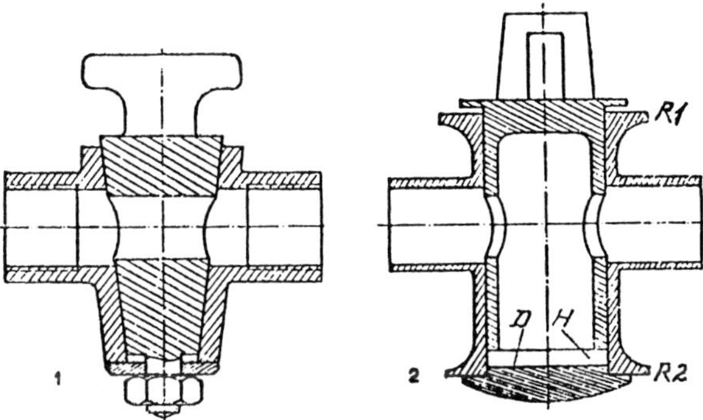

229. Rotary plug tap, in cross section: (1) left, a modern tap; (2) right, a typical ancient tap (Kretzschmer).

engineer would instead unhesitatingly use a gate valve.[66]

The main differences in design from a similar modern tap are shown in Fig. 229. There are three. First, the plug of the modern tap is tapered, resting in a conical housing, while the hollow cylinder that forms its ancient equivalent has parallel sides and the sides of the cylindrical chamber in which it turns are likewise vertical. Technically, the result is that the ancient tap is a lot stiffer and harder to turn, but less liable to leak. This is because, under the static pressure of the water, the plug is pushed upwards. With a tapered plug, even the slightest suspicion of an upward movement tends to lift it clear of its seating, thus reducing friction when it is turned. Conversely, the crack thus opening between plug and housing creates a leak. With an ancient tap, neither occurs. The pressure will still be there, and the central cylinder will tend to move upwards under it, but the move will not open a crack between it and its housing, both being cylindrical. Second, at the base of the modern tap is a large nut, precisely to stop any upward movement of the plug by holding it firmly down. The ancient tap had no equivalent, and this was a cardinal weakness. When the tap was significantly lower than natural water level – in Roman examples, this was usually the level in the neighbouring water tower – the water in it was subject to static pressure generated by the head. Static pressure is exerted in all directions equally. Against this, the thick double side walls of cylinder and housing offered adequate resistance, so lateral thrust was contained. The top of the cylinder, carrying the handle, and the bottom of the housing, covered by a heavy cap, were each strong enough in themselves, but there was nothing to hold them together, and once any serious pressure was applied, the cylinder, not being held down, was free to slide upwards and out of the housing altogether, thus bursting the tap.[67] This was the weak link in the whole Roman water supply system, and explains the need for the water

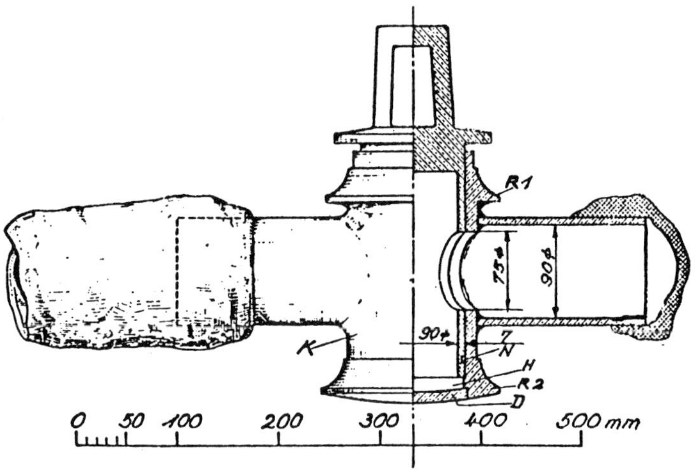

230. Nemi: ancient Roman tap (Kretzschmer).

towers to reduce the pressure. It is ironic that after so many commentators have worried about the ability of Roman pipes to handle high pressures – worries that are largely groundless – it is actually not the pipes but the taps that are the governing factor. It will, indeed, be seen that there were actually two good reasons for avoiding high pressure on a pipe with a Roman tap on it: the static pressure might burst the tap itself, and if the tap were abruptly closed, the water-hammer might harm the pipe. The Romans may not have known or understood both. They may have reduced pressure as a protection against one, and in so doing also covered themselves against the other by accident: but this is guesswork.

The third difference is in the form of the handle, which, in a Roman tap, usually takes the form of the 'square loop' shown in Fig. 230. This is because the tap was turned by an iron wrench or key, much like that used by modern plumbers to turn on street hydrants and other large taps. The cylindrical core of the tap so increased friction that there was no question of turning it with one's fingers. These keys were evidently portable and normally removed once the tap had been turned and, according to Kretzschmer, none has ever been found.[68] However, in the Museo Nazionale at Naples are several taps with the remains of an iron bar stuck in the loop; the interpretation is that somebody lost the proper key and instead used an iron bar that bent under the strain and got jammed in the tap so that it could not be removed.[69]

Inside the house

As the pipe entered the house, it might well, as today, be fitted with a control stopcock. This might supplant the *calix*, if it were used to regulate the supply, but we know too little to draw reliable conclusions. The vast bulk of our knowledge of water entitlement and regulation comes from Frontinus, who bases it entirely on the *calix* system and says nothing at all about taps. Presumably this was the way things were run in Rome, but we must beware of extrapolating from this. Other cities, where taps were more in evidence. and the water supply hence more subject to control, may well have done things differently, and we cannot make too many assumptions based on the practice and policies of metropolitan Rome.[70] When the water entered the house, at Pompeii at least, a further possibility arose. It might be taken directly up to pour into a large lead tank somewhere in the upper parts of the house, around roof level. From the tank branched off further lead pipes, in turn fitted with *calices*, to serve other neighbouring houses. This arrangement evidently means that a group of householders, none of whom wanted to pay the full water tax individually, clubbed together to, as it were, share out a single subscription among themselves. The *calices* meant that consumption and costs could be shared on a pro rata basis.[71] We are here presumably dealing with a different situation from that mentioned by Frontinus, who specifies that a group of individuals can build and use a private *castellum*. Such a *castellum* is built under public authority and each member draws from it, in effect, one official allocation of water. The Pompeii arrangement is something much more modest and more informal. One person, apparently, draws a single allocation from the official *castellum* and then, after the water has entered his house, divides it up among his friends, who reimburse him. The transaction is unofficial, though presumably legal, and the state has nothing to do with it. It is noteworthy that even the well-to-do[72] apparently are to be found joining such a consortium, an indication that getting your own private supply must have been, in relative terms, really quite expensive.

However, given that the supply was there, how was it used? For one thing, if the water was being delivered continuously, twenty-four hours a day, then, although the concession might be expensive, at least you got a great deal of water for it. The smallest pipe, and *calix*, a house could be served by was a *quinaria*, since that was the smallest existing, and that is reckoned to have delivered something around 40 m^3 daily.[73] To set this figure in perspective, the water bill for my own house in Ottawa shows a consumption averaging 80-90 m^3 per four-month period: a Roman house used in one day as much water as a modern house does in two months, and in terms of water consumption per household a Roman house used something like sixty times the modern figure. This does explain why, in Pompeii at least, this monstrous amount could sometimes be divided up among several houses (particularly if they used taps). But assuming it was not, and that, as was usual, this was the amount actually delivered and used by one house, wherever did it all go to?

The short answer, of course, is that most of it went straight into the drains, unused, but it got there by various routes. Within the house the water was fed into an internal system of small-gauge pipes. Though usually hidden, like their modern equivalent, they must sometimes have been surface-mounted and visible, for sometimes in villas, as a sign of conspicuous luxury, and presumably a rare one, they might be made of silver.[74] The arrangement of internal domestic services is again something awaiting further study, but they probably fell into three main categories. First, there must have been somewhere a spout or fountain delivering water for drinking and cooking. The kitchen sink does not seem to have been a standard fitting,[75] despite its popularity at the counters of taverns, and it seems quite likely that in many houses the kitchen did not even have running water. It must be remembered that the two rooms which we today take absolutely for granted in any residential accommodation, and which are most closely dependent on plumbing, the kitchen and the toilet, were often lacking in a Roman dwelling. At Ostia, even large and presumably expensive apartments had neither,[76] the occupants relying on public facilities of various kinds. In the more spacious layouts of Pompeii, where land values were cheaper, houses often have both, but one still cannot rely on it. The spout or fountain, wherever located, presumably ran continuously, with a drain set below it. None the less, discharge taps (though much rarer than stopcocks) do exist and have been found, though it is seldom if ever known what part of the house plumbing system they came from. Designed for horizontal mounting on a wall surface (like a modern tap) they deliver the water either through a horizontal spout (i.e. like a stopcock with the pipe on one side removed), or, more rarely, vertically (i.e. the water channel makes a right-angle bend inside the tap and the delivery spout comes out of the bottom, pointing downwards).[77] In such taps the plug or cylinder could be, and sometimes was, slightly tapered to reduce friction and make it easier to turn; the leakage that normally would result from this was evidently minimal, as the taps were here mounted on a low pressure pipe of small gauge. In consequence, a handle of modern type, which could be turned with the fingers, was sometimes fitted, though such was the force of convention that we also often find the 'square loop' type, turned by a portable key, even where the tap seems to be not stiff or big enough to need it. We may also here notice a tap of highly sophisticated design, rather like a modern mixer tap, which was connected to two separate water lines, hot and cold, discharging both through a single spout (Fig. 231). Two separate examples of this tap have come to light, from the same general area.[78] It must be noted that in both of these taps there is no question of mixing hot and cold water together to produce warm. Only three positions were possible: 'hot', 'cold' and 'off', and the tap was again turned by a separate key.

The presence of hot water raises the second category of use, baths.

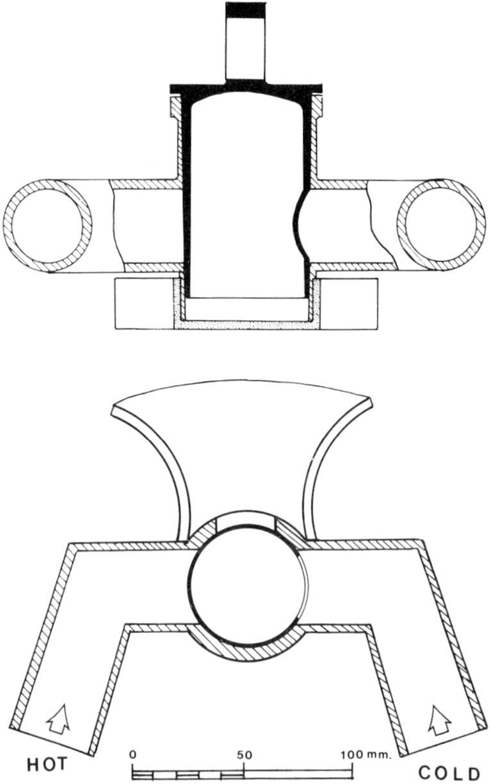

231. Rottweil, South Germany: 'mixer' tap, connected to both cold and hot water lines. It should be noted that the tap had only three positions, 'hot', 'cold', and 'off', so that any mixing of the water to vary the temperature had to be done in the basin below (Kretzschmer).

Baths, of the sauna-like type favoured by the Romans, were an ambitious and sizable installation, and most householders patronised the public baths, either the imperial monsters like the Baths of Caracalla or more modest local establishments. Yet, in spite of the multitude of public baths, one does find private ones as well. Usually they are in country villas or military forts, where there were no urban amenities within convenient reach, but one sometimes finds them installed in private houses by city dwellers who could afford it.[79] Baths notoriously used a lot of water, and in any house that had one would account for a lot of the consumption. The plumbing might be elaborate too. In the private baths at Boscoreale a complex, and very untidy, tangle of lead piping united no less than four stopcocks, all in order to mix hot and cold water to a temperature agreeable to the bather (Fig. 184). In some baths also we see yet another type of fitting – an *epitonium*, a kind of flattened nozzle with no control tap, producing a jet of water spurting out from the wall, rather

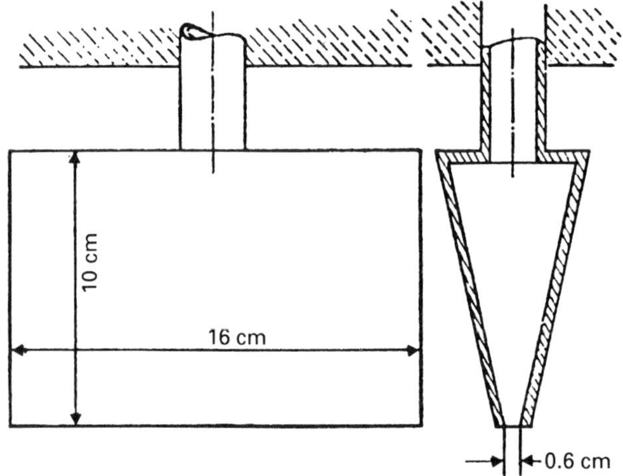

232. Herculaneum: *epitonium*, or shower jet (Kretzschmer).

like a modern bathroom shower-head (Fig. 232).[80]

The third principal category of domestic water use was the garden, if one existed. We have already noted the extensive use of water for watering the plants, and the effect that this had on the design of the garden. As well as that, however, we must also include ornamental fountains. It was only in post-classical times that, with the Villa d'Este and the Alhambra (Granada), water-gardens became an art form in their own right, but already, in the Roman age, fountains were an essential feature of every garden.[81] Given the lower water pressure, there was no possibility of the powerful vertical jets that we so often favour today, but small cascades and basins rather like a modern bird bath were always a possibility. The attraction of such fountains was not only visual. The constant gurgling and splashing made an agreeable background for garden relaxation,[82] and in the hot Mediterranean climate even the psychological associations of trickling water are not to be underrated; it made the garden *sound* cool.[83] As well as the fountains, we find ornamental pools and channels of flowing water, ranging all the way from the pergola and series of basins that formed the centrepieces of the garden of the House of Lorieus Tibertinus at Pompeii, to the Canopus and other grandiose waterworks of Hadrian's Villa at Tivoli (Fig. 233). Sometimes these installations were loosely modelled – and specifically named – after celebrated waterways really existing, the favourites being the Euripos and the Nile: as an ultimate touch in imaginative realism, Lorieus Tibertinus' channel could be made at will to overflow in imitation of the flood of the Nile.[84]

At Bulla Regia in Tunisia, in the Maison de la Pêche, one perhaps finds the use of water in the garden to provide a cool atmosphere carried to its ultimate extreme. Here, the peristyle is mostly occupied by a series of

233. Tivoli: Hadrian's Villa, the Canopus.

intercommunicating pools or shallow tanks. The water enters continuously through a terracotta pipeline, and circulates in turn through the whole complex, so that the pools are always full of cool, fresh water. The remarkable feature here is that in this house there is, quite exceptionally, a full, habitable floor of rooms, including a triclinium, directly under the peristyle.[85] In hot weather, therefore, one could not only retire to the cool of this basement level, but actually dine there with a sheet of cool water circulating directly over one's head. Some sort of downspout then brought the water down to the lower level, there to be enjoyed further, apparently in the form of fountains. This sophisticated hydraulic installation is without parallel, but shows to advantage the ingenuity of the Roman architect and hydraulic engineer in adapting to the hot North African climate. It is perhaps the closest thing to air conditioning the ancient world ever achieved.[86]

12
Drains and Sewers

And so this process must continue in a never-ending cycle.

Aristotle, *Meteorologica* II, 3

We have followed the course of water as it was collected in the catchment area, conveyed by the aqueducts, distributed in the city, and used domestically. It remains only to see how it was then disposed of, and so, by reaching the sea or evaporating, completed the hydrologic cycle that we encountered in Chapter 4.[1]

It is therefore now our task, possibly an unappetising one but scarcely to be avoided, to descend into the sewers and drains.

We must clearly realise that the two words do not mean the same thing, though their functions often overlap. It only needs spelling it out for the distinction to become obvious: sewers are for the conveyance of domestic human waste, and are only found in cities, where they are justified by the density of the population; drains are for the disposal of surplus water, and are found both in cities, to carry off rainfall and overflow from the constantly-running fountains, and in the country, to prevent flooding in the fields. In cities it was often convenient to use one conduit for both, and, perhaps as a result of this, in modern publications and parlance, the terms are often used loosely: 'drain' can mean a conduit for either purpose, while 'sewer' is used exclusively for human waste. Nowadays, familiar with needs of urban hygiene, we perhaps often consider sewage disposal a higher priority than drainage. Impressed by the Roman achievements in water supply, we may even laud their standards in sanitation and attention to public health as one of the high-water marks in ancient social history. It may for a moment, therefore, surprise us to realise that in the ancient world drains came well before sewers, and even in sewage disposal the Romans did not share our sense of priorities. As long ago as 1930, the truth was well put by Alfred Neuburger: 'The hygenic value of drainage was probably less taken into account than considerations of comfort. The larger a town was, the more difficult and the more laborious it became to remove refuse and sewage waters which accumulated in great quantity. For this reason – and probably not for sanitary reasons – drains were built for removing all this matter from the town mechanically and with as little effort as possible.' Lewis Mumford, speaking of ancient water supply generally, but sewers in particular, analyses it as an 'uneconomic combination of refined technical devices and primitive social planning'.[2]

The largest drains were those intended for draining swampland and reclaiming it for agriculture. The greatest, and best known, examples of this are the various schemes for draining Lake Copais in Greece, and in Italy, the Fucine Lake and Lakes Albano and Nemi. They were impressive works of engineering, and the *emissaria* included the longest tunnels in the ancient world, but they are hardly relevant to our study of urban drains.[3] The best known such urban drain was the Cloaca Maxima at Rome (Fig. 240), which probably began as an open ditch draining swampland around the Forum into the Tiber. From its earliest days – tradition ascribes its construction to Tarquinius Priscus, supposedly king of Rome 616–578 BC – it also apparently carried sewage, and it was subsequently covered by stone vaulting. The work took place in stages, reflected by different techniques in the masonry. It is interesting to note that its bottom was also paved, with basalt pavers like a Roman road. This was an all-purpose drain, in that it combined the three functions of sewer, rainwater removal and swamp drainage; and was of great size, in places 4.2 m high by 3.2 m wide. It was large enough, Roman writers agreed, for a wagonload of hay to pass through, and though there is no reason to believe anything so improbable ever happened, Agrippa, in his capacity as water commissioner, did go through it in a boat on a tour of inspection.[4] Quite a number of cities, Greek as well as Roman, if they do not go to these lengths, yet boast their 'great drain', in the form of a large, rectangular conduit, usually lined and covered by stone slabs, and pursuing a strategically chosen course through the centre of the city, perhaps along the main street. This main 'collector drain' was then often served by a number of smaller branches running down side streets.[5] In Greek days, sometimes even in Roman, the stone slabs forming its sides are not mortared together, nor is there any cement lining, so that the drain was, apparently, far from watertight, and more resembles a kind of artificial watercourse or gully, rather than a conduit proper. One conjectures that such channels were often really storm drains for emergency use, a way of directing into a pre-determined and harmless course a sudden torrential run-off of high volume.[6] What use they had in ordinary daily life is a different matter.

For the different types of drain, there were presumably different requirements. Storm drains had to be capable of carrying large amounts of water intermittently, but were usually dry. Drains from swampland would continue in constant flow, as the water that had made the area swampy continued to enter it, from whatever source, and had to be continuously removed; the Cloaca Maxima is a good example. Sewage would be very small in relative volume, but less fluid and harder to transport, unless mixed in with some independent source of flowing water. And, once an aqueduct was built, there was a constant volume of waste water to be carried away after use, whether by overflow from street fountains, from the baths or elsewhere.

Sometimes these requirements could usefully be combined. If there was an aqueduct, the street drains, constantly carrying a stream of waste water, could also be used to carry the sewage, as is recognised by the frequent comments one sees on the necessity of having an extensive overflow to keep the drains flushed. Indeed, the two did go together, and the enormous water surplus provided by the Roman water system, far from being wasteful, was practically a necessity.[7] As in the Cloaca Maxima, sewage could also be run into a swamp or flood drain, provided there would always be water flowing in it. But there were also presumably some of the larger drains that would be of little use for sewage because they were normally more or less dry, and any sewage put into them would not go very far, at least not until the next cloudburst.[8] In such cases, sewage, where it was conveyed away at all, may have been carried in pipes, which would certainly be much more suited to the requirements of the occasion than a large, semi-dry drain. But, as so often in our study, we run into the same difficulty: no really systematic work has ever been done on the question. We may, however, note two points.

First, the simple but ingenious idea of the trap, the U-bend in the drain pipe immediately below most of our plumbing appliances which remains always full of water and prevents odours and gases rising from the drains and back into the house, was unknown to antiquity. Drains and sewers were therefore liable to be smelly, not just in themselves, but for any house or building connected to them. This probably explains the fact that in Jerusalem, around the tenth century BC, two independent systems of street drainage were provided, keeping sewage and waste water quite separate.[9]

Second, and perhaps most important, we must not take sewers for granted. We are liable to do so – particularly, perhaps, in a book on Roman water supply – because of all the laudatory remarks we have so often heard on Roman sanitation. We must therefore realise clearly that many Roman towns had no sewers and/or drains at all, and many more had only a partial network that did not by any means cover the whole town. The arrival of an aqueduct would naturally lead to the building of a drain/sewer, if only to carry off the overflow, but this natural sequel did not always in fact follow, or at least not immediately.[10] Probably no city is better known for its street drains than Pompeii, but, at the time of its destruction in AD 79, they existed only in the area around the forum: in the rest of the town there are no underground drains, and drainage followed the ground surface.[11] This must have been the regular system in many cities, and what it meant was simple.

Drainage of rain and waste water

Rain water was disposed of depending on where it fell. If it fell on a house, the roof (usually of compluviate pattern) funnelled it into a cistern somewhere in the interior. In other words, the function of the roof on an ancient house (other than keeping the interior dry) was not so much to shed water as to collect it. A great deal of the rain falling on a town, therefore, never did get drained away. Instead, it was kept. Rain falling on the streets and open spaces, on the other hand, found its own way downhill, following the lie of the land. It was joined by waste water thrown out of the houses and from any other source. This might not amount to much. But once an aqueduct was brought in, and with it the overflow (of what was, essentially, unused water) from the street fountains and, especially, the baths, then the water to be drained away might amount to a great deal. One has the impression that many small towns, or parts of them, must have been permanently awash in surface water. Where, then, did it all go to? The rough answer is that since absolutely level ground is rare and most towns were built on something of a slope, the water ran down to the lower end until it reached the city wall, from which it escaped by running out through the gates. Where it was necessary, there were also drains through the wall itself, to prevent a pond building up on the inside – a prominent example may be seen alongside the Gate of Nola at Pompeii. This system was not quite as haphazard as it sounds, for where convenient or necessary through the city surface gutters or runnels would be provided to send the water in the right direction. Buildings lying across its natural course, which would have otherwise acted as dams with the water building up behind them, in this way had it led round the sides. Any kind of subterranean accommodation, such as cellars, obviously presented difficulties (cryptoporticos were an especially intractable problem) and required special treatment to stop them simply filling up. One way was to build a second wall around the underground structure, so that it had, in effect, a double wall with a narrow gap of 50 cm or so in between, acting both as a drain and as a barrier to damp seeping through from the sodden earth outside. The problem was particularly severe if the inner face of the wall carried frescoes, and was common enough to evoke prophylactic recommendations from Vitruvius.[12]

In the streets, it was largely this surface water problem that led to the raised sidewalks so familiar from Pompeii and Herculaneum, with stepping stones (*pondera*) at the street corners to enable pedestrians to cross from one side to the other without stepping down. The height of these sidewalks has always been something of a problem, for they sometimes get up to 50 or 60 cm high, which seems to involve, to put it mildly, an excessive safety factor. The streets of Pompeii were washed by the continuous overflow of some 50–60 public fountains (not counting waste water from the baths), each overflow delivering the discharge of a small-gauge lead supply pipe, probably a *quinaria*,[13] not much bigger than a modern garden hose. While the combined output of fifty garden hoses might be enough to maintain a steady trickle or even a small stream along the main streets, it would hardly lend them that Venetian aspect that arises in our mind when we see the height of the stepping stones.

Domestic sewage

Part of the answer may of course come from the fact that what was washing the streets was not just clean water from the fountains. In the absence of sewers, what did private houses do with their sewage? Sometimes it never left the house at all. A toilet was built inside over a cesspit, in which the accumulated discharge simply piled up, rotting away, until there was enough of it to justify a call from the local manure merchant,[14] who came and carted it away for resale as fertiliser. Although this is a practice of which one normally hears little, it was common in all ages down to modern times, often representing a serious factor in local agriculture, or, as a modern authority put it, 'in any society with a sound ecological balance between man and the environment, people have developed the ability to live lightly on the land, contributing to its renewable resources, observing and utilising natural cycles, harmonising actions to give as well as take. One such contribution, part of the traditional symbiosis between town and the countryside, was the provision of night soil as fertiliser for the crops.'[15] It was particularly rich in ammonia, and was good for trees, a healthy growth of which would, among other things, recharge the water table; while 'the air and prevalent sunshine would have quickly killed the anaerobic bacteria, sterilising the material and cancelling the bad smell'.

In the typical Pompeian house, the toilet was often located next to the kitchen. This was a matter of architectural convenience, and if it squares but ill with modern notions of hygiene, we must see in it only another confirmation of Neuburger's dictum with which we began the chapter, that what the Romans had their eye on was comfort rather than health. Or, as Carcopino put it even more forcefully: 'The drainage system of the Roman house is merely a myth begotten of the complacent imagination of modern times. Of all the hardships endured by the inhabitants of ancient Rome, the lack of domestic drainage is the one which would be most severely resented by the Romans of today.'[16]

Another possibility, where space was available, was to connect the toilet to a cesspit or septic tank constructed in the house's own backyard. The connection would be made by a drain, probably terracotta pipes, along which the waste would be flushed by a bucketful of water poured in when needed. The word 'along', rather than 'down' is significant, for it was in all likelihood often impossible to provide the degree of fall, of downward gradient, that a modern plumber would consider essential. In the absence of any anti-odour U-bend trap and the presence of a drain perhaps little better than level, in which things could easily get stuck, one wonders whether the resultant installation represented any great improvement of domestic amenities.[17]

With multi-storey apartments the problem was different. Running water above the ground floor was a rarity, if it ever existed at all, because of the necessity of a head to make it rise to the necessary level. Outgoing, or downgoing, sewage could at least rely on gravity, so one does sometimes find upper-floor apartments with drainage and toilets, but no running water. The drain was usually vertical and consisted of large-gauge terracotta pipes, internal diameter 13–15 cm, embedded in the wall and leading to the cesspit or the sewer, or simply opening into the street.[18] It is assumed that the larger of these pipes are for sewage, and the smaller ones for waste water, but this is apparently conjecture.

In spite of these arrangements – or, more properly, in their absence – a more direct approach was often adopted. It was not at all uncommon for apartments to lack both a toilet and a kitchen. To fill the appropriate needs the occupants expected, and were expected, to go outside. One need was filled by the fast food outlets, *thermopolia*, found on every street corner. To meet the other challenge one had recourse to personal initiative. There were, of course, the public toilets, which we have already considered. Businesses which need urine for industrial use, notably fullers, set out a pot on their front doorsteps to solicit contributions from passers-by on a mutual assistance basis. Others 'betook themselves to some neighbouring dungheap';[19] others evidently did not, and notices were regularly posted on the walls of Pompeian houses warning all and sundry 'Don't do it here – or else!';[20] yet others, in an anticipation of the traditional practice in mediaeval and later cities, simply threw everything into the street from an upper storey window or balcony.[21] One can only generalise by saying that there was no consistent overall picture, and that from one town to another, even in different regions of the same town, and at different periods of time, anything could happen, from one extreme to another.

Street drains and sewers

Where sewers or drains were provided, the simplest form was an open ditch. We do not normally think of anything so primitive in connection with the Romans, but not every town was a Pompeii (see Figs 235, 236). Even at Rome itself, at the height of the empire, the very period when nine or ten aqueducts were daily supplying those vast quantities of water at which we justly marvel, a lot of the city still relied on open sewers down

234. Ostia: reconstruction of water mains and drains under street (A. Pascolini, 1980).

235. Priene: open drain down middle of street.

236. Swarzenacker (Homburg), Germany: street drain, with (restored) wooden props to hold stone side walls in place.

the middle of the streets; passers-by were often knocked into them for the fun of it by Nero, sallying forth incognito on one of his nocturnal escapades.[22] This is the other side of the coin from that represented by the splendid engineering of the Cloaca Maxima, and if things were like this in the imperial capital, what are we to expect in the provinces? In fact, in the outer provinces many towns and settlements, though nominally in the Roman empire, were only half Romanised, and, if there was no substantial Roman army garrison locally stationed who could accelerate and give direction to the process, the progress of the natives, left to their own devices, might not be rapid. Pliny tells us that at Amastris, in Bithynia, a city that he otherwise describes as elegant and beautiful, there was an open sewer running down the main street, which he is proposing to cover over.[23] This must have been quite common practice, and Silchester, in Britain, is merely one more example of it. There the drainage was by open ditches 60 cm wide by 60 cm deep running along the streets, in this case often with their sides reinforced with planking, which carried away both rainwater and household sewage. Silchester also had a bathing establishment of the latest model, prompting a comment that this 'shows a praiseworthy regard for the interests of public health, but the lack of a decent drainage system must have done much to counteract any beneficial results to be gained from regular washing'.[24]

In more developed communities, built stone drains would be provided. It was of course by far and away easiest to put in such drains when the town was still under construction.[25] Otherwise all the streets had to be torn up to install them, unlike the aqueducts and urban water systems, which came in at surface level and caused much less disruption. Street drains therefore tend to be a feature of new towns – provincial *coloniae*, legionary cantonments and the like – and perhaps we may risk a generalisation and suggest that they were commonest in the 'middle provinces' (counting outward geographically from Rome). In the far-flung outposts of empire, such as Britain, Roman civilisation was sometimes thin enough for towns that easily could have had drains not to bother with them, as at Silchester. At the other extreme, old-established Mediterranean cities that were originally built without drains might find it prohibitively difficult to install them afterwards. No doubt this is why Pompeii lacks a full network. An interesting parallel is the use of the typical squared grid street-plan in new Roman construction while the older parts of the cities, built higgledy-piggledy in an era yet innocent of urban planning, often retain their random layout. This is reflected in Pompeii where the 'downtown core' around the forum (ironically the only part that ever did get drains) is clearly distinguishable from the rectangular city blocks of later expansion to the east; and, of course, in the whole of Rome itself.

In Pompeii, a typical example, the drains ran 1 m under the sidewalk, or sometimes the roadway (compare Fig. 234) They were some 50 cm or so wide, with a gabled roof, and provided with inspection manholes, each covered by a round stone lid furnished with a bronze ring in the centre for lifting.[26] They were fed from various sources. Private houses could have a drain from their toilet or kitchen running downhill to empty into the street drain.[27] Baths and public toilets would likewise be connected directly to the system and, given the constant flow of water through each, would have been one of the major contributors. Indeed, baths, located in a town where the drainage was rudimentary, were sometimes provided with their own individual drain independent of the city system.[28] But most of what went into the drains came directly from the streets. This included rainwater, anything thrown out of house windows, the overflow in rainy seasons from cisterns inside the house[29] and, far and away the most copious source, the constant overflow from the street fountains. There were two means by which all of this found its way down into the drains. One was a series of openings in the vertical face of the sidewalk curbstones, sometimes provided with a metal grille to keep out debris but

237. Pompeii: street drain.

often, apparently, simply built into the paving of the roadway at this point, which, running at an angle across the street, would direct the surface water into the hole (Fig. 237), the inconvenience to wheeled traffic evidently being adjudged a sacrifice in a good cause. The other was a series of openings cut in a paving slab set flat in the roadway, forming in effect a stone grating, through which the water falls straight down into the drain below. These differ from the intakes previously mentioned in two ways. First, the grating is laid horizontally in the middle of the street while the apertures in the curbstones are vertical. Second, while the curbstone holes may sometimes have had metal grilles which may have been decorative in form, these have now disappeared, leaving a plainly utilitarian appearance. The holes in the stone gratings, in contrast, are decoratively arranged, usually in curvilinear patterns such as crescents; some typical examples may be seen in Fig. 238.

Once in the street drain, the run-off from the street proceeded downhill along it. There is no telling what the gradient was, for though a whole range of gradients, often widely different and sometimes contradictory, has been published for aqueducts, no figures are available for drains. Problems there must have been, for on a level site there must have been a limit to how far one could go towards establishing an adequate slope by digging the drain ever deeper, at least if one wanted it ever to discharge its contents anywhere by gravity. Presumably there was also less scope than in an aqueduct for achieving a desired gradient by following the terrain: the drain had to follow the streets, or at least go where the houses

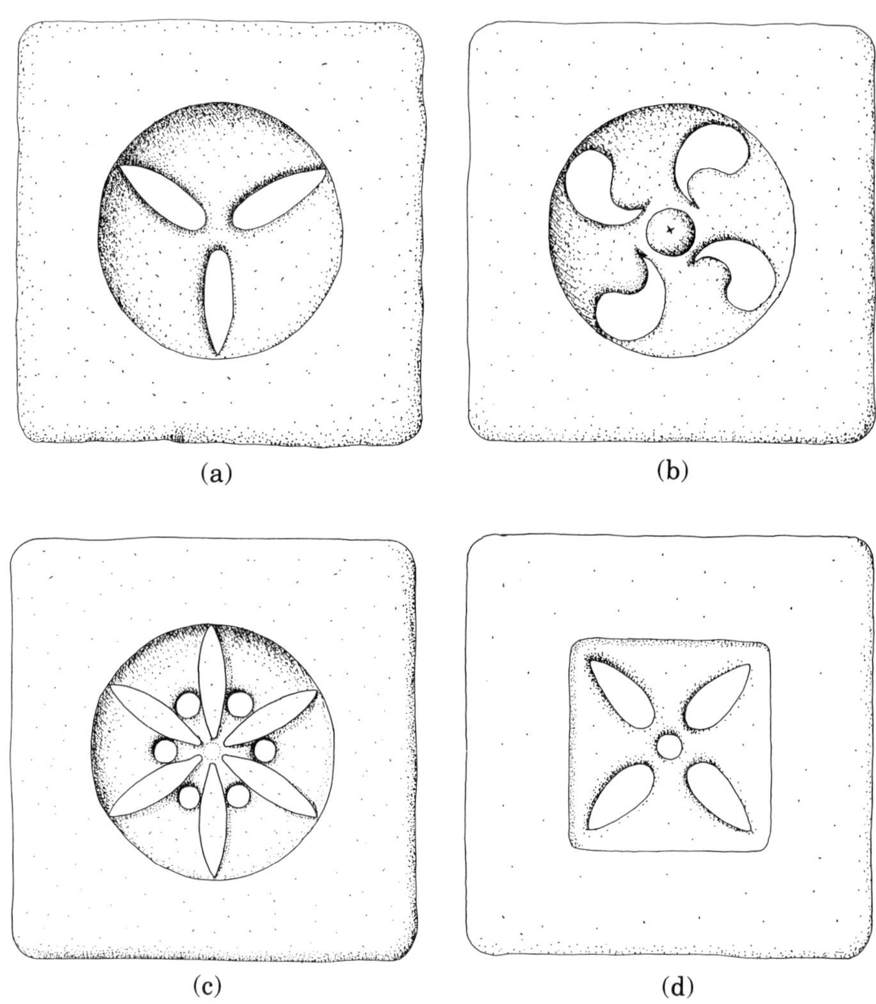

238. Stone drain gratings from (a) Volubilis; (b) Ostia; (c) Side; (d) Housesteads.

were. On the other hand, the whole network of drains was an enterprise on a scale so much smaller than the aqueduct system that such difficulties should not be overemphasised.

The street drains would usually discharge into a large central collector sewer. These, of course, would be much like the Cloaca Maxima – the reader will be aware that on this topic categories often overlap and classification will necessarily be imprecise – except that, in the case of the two best preserved examples, they were apparently planned and built as part of the city from the start. The better known is the long length of drain under the *praetorium*, now the Rathaus, at Cologne (Fig. 239). With dimensions of 2.5 m high by 1.2 wide, a floor 10 m under present street

239. Cologne: main city drain under the *praetorium*, now the Rathaus (photo: K. Grewe, Bonn).

level, and a preserved length of over 100 m, it constitutes an impressive monument.[30] In shape it has a vaulted roof and is more or less of the same proportions as an ordinary aqueduct conduit, though a lot larger. The side walls and vault of the drain are built of cut masonry, squared blocks of tufa, with two preserved vertical inspection shafts of 65 cm^2 section. The other example is the very similar massive drain under the forum at Trier.[31] Both date from the first century AD.

The last stage in drainage came when the drain with its cargo of rainwater, overflow and assorted detritus reached the edge of the city. If the city was built on a river, as so many cities were, there was no further problem. The drain simply emptied into it, and its contents were washed away downstream, to find their way whither they would. There is no record of any sewage-treatment; no doubt it helped that the sewage arrived there already heavily diluted by the fountain overflow coming down the drains. The archway by which the Cloaca Maxima at Rome discharged into the Tiber, and the main drain at Cologne into the Rhine, are still preserved and visible (Fig. 240).

240. Rome: Cloaca Maxima. The exit from the drain into the river Tiber, still existing but now partly hidden by the modern Lungotevere Embankment, is clearly visible (just below the round 'Temple of Vesta') in this nineteenth-century engraving.

Sewage disposal

Cities with no convenient river fell back on other expedients for disposal. The best concise description is still that of Neuburger for the arrangements at Athens: 'The Athenian drainage is worthy of particular notice because it allows the waste water to escape by a method nowadays called "the soakage system". The main drain, consisting partly of false and partly of true vaults, divides, when it has left town, into a number of smaller channels, so that the waters which had first been collected together, again separate into little rivulets. They run along the little drains underground for some distance and thence flow out into the lower-lying plains, where they soak away. It is not known whether plantations were sown after the manner of our irrigation trenches in order to exploit the manures contained in the waters.'[32] Though there is little evidence for or against it, this assuredly must have been a common ancient practice. It is still done in the Mediterranean, and one modern authority notes that at the modern Sicilian town of Aidone, near the ancient site of Morgantina, 'the sewage is spread out on what are called

241. *L'envoi*: in a poster at Gentilly station on the Paris Metro RER line, advertising ingenuity suggests a productive future for the Pont du Gard, converted to a railway viaduct (compare Fig. 87).

sewage farms or sanitary fields. The steep slopes of the Serra Orlando ridge [Morgantina] could have been used in this manner in antiquity, handling sewage that would flow out through the drain pipes and channels.'[33] No doubt it would be more reassuring to quote examples from more typically Roman sites, but in a field where evidence is scanty we may yet with some confidence accept the above as reflecting what must have been common practice, widespread throughout the Roman empire. So was the hydrologic cycle brought full circle.

APPENDIX

Facts, Figures & Formulae

Few engineering data are sufficiently precise to justify the use of more than three significant figures.

R.W. Fox and A.T. McDonald, *Introduction to Fluid Mechanics* (New York, 1985), 2

In this appendix will be found grouped together some of the facts and figures to which the reader will most commonly wish to refer. It has not been possible to arrange them in any logical sequence or consistent pattern. Nor has it always been easy to establish even 'records' – the longest aqueduct, for example. What length does one give for Carthage? Does one count only the original line, or also later additions? Does one add in branches to the total mileage? Nevertheless, listed below are some outstanding features.

Long aqueducts:	Aqua Marcia (Rome)	91 km
	Anio Novus (Rome)	87 km
	Carthage (total)	132 km
	(Zaghouan – Carthage)	90 km
	Cologne (Eifel)	95 km
Large discharge:*	Anio Novus (Rome)	189,500 m³ (Adam)
	Marcia	187,600 m³
Total length of aqueducts (11) of Rome:		502 km
Largest bridge:	Pont du Gard	275 m long
		48.77 m high
		24.52 m max. span
Largest siphons:	Pergamon (Madradag)	3,000 m long
		190 m deep
		4,000 m³ discharge*
		1 pipe
	Beaunant (Gier, Lyon)	2,600 m long
		123 m deep
		25,000 m³ discharge*
		9 pipes
Longest tunnels:	Valle Barberini (Anio Novus)	2.25 km
	Mornant (Gier, Lyon).	825 m

* per 24 hours.

Maximum slope:

(in short stretches)	Carthage (over 6 km, Zaghouan-Moghrane).	28 m per km (2.8%)
(overall)	Segovia	16.4 m per km
	Aqua Julia	12.4 m per km
Minimum slope:	Nîmes (for 10 km Pont du Gard to St Bonnet)	0.07 m per km
	Nîmes (overall)	0.34 m per km
	Pergamon, Kaikos	0.3 m per km

* per 24 hours.

Aqueduct statistics

Location	*Length of channel (km)*	*Average slope (m per km)*	*Discharge (m^3/24 hrs)*	*Source, remarks*
Rome				
Appia	16	0.6	73,000	Total given here
Anio Vetus	64	3.6	175,920	estimated by
Marcia	91	2.7	187,600	P. Grimal.
Tepula	18	5.0	17,800	Alternative
Julia	23	12.4	48,240	estimate by
Virgo	21	0.2	100,160	H. Fahlbusch
Alsietina	33	6.0	15,680	520,000-
Claudia	69	3.8	184,220	635,000 (*WAS* 1,
Anio Novus	87	3.8	189,520	137).
Traiana	58	3.8	113,920	
Alexandrina	22	1.0	21,160	
Rome total			1,127,220	
Lyon				
Mont d'Or	28	3.27	10,000	
Craponne	25	16.8	13,000	Cascades
Brévenne	66	5.30	28,000	Cascades
Gier	75	1.47	25,000	
Lyon total			76,000	Lyon figures based on J. Burdy, *Doss. Arch* 38 (1979), 65
France				
Arles	48	0.625	8,000	Stübinger
Metz	22	0.56	22,000	Adam
Nîmes	50	0.34	20,000	Esperandieu (see Hauck, 169)
Saintes	7.5	0.87	19,000	Adam
Sens	17	0.95	40,760	Adam
Strasbourg (Kuttolsheim)	20	3.13	2,160	*JEAR* 176

Roman Aqueducts

Location	*Length of channel (km)*	*Average slope (m per km)*	*Discharge (m^3/24 hrs)*	*Source, remarks*
Germany				
Cologne (Eifel)	95.4	3.89	21,600	*WAS* 3,89
Trier	13	0.6	25,450	Neyses, 6
Switzerland				
Aventicum	8	4.4	2,880	Olivier, 10-11
Geneva	11	0.55	8,640	Blondel, 55
Spain				
Segovia	15	16.4	1,728	Gallardo, 19
Africa				
Carthage	84	1.5	17,280	Moghrane-Ctge. section only
Cherchel	45	2.0	40,000	Cherchel, 154; 148
Italy				
Bologna	20	1.0	35,000	*WAS* 3, 185
Turkey				
Pergamon, Kaikos	50	0.3	20,000	*WAS* 2, 35
Pergamon, Madradag	42	7.0?*	2,700	*WAS* 2, 26

* level of source uncertain, and line includes major siphon.

It must be emphasised that since these figures come from a variety of different sources, and – particularly estimates of discharge – may be calculated in different ways, they are not necessarily internally consistent, and should be used for comparison only with caution.

Useful Roman measurements

1 Roman foot = 0.296 m
1 *actus* = 36 m
1 Roman mile = 5,000 Roman feet, = 1.48 km
(*mille passuum*)

Weight of standard lead pipes

(Estimated by A. Kottman and O. Halter, *Schriftenreihe der Frontinus Gesellschaft*, Heft 7 (1984), 75. For an alternative estimate, see ch.11 n.20 p. 467.)

The gauge of the pipe is estimated by the breadth of the flat plate used to form the pipe, as measured in dactyls (finger breadths), i.e. a '5-finger pipe' is a pipe of five-fingers' breadth circumference, a *quinaria*. Note that the pipes are measured by circumference, not diameter, as is usual today.

Weight in kg per ten-foot length

Circumference	*Weight*
5 fingers	27.24 kg
8	45.40
10	54.48
15	81.72
30	163.44
40	217.92
50	272.40
80	435.84
100	544.80

Total weight of lead in all the siphons at Lyon: 10,000–15,000 tons

Total length of all Lyon siphon pipes: 150 km (average weight of pipe
(*reckoning 9 per siphon*) = 100 kg per m
= 230 kg per 10 ft)

Formulae

For those unfamiliar with the use of formulae – a category in which I most definitely include myself – several difficulties present themselves, and there are several points that should be made. To the non-mathematician and non-engineer, all but the most basic arithmetic and certainly the whole of algebra are usually little more than a dim memory of a lost skill, echoing down the years from schooldays long ago. The whole idea of solving an equation smacks of an adventurous safari into *terra incognita*, and the very appearance of a formula looks like a foreign language. Paradoxically, this should be a great comfort to the orthodox classicist, historian or philologist, for that is exactly what a formula is, and a foreign language, even if unknown, is never something he has felt uncomfortable with. If the vital book on a subject is in a language one does not know, it has always been an accepted responsibility of the academic discipline that one gets out a dictionary and does one's best. And a foreign language, with its own rules and conventions, is what formulae are – a language, indeed, in which engineers and others converse with great fluency. The non-mathematical reader, therefore, while he may not be able to master formulaic communications, has little excuse for throwing up his hands and not even making the attempt.

But there are things he should know. First of all, a formula can be solved for any one of the terms in it. Usually it is expressed in the form A = so-and-so, the 'equals' sign being followed by an array of figures of greater or lesser complexity, and meaning, in effect, that if these figures are manipulated in the prescribed fashion, the result will be A. This presumes that A is unknown, but if it is known, the equation can be rearranged to find the value of one (but only one) of the other terms. In effect, this is not much more than saying that one usually adds up a string of figures to find out what is the total, but if the total is already known but one of the figures is not, it is easy enough to see what is missing to bring the others up to the known result. Thus, in aqueducts the quantity or discharge of water carried (Q) is found by multiplying the average velocity of flow (V) by the cross-sectional area (A): Q=VA. But likewise the velocity is the quantity divided by the area ($\bar{V}$=Q/A), and the area the quantity divided by the velocity (A=Q/$\bar{V}$). So, provided two out of these three terms are known, the equation can be rearranged to find the third. Likewise, even if one of the two is not known, it can be assumed, and the answer based on the assumption: '*assuming* a velocity of ..., the quantity will then be ...'

The next point to observe is that there is no uniform convention on the meaning of the letters in the various formulae. Properly, when a formula is to be applied, the author ought to explain what the various letters stand for: 'where Q = quantity of discharge, $\bar{V}$ = average velocity ...' etc. But a quick glance through the publications, including standard engineering handbooks, will show that this is not so. Usually the letter is an acronym for the value concerned – Q for quantity, A for cross-sectional area, but this is not to be relied on. The same formula may appear in different works featuring in one Q for Quantity of discharge, and in another D for quantity of Discharge, A for Area of cross-section, and S for area of cross-Section, and so forth. In most textbooks on hydraulics Q means 'total' discharge, while q means 'unit-width' discharge. The same letter may also stand for different things, S being either cross-Section or Slope. There are also other complications. The letters sometimes change on a linguistic basis, depending on whether the particular book we are dealing with is written in English, French or another language. In an English-language publication, H is liable to stand for Head, the distance of a pipe below the water surface in the header tank serving it, and in a French one for Hauteur, the depth of the water in an open channel. In Haberey (p. 97) U stands for *benetzter Umfang*, Perimeter, while in Germain de Montauzan (p. 345) U stands for velocity of flow. Another problem is caused by the use of both Imperial and metric (often abbreviated to SI – *Système International d'Unités*) figures. Sometimes this does not matter. A gradient of 3 feet in 1,000 feet and one 3 metres in 1,000 metres come to the same thing, and if '3' is entered into the equation, it will work, no matter which measurement standard it is based on. Sometimes it does matter. Flow velocities of 2 feet per second and 2 metres per second are obviously very different things. One must therefore be careful. Nearly all modern archaeological and scientific books use exclusively metric figures. So do a lot of the engineering handbooks, but not all of them, particularly the older ones, for there are many materials that engineers often deal with, such as planks or pipes, that are still commercially produced in many countries to non-metric standards. One therefore has to be careful whether one is dealing with metres or feet. Unfortunately, sometimes care is not enough, for the book simply may not say, entailing a lot of hunting back and forth in an attempt to find what standard is being used.

Yet another complication is to be found in the fact that the same equation may be represented in two (or more) forms that come, mathematically, to the same thing. Thus, a number to the power of one-half is just another way of saying its square root, so $a^{1/2}$ and $\sqrt{a}$ are the same thing, and, given the different symbols sometimes used, $V^{1/2}$ and $\sqrt{u}$ might well be the same thing also.

By now the non-technical reader is no doubt beginning to despair of ever finding his way along a trail so ill-blazed (and one that was supposed to be scientifically precise, at that!), so I will add but one more worry. Sometimes the author will spell out in the equation all the various steps that have to be gone through to get the answer; sometimes he will combine several of these steps to produce an equation that is simpler and a better exposition of the essentials – but which, in appearance, looks very different from the other one. Thus, for example, de Montauzan (p. 345) gives as the formula for working out velocity of flow in an aqueduct channel

$$U = C\sqrt{RI}$$

Ten pages previously, on p. 335, he gives

$$V = 63.25 \sqrt{\frac{LH\,I}{L + 2H}}$$

The two formulae are the same. U and V both stand for the mean velocity of flow. C is a 'resistance' coefficient that, in this particular case, has the value 63.25. For the rest, the trouble is the hydraulic radius. This figure is produced by dividing the cross-sectional area of the part of the channel occupied by the water by the 'wetted perimeter'. The area is calculated by multiplying the width (L, largeur) by the depth (H, hauteur): L × H or LH. The 'wetted perimeter' is calculated by adding together the width (L) of the bottom and the height (H) of the two sides, giving us L + 2H. The entire calculation for the hydraulic radius (R) is therefore

$$R = \frac{LH}{L+2H}$$

Multiplying this by the slope (I, inclinaison), gives us

$$RI, \text{ or } \frac{LH \times I}{L + 2H}$$

The square root of this is thus $\sqrt{RI}$, or

$$\sqrt{\frac{LH\,I}{L + 2H}}$$

and so $U = C\sqrt{RI}$ is the same thing as

$$V = 63.25\sqrt{\frac{LH\,I}{L + 2H}},$$

the only real difference being that in the first version the assumption is made that R has already been calculated, and in the second the various steps for doing so are all spelled out. We may note that all this concerns only the different ways of representing the formula itself. No actual figures from a real aqueduct have been applied to it, this being the next step that de Montauzan takes, after the example printed above. The reader, if he has subconsciously thought of 'the formula' as a kind of obscure, magic, but infallible device that will 'give you the answer', may find all the above daunting. Engineers and those accustomed to handling formulae experience no difficulty, being used to the situation, but it is hoped that the above caveats and comments may be of use to the inexperienced.

I give below some of the formulae most useful in the context of this book, with comments.

Siphons: static pressure

Static pressure in the pipes is generated by the distance the siphon (or the particular part of it under consideration) lies below the free surface water level

(H, the head). This pressure is normally expressed in terms of the vertical column of water supported, in metres (e.g. a head of 12 m).

Another way is to express it in terms of atmospheres. Roughly, the pressure increases by one atmosphere for every 10 m, so a siphon that dips 50 m below the header tank has, at the bottom, a pressure of 5 atmospheres in the pipes. More precisely, a pressure of one atmosphere is the equivalent to the pressure produced by a column of water 10.33 m high (pressure, 101.3 KiloPascals; 14.7 lb per sq. inch): conventionally, this is atmospheric pressure measured at sea level.

Resistance of pipes to static pressure

The formula is

$$2R\varepsilon = 1000\ DH; \text{ or } H = \frac{2R\varepsilon}{1000\ D}$$

where ε = thickness of the pipe (m).
D = internal pipe diameter (m).
H = pressure (measured in metres of water column) at which pipe will burst.
R = coefficient of resistance of material from which pipe is made (for lead, = 1.35 kg per mm^2, or 1,350,000 kg per m^2

Example: for pipe made of lead, of internal diameter 0.579 m, and thickness of sidewalls 0.006 m, to find the bursting pressure, H:

$$H = \frac{2 \times 1,350,000 \times 0.006}{1000 \times 0.579}$$
$$= 27.9 \text{ m.}$$

Thus the pipe will sustain pressures of up to 28 m below the free surface, or just under 3 atmospheres, before bursting. (GdM 182: the very thin walls of the pipe are to be noted; the pipe concerned is that recommended by Vitruvius for siphons.)

Velocity of flow in submerged pipes or orifices

The mean velocity of flow in a pipe is, for our purposes, governed by the head under which it operates. The formula is

$$\bar{V} = \sqrt{2gh}$$

where $\bar{V}$ = mean velocity, g = acceleration by gravity [= 9.81 (m/sec^2)] and h = the head.

To give the quantity of discharge, one multiplies mean velocity by area (A) of cross-section:

$$Q = A\sqrt{2gh}$$ (Grimal, 82; Herschel, 216-19)

Example: for a pipe drawing water from a header tank, at a level 1.5 m. below the water surface, the velocity will be

$$\bar{V} = \sqrt{2gh};\ \bar{V} = \sqrt{2 \times 9.81 \times 1.5};\ = 5.42\ \text{m/sec}$$

and supposing the pipe to have an interior diameter of 25 cm, that gives a cross-sectional area (πr^2) of 0.049 m^2, and a discharge ($Q = A\bar{V}$) of

$$0.049 \times 5.42 = 0.266\ \text{m}^3\text{/sec.}$$

Velocity of flow in open channels
(The Chezy/Bazin/Manning formulae)

This is probably the most complicated formula which readers of this book will encounter. The figure with which they will most often be concerned is the discharge, the volume of water provided daily by a given aqueduct, and the formula for finding that is simplicity itself: $Q = A\bar{V}$. To get the quantity, multiply mean velocity of flow by cross-sectional area (always remembering that what counts is not the dimensions of the dry conduit as excavated by an archaeologist, but the part actually occupied by water, depending on its depth – indicated, usually, by traces of incrustation).

Example: Conduit is 1.0 m wide, with water running at a depth of 0.30 m and a mean velocity of 1.5 m/sec.

$$\begin{aligned} Q &= A\bar{V} \\ &= 1.0 \times 0.30 \times 1.5 \\ &= 0.45\ \text{m}^3\text{/sec.} = 27\ \text{m}^3\text{/min} = 1{,}620\ \text{m}^3\text{/hour} \\ &= 38{,}880\ \text{m}^3 \text{ per 24 hour day} \end{aligned}$$

This is easy enough, and finding A, the cross-sectional area, is merely a matter of measuring bottom and sides, and then multiplying. But how does one find the mean velocity of flow?

The Chezy formula

The formula for this was established by the French engineer A. Chezy (1718-1782):

$$\bar{V} = C\sqrt{R \times S}$$

Here R is the hydraulic radius of the conduit, as defined above (i.e. R = A/P, where A is the cross-sectional area, and P the wetted perimeter); S is the slope (i.e. the aqueduct gradient, expressed in metres per 1,000 m. If a vertical drop of 0.46 m has been measured over a 800 m length, then 0.46/800 inserted into the equation will give the right answer); and C is the problem. C is a coefficient. That is, it is not anything measured on the aqueduct itself, but an arbitrarily chosen number ensuring that, when the others are multiplied by it, it gives a result that forms the information desired. What then, is this number? Chezy thought it was a constant, but his work has been refined by two later engineers, the Irishman R. Manning (1816-1897) and the Frenchman H. Bazin (1829-1917), both of whom realised that the value of C varies with the degree of roughness of the channel surface (and the degree of friction it causes), to say nothing of the influence of other factors as well.

To find what this coefficient is, one therefore has to look up sets of tables, as published in any technical handbook, in which all possible channel linings are arranged in order of smoothness, from plastic to brickwork to weedy rivers. Although the principle is the same in the formulae of Manning and of Bazin, it is important to realise that their work is in parallel, resulting in two alternative formulae, both based on Chezy, and either capable of giving the required answer, though by a different procedure. Engineers, especially in America, seem to find the Manning formula easiest to handle, others prefer the Bazin version.

The Bazin formula

This is basically still the Chezy formula:

$$\bar{V} = C\sqrt{R \times S}$$

except that we now have a set of values for C. In theory one should look them up in the tables, but since the tables exist simply out of a need to cover all conceivable technical eventualities, while all we need is the figure for smoothly-plastered concrete (which is what just about all aqueducts were made of), the reader may take it as a rough guide that the coefficient to use is something between 60 and 80, the higher figure being for smoother, new work.

Example: To go back to the calculation previously quoted for the Marcia, we may now substitute values in the formula

$$\bar{V} = 63.25 \sqrt{\frac{LH \times I}{L + 2H}} \qquad \text{(GdM 335)}$$

$\bar{V}$ = mean or average velocity of flow. 63.25 is the coefficient (as here established by Blumensthil for this particular aqueduct). L = largeur (width of channel), or 1.50 m, and H = hauteur (depth of water), or 1.0 m. I, inclinaison (slope) has been measured at 0.46 m over a horizontal distance of 800 m. The entire formula thus works out as

$$\bar{V} = 63.25 \sqrt{\frac{1.50 \times 1.0}{3.50} \times \frac{0.46}{800}}$$

$$= 63.25 \times 0.0155$$
$$= 0.987 \text{ m/sec}$$

For further use of the Bazin formula, see GdM 172, Cherchel 87, Haberey 97, and Burdy (ch.8 n. 23 p. 442), 36-7.

The Manning formula

The Manning formula differs from Bazin in two main ways. First, it avoids some of the steps in the Bazin formula by using the hydraulic radius to the power of two-thirds (i.e. the cube root squared). Second, it uses a different set of coefficients, ranging from 0.013 to 0.015 (in increasing order of roughness of the concrete) and conventionally indicated by the symbol n (Horton's values).* As applied to the formula, these are transformed in the SI (metric) system of units by dividing them into 1, the result being referred to conventionally as K (for Kutter,

* Full table in L.C. Urquhart, *Civil Engineering Handbook*[2] (New York, 1940), 322-3.

the author of this practice).

Thus K = 1/n, and where n = 0.013, K = 76.9; where n = 0.015, K = 66.6. It will be noted that n numbers get higher as the surface gets rougher, and K numbers lower. Hauck gives the K coefficient as ranging from 100 for very smooth concrete to 55 for very rough. The Manning formula itself is

$$\bar{V} = K R^{2/3} S^{1/2} \text{ m/sec.}$$

Where $\bar{V}$ is average velocity, R the hydraulic radius, and S the slope ($R^{2/3}$ is the cube root of R squared, and $S^{1/2}$ is the square root of S).

The equivalent form in Imperial units is

$$\bar{V} = \frac{1.486}{n} R^{2/3} S^{1/2} \text{ ft/sec.}$$

However, although the Manning formula is more popular with engineers, the non-technical reader will probably find Bazin easier to handle.

For further details on all three – Chezy, Bazin and Manning – see:

L.C. Urquhart, *Civil Engineering Handbook*[2] (New York, 1940), 319, 321.
P.S. Barna, *Fluid Mechanics for Engineers* (London, 1957), 78.
J.R.D. Francis, *A Textbook of Fluid Mechanics for Engineering Students* (London, 1958), 223, 255.
R.W. Fox and A.T. McDonald, *Introduction to Fluid Mechanics* (New York, 1985), 522-3.

In these works, as in others I have quoted, the reader will find the relevant facts and figures often different from what I have stated, as other minor factors are taken into consideration; and if, having expected to find himself in a world of scientific precision and clear uniformity, he has the impression of being instead in something looking uncommonly like a jungle, he will have at least my sympathy.*

* My thanks are due to Prof. D.R. Townsend of the Faculty of Engineering, University of Ottawa, who very kindly vetted the text of this appendix to ensure technical accuracy. Prof. Townsend points out that the Chezy formula was found to be inaccurate for slow velocities in smooth conduits, such as aqueducts, and this led to the development of the Manning equation. I include the Chezy (Bazin) version here as it has been much used (and still is) in aqueduct studies, and in any case I suspect it is sufficiently accurate for the purposes of most readers of this book.

Abbreviations

Abbreviations used are as listed in *L'Année Philologique*, plus the following:

Ashby	Ashby, T., *The Aqueducts of Ancient Rome* (Oxford, 1935).
Birebent	Birebent, J., *Aquae Romanae* (Algeria, 1962).
Callebat	Callebat, L. (ed.), *Vitruve VIII* (Paris [editions Budé] 1973).
Cherchel	Leveau, P. & Paillet, J-L., *L'alimentation en eau de Caesarea de Mauretanie et l'aqueduc de Cherchel* (Paris, 1976).
Dolaucothi	Jones, G.B.D., Blakeney, I.J. & MacPherson, E.C.F., 'Dolaucothi: the Roman aqueduct', *Bulletin of the Board of Celtic Studies, Univ. of Wales* 19 (1960), 71-80.
Esperandieu	Esperandieu, E., *Le Pont du Gard* (repr. Paris, 1979).
Fahlbusch	Fahlbusch, H., 'Vergleich antiker griechischer und römischer Wasserversorgungsanlagen', *Leichtweiss-Institut für Wasserbau, Mitteilungen*, Heft 73 (1982).
FCAS	Hodge, A. Trevor (ed.), *Future Currents in Aqueduct Studies* (Leeds, 1991).
GdM	de Montauzan, G., *Les aqueducs antiques de Lyon* (Paris, 1909).
Grewe	Grewe, K., *Planung und Trassierung römischer Wasserleitungen* (Scriftenreihe der Frontinus-Gesellschaft, Supplementband I; Verlag Chmielorz GmbH., Wiesbaden, 1985).
Grewe, *Atlas*	Grewe, K., *Atlas der römischen Wasserleitungen nach Köln* (Rheinland-Verlag GmbH., Köln, 1986).
Grimal	Grimal, P. (ed.), *Frontinus: les aqueducs de la ville de Rome* (Paris [editions Budé] 1961).
Haberey	Haberey, W., *Die römischen Wasserleitungen nach Köln* (Bonn, 1972).
Hanson	Hanson, J., *Municipal and Military Water Supply and Drainage in Roman Britain* (University of London, Institute of Archaeology, 1970), unpublished dissertation, available on microfilm.
Hauck	Hauck, G., *The Aqueduct of Nemausus* (Jefferson, N.C./ London, 1988).
HD	Smith, N.A.F., *A History of Dams* (London, 1971).
Herschel	Herschel, C., *Frontinus: the two books on the water supply of Rome* (New England Water Works Association: Boston, 1973 [repr. of 1899 ed.]).
JEAR	Boucher, J.-P. (ed.), *Journées d'études sur les aqueducs romains* (Paris, 1983).

Kretzschmer	Kretzschmer, F. *La technique romaine* (Brussels, 1966).
Landels	Landels, J.G., *Engineering in the Ancient World* (London, 1978).
MAGR	Grenier, A., 'Les monuments des eaux', *Manuel d'archéologie gallo-romaine* 4, 1 (Paris, 1960).
Mitt. L. Inst.	*Leichtweiss-Institut für Wasserbau der Technischen Universität Braunschweig: Mitteilungen.*
Pace	Pace, P., *Gli Acquedotti di Roma* (Rome, 1983).
Samesreuther	Samesreuther, E., *Römischen Wasserleitungen in den Rheinlanden*: 26. *Bericht der Romisch-Germanischen Kommission* (Berlin, 1938).
SAT	Forbes, R.J., *Studies in Ancient Technology*, vol. 1 (Leiden, 1955) (Note: The abbreviation *SAT* is used to refer to vol. 1. Other volumes are shown as *SAT* 2, etc.)
Smith	Smith, N.A.F., 'Attitudes to Roman engineering and the question of the inverted siphon', *History of Technology*, vol. 1 (1976), 45-71.
SRA	Hodge, A. Trevor, 'Siphons in Roman aqueducts', *PBSR* 51 (1983).
Van Deman	Van Deman, E., *The Building of the Roman Aqueducts* (Washington, 1934).
Ward-Perkins	Ward-Perkins, J.B., 'The aqueduct of Aspendos', *PBSR* 23 (1955), 115–23.
WAS	*Die Wasserversorgung antiker Städte*, vol. 1 (ed. G. Garbrecht; 1986 (repr.), R. Oldenbourg Verlag: Munich; title, *Wasserversorgung im antiken Rom*); vol. 2 (Verlag Philipp von Zabern: Mainz, 1987); vol. 3 (von Zabern: Mainz, 1988). All three volumes form part of a *Geschichte der Wasserversorgung* (ongoing), published by the Frontinus Gesellschaft.
White	White, K.D., *Greek and Roman Technology* (London, 1984).

Supplementary Bibliography for Second Edition, 2002

For further and fuller bibliographies the reader is particularly referred to titles marked *.

Aicher, P.J., *Guide to the Aqueducts of Ancient Rome* (Wauconda, Ill., 1995), biblio., pp. 171-4.

Andrieu, J.-L., *Béziers, l'aqueduc romain* (Besançon, 1990).

Blackman, D.R. & Hodge, A. Trevor, *Frontinus' Legacy* (Ann Arbor, 2000).

*Bruun, Ch., *The Water Supply of Ancient Rome* (Helsinki, 1991).

Burdy, J., *Lyon: l'aqueduc romain du Gier* (Lyon, 1996).

Burés Vilaseca, L., *Les estructures hidràuliques a la ciutat antiga: l'exemples d'Empúriés* (Barcelona, 1998).

Chanson, H., 'Hydraulics of Roman aqueducts … Cascades …', *AJA* 104 (2001), 47-92.

Clamagiraud, E. et al., 'L'aqueduc de Carthage', *Journal de l'Huile Blanche* 6 (1990), 423-31.

Crouch, D.P., *Water Management in Ancient Greek Cities* (Oxford, 1993), biblio., pp. 347-66.

De Laine, J., *The Baths of Caracalla* (*JRA*, Portsmouth, RI, 1997).

*Evans, H.B., *Water Distribution in Ancient Rome* (Ann Arbor, 1994).

Fabre, G. et al., *L'Aqueduc de Nîmes et le Pont du Gard* (Nîmes, 1991).

Grewe, K., *Licht am Ende des Tunnels* (Mainz, 1998), biblio., pp. 203-11.

Jansen, C.M.J. (ed.), *Cura Aquarum in Sicilia* (Transactions; Leiden, 2000).

Kessener, P., 'The aqueduct at Aspendos and its inverted siphon', *JRA* 13 (2000), 104-32.

Kolowski-Ostrow, A. et al., 'Water in the Roman town', *JRA* 10 (1997), 181-91.

Leveau, Ph., 'The Barbegal water mill', *JRA* 9 (1996), 157-63.

Lewis, M.J.T., 'Vitruvius and Greek aqueducts', *PBSR* 67 (1999), 137-53.

Liberati Silverio, A.M. & Pisani Sartorio, G. (eds), *Il Trionfo dell'Aqua* (Transactions; Rome, 1992).

Lombardi, L. & Corazza, A., *Le Terme di Caracalla* (Rome, 1995).

Malissard, A., *Les Romains et l'eau* (Paris, 1994).

Meshel, Z. et al., *The Water Supply of Susita* (Tel Aviv/ Lubeck, 1996).

Smith, N.A.F., 'The Pont du Gard and the aqueduct of Nîmes', *Trans. Newcomen Soc.* 62 (1990-1), 67-84.

Taylor, R., 'Torrent or trickle? The Aqua Alsietina …', *AJA* 101 (1997), 465-93.

*Wikander, O. (ed.), *Handbook of Ancient Water Technology* (series: 'Technology and Change in History'; Leiden, 2000) (easily the fullest account), biblio., pp. 661-702.

Wilson, A., 'Deliveries *extra urbem* … aqueducts and countryside', *JRA* 12 (1999), 314-30.

Wilson, A., 'The aqueducts of Italy and Gaul' (rev.), *JRA* 13 (2000), 597-604.

*Wilson, R.J.A., '*Tot aquarum* … recent studies on aqueducts and water supply', *JRA* 9 (1996), 5-29.

Bibliography

1. General reference works

Ancient sources

2. Frontinus and commentaries
3. Vitruvius and commentaries

General works on hydraulics and aqueducts

4. General hydraulics
5. Modern handbooks on hydraulics
6. Hydrology
7. General works on aqueducts
8. General works on technology and architecture with material on aqueducts
9. Dams
10. Wells
11. Mines
12. Pulleys

Non-Roman waterworks

13. Qanats
14. Early Greek
15. Eupalinos' tunnel, Samos
16. Hellenistic
17. Etruscan
18. Irrigation

Roman waterworks

19. Roman waterworks in general
20. Surveying and construction
21. Construction materials and techniques
22. Siphons
23. Discharge and lime incrustation (*sinter*)
24. Pipes and pipelines
25. Health and lead poisoning
26. Filtration plants/cisterns
27. Architectural features
28. Taps
29. Drainage

Industrial aqueducts

Roman waterworks: a geographical survey

For abbreviations see pp. 357-8.

1. General reference works

1.1 Carcopino, J., *Daily Life in Ancient Rome* (Harmondsworth, 1956).
1.2 Daremberg, C. & Saglio, E., *Dictionnaire des antiquités grecques et romaines* (Hachette: Paris, 1873-1917).
1.3 Dinsmoor, W.B., *Architecture of Ancient Greece* (London, 1952 [reprint]).
1.4 Hammond, N.G.L., *Atlas of the Greek and Roman World in Antiquity* (Noyes: Park Ridge, N.J., 1981).
1.5 Oleson, J., *Bronze Age, Greek and Roman Technology: a select annotated bibliography* (New York/London, 1986).
1.6 Pauly-Wissowa, *Realencyclopädie der klassischen Altertumswissenschaft*.
1.7 Stillwell, R., *Princeton Encyclopedia of Classical Sites* (Princeton, 1976).
1.8 Ward-Perkins, J.B., *Roman Imperial Architecture* (Harmondsworth, 1981).

Ancient sources

2. Frontinus, De Aquaeductu: commentaries

2.1 Bennett, C.E. (ed.), *Frontinus: The Stratagems, and the aqueducts of Rome* (Loeb Classical Library, Cambridge, Mass. & London, 1925).

2.2 Eck, W., 'Die Gestalt Frontins in ihrer politischen und sozialen Umwelt', *Wasserversorgung im antiken Rom* (1986), 47-62.
2.3 Grimal, P. (ed.), *Frontinus: Les aqueducs de la ville de Rome* (Paris [editions Budé] 1961). [abbrev. Grimal].
2.4 Herschel, C., *Frontinus: the two books on the water supply of Rome* (New England Water Works Association: Boston, 1973 [repr. of 1899 edition]) [abbrev. Herschel].
2.5 Hodge, A. Trevor, 'Frontinus', in H. Temporini & W. Haase (eds.), *Aufstieg und Niedergang der römischen Welt* (Berlin, forthcoming).
2.6 Kunderewicz, C. (ed.), *Iulii Frontini de Aquaeductu Urbis Romae.* (Teubner: Leipzig, 1973).
2.7 Rodgers, R.H., 'An administrator's hydraulics: Frontinus *Aq*. 35-36', *FCAS*, 15-20.
2.8 Tannery, P., 'Frontin et Vitruve', *R.Ph.* 21 (1897).

3. Vitruvius, De Architectura: commentaries

3.1 Callebat, L., 'Le vocabulaire de l'hydraulique dans le livre VIII du de Architectura de Vitruve', *Revue de Philologie* 47 (1973), 313-29.
3.2 Callebat, L., (ed.), *Vitruve VIII* (Paris [editions Budé] 1973). [abbrev. Callebat]
3.3 Fahlbusch, H., 'Remarks on Vitruvius' colluviariae', *IAHR, XVIII Congress Proceedings*, vol. 6 (1979) 669-72. [See 16.5.]
3.4 Fahlbusch, H., 'Vitruvius and Frontinus: hydraulics in the Roman period', in *Hydraulics and Hydraulic Research*, ed. G. Garbrecht (Rotterdam/ Boston, 1987), 23-31.
3.5 Granger, F. (ed.), *Vitruvius on Architecture* (2 vols, Loeb Classical Library, Cambridge, Mass. and London, 1931).
3.6 Grimal, P., 'Vitruve et la technique des aqueducs', *Revue de Philologie* 29 (1945), 162-74.
3.7 Morgan, M.H. (tr.), *Vitruvius: the ten books on architecture* (New York, 1960 [republication of 1914 edition]).
3.8 Mortet, V. 'Recherches critiques sur Vitruve et son oeuvre: IV, Vitruve et l'hydraulique romaine', *Revue archéologique* (1907), 75-83.
3.9 Plommer, W.H., *Vitruvius and Later Roman Building Manuals* (Cambridge, 1973).
3.10 Praeger, F.D., 'Vitruvius and the elevated aqueducts', *History of Technology* 3 (1978), 105-201.

General works on hydraulics and aqueducts

4. General hydraulics

4.1 Argyropoulos, P., 'Ancient Greek hydraulic engineering experience and technique', *IAHR, XVIII Congress Proceedings*, vol. 6 (1979), 688-9. [See 16.5.]
4.2 Bethemont, J., 'Sur les origines de l'agriculture hydraulique', *ICID, XI Congress Proceedings: special session on the history of irrigation, drainage and flood control* (1981), 40-58.
4.3 Bonnin, A., *L'eau dans l'antiquité: l'hydraulique avant notre ère* (Eyrolles: Paris, 1984).
4.4 Camp II, J.McK., 'Water supply and its historical context', *FCAS*, 105-11.

4.5 Crouch, D., 'Modern insights from the study of ancient Greek water management', *FCAS*, 93-103.
4.6 Eck, W., 'Die Wasserversorgung im römischen Reich: Soziopolitische Bedingungen, Recht und Administration', *WAS* 2 (1987), 49-102.
4.7 Garbrecht, G., 'Griechische Beiträge zur Hydrologie und Hydraulik', *Leichtweiss-Institut für Wasserbau, Mitteilungen*, Heft 71 (1981), 19-41.
4.8 Garbrecht, G., 'Interdisciplinary co-operation in archaeological investigations', *FCAS*, 1-6.
4.9 Hodge, A. Trevor, 'A plain man's guide to Roman plumbing', *EMC/Classical Views* (University of Calgary), 27 N.S. 2 (1983), 311-28.
4.10 Hodge, A. Trevor, 'Conclusion', *FCAS*, 163-73.
4.11 Leveau, Ph., 'Research on Roman aqueducts in the past 10 years', *FCAS*, 149-62.

5. *Modern handbooks on hydraulics*

5.1 Babitt, H.E. & Doland, J.J., *Water Supply Engineering* (New York, 1955).
5.2 Barna, P.S., *Fluid Mechanics for Engineers* (London, 1957).
5.3 Fox, R.W. & McDonald, A.T., *Introduction to Fluid Mechanics* (New York, 1985).
5.4 Francis, J.R.D., *A Textbook of Fluid Mechanics for Engineering Students* (London, 1958).
5.5 Gordon, J.E., *Structures* (Harmondsworth, 1987).
5.6 Hobbs, A.T. (ed.), *Manual of British Water Supply Practice* (Cambridge, 1950).
5.7 Merritt, F.S., *Standard Handbook for Civil Engineers*, 2nd ed. (New York, 1976).
5.8 Urquhart, L.C., *Civil Engineering Handbook*, 2nd ed. (New York, 1940).
5.9 Vollmer, E., *Encyclopedia of Hydraulics, Soils and Foundation Engineering* (Amsterdam/London/New York, 1967).

6. *Hydrology*

6.1 Biswas, A.K., *History of Hydrology* (Amsterdam/London, 1970).
6.2 Meinzer, O.E., *U.S. Geological Survey, Water Supply Paper* 494 (1923), 50-4.
6.3 Rouse, H. & Ince, S., *History of Hydraulics* (Iowa, 1957).
6.4 Ven Te Chow (ed.), *Handbook of Applied Hydrology* (New York, 1964).
6.5 Wisler, C.O. & Brater, E.F., *Hydrology* (New York & London, 1959).
See also: 40.13 (Haguenhauer)

7. *General works on aqueducts*

7.1 Ashby, T., *The Aqueducts of Ancient Rome* (Oxford, 1935).
7.2 Belgrand, M., *Les aqueducs romains* (Paris, 1875).
7.3 Fahlbusch, H., 'Vergleich antiker griechischer und römischer Wasserversorgungsanlagen', *Leichtweiss-Institut für Wasserbau, Mitteilungen*, Heft 73 (1982). [abbrev. Fahlbusch].
7.4 Fahlbusch, H., 'Vergleich griechischer und römischer Wasserversorgungsanlagen', *Schriftenreihe der Frontinus-Gesellschaft*, Heft 6 (1982), 45-86.
7.5 Fahlbusch, H., 'Elemente griechisher und römischer Wasserversorgungsanlagen', *WAS* 2 (1987) 133-63.

7.6 Garbrecht, G., 'Mensch und Wasser im Altertum', *WAS* 3 (1988), 13-42.
7.7 Grenier, A., *Manuel d'archéologie gallo-romaine* 4, 1, 'Les monuments des eaux' (Paris, 1960). [abbrev. *MAGR*]
7.8 Haberey, W., *Die römischen Wasserleitungen nach Köln* (Bonn, 1972). [abbrev. Haberey]
7.9 Hodge, A. Trevor, 'Aqueducts', in *Roman Public Buildings*, ed. I.M. Barton (Exeter, 1989), 127-49.
7.10 Hodge, A. Trevor (ed.), *Future Currents in Aqueduct Studies* (Leeds, 1990). [abbrev. *FCAS*]
7.11 *Journées d'étude sur les aqueducs romains*, ed. J.-P. Boucher (Paris, 1983). [abbrev. *JEAR*]
7.12 Leveau, Ph., 'Research on Roman aqueducts in the past 10 years', *FCAS*, 149f.
7.13 Morgan, M.H., 'Remarks on the water supply of ancient Rome', *Proceedings of the American Philological Association* 33 (1902), 30-7.
7.14 Smith, N.A.F., *Man and Water* (London, 1976).
7.15 Squassi, F., *L'Arte Idro-Sanitaria degli Antichi* (Tolentino: Filelefo, 1954).
7.16 Van Buren, 'Wasserleitungen', *Paulys Realencyclopädie* VIII A (1955), 472-5.
7.17 *Wasserversorgung im antiken Rom*. Herausgeber: Frontinus-Gesellschaft e.V. (R. Oldenbourg Verlag: Munich, 1986). [abbrev. *WAS* 1]
7.18 *Die Wasserversorgung antiker Städte* (vol. 2). Herausgeber: Frontinus-Gesellschaft e.V. (Philipp von Zabern: Mainz am Rhein, 1987). [abbrev. *WAS* 2]
7.19 *Die Wasserversorgung antiker Städte* (vol. 3). Herausgeber: Frontinus-Gesellschaft e.V. (Philipp von Zabern: Mainz am Rhein, 1988). [abbrev. *WAS* 3]
See also: 45.26, 45.27, 45.28, 45.31.

8. General works on technology and archaeology with material on aqueducts

8.1 Adam, J.-P., *La construction romaine* (Paris, 1984), ch. 10, 257-86.
8.2 Blake, M., *Roman Building Construction in Italy from Tiberius through the Flavians* (Washington, 1959).
8.3 Drower, M.S., 'Water supply, irrigation and agriculture' in *A History of Technology*, ed. C. Singer et al. (Oxford, 1954-58), vol. 1, 525.
8.4 Forbes, R.J., *Studies in Ancient Technology* (Leiden, 1955), vol. 1, 1-94. [abbrev. *SAT*]
8.5 Gordon, J.E., *Structures* (Harmondsworth, 1978).
8.6 Healey, J.F., *Mining and Metallurgy in the Greek and Roman World* (London, 1975). [for tunnelling and geology]
8.7 Hodges, H., *Technology in the Ancient World* (London, 1970).
8.8 Johnson, A., *Roman Forts* (London, 1983).
8.9 Kretzschmer, F., *La technique romaine*, tr. J. Breuer & F. Ulrix (Brussels, 1966), 47-63. [abbrev. Kretzschmer]
8.10 Lamprecht, H.O., *Caementitium Opus* (Dusseldorf, 1984).
8.11 Landels, J.G., *Engineering in the Ancient World* (London, 1978), esp. 34-57. [abbrev. Landels]
8.12 McKay, A.G., *Houses, Villas and Palaces in the Roman World* (London, 1975). [baths]

8.13 Merckel, C., *Die Ingenieurtechnik im Altertum* (Berlin, 1899).
8.14 Nash, E., *Pictorial Dictionary of Rome*, rev. ed. (London, 1986).
8.15 Neuburger, A., *Technical Arts of the Ancients* (New York, 1930), esp. 409-50.
8.16 Singer, C., Holmyard, E.J., Hall, A.R. & Williams, T.I., *History of Technology*, vol. 1 (Oxford, 1954) 520-55; vol. 2 (Oxford, 1956), 663-92.
8.17 Ward-Perkins, J.B., *Etruscan and Roman Architecture* (Harmondsworth, 1970).
8.18 White, K.D., *Greek and Roman Technology* (London, 1984). [abbrev. White]
8.19 White, K.D., *Roman Farming* (London, 1970). [irrigation]

9. Dams

9.1 Benoît, F., 'Le barrage et l'aqueduc romain de Saint-Rémy en Provence', *REA* 37 (1935), 331-9.
9.2 Casado, C.F., 'Las Presas Romanas en España', *Revista de Obras Publicas*, VII Congreso Internacional de Grandes Presas (1961), 361-3.
9.3 de Castro Gil, J., 'El Pantano de Proserpina', *Revista de Obras Publicas* (1933), 449-54.
9.4 Garbrecht, G., *Historische Talsperren* (Stuttgart, 1991).
9.5 Giovannoni, G., *L'Architettura nei Monasteri Sublacensi* (Rome, 1904).
9.6 Gomez, R.C., 'Chronologia des las Fabricas no Romanas del Pantano de Proserpina', *Revista de Obras Publicas* (1943), 558-61.
9.7 Grewe, K., 'Merida', *WAS* 3 (1988), 204-6. [Proserpina and Cornalvo]
9.8 Hathaway, G.A., 'Dams: their effect on some ancient civilizations', *Civil Engineering*, vol. 28, no. 1 (January, 1958), 58-63.
9.9 Murray, W.M., 'The ancient dam of the Mytikas valley', *AJA* 88 (1984), 195-203.
9.10 Murray, W.M., 'A dam below Glosses', *AJA* 86 (1982), 279.
9.11 Rolland, H., 'Fouilles de Glanum', *Gallia*, Suppl. 1 (1946), 45.
9.12 Schnitter, N., 'A short history of dam engineering', *Water Power* (April, 1967), 142-8.
9.13 Schnitter, N., 'The evolution of the arch dam', *Int. of Water Power and Dam Construction*, vol. 28 (1976), no. 10, 34-40; no. 11, 19-21.
9.14 Schnitter, N., 'Römische Talsperren', *Antike Welt* 2 (1978), 25-32.
9.15 Schnitter, N., 'Antike Talsperren in Anatolien', *Leichtweiss-Institut für Wasserbau, Mitteilungen*, Heft 64 (1979), 8.
9.16 Schnitter Reinhardt, N., 'Les barrages romains', *Doss. Arch.*, no. 38 (Oct./Nov., 1979), 20-5.
9.17 Schnitter, N., 'Altgriechischer Wasserbau', *Schweizer Ingenieur und Architect*, Heft 24 (1984), 479-86.
9.18 Smith, N.A.F., *A History of Dams* (London, 1971). [abbrev. *HD*]
9.19 Smith, N.A.F., 'The Roman dams of Subiaco', *Technology and Culture* 22 (1970), 58-68.
9.20 Vita-Finzi, C., 'Roman dams in Tripolitania', *Antiquity* 35 (1961), 14-20.
9.21 Vita-Finzi, C. & Brogan, O., 'Roman dams on the Wadi Megenin', *Libya Antiqua* 2 (1965), 65-71.

10. Wells

10.1 Dauvergne, R., *Sources minerales, thermes gallo-romains et occupation du sol aux Fontaines-Salées* (Paris, 1944).

10.2 Hawkes, C.F.C. & Hull, M.R., *Camulodunum* (Oxford, 1947).
10.3 *Hesp.* 23 (1954), 87-101.
10.4 Jacobi, H., 'Die Be- und Ent-wasserung unserer Limeskastelle', *Saalburg Jhb.* 8 (1934), 32-60.
10.5 Maiuri, A., *Pompeii: Pozzi e Condotture d'Aqua nell'Antica Città, N.Sc.* (1931), 546-76.
10.6 Mertens, J., 'Puits antiques à Elewijt et les puits romains en bois', *Antiquité Classique* 20 (1951), 97.
10.7 Miller, S.G., 'Menon's cistern', *Hesp.* 43 (1974), 194-9, 228-9.
10.8 Oswald, F., 'The pottery of a Claudian well at Margidunum', *JRS* 13 (1923), 114-26.
10.9 Pimienta, J., *Le captage des eaux souterraines* (Paris, 1972).
See also: 7.14, 18.26, 37.9, 37.10, 46.2.

11. Mines

11.1 Ardaillon, E., *Les mines du Laurion* (Paris, 1897).
11.2 Davies, O., *Roman Mines in Europe* (Oxford, 1935), 24-5.
11.3 Healy, J.F., *Mining and Metallurgy in the Greek and Roman World* (London, 1975).
11.4 Kalcyk, H., 'Water in ancient mining', *Leichtweiss-Institut für Wasserbau, Mitteilungen*, Heft 82 (1984), 20f.
11.5 Leger, A., *Les travaux publics, les mines et la metallurgie au temps des Romains* (Paris, 1875).
11.6 Murray, G.W., 'Gold mines of the Turin Papyrus', in Ball, J., *Egypt in the Classical Geographers* (Cairo, 1942), 180-2.
11.7 Nougier, L.R., *Géographie humaine préhistorique* (Paris, 1959).
11.8 *Thonglushan: a pearl among ancient mines*, ed. Huangshi Museum, Hubei; Chinese Society of Metals, Publication Committee; and Archaeometallurgy Group, Beijing University of Iron and Steel Technology (Cultural Relics Publishing House: Beijing, 1980).
11.9 Tylecote, R.F., *Metallurgy in Archaeology* (London, 1962).
11.10 Wertime, T.A., 'A metallurgical exploration through the Persian desert', *Science* 159 (1968), 927-35.

12. Pulleys

12.1 *Expedition archéologique de Délos* 8, 2, 348.
12.2 Shaw, J.W., 'A double-sheaved pulley block from Kenchreai', *Hesp.* 35 (1967), 389-401.
12.3 Sparkes, B.A., 'Illustrating Aristophanes', *JHS* 95 (1975), 130-1.

Non-Roman waterworks

13. Qanats

13.1 Bemont, F., 'L'irrigation en Iran', *Annales de Géographie* 52 (1961), 597-620.
13.2 Butler, M.A., 'Irrigation in Persia by kanats', *Civil Engineering* 3 (1933), 69-73.
13.3 Cornet, A., 'Essai sur l'hydrologie du Grand Erg occidental et des régions

limitrophies: les foggaras', *Travaux de l'Institut de Recherches Sahariennes* (1952), 71-122.

13.4 English, P.W., 'The origin and spread of qanats in the Old World', *Proc. of the American Philosoph. Soc.* 112 (1968), 170-81.

13.5 Feylessoufi, E., *Eaux souterraines, kanats et puits profonds en Iran* (Teheran Univ. Press, 1959).

13.6 Goblot, H., *Les qanats: une technique d'acquisition de l'eau*, ed. Mouton/École des Hautes Études en Sciences Sociales (Paris, 1979).

13.7 Krenkov, 'The construction of subterraneous water supply during the Abbasid Kaliphate', *Transactions of the Scottish Oriental Society* (Univ. of Glasgow, 1951), 23-32.

13.8 Kuros, G.R., *Irans Kampf um Wasser* (Berlin, 1943).

13.9 Mazaheri, M.A., 'Le traité d'exploitation des eaux souterraines d'al Kardji', *Archives Internationales d'Histoire des Sciences* 18 (juin-dec, 1965), 300-1.

13.10 Mazaheri, Aly, *al Karagi, La civilisation des eaux cachées: traité de l'exploitation des eaux souterraines (composé en 1017), texte établi, traduit et commenté* (IDERIC, Université de Nice: Nice, 1973).

13.11 Michel, G., 'Kanate in der Volksrepublik China', *Schriftenreihe der Frontinus-Gesell.*, Heft 11 (1988), 119-30.

13.12 Noel, Col. E., 'Qanats', *Journal of the Royal Central Asian Society* 41 (1944), 191-202.

13.13 Wilson, Sir A.T., *The Persian Gulf* (London, 1954).

13.14 Wolski, K., 'Les Karez', *Folia Orientalia* (Univ. of Cracow) 6b (1964), 179-204.

13.15 Wulff, H.E., 'The qanats of Iran', *Scientific American* 218, 4 (1968), 94-107.

14. Early Greek waterworks construction

14.1 Burns, A., 'Ancient Greek water supply and city planning: a study of Syracuse and Acragas', *Technology and Culture* 15.3 (July, 1974), 389-412.

14.2 Camp, J.M., *The Water Supply of Ancient Athens from 3000 to 86 B.C.*, Ph.D. thesis (Princeton, Jan. 1977; University Microfilms International).

14.3 Dellbruck, R. & Volkmuller, K.G., 'Das Brunnenhaus des Theagenes', *AM* 25 (1900), 23-33. [Megara]

14.4 Dorpfeld, W., 'Die Ausgrabungen an der Enneakrounos', *AM* 19 (1894), 144-6.

14.5 Dunkley, B., 'Greek fountain-buildings before 300 B.C.', *BSA* 36, Session 1935-36 (1939), 142-204.

14.6 Fahlbusch, H., 'Vergleich griechischer und römischer Wasserversorgungs-anlagen', *Schriftenreihe der Frontinus-Gesellschaft*, Heft 6, *Symposium über die historische Entwicklung der Wasserversorgungstechnik* (Köln, 1982).

14.7 Glaser, F., 'Griechische Brunnenhäuser', *Leichtweiss-Institut für Wasserbau, Mitteilungen*, Heft 71 (1918), 177-203.

14.8 Graber, F., 'Die Enneakrounos', *AM* 30 (1905), 1-64.

14.9 Graber, F., 'Die Wasserleitung des Peisistratos und die Wasserversorgung des alten Athen', *Zentralblatt der Bauverwaltung* (1905).

14.10 Gruben, G., 'Das Quellhaus von Megara', *AD* 19 (1964), 37-41.

14.11 Kienast, H., 'Zur Wasserversorgung griechischer Staate', *Wohnungsbau im Altertum* (Berlin, 1979).

14.12 Knauss, J., Heinrich, B. & Kalcyk, H., *Die Wasserbauten der Minyer – die*

älteste Flussregulierung Europas, Institut für Wassermengenwirtschaft und Versuchsanstalt für Wasserbau, Munich, Berichte, no. 50 (1984).
14.13 Lang, M., *Waterworks in the Athenian Agora* (Princeton, 1968).
14.14 Pfriemer, U., 'Das grosse Brunnenhaus in Megara', *Leichtweiss-Institut für Wasserbau, Mitteilungen*, Heft 71 (1981), 269-74.
14.15 Thompson, H.A. & Wycherley, R.E., *The Agora of Athens (The Athenian Agora)*, vol. 14 (Princeton, 1972), 198.
14.16 Tomlinson, R.A., 'Perachora: the remains outside the two sanctuaries', *BSA* 64 (1969), 195-232.
14.17 Tomlinson, R.A., 'The Perachora waterworks: addenda', *BSA* 71 (1976), 147-8.
14.18 Travlos, J., *Pictorial Dictionary of Athens* (London, 1971).
See also: 4.4.

15. Eupalinos' tunnel, Samos

15.1 Burns, A., 'The tunnel of Eupalinos and the tunnel problem of Hero of Alexandria', *Isis* 62 (1971), 172-85.
15.2 Fabricius, E., *RE* 6, 1159, s.v. 'Eupalinos'.
15.3 Fabricius, E., 'Altertumer auf der insel Samos', *AM* (1884), 171ff.
15.4 Goodfield, J., 'The tunnel of Eupalinos', *Scientific American* 210 (June, 1964), 104-10.
15.5 Goodfield, J. & Toulmin, S., 'How was Eupalinos' tunnel aligned?', *Isis* 56 (1965), 45-56.
15.6 Jantzen, U., Kienast, H. & Felsch, R.S.C., 'Die Wasserleitung des Eupalinos (Samos 1971)', *AA* 88 (1973), 72-89; 401-4.
15.7 Jantzen, U., Kienast, H. & Felsch, R.S.C. 'Die Wasserleitung des Eupalinos (Samos 1973/74)', *AA* 90 (1975), 19-35.
15.8 Kastenbein, W., 'Untersuchungen am Stollen des Eupalinos auf Samos', *AA* (1960), 178-98.
15.9 Kienast, H., 'Die Wasserleitung des Eupalinos auf Samos', *Leichtweiss-Institut für Wasserbau, Mitteilungen*, Heft 64 (1979), 31f.
15.10 Kienast, H., 'Bauelemente griechischer Wasserversorgungsanlagen', *Leichtweiss-Institut für Wasserbau, Mitteilungen*, Heft 71 (1981), 45-67.
15.11 Kienast, H., 'Hezekiah's water tunnel and the Eupalinos tunnel of Samos, a comparison', *Leichtweiss-Institut für Wasserbau, Mitteilungen*, Heft 82 (1984), 20f.
15.12 Kienast, H., 'Samos', *WAS* 2 (1987), 214-17.
15.13 van Waerden, B.L., 'Eupalinos and his tunnel', *Isis* 59 (1968), 82-3.

16. Hellenistic waterworks

16.1 Bammer, A., 'Ephesos', *AA* (1972), 714-28.
16.2 Bucholz, H.G., *Methymna* (Mainz, 1975).
16.3 Crouch, D.P., 'The Hellenistic water system of Morgantina, Sicily', *AJA* 88 (1984), 353-65.
16.4 Fahlbusch, H., 'Gedanken zur Entsinterung von Druckrohren im Verlauf römischer Fernwasserleitungen', *3R International*, 25 Jahrgang, Heft 1/2 (Jan./Feb. 1986), 73-9.

Pergamon

16.5 Brinker, W., 'The effects of the Smyrna earthquake of 178 AD on the water supply of Pergamum', *Proceedings of the 17th Congress of the International Association of Hydraulic Research (IAHR)* (Baden-Baden, 1977), 768-9.
16.6 Fahlbusch, H., 'The development of the Pergamon water supply between 200 BC and 300 AD', *Proceedings of the 17th Congress of the International Association of Hydraulic Research (IAHR)* (Baden-Baden, 1977), 758-62.
16.7 Garbrecht, G. & Holtorf, G., 'Wasserwirtschaftliche Anlagen des antiken Pergamon: Die Madradag-Leitung', *Leichtweiss Institut für Wasserbau, Mitteilungen,* Heft 37 (1973).
16.8 Garbrecht, G. & Fahlbusch, H., 'Neue erkenntnisse über die Druckrohrleitung von Pergamon: Die Kaikos-Leitung', *Leichtweiss Institut für Wasserbau, Mitteilungen,* Heft 50 (1976).
16.9 Garbrecht, G., 'Wasserwirtschaftliche Anlagen des antiken Pergamon: Die Druckleitung', *Leichtweiss Institut für Wasserbau, Mitteilungen,* Heft 60 (1978).
16.10 Garbrecht, G. & Fahlbusch, H., 'Wasserwirtschaftliche Anlagen des antiken Pergamon: Das hellenistische Leitungssystem', *Leichtweiss Institut für Wasserbau, Mitteilungen,* (1981).
16.11 Garbrecht, G., 'L'alimentation en eau de Pergame', *Doss. Arch.* 38 (Oct./Nov., 1979), 26-33.
16.12 Garbrecht, G. & Fahlbusch, H., 'The pipes of the pressure conduit of Pergamon', *Proceedings of the 17th Congress of the International Association of Hydraulic Research (IAHR)* (Baden-Baden, 1977), 763-7.
16.13 Garbrecht, G. & Holtorf, G., 'Wass. An. Ant. Perg: die Madradag-Leitung', *Leichtweiss Institut für Wasserbau, Mitteilungen,* Heft 37 (1973).
16.14 Graber, H., 'Die Wasserleitungen von Pergamon', *Abhandlungen der Deutschen Akademie der Wissenschaften zu Berlin* (1888), 1-19; 26-31.
16.15 Graber, F., *Altertumer von Pergamon 1,3: Die Wasserleitungen* (Berlin, 1913).
See also: 53.35.

17. *Etruscan waterworks*

17.1 Ashby, T., *The Roman Campagna in Classical Times* (London, 1927).
17.2 Bonnin, J., 'Les hydrauliciens étrusques, des précurseurs', *La houille blanche* (Grenoble, 1973), 641-9.
17.3 Fraccaro, P., 'Di alcuni antichissimi lavori idraulici di Roma e della Campagna', *Opuscola* (Pavia, 1957), vol. 3, 1-36 (repr. from *Bulletino della Società Geografica Italiana*, 5 (1919), 186-215).
17.4 de la Blanchère, R., 'Cuniculus', *Dictionnaire des Antiquités*, ed. Daremberg & Saglio (Paris, 1887), vol. I, 2, 1589-94.
17.5 Judson, S. & Kahane, A., 'Underground drainageways in Southern Etruria', *PBSR* 31 (1963), 74-99.
17.6 Ward-Perkins, J.B., 'Veii: the historical topography of the ancient city', *PBSR* 29 (1961), 47-52.
17.7 Ward-Perkins, J.B., 'Etruscan engineering: road-building, water supply and drainage', *Hommages à Albert Grenier*, vol. 3 (Latomus, 1962), 1636-43.

18. Irrigation systems

18.1 Abercrombie, T.J., 'Behind the veil of troubled Yemen', *National Geographic Magazine* 125 (1964), 423.

18.2 Biswas, M. & A., *Food, Climate and Man* (John Wiley & Sons: New York, 1979).

18.3 Carton, L., 'Hydraulique dans l'antiquité', *Revue Tunisienne* 19 (1912), 221-30.

18.4 Caton-Thomas, G. & Gardener, E.W., 'The prehistoric geography of Kharga oasis', *Geographical Journal* 80 (1932), 369-406.

18.5 Caton-Thomas, G. & Gardener, E.W., *Kharga Oasis in Prehistory* (Univ. of London, 1952).

18.6 Costa, P.M., 'Notes on traditional hydraulics and agriculture in Oman', *World Archaeology* 14.3 (1982).

18.7 Evenari, M., Shanan, L. & Tadmor, N. *The Negev: the challenge of a desert* (Cambridge, Mass., 1982).

18.8 d'Escurac, P.H., 'Irrigation et la vie paysanne dans l'Afrique du Nord antique', *Ktema* 5 (1980), 177-91.

18.9 Fahlbusch, H., 'Roman long distance water conduits for irrigation and water supply', *ICID XI Proceedings: special sessions on the history of irrigation, drainage and flood control* (Grenoble, 1981), 61-75. [See p. 472, n. 55.]

18.10 Falkenmark, M. & Lindh, G., *Water for a Starving World* (Boulder, Colorado, 1976), tr. from the Swedish edition (Stockholm, 1975), Unesco and UN Water Conference.

18.11 Garbrecht, G., 'Irrigation throughout history – problems and solutions', *Water Resources Developments in Perspective*, IAHR, International Symposium on Water for the Future (Rome, 1986), 3-18.

18.12 Glick, T.F., *Irrigation and Society in Mediaeval Valencia* (Cambridge, Mass., 1972).

18.13 Glueck, N., *Rivers in the Desert* (New York, 1959).

18.14 Heitland, W.E., *Agricola* (Cambridge, 1921).

18.15 Hammond, P., 'Desert waterworks of the ancient Nabataeans', *Natural History* 76 (June-July 1967), 38ff.

18.16 Kedar, Y., 'Water and soil from the desert', *Geogr. Journal* 123 (1957).

18.17 Knapp, C., 'Irrigation among the Greeks and Romans', *Classical Weekly* 12 (1919), 73-4; 81-2.

18.18 Lassoe, J., 'The irrigation system at Ulhu (eighth century BC)', *Journal of Cuneiform Studies* 5 (1951), 1 21-32.

18.19 Nilsson, M., *Eranos* 43 (1945), 301-3.

18.20 de Patchere, F., 'Le règlement d'irrigation à Lamasba', *Mélanges de l'École Française à Rome* 28 (1908), 373-405.

18.21 *RE* 1, 1 (s.v. 'Ackerbau') F. Olck (1893).

18.22 *SAT* 2, 1-44.

18.23 Semple, E.C., *The Geography of the Mediterranean Region in Relation to Ancient History* (London, 1932), 385-8.

18.24 Shaw, B.D., 'Lamasba: an ancient irrigation community', *Antiquités Africaines* 18 (1982), 61-103.

18.25 Shaw, B.D., 'Water and society in the ancient Maghrib', *Antiquités Africaines* 20 (1984), 121-73.

18.26 Shaw, B.D. 'The noblest monuments and the smallest things: wells, walls and aqueducts in the making of Roman Africa', *FCAS*, 63-91.

18.27 Wesyermann, W.L., 'Aelius Gallus and the reorganization of the irrigation system of Egypt under Augustus', *Cl. Ph* 12 (July, 1917), 237-43.
See also: 6.1, 9.19.

Roman waterworks

19. Roman waterworks in general

19.1 van Buren, A.W., 'Wasserleitungen', *RE* 2, Reihe 15 (Stuttgart, 1955).
19.2 Eck, W., 'Organisation und administration der Wasserversorgung Roms', *Wasserversorgung im antiken Rom* (1986), 63-77.
19.3 Eck, W., 'Die Wasserversorgung im römischen Reich: sozio-politische Bedingungen, Recht und Administration', *WAS* 2, 49-101.
19.4 Février, P-A., 'Armé et aqueducs', *JEAR*, 133-40.
19.5 Garbrecht, G., 'Wasserversorgungstechnik in römischer Zeit', *Wasserversorgung im antiken Rom* (1986), 11-43.
19.6 Grewe, K., 'Römische Wasserleitungen nördlich der Alpen', *WAS* 3 (1988), 45-97.
19.7 Gross, W.H., 'Artikel, Kanal ... Kanalisation', *Der Kleine Pauly*, Bd. 3 (Stuttgart, 1969), 104-6.
19.8 Hainzmann, M., *Untersuchungen zur Geschichte und Ertwaltung der städtrömischen Wasserleitungen* (Graz, 1973).
19.9 [various authors], *Les aqueducs romains, Les dossiers d'archéologie* (Dijon) 38 (Oct.-Nov. 1979).
19.10 Volkmann, H., 'Die Wasserversorgung einer Römerstadt', *AA* (1963), 602-22.

20. Surveying and construction

20.1 Adam, J.-P., 'Groma et Chorobates, exercises de topographie antique', *Mefra* 94 (1982).
20.2 Adam, J.-P., *La construction romaine* (Paris, 1984), ch. 1, 9-22.
20.3 Boardman, J. et al., *Greek Art and Architecture* (New York, 1967).
20.4 Blume, F., Lachmann, K. & Rudorff, A., *Die Schriften der römischen Feldmesser* (Berlin, 1848-52; repr. 1962).
20.5 Burford, A., *The Greek Temple Builders at Epidauros* (Liverpool, 1969). [contracts and work force]
20.6 Carettoni, G., *S.P.Q.R.: la pianta marmorea di Roma Antica* (Rome, 1960).
20.7 Cohen, M.R. & Drabkin, I.E., *Source Book in Greek Science* (Cambridge, Mass. 1958).
20.8 *Corpus Agrimensorum Romanorum*, ed. C. Thulin (Teubner: Leipzig, 1913).
20.9 Dilke, O.A.W., *The Roman Land Surveyors* (Newton Abbot, England, 1971).
20.10 Dilke, O.A.W., *Greek and Roman Maps* (London, 1985).
20.11 Evans, H.B. 'Agrippa's water plan', *AJA* 86 (1982), 401-11.
20.12 Giorgetti, D., IV, 'L'acquedotto Romano di Bologna: l'antico cunicolo ed i sistemi di avanzamento in cavo cieco', *Acquedotto 2000: Bologna, l'Acqua del Duemila La Duemila Anni* (Grafis Edizioni: Casalecchio di Reno, 1985), 37-107.
20.13 Grewe, K., 'Über die Rekonstruktionversuche des Chorobates', *Allgem. Vermess.-Nachr.* 88 (1981), 205.

20.14 Grewe, K., *Planung und Trassierung römischer Wasserleitungen*, Schriftenreihe der Frontinus-Gesellschaft, Supplementband 1 (Verlag Chmielorz GmbH: Wiesbaden, 1985). [abbrev. Grewe]
20.15 Grewe, K., *Atlas der römischen Wasserleitungen nach Köln* (Rheinland-Verlag GmbH: Köln, 1986). ('Planung, Bemessung und Bau römischer Wasserversorgungsanlagen' – G. Garbrecht, 215-24; ('Vermessungsmethoden beim Bau von Fernwasserleitungen' – K. Grewe, 225-34). [abbrev. Grewe, *Atlas*]
20.16 Hauck, G., *The Aqueduct of Nemausus* (Jefferson, N.C./London, 1988). [abbrev. Hauck]
20.17 Hultsch, F.O., 'Chorobates', *RE* 3, 2, (1899), 2439-40.
20.18 al Karagi, M., *La civilisation des eaux cachées*, ed. & tr. Aly Mazaheri (Université de Nice, Institut d'Études et de Recherches Interethniques et Interculturelles: Nice, 1973).
20.19 Keller, F.J., 'Der römische Wasserleitungenstollen am Halberg bei Saarbrücken', *Bericht der Staatlichen Denkmalerpflege im Saarland* 12 (1965), 66-77.
20.20 McMullen, R., 'Roman imperial building in the provinces', *HCSP* (1959), 207-35. [esp. the role of the Roman army]
20.21 Piganiol, A., *Les documents cadastraux de la colonie romaine d'Orange, Gallia*, Suppl. 16 (Paris, 1962).
20.22 Sherk, R.K., 'Roman geographical exploration and military maps', *ANRW* 2, 1 (Berlin, 1974), 534-62.
20.23 Smith, N.A.F., 'Problems of design and analysis', *FCAS*, 113-28.
20.24 Thornton, M.K., 'Julio-Claudian building programs: eat, drink and be merry', *Historia* 35 (1986), 28-44.

21. Construction materials and techniques

21.1 Grimal, P., 'Vitruve et la technique des aqueducs', *Rev. Phil.* 19 (1945), 169.
21.2 Hauck, G.F., *The Aqueduct of Nemausus* (Jefferson, N.C./London, 1988).
21.3 Lamprecht, H.O., *Caementitium Opus* (Dusseldorf, 1984).
21.4 Lamprecht, H.O., 'Bau- und Materialtechnik bei antiken Wasserversorgungsanlagen', *WAS* 3 (1988), 129-55.
21.5 Leveau, Ph., 'La construction des aqueducs', *Doss. Arch.* 38 (Oct./Nov., 1979).
21.6 Malinowski, R., 'Concretes and mortars in ancient aqueducts' *Concrete International*, vol. 1, no. 1 (Jan., 1979), 66-76.
21.7 Malinowski, R., 'Einige Baustoffe probleme der antiken Aquedukten', *JEAR*, 245-74.
21.8 Malinowski, R., 'Ancient mortars and concretes: aspects of their durability', *History of Technology* (1982), 89-100.
21.9 Matthews, K.D., 'Roman aqueducts: technical aspects of their construction', *Expedition* (Journal of the University Museum, University of Pennsylvania; Fall, 1970).
21.10 Orlandos, A., *Hulika Domês tôn Archaiôn Hellênôn* (Athens, 1959-60).
See also: 1.8.

22. Siphons

22.1 Burns, A., 'Greek water supply and city planning', *Technology and Culture* 15, 3 (1974), 408-9.

22.2 Bassel, R., *N.Sc.A.* (1882), 418. [Alatri]

22.3 Carton, L., 'Hydraulique dans l'antiquité en Barbarie', *Revue Tunisienne* 19 (1912), 221-33.

22.4 Garbrecht, G., 'Hydraulic towers in Roman pressure conduits', *Proceedings of the 17th Congress of the International Association of Hydraulic Research (IAHR)* (Baden-Baden, 1977), 778-80.

22.5 Hodge, A. Trevor, 'Siphons in Roman aqueducts', *PBSR* 51 (1983). [abbrev. SRA]

22.6 Hodge, A. Trevor, 'Siphons in Roman aqueducts', *Scientific American* (June, 1985).

22.7 Lanckoronski, K.G., *Die Städte Pamphyliens und Pisidiens* 1 (Vienna, 1890). [Aspendos]

22.8 Malinowski, R. & Peleg, Y., 'Some pressure pipelines in Israel', *Leichtweiss-Institut für Wasserbau, Mitteilungen*, Heft 71 (1984), 239-65.

22.9 Smith, N.A.F., 'Attitudes to Roman engineering and the question of the inverted siphon', *History of Technology*, vol. 1 (1976), 45-71. [abbrev. Smith]

22.10 Stehlin, K., 'Über die Colliviaria oder Colliquiaria der römischen Wasserleitungen', *Anzeiger für schweizerische Altertumskunde* 20 (1918), 167-75.

23. Discharge and lime incrustation (sinter)

23.1 Baatz, D., 'Bemerkungen zu Kalksinter im Römerkanal', *Das Rheinische Landesmuseum* 6/78 (1978), 90.

23.2 Bailhache, M., 'Le débit des aqueducs gallo-romains', *JEAR* 19-49, esp. 40-1. [table of output/quality]

23.4 Blackman, D.R., 'The volume of water delivered by the four great aqueducts of Rome', *PBSR* 46 (1978), 52-72.

23.6 Committee 861D, Water meters – selection, installation, testing and maintenance: Chapter One, 'Early history of water measurement and the development of meters', *Journal of the American Water Works Association* 51 (1959), 791-9.

23.7 Fahlbusch, H., 'Maintenance problems in ancient aqueducts', *FCAS*, 7-14.

23.8 Fahlbusch, H., 'Über Abflussmessung und Standardisierung bei den Wasserversorgungsanlagen Roms', *Wasserversorgung im antiken Rom* (1986), 129-44.

23.9 Gilly, I.-C., 'Les dépots calcaires de l'aqueduc antique de Nîmes', *École Antique de Nîmes, Bulletin Annuel* 6-7 (1971-72), 61-72.

23.10 Grohmann, A., 'Leserbrief zu Kalksinter im Römerkanal', *Das Rheinische Landesmuseum Bonn* 6/78 (1978), 91.

23.11 Porath, Y., 'Lime plaster in aqueducts: a new chronological indicator', *Leichtweiss-Institut für Wasserbau, Mitteilungen*, Heft 82 (1984), 16.

23.12 Richardson, C.G., 'The measurement of flowing water', *Water and Sewage Works* 102 (1955), 379-85.

23.13 Schmitz, W., 'Kalksinter in Römerkanal', *Das Rheinische Landesmuseum Bonn* 4/78 (1978), 55-7.

23.14 Wilber, D.N., 'The plateau of Daphne', *Antioch on the Orontes II*, ed. R. Stillwell (Princeton, 1938) [saline deposit]

24. Pipes and pipelines

24.1 Andreossy, Count A.F., *Constantinople et le Bosphore de Thrace* (Paris, 1828). [*suterazi*]
24.3 *Archaeologia Aeliana* 4, 37 (1959). [lead pipes]
24.3 *Les Archives de la Commission des Monuments Historiques* (Paris, 1855-72), ed. Gide et J. Beaudry, vol. 1, 3.
24.4 di Fenizio, Cl., 'Sulla portata degli acquedotti romani e determinazione della quinaria', *Giornale del Genio Civile* 54 (1916), 227-331.
24.5 di Fenizio, Cl., 'Studio sulla portata degli antichi acquedotti romani e determinazione della quinaria', *Nuova Appendice* (Stabilimento Tipo-Lithografico del Genio Civile, Roma, 20 Maggio 1930).
24.6 Hodge, A. Trevor, 'How did Frontinus measure the quinaria?', *AJA* 88 (1984), 205-16.
24.7 Jacono, L., 'La misura delle antiche fistule plumbee', *Rivista di Studi Pompeiani* 1, 2 (Napoli, 1934-5).
24.8 Mahol, J., 'Les tuyaux de plomb. Histoire et progrès de leur fabrication', *La Nature* (1 Dec., 1937), 503-7.
24.9 Peleg, Y., 'Ancient pipelines in Israel', *FCAS*, 129-40.
24.10 Penn, W.S., 'Roman lead pipes', *The Plumbing Trade Journal* (Great Britain) (Sept., 1961), 62-4.
24.11 Schmidt, J. & Gieger, H.J., 'About the sealing of antique clay pipe conduits', International Association of Hydraulic Research, *XVII Congress Proceedings*, vol. 6 (1977), 775-7.
See also: 14.13, 40.37, 41.1.

25. Health and lead poisoning

25.1 Amulree, Lord, 'Hygenic conditions in ancient Rome and modern London', *Medical History* 17 (1973), 244-55.
25.2 Hodge, A. Trevor, 'Vitruvius, lead pipes and lead poisoning', *AJA* 85 (1981), 486-91.
25.3 Kobert, R., *Chronische Bleivergiftung im Klassische Altertum* (in Diergard (ed.), *Beiträge aus der Geschichte der Chemie* (Leipzig, 1909)).
25.4 Mackie, A., Townshend, A. & Waldron, H.A., 'Lead concentrations in bones from Roman York', *Journal of Archaeological Science* 2 (1975), 235-7.
25.5 Mackie, A., Townshend, A. & Waldron, H.A., 'Lead content of some Romano-British bones', *Archaeometry* 18 (1976), 221-7.
25.6 Needleman, L. & D., 'Lead poisoning and the decline of the Roman aristocracy', *Classical Views* (publ. by the University of Calgary) 29, N.S. 4 (1985), 63-94.
25.7 Schoenen, D., 'Antike Wasserversorgungen aus der Sicht moderner Trinkwasserhygiene', *Leichtweiss-Institut für Wasserbau, Mitteilungen*, Heft 82 (1984), 15.
25.8 Winkelmann, O., 'Hygienische Aspekte der Wasserversorgung antiker Städte', *WAS* 3 (1988), 157-70.

26. Filtration plants/cisterns

26.1 Berthier, A., 'Note sur un filtre romain découvert à Constantine', *Receuil des Notices et Mémoires de la Société Archéologique de Constantine* 69 (1960-61), 175-8.

26.2 Crasta, F.M., Fasso, C.A., Patta, F. & Putzu, G., *Carthaginian Roman Cisterns in Sardinia* (University of Hawaii: Honolulu, 1982).
26.3 Nash, E., *Pictorial Dictionary of Ancient Rome* (London, 1961), vol. 1, 37. [engraving of a terminal reservoir – Claudia]
26.4 Perello, E.R., *Ampurias* (Barcelona, 1979).
See also: 7.12, 40.20.

27. *Architectural features*

27.1 Glaser, Franz, 'Brunnen und Nymphäen', *WAS* 2 (1987), 103-1.
27.2 Hülsen, J., 'Zur Entwicklung der antiken Brunnenarchitektur', in Milet 1, 5, *Das Nymphäum* (1919), 73ff.
27.3 Neuerburg, N., *L'archittetura delle fontane e dei ninfei nell'Italia antica*, Memorie della Accademia di Archeologia, Lettere e Belle Arti, Napoli, N.S. 5 (Naples, 1965).
27.4 Richmond, I.A., 'Commemorative arches and city gates in the Augustan age', *JRS* 23 (1933), 149-74.
27.5 Wiegand, T. & Schrader, 'Wasseranlagen', *Priene*, vol. 4, Berlin (1904), 68-80.

28. *Taps*

28.1 Kretzchmer, F., 'La robinetterie romaine', *RAE* 11 (1960), 89-113.
28.2 Kretzschmer, F., 'Römische Wasserhähne', *Jb. Schweiz, Gesel. Urg.* 48 (1960-61), 50-62.

29. *Drainage*

29.1 Adam, J-P., *La construction romaine* (Paris, 1984), 283-6.
29.2 Adam, J-P. & Bourgeois, Cl., 'Un ensemble monumentale gallo-romain en terre dans le sous-sol de Bourges', *Gallia* 35 (1977).
29.3 Corbin, A., *The Foul and the Fragrant: odor and the French social imagination* (Harvard University Press, 1986).
29.4 Crouch, D., 'The Hellenistic water system of Morgantina, Sicily', *AJA* 88 (1984), 353-65.
29.5 Forbes, R.J., 'Hydraulic engineering and sanitation', in Singer et al. (see 8.16) vol. 2, 663-94.
29.6 Friedlander, L., *Roman Life and Manners*, tr. L.A. Magnus (Barnes & Noble: New York, 1965; repr. of 1913 ed.).
29.7 Frezouls, E., *Le cryptoportique de Rheims, les cryptoportiques dans l'architecture romaine, École française de Rome* 14 (1974), 293f.
29.8 *Gallia* 26 (1968), 580-1. [St. Romain-en-Gal]
29.9 Kahrstedt, U., 'Der Kopaissee im Altertum und die Minyschen Kanale', *JDI* 52 (1937), 1-20.
29.10 Kenny, E.J.A., 'The ancient drainage of the Copais', *Ann. Archaeol. Anthr.* (1935), 189-206.
29.11 Owens, E.J., 'The koprologoi at Athens in the fifth and fourth centuries BC', *CQ* 33 (1983), 44-50.
29.12 'Roman Britain in 1934', *JRS* 25 (1935), 213. [Caister]
29.13 Scobie, A., 'Slums, sanitation and morality in the ancient world', *Klio* 68 (1986), 399-434.

29.14 Will, E., 'Le cryptoportique de Bavay', *Revue du Nord* 40 (1958), 42 (1960), 46 (1964).

Industrial aqueducts

30. Water in industry

30.1 Ausonius, *Mosella* 5, 362-4.
30.2 Bennett, R. & Elton, J., *History of Corn Milling* 2 (London, 1899).
30.3 Lewis, P.R. & Jones, J.D.B., 'The Dolaucothi gold mines, 1: the surface evidence', *Antiquaries Journal* 49 (1969), 244-72.
30.4 Lewis, P.R. & Jones, J.D.B., 'Roman gold-mining in north-west Spain', *JRS* 60 (1970), 169-85.
30.5 Rakob, F., 'Wasser als Element römischer Infrastruktur', *D U Die Kunstzeitschrift* 3 (1979), 66.
30.6 Simms, D.L., 'Water-driven saws, Ausonius and the authenticity of the *mosella*', *Technology and Culture* 24 (1983), 635-43.

31. Water mills

31.1 Curwen, C.E., 'The problem of early water mills', *Antiquity* 18 (1944), 130-46.
31.2 Moritz, L.A., *Grain-Mills and Flour in Classical Antiquity* (Oxford, 1958).
31.3 Oleson, J.P., 'A Roman water-mill on the crocodilian river near Caesarea', *Zeit. des Deut. Palastina-Vereins*, vol. 100, 137-52.
31.4 Parsons, A.W., 'A Roman water-mill in the Athenian Agora', *Hesp.* 5 (1936), 70-90.
31.5 Reynolds, T.S., *Stronger than a Hundred Men* (Baltimore/London, 1983). [history of water-wheels]
31.6 Schiøler, T. & Wikander, Ö., 'A Roman watermill in the Baths of Caracalla', *Opuscula Romana* 14.4 (Stockholm, 1983), 47-64.
31.7 Schiøler, T., 'Vandmollerne ved Krokodillerfloden', *Sfinx* (a journal of Aarhus University, Denmark) 8 (1985), 12-14.
31.8 Smith, N.A.F., 'The origins of water power: a problem of evidence and expectations', *Transactions of the Newcomen Society* 55 (1983-84), 67-84.
31.9 Spain, R.J., 'The second-century Romano-British watermill at Ickham, Kent', *History of Technology* 9 (1984), 143-80.
31.10 Van Buren, A.W. & Stevens, G.P., 'The Aqua Traiana and the mills on the Janiculum', *Memoirs of the American Academy in Rome* 1 (1915-16), 59-62.
31.11 Van Buren, A.W. & Stevens, G.P., 'The Aqua Traiana and the mills on the Janiculum', *Memoirs of the American Academy in Rome* 6 (1927), 137-46.
31.12 Wikander, Ö., *Exploitation of Water-Power or Technical Stagnation?* (Scripta Minora 1983-84:3) (C.K.W. Glerup: Lund, Sweden, 1984).
31.13 Wikander, Ö., 'Water-mills in ancient Rome', *Opuscula Romana*, series 4, year 26, no. 12 (1979).
31.14 Wikander, Ö., 'Water mills and aqueducts', *FCAS*, 141-8.

32. Barbegal

32.1 Benoît, F., 'L'usine de meunerie hydraulique de Barbegal, Arles', *RA* 15 (1940), 19-80.

32.2 Fleming, S., 'Gallic waterpower: the mills of Barbegal', *Archaeology* 36.6 (Nov./Dec., 1983), 68.
32.3 Hodge, A. Trevor, 'A Roman factory', *Scientific American* 262-11 (Nov. 1990), 106-11.
32.4 Sagui, C.L., 'La meunerie de Barbegal (France) et les roues hydrauliques chez les anciens et au Moyen Age', *Isis* 38 (1947), 224-31.
32.5 Sellin, R.H.J., 'The large Roman water mill at Barbegal (France)', *History of Technology* 8 (1983), 91-109.

33. *Machines and allied topics*

33.1 Brumbaugh, R.S., *Ancient Greek gadgets and machines* (New York, 1966).
33.2 Burford, A., 'Heavy transport in classical antiquity', *Economic History Review* 13 (1960), 1-18.
33.3 Burstall, A., *Simple Working Models of Historic Machines* (London, 1968).
33.4 Cohen, M.R. & Drabkin, I.E., *Source Book in Greek Science* (Cambridge, Mass., 1958), 326-9.
33.5 Finley, M.I., *Economy and Society in Ancient Greece* (Harmondsworth, 1983), 184.
33.6 Oleson, J.P., *Greek and Roman Mechanical Water-Lifting Devices* (Toronto, 1984).
33.7 White, L., *Mediaeval Technology and Social Change* (Oxford, 1962).

34. *Recreational use*

34.1 Banti, L., 'Piscina' in *RE* 20 (1950).
34.2 Brodner, E., *Untersuchungen an den Caracallathermen* (Berlin, 1951).
34.3 Cunliffe, B.W., *Roman Bath* (London, 1970).
34.4 Cunliffe, B.W., 'The temple of Sulis Minerva at Bath', *Archaeology* 36.6 (Nov.-Dec., 1983), 16-23.
34.5 Cunliffe, B., *Roman Bath Discovered* (London, 1984).
34.6 Grimal, P., *Roman Cities*, tr. & ed. G. Michael Woloch (Madison, Wisc., 1983), 182.
34.7 Heinz, W., *Römische Thermen* (Zürich/München, 1983).
34.8 Johnson, A., *Roman Forts* (London, 1983). [baths]
34.9 Krencker, D. & Kruger, E., *Die Trierer Kaiserthermen (mit einer Übersicht über die wichtigsten Thermenanlagen des römischen Reiches)* (Augsburg, 1929).
34.10 *MAGR* 231-384.
34.11 Manderscheid, H., 'Römische Thermen. Aspekte von Architektur, Technik und Ausstattung', *WAS* 3 (1988), 99-125.
34.12 Robertson, D.S., *Handbook of Greek and Roman Architecture*, 2nd ed. (Cambridge, 1943). [baths]
34.13 Sear, F., *Roman Architecture* (Cornell Univ. Press: Ithaca, N.Y., 1983). [baths]

Roman waterworks: a geographical survey

35. *Algeria*

35.1 Ballu, A., *Les ruines de Timgad* (Paris, 1897-1911).

35.2 Birebent, J., *Aquae Romanae* (Algeria, 1962) [abbrev. Birebent]
35.3 Godet, R., 'Le ravitaillement de Timgad en eau potable', *Lybica. Archaeologie, Epigraphie* 2, 1er sem. (1954), 65-72.
35.4 Grewe, K., *Planung und Trassierung römischer Wasserleitungen*, Schriftenreihe der Frontinus-Gesellschaft, Supplementband 1 (Verlag Chmielorz GmbH.: Wiesbaden, 1985). [abbrev. Grewe]
35.5 Gsell, S., *Monuments antiques de l'Algérie* (Paris, 1901).
35.6 Gsell, S., *Enquête administrative sur les travaux hydrauliques anciens en Algérie* (Paris, 1902).
35.7 Leveau, P. & Paillet, J-L., *L'alimentation en eau de Caesarea de Mauretanie et l'aqueduc de Cherchel* (Paris, 1976). [abbrev. Cherchel]
35.8 Leveau, Ph., 'Saldae', *WAS* 3 (1988), 215-18.
35.9 Vertet, H., 'Aqueducs de Rusicade', *JEAR*, 349-69.

36. Austria

36.1 Grewe, K., 'Die Wasserversorgung militärischer Lager und Kastelle: Vindobona/Wien', *WAS* 3 (1988), 56-9.
36.2 Kling, A., 'Die römischen und mittelalterlichen Wasserversorgungen in Wien', *Symposium: Historische Entwickling der Wasserwirtschaft und der Wasserversorgung* (1981), 764-9.

37. Britain

General

37.1 Hanson, J., *Municipal and Military Water Supply and Drainage in Roman Britain* (University of London: Institute of Archaeology, 1970), unpublished dissertation, available on microfilm. [abbrev. Hansen]
37.2 Stephens, G.R., 'Civic aqueducts in Britain', *Britannia* 16 (1985) 197-208.

Individual

37.3 Atkinson, D., *Wroxeter* (Oxford, 1942).
37.4 Boon, G.C., *Roman Silchester* (London, 1957).
37.5 Grewe, K., 'Die Wasserversorgung von Badeorten: Aquae Sulis/Bath', *WAS* 3 (1988), 63-5.
37.6 Jones, G.B.D., Blakeney, I.J. & MacPherson, E.C.F., 'Dolaucothi: the Roman aqueduct', *Bulletin of the Board of Celtic Studies, University of Wales* 19 (1960), 71-80.
37.7 Richmond, I.A., note on Greatchesters, *Journal of Roman Studies* 35 (1945), 80.
37.8 Thompson, F.H., 'The Roman aqueduct at Lincoln', *Archaeological Journal* 111 (July, 1955), 106ff.
37.9 *York: Eburacum* (1962), 1, 38.

London

37.10 Marsden, P., *Roman London* (London, 1980).
37.11 Merrifield, R., *The Roman City of London* (London, 1965).
37.12 Wheeler, E.M., *London in Roman Times* (London, 1930).
See also: 34.3, 34.4, 34.5.

38. Bulgaria

38.1 Biernacka-Lubanska, M., 'The water-supply of Novae', *Archeologia* 30 (1979), 57-67.

39. Cyprus

39.1 Last, J.S., 'Kourion: the ancient water supply', *Proceedings of the American Philosophical Society* 119 (1975), 39-72.

40. France

General

40.1 Bailhache, M., 'Le vieillissement des aqueducs et la baisse de leur débit', *Les aqueducs romains. Les dossiers d'archéologie* 38 (1979), 62-71. [See 23.2.]
40.2 Blanchet, A., *Recherches sur les aqueducs et les cloaques de la Gaule romaine* (Paris, 1908).
40.3 Boucher, J.P., (ed.), *Journées d'études sur les aqueducs romains, Lyons, 26-28 mai, 1977* (Paris, 1983). [abbrev. *JEAR*]
40.4 *Forma Orbis Romani: Carte archaeologique de la Gaule romaine, Département du Gard*, ed. A. Blanchet (Paris, 1941).
40.5 Grenier, A., 'Les monuments des eaux', *Manuel d'archéologie gallo-romaine* 4,1 (Paris, 1960). [abbrev. *MAGR*]
40.6 MacKendrick, P., *Roman France* (London, 1971).
40.7 [various authors], *Les aqueducs romains. Les dossiers d'archéologie* (Dijon) 38 (Oct.-Nov. 1979).
For most individual locations see *MAGR*.

Aix-en-Provence

40.8 Boiron, R. & Moliner, M., 'Aix-en-Provence' in *WAS* 3 (1988), 173-6.

Arles

40.9 Auvergne, J., 'Fontvieille inédit', *Bulletin de la Société des amis du vieil Arles*, vol. 2.
See Nîmes: Stübinger

Barbegal

See Water mills (31)

Fontenay

40.10 Liot, C., 'Contribution à l'étude des aqueducs romains en Touraine ... l'aqueduc de Fontenay', *RA* 2 (1963), 293-310; 3 (1964) 3-17 and 125-37.

Fréjus

40.11 Donnadieu, A., *La Pompeii de la Provence, Fréjus, Forum Julii* (Paris, 1927). [wind and buttresses]

40.12 Février, P.A., *Fréjus, Itineraires Ligures* (Institut International d'Études Ligures, 1977).

Grand

40.13 Bertaux, J-P., 'Les galeries souterraines'; André, O. and Bertaux, V., 'Les aqueducs souterrains construits par les Romains'; Haguenhauer, B. and Deletie, P., 'Mythe ou réalité de la ressource en eau du site'; André, O., Bertaux, J-P. and Nion, S., 'Les ressources hydrauliques du site'; *Dossiers d'Archéologie* 162 (juil.–août, 1991), 28-33, 34-7, 67-72; 82-3.

Lyon

40.14 Audin, A., 'Le réservoir terminal de l'aqueduc du Gier à Lyon', *JEAR*, 13-18.
40.15 Burdy, J., 'Lyon', *WAS* 3 (1988), 190-8.
40.16 Burdy, J., 'Some directions of future research for the aqueduct of Lugdunum (Lyon)', *FCAS*, 29-44.
40.17 Jeancolas, L., 'Aqueducs antiques de Lyon', *JEAR*, 179-205.
40.18 de Montauzan, G., *Les aqueducs antiques de Lyon* (Paris, 1909). [abbrev. GdM]

Metz

40.19 Grewe, K., 'Die Wasserversorgung von zivilen Siedlungsplätzen und Kolonien: Divodurum/Metz', *WAS* 3 (1988), 76-8.

Nîmes

40.20 *Les Archives de la Commission des Monuments Historiques*, ed. Gide & Beaudry (Paris, 1855-72).
40.21 Esperandieu, E., *Le Pont du Gard* (repr. Paris, 1979).
40.22 Fiches, J.L. & Paillet, J.L. 'Nîmes', *WAS* 3 (1988).
40.23 Hauck, G.F.W., *The Aqueduct of Nemausus* (Jefferson, N.C./London, 1988).
40.24 Hauck, G.F.W., 'Structural designs of the Pont du Gard', *Journal of Structural Engineering*, Am. Soc. of Civil Engineers 112, no. 1 (Jan., 1986).
40.25 Hauck, G.F.W., & Novak, R.A., 'Interaction of flow and incrustation in the Roman aqueduct of Nîmes', *Journal of Hydraulic Engineering*, Am. Soc. of Civil Engineers 113, no. 2 (Feb., 1987).
40.25 (a) Hauck, G.F.W. & Novak, R.A., 'Water flow in the castellum at Nîmes', *AJA* 92 (1988), 393-407.
40.26 Hauck, G., 'The Roman aqueduct of Nîmes', *Scientific American* (March, 1989), 98-104.
40.27 Lassalle, V., 'Le Pont du Gard et l'aqueduc de Nîmes', *Doss. Arch.* 38 (1979), 52-61.
40.28 Naumann, R., *Der Quelbezirk von Nîmes* (Berlin-Leipzig, 1937).
40.29 Pelet, A., *Mémoires Acad. Gard.* (1845-46).
40.30 Pelet, A., *Descriptions des monuments grecs et romains* (Nîmes, 1876).
40.31 Smith, N.A.F., 'The Pont du Gard and the aqueduct of Nîmes', *Transactions of the Newcomen Society* 62 (1990-91).
40.32 Stübinger, O., *Die Wasserleitungen von Nîmes und Arles* (Heidelberg, 1909).

Paris

40.33 Grewe, K., 'Die Wasserversorgung von zivilen Siedlungsplätzen und Kolonien: Lutetia Parisiorum/Paris', *WAS* 3 (1988), 78-9.

Saintes

40.34 Triou, A., 'Les aqueducs gallo-romaines de Saintes', *Gallia* 26 (1968), 119-44.

Strasbourg/Kuttolsheim

40.35 Grewe, K., 'Die Wasserversorgung von zivilen Siedlungsplätzen und Kolonien: Argentorate/Strasbourg', *WAS* 3 (1988), 72-6.

40.36 Hatt, J.J., 'La conduite d'eau romaine de Kuttolsheim à Strasbourg', *JEAR*, 175-7.

40.37 Stieber, A., 'Observations concernant la conduite de l'eau romaine de Kuttolsheim à Strasbourg', *Cahiers alsaciens d'archéologie, d'art et d'histoire* 4 (1960), 45-52.

Vienne

40.38 Pelletier, A., 'L'alimentation en eau à Vienne dans l'antiquité', *JEAR*, 293-307.

41. Germany

General

41.1 Samesreuther, E., 'Römische Wasserleitungen in den Rheinlanden', 26. *Bericht der römisch-germanischen Kommission* (Berlin, 1938).

Individual

41.2 Engelhardt, R., *Die römische Wasserleitung in Bingen* (Bingen, 1978).

41.3 Grewe, K., 'Der Aquaedukttunnel, durch den Drover Berg bei Vettweiss-Soller, Kreis Düren', *Ausgrabungen im Rheinland 1981/82* (Bonn, 1983), 159.

41.4 Grewe, K., 'Römische Tunnelabsteckung am Beispiel des Drover-Berg-Tunnels', *Allgem. Vermess.-Nachr.* 91 (1984), 246.

41.4a Grewe, K., *Atlas der römischen Wasserleitungen nach Köln* (Rheinland-Verlag GmbH., Köln, 1986) (abbrev. Grewe, *Atlas*).

41.5 Grewe, K., 'Die Wasserversorgung von zivilen Siedlungsplätzen und Kolonien: Colonia Augusta Treverorum/Trier', *WAS* 3 (1988), 79-83.

41.6 Grewe, K., 'Die Wasserversorgung von zivilen Siedlungspläten und Kolonien: Colonia Ulpia Traiana', *WAS* 3 (1988), 83-4.

41.7 Grewe, K., 'Die Wasserversorgung von zivilen Siedlungsplätzen und Kolonien: Colonia Claudia Ara Aggrippinensium/Köln', *WAS* 3 (1988), 84-9.

41.8 Grewe, K., 'Die Wasserversorgung militärischer Lager und Kastelle: Bonna/Bonn', *WAS* 3 (1988), 50-2.

41.9 Grewe, K., 'Die Wasserversorgung militärischer Lager und Kastelle: Castellum Mattiacorum/(Mainz-)Kastel', *WAS* 3 (1988), 52-3.

41.10 Grewe, K., 'Die Wasserversorgung militärischer Lager und Kastelle: Regina Castra/Regensburg', *WAS* 3 (1988), 59.

41.11 Grewe, K., 'Die Wasserversorgung militärischer Lager und Kastelle: Aquae Mattaicorum/Wiesbaden', *WAS* 3 (1988), 59-60.
41.12 Grewe, K., 'Die Wasserversorgung von Badeorten: Aquae Granni/Aachen', *WAS* 3 (1988), 61-3.
41.13 Grewe, K., 'Die Wasserversorgung von zivilen Siedlungsplätzen und Kolonien: Augusta Vindelicum/Augsburg', *WAS* 3 (1988), 70-1.
41.14 Grewe, K., 'Tunnelbauten für Wasserleitungen: Halberg-Tunnel bei Düren', *WAS* 3 (1988), 92-3.
41.15 Haberey, W., *Die römischen Wasserleitungen nach Köln* (Köln, 1972). [abbrev. Haberey]
41.16 von Kaphengst, C. & Rupprecht, G., 'Mainz', *WAS* 3 (1988), 199-203.
41.17 Keller, F.J., 'Der römische Wasserleitungenstollen am Halberg bei Saarbrücken', *Bericht des staatl. Denkmalspflege im Saarland* 12 (1965), 67-77.
41.18 Neyses, A., 'Die Ruwer-Wasserleitung des römischen Trier', *JEAR*, 173-283.
See also: 34.9.

42. Greece

42.1 Case, J.F., 'The Ancient Roman Aqueduct at Athens', *ASCE Transactions*, Paper no. 1627 (1927) 281-290.
42.2 Gräber, F., 'Die Enneakrunos', *Mitteilungen des Deutschen Archäol. Instituts in Athen*, 30 (1905) 1-64.
42.3 Kienast, H. 'Athen', *WAS* 2 (1987), 167-71.
42.4 Robinson, D.M., 'Excavations at Olynthos', no. 8. *Water Supply and Drainage* (Baltimore, 1938), 307-11.
42.5 Ziller, E., 'Untersuchungen über die antiken Wasserleitungen Athen', *Mitteilungen d. deutschen archäol. Instituts Athen* 2 (1877), 107-31.

43. Iran

43.1 Hartung, F., 'Altiranische Grosswasserbauten', *Wasser und Energiewirtschaft/Cours d'eau et energie*, no. 4 (1972), 117-32.
43.2 Stein, Sir A., 'Surveys on the Roman frontier in Iran and Transjordan', *The Geographic Journal* 95 (June, 1940), 430.

44. Israel

44.1 Malinowski, R. & Peleg, Y., 'Some pressure pipelines in Israel', *Papers at the Symposium of Historical Water Development Projects in the Eastern Mediterranean: Leichtweiss-Institut für Wasserbau, Mitteilungen*, Heft 82 (1984).
44.2 Mazar, A., 'Survey of the Jerusalem aqueducts', *Leichtweiss-Institut für Wasserbau, Mitteilungen*, Heft 82 (1984), 7f.
44.3 Mazar, A., 'Jerusalem', *WAS* 2 (1987), 185-8.
44.4 Netzer, E., 'Masada', *WAS* 2 (1987), 189-92.
44.5 Olami, Y. & Peleg, Y., 'The water system of Caesarea Maritima', *Israel Exploration Journal* 27 (1977), 127-37.
44.6 Peleg, Y., 'The water supply system of Caesarea', *Leichtweiss-Institut für Wasserbau, Mitteilungen*, Heft 82 (1984) 4.

44.7 Peleg, Y., 'The aqueducts to Dor', *Leichtweiss-Institut für Wasserbau, Mitteilungen*, Heft 82 (1984), 7.
44.8 Peleg, Y., 'Die Wasseranlagen von Megiddo', *Leichtweiss-Institut für Wasserbau, Mitteilungen*, Heft 82 (1984), 1.
44.9 Peleg, Y., 'Wasserwerke aus der israelischen Zeit', *Leichtweiss-Institut für Wasserbau, Mitteilungen*, Heft 82 (1984), 2.
44.10 Peleg, Y., 'Das Stauwerk für die untere Wasserleitung nach Caesarea', *Leichtweiss-Institut für Wasserbau, Mitteilungen*, Heft 89 (1986), 15.
44.11 Peleg, Y., 'Caesarea Maritima', *WAS* 2 (1987), 176-9.
44.12 Peleg, Y., 'Ancient pipelines in Israel', *FCAS*, 129-40.
44.13 Reich, R., 'Domestic water installations in Jerusalem of the Second Temple', *Leichtweiss-Institut für Wasserbau, Mitteilungen*, Heft 82 (1984), 13.
See also: 24.9.

45. Italy

General

45.1 di Fenizio, C., 'Sulla ubicazione della piscina dell'acquedotto Marcio e della misura di portata in essa eseguita da Frontino', *Tipografia del Genio Civile* (Rome, 1931).
45.2 di Fenizio, C., 'L'acqua Appia: La misura delle acque "more romano" e la technica delle condotte nel primo secolo dell'era volgare', *Giornale del Genio Civile*, fasc. 9-10 and 11-12 (1947).

Angitia

45.3 Giovanni, G., 'L'aquedotto romano di Angitia', *Rendiconti della Pontificia Accademia romana di Archaeologia* 9 (1935), 70ff.

Bologna

45.4 Giorgetti, D., IV, 'L'acquedotto romano di Bologna: l'antico cunicolo ed i sistemi di avanzamento in cavo cieco', *Acquedotto 2000: Bologna, l'Acqua del Duemila La Duemila Anni* (Grafis Edizioni: Casalecchio di Reno, 1985), 37-107.
45.5 Giorgetti, D., 'Bologna', *WAS* 3 (1988), 180-5.

Minturnae

45.6 Butler, H.C. 'The aqueduct of Minturnae', *AJA* 5 (1901), 187-92.
45.7 Richmond, I.A., 'Commemorative arches and city gates in the Augustan age', *JRS* 23 (1933), 149-74.

Ostia

45.8 Becatti, G., *Ostia* (Rome, 1970).
45.9 Hermansen, G., *Ostia: aspects of Roman city life* (Edmonton, 1982).
45.10 Meiggs, R., *Roman Ostia* (Oxford, 1973).

Pompeii

45.11 Eschebach, H., 'Pompeii: la distribution des eaux dans une grande ville romaine', *Doss. Arch.* 38 (Oct.-Nov., 1979).
45.12 Eschebach, H., 'Die innerstädtsche Gebrauchwasserversorgung dargestellt am Beispiel Pompejis', *JEAR* (1983), 81-132.
45.13 Eschebach, L., 'Pompeji', *WAS* 2 (1987), 202-7.
45.14 Jashemski, W.F., *The Gardens of Pompeii* (New York, 1979).
45.15 Maiuri, A., 'Pompeii: pozzi e condotture d'aqua nell'antica città', *N. Sc.* (1931-), 546-76.
45.16 Paribeni, R., 'Pompeii – relazione degli scavi eseguiti durante il mese di novembre', *Les nouvelles archives des missions*, vol. 15, fasc. 2 (excavation of the Pompeii castellum) *N.Sc.* (1903), 25-33.
45.17 Sgorbbo, I., 'Serino. L'aquedotto romano della Campania "Fontis Augustei Aquaeductus" ', *N.Sc.* (1938), 75f.

Rome

45.18 Ashby, T., *The Aqueducts of Ancient Rome* (Oxford, 1935).
45.19 Astin, A.E., 'Water to the Capitol: a note on Frontinus *De Aquis* I, 7, 5', *Latomus*, vol. 20 (1961), 541-8.
45.20 Bieber, M., 'The Aqua Marcia in coins and in ruins', *Archaeology*, vol. 20 (1967), no. 3, 194-6.
45.21 Fabretti, R.G., *De aquis et aquaductibus veteris Romae dissertationes tres* (Giov. Batt. Bussotti: Rome, 1680; repr. Collegium Graphicum, Portland, Oreg., 1972).
45.22 Garbrecht, G. 'Rom', *WAS* 2 (1987), 208-13.
45.23 Kühne, G., 'Die Wasserversorgung der antiken Stadt Rom', *Wasserversorgung im antiken Rom* (1986), 79-128.
45.24 Lanciani, R., *Le aque e gli aquedotti di Roma antica* (Rome, 1881).
45.25 Lanciani, R., *I commentari di Frontino intorno le acque e gli acquedotti* (Rome, 1881).
45.26 Pace, P., *Gli acquedotti di Roma* (Rome, 1983).
45.27 Panimolle, G., *Gli acquedotti di Roma antica* (Rome, 1968).
45.28 Parker, J.H., *The Aqueducts of Ancient Rome* (Oxford, 1876).
45.29 Van Buren, A.W., 'Come fu condotta l'acqua al monte Capitolino?' *Rendiconti della Pontificia Accademia Romana di Archaeologia*, vol. 18 (1941-42).
45.30 Van Deman, E., *The Building of the Roman Aqueducts* (Washington, 1934).
45.31 Winslow, E.M., *A Libation to the Gods* (London, 1963).

Subiaco

45.32 Smith, N.A.F., 'The Roman dams of Subiaco', *Technology and Culture* 22 (1970), 58-68.

Syracuse

45.33 Burns, A., 'Ancient Greek water supply and city planning: a study of Syracuse and Acragas', *Technology and Culture* 15.3 (July, 1974), 389-412.
45.34 Schubring, J., 'Die Bewässerung von Syrakus', *Philologus, Zeitschr. f. d. klassische Altertum*, no. 22 (1985), 577-638.

46. Jordan

46.1 Lindner, M., 'Petra', *WAS* 2 (1987), 196-201.
46.2 Oleson, J.P., 'Aqueducts, cisterns and the strategy of water-supply at Nabataean and Roman Avara', *FCAS*, 45-62.
46.3 Oleson, J.P., 'The Humayma hydraulic survey: preliminary report', *Annual of the Department of Antiquities of Jordan* 30 (1986), 253-60.
46.4 Oleson, J.P., 'Nabataean and Roman water use in Edom: the Humayma hydraulic survey, 1987', *EMC/Classical Views* (Calgary) 32 N.S. 7 (1988), 117-29.

47. Morocco

47.1 Etienne, R., *Le quartier nord-est de Volubilis* (Paris, 1960).

48. North Africa

48.1 MacKendrick, P., *The North African Stones Speak* (London, 1980).
48.2 Romanelli, P., *La politica Romana delle acque* in *Tripolitania*, in *La Rinascita della Tripolitania* (Milan, 1926).
48.3 Saladin, M.H., [report untitled] *Archives des missions scientifiques et littéraires*, vol. 13 (Paris, 1887), 1-225.
See also: 18.26, 20.20.

49. Spain

General

49.1 Casado, C.F., *Acueductos Romanos en España* (Madrid, 1972).
49.2 Grewe, K., 'Römische Wasserleitungen in Spanien', *Schriftenreihe der Frontinus-Gesellsch.* 7 (Köln, 1984).
49.3 Wiseman, F.M., *Roman Spain* (London, 1956).

Almuñécar

49.4 Grewe, K., 'Almuñécar', *WAS* 3 (1988), 177-9.

Mérida

49.5 Jiménez, A., 'Los Acueductos de Mérida', *Augusta Emerita: Actas del Simposio Internacional Commemorativo del Bimilenario de Mérida* (Madrid, 1976), 111-25.
49.6 Jiménez, A., 'Los Acueductos de Emerita', *Actas del Bimilenario de Mérida* (Madrid, 1969), 103.
See also Dams (9): Proserpina, Cornalvo

Segovia

49.7 Casado, C.F., *El Acueducto de Segovia* (Barcelona, 1973).
49.8 Gallardo, A.R. *Supervivencia de una obra Hidraulica: el Aqueducto de Segovia* (Segovia, 1975).

49.9 Grewe, K., 'Segovia', *WAS* 3 (1988), 219-23.

50. Switzerland

50.1 Blondel, L., 'L'aqueduc antique de Genève', *Genava* (1928), 21ff.
50.2 Olivier, E., *L'alimentation d'Aventicum en eau* (Neuchatel, 1942), repr. from *Bulletin de la Société Neuchateloise de Géographie* 48 (1942).
50.3 Olivier, E., *Médecine et santé du Pays de Vaud*, in Bibliothèque Historique Vaudoise 29-31 (Lausanne, 1962).
50.4 Tissieres, P., 'Quelques problèmes de l'eau a forum Claudii Vallnesium' (Octodurus) *Annales Valaisannes* 2e serie, 53e année (1978), 182.

51. Syria

51.1 Crouch, D., 'The water system of Palmyra', *Studia Palmyrenskie*, vol. 7 (1975), 151-86.

52. Tunisia

52.1 Baur, A., 'Wasserversorgung in Tunisien', *Schriftenreihe der Frontinus-Gesellschaft*, Heft 10, Symposium (Köln, 1986), 163-73.
52.2 Gauckler, P., *Enquête sur les installations hydrauliques romaines en Tunisie* (Tunis, 1897-1912).

Carthage

52.3 Rakob, F., 'Das Quellenheiligtum in Zaghouan und die römische Wasserleitung nach Karthago', *Mitteilungen des deutschen archäol. Instituts, römische Abteilung* (1974), 41-88.
52.4 Rakob, F., 'L'aqueduc de Carthage', *Doss. Arch.* 38 (Oct.-Nov. 1979).

53. Turkey

General

53.1 Bean, G.E., *Aegean Turkey* (London, 1966).
53.2 Bean, G.E., *Turkey's Southern Shore* (London, 1968).
53.3 Bean, G.E., *Lycian Turkey* (London, 1978).
53.4 Bean, G.E., *Turkey Beyond the Maeander* (London, 1980).
53.5 Lanckoronski, C., *Städte Pamphyliens und Pisidiens*, vol. 1 (Vienna, 1890).
53.6 Özis, Ü., 'Su mühendisligi tarihi açisindin Anadoludaki eski su yapilari (Ancient water works in Anatolia with regard to hydraulics engineering history)' VI. *Bilim Kongresi, Mühendislik Arastirma Grubu Tebligleri*, n. 384/MAG. 49 (1977) 1-21, repr. (Izmir, 1987). [in Turkish; numerous illustrations and plans]
53.7 Özis, Ü., 'Ancient water works in Anatolia', *London Water Resources Development* 3, no. 1 (1987), 55-62.
53.8 Texier, C., *Description de l'Asie Mineure* (Paris, 1839-49).
53.9 Weber, G., 'Wasserleitungen in kleinasiatischen Städten I und II', *JDI* 19 (1904), 86-101; 20 (1905), 202-10.

Alabanda

53.10 Özis, Ü., Atalay, A., Hasal, M. & Atalay (Utku), V., 'Antike Fernwasserleitungen von Alabanda und Gerga', *Leichtweiss-Institut für Wasserbau, Mitteilungen*, Heft 64 (1979), 8f.

Antioch (Orontes)

53.11 Lassus, J., 'L'eau courante à Antioch', *JEAR*, 207-29.
53.12 Lassus, J., 'Das fliessende Wasser von Antiochia', *Leichtweiss-Institut für Wasserbau, Mitteilungen, Tagung über römische Wasserversorgungsanlagen*, B. 3 (1977), 27.
53.13 Lassus, J., 'Une villa de plaisance a Daphne-Yakto', *Antioch on the Orontes*, ed. R. Stillwell (Princeton, 1938).
53.14 Weulersse J., 'Antioche, Essai de géographie urbaine', *Bullétin des Études Orientales* 4 (1934), 27-9.

Aphrodisias

53.15 Cakir, M., Verim, Ö. & Afsar, R., *Antik Afrodisias kenti su yapilari* (Water works of the antique city Aphrodisias) (Izmir, 1978).

Aspendos

53.16 Fahlbusch, H., 'Aspendos', *WAS* 2 (1987), 172-5.
53.17 Ward-Perkins, J.B., 'The aqueduct of Aspendos', *PBSR* 23 (1955), 115-23. [abbrev. Ward-Perkins]

Cilicia

53.18 Regler, R.M., 'Siedlungswasserbau der antiken Kustenstädte Westkilikiens', in E. Rosenbaum et al., *A Survey of Coastal Cities in Western Cilicia* (Ankara, 1967), 87-94.

Ephesos

53.19 Alzinger, W., 'Ephesos', *WAS* 2 (1987) 180-4.
53.20 Linguri, M., Tulgar, T. & Samli, F., *Antik Efes sehri su getirme ve kanalizasyonu* (Water supply and sewerage of the antique city Ephesus) (Izmir, 1974).
53.21 Wieberg, W., 'Der Aquädukt des C. Sextus Pollio', *Forschungen in Ephesus*, Bd. 3 (Vienna, 1923), 256-65.

Foça

53.22 Önen, N., Özyurt, S. & YaGci, G., *Foça tarihsel su iletimi* (Historical water conveyance to Phocea) (Izmir, 1975).

Istanbul

53.23 Dalman, K.O., *Der Valens-Aquädukt im Konstantinopel, Istanbuler Forschungen des deutschen Archäologischen Instituts*, no. 3 (Berlin, 1933).

53.24 Eyice, S., 'Byzantinische Wasserversorgungsanlagen in Istanbul', *Leichtweiss Institut für Wasserbau, Mitteilungen*, Heft 64 (1979), 31.
53.25 Forchheimer, P. & Strzygowski, J., *Die byzantinischen Wasserbehälter von Konstantinopel* (Vienna, 1983).
53.26 Özis, Ü., 'Historical parallels [between] the water supply development of Roma and Istanbul', IAHR, *Water Resources Developments in Perspective* (1987), 35-44. [See 16.5.]

Laodicea

53.27 Weber, G., 'Die Flüsse von Laodicea', *Mitteilungen des deutschen archäologischen Instituts, athenische Abteilung* 23 (1898), 178-95.
53.28 Weber, G., 'Die Hochdruck-Wasserleitung von Laodicea ad Lycum', *JDI* 13 (1898), 1-13.

Miletos

53.29 Sertöz, A., *Antik Milet sehrinde su getirme* (Water supply to the antique city Miletus) (Izmir, 1974).
53.30 Wiegand, T., 'Zur Entwicklung der antiken Brunnenarchitektur', *Milet I, 5: Das Nymphaeum* (Berlin, 1919), 73-88.

Pergamon

53.31 Fahlbusch, H., 'Wasserversorgung griechischer Städte, dargestellt am Beispiel Pergamon', *Leichtweiss-Institut für Wasserbau, Mitteilungen*, Heft 71 (1981), 137-73.
53.32 Garbrecht, G. & Fahlbusch, H., 'Wasserwirtschaftliche Anlagen des antiken Pergamon: Die Kaikos-Leitung', *Leichtweiss-Institut für Wasserbau, Mitteilungen*, Heft 44 (1975).
53.33 Garbrecht, G. & Fahlbusch, H., 'Wasserwirtschaftliche Anlagen des antiken Pergamon: Umbau und Neubau der Kaikos-Leitung', *Leichtweiss-Institut für Wasserbau, Mitteilungen*, Heft 60 (1978).
53.34 Garbrecht, G., 'Die Entwicklung der Wasserwirtschaft Pergamons bis zur römischen Kaiserzeit', *Leichtweiss-Institut für Wasserbau, Mitteilungen, Tagung über römische Wasserversorgungsanlagen*, Bd.2 (1977), 29.
53.35 Garbrecht, G., 'Die Wasserversorgung des antiken Pergamon', *WAS* 2 (1987), 11-47.
53.36 Hecht, K., 'Wasserwirtschaftliche Anlagen des antiken Pergamon: Die Aquädukte der Madradag-Kanalleitung sowie die Aquädukte XLII und XLIII der Aksu-Leitung', *Leichtweiss-Institut für Wasserbau, Mitteilungen*, Heft 78 (1983), 134.
See also: 16.14, 16.15.

Perge

53.37 Fahlbusch, H., 'Perge', *WAS* 2 (1987), 193-5.

Priene

53.38 Tanriöver, A., *Prienedeki tarihi su yapilari* (Historical waterworks in Priene) (Izmir, 1974).

53.39 Wiegand, T. & Schrader, H., 'Wasseranlagen', *Priene*, vol. 4 (Berlin, 1904), 68-80.

Side

53.40 Fahlbusch, H., 'Side', *WAS* 2 (1987), 218-21.
53.41 Izmirligil, Ü., 'Die Wasserversorgungsanlagen von Side', *Leichtweiss-Institut für Wasserbau, Mitteilungen*, Heft 64 (1979).
53.42 Karanjac, J. & Günay, G., 'Dumanli spring, Turkey – the largest spring in the world?', *Journal of Hydrology* 45 (1980), 19-35.
53.43 Mansel, A.M., *Die Ruinen von Side* (Berlin, 1963).
53.44 Mansel, A.M., *Side: 1947-1966 Yillari Kazilari ve Arastirmalarin Sonuçlari* (Ankara, 1967), 87-94.

Smyrna (Izmir)

53.45 Weber, G., 'Die Wasserleitungen von Smyrna I und II', *JDI* 14 (1899), 4-25 & 167-88.

Notes

For abbreviations see pp. 357-8.

Chapter 1

1. *MAGR* 110 n. 3.

2. Compare the ancient Greeks' description of the massive masonry of Tiryns and Mycenae as 'Cyclopean', on the grounds that only giants such as the mythical Cyclopes could have lifted the blocks. For Arab reactions to the Carthage aqueduct, see *JEAR* 312: the seventeenth-century writer El Kairouani maintains that its arches are the Africans' answer to the Egyptians' pyramids.

3. Herschel xvii: 'We laymen have been waiting a long time for Latin scholars to do this thing for us, and they have not responded.' Herschel's complaint was not so much that classicists did not try to master the hydraulics of Frontinus (evidently accepting their unspoken position that this would be quite beyond their powers – though it was not beyond Frontinus'), as that they refused even to translate him, thereby leaving 'very many kinds of people, professional men and others, ... shut out from the pleasure of reading our author so long as he has been immured in the Latin tongue'. Worse than that, some even have 'spoken rather scornfully of our author as though he were not worth the reading in any language', an idea well summed up in Duruy's *History of Rome* (as quoted by Herschel): 'Columella, Pomponius Mela and Frontinus have left some valuable remarks on agriculture, geography, tactics, and aqueducts; but their books belong to the class which furnish facts without giving ideas.'

For further discussion of some of the points raised, see *FCAS passim*, esp. A. Trevor Hodge, 'Conclusion', 163-72.

4. See pp. 48-9. below.

5. e.g. Nicomedia, in Pliny, *Ep.* 10, 37. In *FCAS*, 153-4, there is a fine analysis by Philippe Leveau of the relative costs of aqueduct-building, which makes it plain that, unlike such things as theatres, even millionaires could not afford to finance one single-handed. The 3.5 million sesterces lost by the people of Nicomedia on their aqueduct would have been enough to build only a conduit less than 2 km long.

6. N.J. Schnitter-Reinhardt, 'Koloniale Aquädukte in Mexico', *Schriftenreihe der Frontinus-Gesellschaft*, Heft 5 (Symposium, Berlin, 1981), 71-81 (with full table, p. 81, of the statistics of forty Mexican aqueducts).

7. See pp. 284-5 below.

8. The celebrated fountain of Trevi, at Rome, is a survival, from a later date, of the same thing, being constructed as a grandiose city terminal of what was originally the Aqua Virgo (19 BC), rebuilt and restored to service by the Popes in the fifteenth and sixteenth centuries. Agrippa's 300 statues are attested by Pliny, *NH* XXXVI, 121.

9. The most comprehensive list I have seen is that formulated by G. Garbrecht (private letter): 'history, political administration, legal aspects, sociology, water demand, engineering, hydrology, hydraulics, stability analysis, construction materials and techniques, literary evidence, archaeological excavation, and ecology.' On most of these aspects little or nothing has been done, though one must make particular mention of Werner Eck, 'Die Wasserversorgung im römischen Reich: Sozio-politische Bedingungen, Recht und Administration', *WAS* 2 (1987), 51-101.

10. The widest and most concentrated treatment of this particular aspect, though limited to one single aqueduct, is George W. Hauck, *The Aqueduct of Nemausus* (Jefferson, N.C./London 1988) (henceforward 'Hauck'). The semi-fictionalised form of this work, an account of the building of the aqueduct of Nîmes, as seen through the eyes of the work force and the engineer in charge, while sacrificing the orthodox format of traditional scholarship, ensures that no point in the building process is omitted or glossed over. The weakness is that since, at doubtful points, the author has either to select arbitrarily which interpretation to follow, or to fall back on downright invention so as to complete his narrative, the unwary or uninstructed reader may be misled into thinking that whenever the author describes something as happening, then that must be what really did happen. This caveat apart, many readers, and students of engineering in particular, will find Hauck's work of great interest.

11. Wooden pipes, pp. 111-13. below; stone pipes, p. 110; dams, p. 79f; reservoirs, p. 279.

12. Herschel 259: 'The truth is, that these much vaunted works, whose *ruins* stand so long, were very poorly designed to *contain water*.' Note: the pagination in the 1973 reprint (see bibliography) is different, all pages being 28 less than the original edition. All page numbers here given are the original figures.

13. *CIL* VIII, 18587 (Lamasba); XIV, 7696 (Crabra); XIII, 1623 (Chagnon); VIII, 2728 (Nonius Datus; his tunnel was at Saldae, the modern Bejaia (Fr. Bougie). For further commentary on his works see ch. 6 n. 6 p. 423; also Hauck 42-3).

14. Smith 58.

15. For a particularly flagrant example, see pp. 280-1 and esp. ch. 10, n. 14, p. 456.

16. Grimal XV: 'le *De Aquae Ductu* est donc, d'abord, un écrit politique.' When one thinks of Frontinus' publication of his successes in technical reforms in a public service and in a political context, one cannot but be reminded of the proud boast of Mussolini and the Italian Fascists that they made the trains run on time.

17. '*Nosse quod suscepi*' – Frontinus, *Aq.* 1 (Preface). The extent to which Frontinus' work rests on his own observation and how far on the archives of the Water Office, is fully discussed by Grimal X-XIV.

18. Frontinus wrote a book on surveying, now surviving only in fragmentary form, in the *Corpus Agrimensorum Romanorum*, ed. C. Thulin (Teubner ed., Stuttgart, 1971), 50, 70. He speaks of the '*mensor, bonus vir et iustus*', quoted with approval by Herschel (201), who translates *mensor* as 'engineer'. Frontinus also wrote books on Warfare (*De Re Militari*), to which an appendix, 'The Stratagems', survives intact, and on the Tactics of Homeric War; he is also reported to have written a book on agriculture, but one with the title *De Coloniis* attributed to him is apparently not his work (A. Kappelmacher, *RE* X, 1, 597; 603, s.v. *Iulius* (*sc.* Frontinus)). Aelian, *De Instruendis Aciebus, praef.*, found him both informed and enthusiastic on Greek scholarly works. The poet Martial (X, 58, 1-6) recounts passing literary *soirées* with him while he was relaxing off-duty at his villa near Anxur (= Terracina) – '*doctas tecum celebrare vacabat/Pieridas*'. He seems to have been a man of wide interests and abilities, and we must beware of assuming, by a kind of *banausis* in reverse, that being a down-to-earth administrator or engineer disqualifies a man from having any appreciation of art and literature.

19. Vitr. VIII, 5, 3; Archim. *Fluit.* 1, 2. Callebat, (142) notes that the quotation is inappropriate to Vitruvius' context, 'et notre auteur semble surtout avoir cédé ici à ses "tentations érudites" ', but the point is that Vitruvius, unlike Frontinus, realises the need to express a general theory of hydraulics, and Archimedes' dictum is itself accurate.

20. It has been well put by C.E. Bennett, *Frontinus* (Loeb ed., London, 1925), xiv: 'Were one asked to point out, in all Roman history, another such example of civic pride and conscientious performance of civic duty, it would be difficult to know where to find it. Men of genius, courage, patriotism are not lacking, but examples are few of men who laboured with such whole-souled devotion in the performance of a homely duty, the reward for which could not possibly be large, and might possibly not exceed the approval of one's own conscience.'

21. Pliny, *Ep.* 9, 19, 1: '*impensa monumenti supervacua est; memoria nostri durabit, si vita meruerimus.*'

Chapter 2

1. The best known was at the city of Ulhu, which is described in detail, on an Akkadian cuneiform tablet now in the Louvre, by the Assyrian king Sargon II, who in 714 BC destroyed it. *SAT* 157, 183; J. Lassoe, 'The irrigation system at Ulhu (eighth century BC)', *Journ. of Cuneiform Studies* 5 (1951), 1, 21-32. Goblot (n. 4 below), 67, n. 6.

2. *SAT* 159-63; T. Jacobsen & Seton Lloyd, *Sennacherib's Aqueduct at Jerwan* (Chicago, 1935).

3. *SAT* 183; II Chron. 32:30.

4. The definitive study of the qanat is Henri Goblot, *Les qanats: une technique d'acquisition de l'eau* (ed. Mouton/École des Hautes Études en Sciences Sociales: Paris, 1979); hereafter in this chapter referred to as 'Goblot'. He gives an exhaustive bibliography (199-224). Qanats are also considered in *SAT* 156-8. We may further mention: M.A. Butler, 'Irrigation in Persia by kanats', *Civil Engineering* 3 (1933), 69-73; Col. E. Noel, 'Qanats', *Journal of the Royal Central Asian Society* 41 (1944), 191-202; P.W. English, 'The origin and spread of qanats in the Old World', *Proc. Amer. Philosoph. Soc.* 112 (1968), 170-81; H.E. Wulff, 'The qanats of Iran', *Scientific American* 218 (4), (1968), 94-105; F. Bemont, 'L'irrigation en Iran', *Annales de Géographie* 52 (1961), 597-620; E. Feylessoufi, *Eaux souterraines, kanats et puits profonds en Iran* (Teheran, 1959); G.R. Kuros, *Irans Kampf um Wasser* (Berlin, 1943); Gert Michel, 'Kanate in der Volksrepublik China', *Schriftenreihe der Frontinus-Gesellschaft*, Heft 11 (1988). For an evaluation of some of these articles, see Goblot 13-15. My own following account of qanats is largely based on Goblot, to whom I here cheerfully acknowledge my debt; its magnitude will be fully apparent only to the reader who refers to Goblot's work.

5. For a description of these see p. 198f. below.

6. See p. 208f. below. The only real contemporary written description of early qanat-building is a treatise in Arabic by Mohammad al Kardji dating to about AD 1010. It considers the matter in detail, including sections on tunnelling, surveying instruments and their operation, and maintaining a gradient. An edition in French was published in 1973 (Aly Mazaheri, *al Karagi, La civilisation* *des eaux cachées: traité de l'exploitation des eaux souterraines (composé en 1017), texte établi, traduit et commenté* (IDERIC, Université de Nice, 1973). See also summaries and references in F. Krenkov, 'The construction of subterraneous water supply during the Abbasid Caliphate', *Transactions of the Scottish Oriental Society* (Univ. of Glasgow, 1951), 23-32; K. Wolski, 'Les Karez', *Folia Orientalia* 6 (Univ. of Cracow, 1964), 179-204; M.A. Mazaheri, 'Le traité d'exploitation des eaux souterraines d'al Kardji', *Archives Internationales d'Histoire des Sciences* 18 (juin-dec., 1965), 300-1. For a discussion of the identity of this 'Iranian Frontinus', see Goblot 74-7 (esp. his n. 33), who also gives a full listing of the thirty chapter headings of the work.

7. Num. 32: 42; I Chron. 2:23. In the King James Version it is spelled 'Kenath'.

8. Goblot 19-22: Madrid = *madjira* + Lat. suffix *-etum*, = 'place of qanats' (Goblot 137). In Cyprus the current term, sanctified by use in official government documents, is 'chain of wells'. This is a misnomer, failing to recognise that the vital part of a qanat is the horizontal gallery that carries the water, the vertical shafts being there purely to assist its construction: it is rather like calling a railway tunnel a chain of wells because of its ventilation shafts.

9. Sir A.T. Wilson, *The Persian Gulf* (London, 1954). Hdt III, 9. The trouble is that Herodotus reports an impossibly long piped aqueduct made, 'they say', from the skins of oxen. The imagination does somewhat balk at oxhide pipes, especially stretching over 'a twelve-days' journey', while a linguistic confusion in translation between 'qanat' and 'reed, pipe' (see above) is quite likely. See Goblot 111.

10. Egypt: *SAT* 158; Goblot 112-15; G. Caton-Thompson & E.W. Gardner, 'The prehistoric geography of Kharga oasis', *Geographical Journal* 80 (1932), 369-406; G. Caton-Thompson & E.W. Gardner, *Kharga Oasis in Prehistory* (Univ. of London, 1952). Etruscans and *cuniculi*: Goblot 192. For Etruscan water supply see p. 45 below. Carthage: Goblot 120-1.

11. Timgad: aqueducts fed by 'des qanats, que l'auteur de la monographie sur l'hydraulique de Timgad ne savait ni nommer, ni décrire' – Goblot 123. The author is R. Godet, 'Le ravitaillement de Timgad en eau potable', *Lybica. Archéologie, epigraphie* 2, 1e sem. (1954), 65-72. The Algerian drainage tunnels disputed by Goblot (124) were so classified by Birebent.

12. Morocco: Goblot 147-58, esp. 152; Madrid: Goblot 136-39; Los Angeles: Goblot 29 n. 7. The Los Angeles qanats are all gone today. Goblot also (176-9) outlines the evidence for the qanat's penetration of Mongolia and China. His map showing the history of its expansion through the world (fig. 15, p. 176) repays study. See also G. Michel, n. 4 above.

13. Diod. Sic. V, 37 (drainage galleries in mines of Baetica; his description is of Augustan date, but the workings are of Carthaginian origin). See also O. Davies, *Roman Mines in Europe* (Oxford, 1935), 24-5. Goblot 62.

14. The papyrus of Turin, dated to 1500 BC. G.W. Murray, 'Gold mines of the Turin papyrus', in J. Ball, *Egypt in the Classical Geographers* (Cairo, 1942), 180-2. Goblot 61. *SAT* 7, 126 (dated to 1250). O.A.W. Dilke, *Greek and Roman Maps* (London, 1985), 15, fig. 1.

15. The earliest mines known were for mining flint; some have been dated by Carbon-14 to around 3350 (L.R. Nougier, *Géographie humaine préhistorique* (Paris, 1959). Mining for metal, mostly copper but possibly tin, can be traced back in the Zagros mountains (on the borders of Iran and Iraq) and on the Iranian plateau to the second millennium BC. T.A. Wertime, 'A metallurgical exploration through the Persian desert', *Science* 159 (1968), 927-35. Goblot 61. *SAT* 9, 88 (and pp. 1-117 for early copper metallurgy in general, with bibliography).

16. Goblot 27-8.

17. *SAT* 158. Goblot surprisingly gives no tunnel dimensions but this seems about right. However, some of Forbes's other figures are at variance with Goblot's: he gives the average tunnel gradient as 1-3%, which seems impossibly steep (= 10-30 m per km) and 20-60 times steeper than Goblot puts it. Goblot is beyond doubt the more reliable source.

18. Goblot 35. One may compare the construction technique of the deep-level tubes on the London underground railway, which, cutting through clay, runs in a tunnel of circular iron segments bolted together.

19. They also much resemble the craters from a stick of bombs dropped from a bomber, and have sometimes been mistaken for this.

20. 10-15 km, Gonabad: Goblot 39; 35-6. At Gonabad there is a series of three separate qanats side by side, of thirteenth-century AD date. All are enormous, with a tunnel 35 km long and a maximum well depth of 300 m. Spacing: Goblot pl. 2 and p. 36. A study of modern qanats (*foggaras*) in the Sahara notes that 'les puits d'evacuation et d'aération sont en très grand nombre, ce qui exprime la pauvreté technique des constructeurs, incapable de forer plus de 5 ou 8 mètres de galerie sans un orifice d'évacuation' (A. Cornet, 'Essai sur l'hydrologie du Grand Erg occidental et des régions limitrophes: les foggaras', *Travaux de l'Institut de Recherches Sahariennes* 8 (1952), 71-122). The same author notes the excessive slope of the tunnel – an average of 5 to 6%, but rising sometimes as high as 26 or even 45%, causing excessive erosion. He sums up: 'Si cette décadence se poursuit, bientôt plus personne ne sera capable de construire un *foggara*.' For a full exposition of the actual techniques in planning and digging a qanat, see Goblot 30-4; also my pp. 205ff. below.

21. Vitruv. VIII, 6, 3; but see ch. 5, n.19 p. 415. Callebat 160.

22. It also implies a very large work force with no need to spare the labour. A 100 metre-long stretch of tunnel will be served by, say, five shafts. Where the tunnel is 100 m below the surface, as it often is, each shaft will itself be 100 m deep, giving a total of 500 m of largely useless vertical shafting to be dug just so as to give access to digging the 100 m horizontally that really counted. But all five shafts could be sunk together. They therefore took five times as much work but the same amount of time, and enabled the 100 m of tunnel to be dug in one-fifth of the time that would be needed if there were only one access point. It therefore seems to me that this labour-intensive approach may illustrate the relative priorities of the qanat builders. A very great amount of extra work, provided it could all be done simultaneously, was accepted as a worthwhile price to pay for accelerating the total job schedule. This would seem to indicate a large, cheap labour force, and suggests that a close spacing of the shafts, as complained of in the Sahara by Cornet, may reflect economic considerations and the availability of labour rather than engineering incompetence. Goblot seems not to have considered this.

23. It is not clear how much time had to be spent cleaning the qanats. Goblot speaks of it as a never-ending task, like the removal of the incrustation in Roman aqueducts. Speaking of the analogous Etruscan *cuniculi*, Sheldon Judson and Anne Kahane (n. 68 below) comment that a lot of them are still working satisfactorily today, without any cleaning or maintenance at all over a period of 2,000 years. Compare also GdM 287.

24. Goblot 41-2. It is estimated (ibid.) that in 1961 there were 22,000 qanats still operating in Iran, with a total discharge of around 48,384,000 m^3 per day (560,000 l/sec). This is the equivalent of a large river; in fact, it is about the same discharge as that of the Garonne at Bordeaux.

25. The Clepsydra is published by A.W. Parsons, 'Klepsydra and the paved court of Pythion', *Hesp.* 12 (1943), 191-267. From calculations made by Parsons (who drained it and let it fill again), in the arid days of late summer its average daily discharge is around 2.5 m^3 (compare p. 305 and ch. 11 n. 4 p. 464). We may also notice (op. cit., 223) that 2,000 people besieged in the Acropolis from July 1826, till the end of May 1827, managed to subsist on this supply, which comes out to around 1.5-2.0 litres per person daily.

26. The Greeks differentiated by calling a natural spring *pêgê*, a fountain *krênê* (R.E. Wycherley, *CR* 51 (1937), 2-3).

27. The replacement of Kallirhoe by the artificial Enneakrounos is due, typically, to the 'tyrannic benificence' of Peisistratos. The actual location of the Enneakrounos has been for long disputed. Pausanias (I, 14, 1) clearly puts it in the Athenian Agora, at the south-east corner, next to the Odeion of Agrippa, and the excavators of the Agora agree: 'The Southeast Fountain House is the strongest candidate now known for the name Enneakrounos' – Homer A. Thompson & R.E. Wycherley, *The Agora of Athens* (*The Athenian Agora*, vol. 14; Princeton, 1972), 198; so also R.E. Wycherley, *The Stones of Athens* (Princeton, 1978), 248. The other possible location is on the other side of the Acropolis altogether, near the Olympieion and alongside the Ilissos. Travlos (J. Travlos, *Pictorial Dictionary of Athens* (London, 1971), 264) supports this identification, though his phrasing is somewhat ambiguous, making it appear that he supports both locations: he apparently means that the real Enneakrounos was on the Ilissos, but the Agora one was the one Pausanias, wrongly, so identified. He repeats the Ilissos identification in his article (s.v. 'Athens') in the *Princeton Encylopedia of Classical Sites* (Princeton, 1976), 109. See also F. Graber, 'Die Enneakrounos', *AM* 30 (1905), 1-64.

28. The reader needs to be careful of the terminology. What I have here called pipelines are sometimes referred to as aqueducts, and the Agora picture book by Mabel Lang, *Waterworks in the Athenian Agora* (Princeton, 1968), fig. 20 and accompanying text, describes 'a large stone aqueduct'. This could be misleading, since the actual water ran only in a 20 cm deep channel sunk in the middle of the floor, and in three terracotta ducts slung half-way up the walls; the actual stone structure inside which these four water conduits run is thus something rather more like a maintenance gallery.

29. op. cit. below fig. 20.

30. As, e.g., at Kuttolsheim and Caesarea (pp. 116-17 below).

31 Paus. I, 40. G. Gruben, 'Das Quellhaus von Megara', *AD* 19 (1964) 37-41; R. Dellbruck & K.G. Volkmuller, 'Das Brunnenhaus des Theagenes', *AM* 25 (1900), 23-33.

32 H. Fahlbusch, 'Vergleich griechischer und römischer Wasserversorgungs-anlagen', *Schriftenreihe der Frontinus-Gesellschaft,* Heft 6, *Symposium über die historische Entwicklung der Wasserversorgungstechnik* (Cologne, 1982), 78.

33. On the Samos aqueduct see E. Fabricius, 'Altertumer auf der insel Samos', *AM* (1884), 171ff.; H. Kienast, 'Zur Wasserversorgung griechischer Staate', *Wohnungsbau im Altertum* (Berlin, 1979); H. Fahlbusch (n. 32 above), 50-1. Fabricius noted that not only were the springs still flowing, but that the reservoir was still acting as a settling tank, its floor being covered with a layer of mud.

34 Hdt. III, 60. The fact that Eupalinos came from Megara, the site of the fountain of Theagenes already described, led F. Graber ('Die Wasserleitung des Peisistratos und die Wasserversorgung des alten Athen', *Zentralblatt der*

Bauverwaltung (1905), 557), to suggest that at this period Megara was a recognised leader in hydraulic engineering.

35. Most surprisingly, the figure for the length is given quite wrongly in the standard commentary on Herodotus, the authors of which even rebuke him for his inaccuracy: 'he exaggerates the length of the tunnel, which is really about 1,100 feet' (W.W. How & J. Wells, *A Commentary on Herodotus* (Oxford, 1912), vol. 2, 237). The real length is 1,010 m. Herodotus considers it one of the three engineering wonders of the Greek world.

36 'horizontal ...' – White 160. Despite White's clear statement, his own drawing (fig. 164, p. 159) shows the tunnel as sloping. What has caused all the confusion is the fact that one end of the tunnel is 2 m lower than the other, leading to the natural supposition that the tunnel inside was at a downhill slope connecting the two. This is not so. From each end the tunnel runs level, and when the two halves meet they are still some distance apart vertically, a discrepancy that has had to be resolved by some juggling around the junction point in the middle of the mountain. See W. Kastenbein, 'Untersuchungen am Stollen des Eupalinos auf Samos', *AA* (1960), 178-98; A. Burns, 'The tunnel of Eupalinos and the tunnel problem of Hero of Alexandria', *Isis* 62 (1971), 172-85, esp. p. 175. On the tunnel see also U. Jantzen, H. Kienast & R.S.C. Felsch, 'Die Wasserleitung des Eupalinos (Samos, 1971)', *AA* (1973), 72-89; U. Jantzen, R.S.C. Felsch & H. Kienast, 'Die Wasserleitung des Eupalinos (Samos, 1973/74)', *AA* (1975), 19-35; J. Goodfield, 'The tunnel of Eupalinus', *Scientific American* 210 (June, 1964), 104-10 (good photographs); Fahlbusch (n. 32 above), 59-61; H.J. Kienast, 'Samos' *WAS* 2 (1987), 214-17.

37. 'Er ist vielleicht als Arbeitsplattform für die eigentliche Wasserleitung zu deuten' – Fahlbusch (n. 32 above), 59. White 160.

38. Compare Hdt III, 146.

39. Athens, Syracuse, Acragas: Burns (n. 36 above), 171, 174; A. Burns, 'Greek water supply and city planning', *Technology and Culture* (1974), 393-4. For these 'double tunnels', see also W. Dorpfeld, 'Die Ausgrabungen an der Enneakrounos', *AM* 19 (1894), 144-5, and E. Fabricius, *RE* VI, 1159, s.v. 'Eupalinos'. The suggestion was made by Curtius that in tunnels without vertical shafts to the surface the second horizontal gallery was cut to facilitate ventilation.

40. Burns (n. 36 above), 174, suggests three possible reasons for the irregularities; (1) the miners were directing the tunnel by the sound of each other's picks, and the passage of sound underground is often distorted by various geological features; (2) the builder erroneously calculated that they should have already met, and 'consequently the two groups started looking for each other'; (3) by both diverging on to what was, in fact, known to be a collision course, they made an eventual junction certain, provided they could be sure that both galleries were at the same depth – there was no possibility of missing each other in an overlap, as happened at Saldae (p. 128 below).

41. J. Goodfield & S. Toulmin, 'How was Eupalinos' tunnel aligned?', *Isis* 56 (1965), 45-56; B.L. van Waerden, 'Eupalinos and his tunnel', *Isis* 59 (1968), 82-3; Burns loc. cit. (n. 40 above).

42. One of the more interesting of the rare exceptions is the pipeline from the spring in the late classical altar complex at Ephesos, which was formed of lengths of lead pipe with their joints embedded in a series of round marble blocks (the *Marmormuffen* of the German excavators; A. Bammer, *AA* (1972), 724, figs 16-17); *WAS* 2 (1987), 180, abb. 1.

43. B. Dunkley, 'Greek fountain-buildings before 300 BC', *BSA* 36 Session 1935-6

(1939), 142-204. Pipes: Fahlbusch (n. 32 above), 53-9 (see fig. 6, p. 54, for profiles of pipe sections in chronological order).

44. Burns (n. 38 above), 405; Hippocrates, *Airs, Waters and Places passim*; Ar. *Pol.* VII, 1330a39-1330b14. For the military view, F.E. Winter, *Greek Fortifications* (Toronto, 1971), 47: 'It must surely have been military necessity that led post-Archaic town-planners and engineers to keep their water-lines as far as possible below ground, that is, out of enemy reach.'

45. R.A. Tomlinson, 'Perachora: the remains outside the two sanctuaries', *BSA* 64 (1969), 155-258 (for the waterworks, 157-64, 195-232): 'The width of the runnel seems to have been determined by the size of the available ridge tiles' 202); this does not suggest that the runnel itself was considered a work of great importance.

46. Bucket-chain: Tomlinson (n. 45 above), 225-31. J. Oleson, *Greek and Roman Mechanical Water-Lifting Devices* (Toronto, 1984), 237-40.

47. D.S. Robertson, *Handbook of Greek and Roman Architecture* (Cambridge, 1943), 190, fig. 85.

48. 'Inverted siphon' – see the comments of Smith 51. GdM 176; Callebat 169. 'Sag pipe': Frederick S. Merritt, *Standard Handbook for Civil Engineers*, 2nd ed (New York, 1976), 21-31.

49. This list is based upon Fahlbusch (see Abbreviations p. 357), table 4, p. 65. See *WAS* 1 (1986), 181-3. Ephesos: A. Bammer, *AA* (1972), 714-28. Methymna: H.G. Bucholz, *Methymna* (Mainz, 1975), 57-8. Smyrna: G. Weber, 'Die Wasserleitungen von Smyrna I und II', *JDI* 14 (1899), 4-25 and 167-88. Laodicea: G. Weber, 'Die Hochdruck-Wasserleitung von Laodicea ad Lycum', *JDI* 13 (1898), 1-13. Patara: C. Merckel, *Die Ingenieurtechnik im Altertum* (Berlin, 1899), 504-6. See my Fig. 12. For long this was thought to be an early siphon because it has as a *venter* a kind of embankment of Cyclopean masonry. This is now questioned by J.J. Coulton (private communication) who, on the basis of a Vespasianic inscription on the *venter*, supports a Roman dating for the structure in its present form. Fahlbusch (op. cit. above), 85, dates it 'wohl in vorrömischer Zeit' and, in a more recent article (H. Fahlbusch, 'Gedanken zur Entsinterung von Druckrohren im Verlauf römischer Fernwasserleitungen', *3R International*, 24 Jahrgang, Heft 1/2 (Jan./Feb. 1986), 73-9) summarises: 'Auf der Ostseite über einem Durchgang ist eine Inschrift angebracht, derzufolge die gesamte Anlage zur Zeit Vespasians (69 bis 79 n.Chr.) von Grunde auf erneuert wurde.' For Pergamon, see below. All the others are to be found in G. Weber, 'Wasserleitungen in kleinasiatischen Städten I und II', *JDI* 19 (1904), 86-101 and 202-10. His dating (p. 209) is 'Alle Hochdruckleitungen in Pergamon, Smyrna, Trapezopolis, Tralleis, Laodikeia, Apameia Kibotos, Magnesia ad Sipylum, Philadelphia, Blaundos, Akmonia und den beiden Antiocheia sind in die hellenistische Epoche zurückzudatieren'. Perhaps more realistically Fahlbusch lists them all as 'frührömisch(?)'. Some of these (e.g. Antioch in Pisidia, Tralleis, Blaundos) are for part of their length carried on an arched arcade, in contradiction to my generalisation in the main text above that Hellenistic engineers did not do this. One must remember the disputed and unreliable (?) dating. If they really are Hellenistic and engineers of that age, when they were not building orthodox temples, did know and often used the arch, then we are faced with a basic re-evaluation of the accepted history of Greek architecture.

For other siphons outside Asia Minor, see Smith 52; SRA 177 n. 9; Merckel (above), 506ff.; Van Buren, *RE* VIIIA, 473. They have been listed in such places as Catania, Syracuse, and Selinus, usually on apparently slight evidence. This, however, is not to be despised. Siphons have always been under-rated, rather than the reverse. Archaeologists have always been more likely to overlook siphons that actually exist than to go around seeing them everywhere, so even slight evidence is to be respected.

50. G. Weber, 'Die Wasserleitungen von Smyrna I und II', *JDI* 14 (1899), 18-20. Ordinary stone pipe blocks ('excluding stone junction blocks') from 84 sites are listed, with bibliography and table of dimensions, by J.J. Coulton, 'Oinoanda: the water supply and aqueduct', *Anatolian Studies* 26 (1986), 46 n. 104. At Oinoanda alone, 33 blocks have been found (p. 34).

51. Smith 57; SRA 197-9.

52. It is worth recounting what happened. In June 1985, I published an article, 'Siphons in Roman aqueducts', in the *Scientific American*. Reaction from readers and other scholars led to an extensive continuing correspondence. The chief product of all this was a large and heterogeneous mass of uncoordinated and sometimes contradictory but very valuable material from a source of professional expertise that classicists are rarely privileged to tap. Much of my comments on the practicalities of siphon operation are derived from this source, and I am happy here to acknowledge my grateful debt to all who contributed.

53. K.G. Lanckoronski, *Die Städte Pamphyliens und Pisidiens* 1 (Vienna, 1890), 124. He does not discuss its purpose. J.B. Ward-Perkins, 'The aqueduct of Aspendos', *PBSR* 23 (1955), 119, fig. 2, reproduces essentially the same drawing, 'after Lanckoronski', but omits from it the vertical hole, without explanation.

54. G. Weber, 'Die Hochdruck-Wasserleitung von Laodicea ad Lycum', *JDI* 13 (1898), 6: 'noch einen runden Stein in das Loch eingepafst und die Fuge mit Kalkmortel vergossen'.

55. Susita, or Hippos, was one of the towns of the Decapolis, near the Golan Heights, and was served by a siphon 500 m long, 60 m head, carried in a pipe made of basalt blocks (bore diam. 30 cm). A length of 24 pipe sections has been uncovered *in situ*, and 'some of the pipes have circular holes cut in their top, a device which is very hard to explain' (R. Malinowski & Y. Peleg, 'Some pressure pipelines in Israel', *Papers at the Symposium of Historical Water Development Projects in the Eastern Mediterranean: Mitt. L. Inst.*, Heft 82 (1984), p. 22; also Y. Peleg, 'Die Wasserleitung nach Hippos und ihr Stein-düker' *Mitt. L. Inst.*, Heft 103 (1989), 325-36). I am grateful to Jehuda Peleg for bringing this to my attention: he writes (private communication) that the pipes where excavated were 'about 60 cm below ground level', which does complicate things if the holes were to be used as some sort of vent. See also *FCAS*, 132-3 (w. illustration).

56. SRA 216-17. In modern practice pipes are sometimes cleaned by pulling through them a large plastic sphere slightly smaller than the bore of the pipe, while keeping the water running; the increased velocity as it squeezes past the circumference of the sphere scours clear the inner surfaces of the pipe. Mr. R. Carrouché, a Los Angeles hydraulic engineer, also tells me that in the oilfields pipes are cleaned by inserting into them a device known as a 'go-devil' which cleans them out as it is propelled along by the current behind it, without being pulled. See ch. 6 n. 57 p. 430.

57. At both Laodicea and Smyrna these small vent holes are published by Weber (n. 49 above): p. 8, fig. 12 (Laodicea); p. 13, fig. 11 (Smyrna). Apart from suggesting that their purpose was 'zu Controllierung', he does not explain further. The shape of the holes, a sort of bell-shaped funnel, strongly suggests that they were corked with some sort of stopper. On the principle of 'safety valves' outlined above, Prof. Thorkhild Schiøler suggests to me (private communication) that they were blocked up by a round stone, embedded in plaster or mud and designed to blow free in the event of water hammer. Their shape certainly would be very suitable.

58. A modern plumber, Mr K. Puma, suggests to me that this accounts for the number of 'vent holes' in the Laodicea siphon: they were cut one after the other to free some obstruction that kept moving further on down the pipe. He has often seen rows of holes in modern terracotta pipelines that resulted from this, the commonest form of blockage being plaster or grouting that had come loose from the joints in the pipes, where it had been imperfectly applied. To me, however, the 'vent holes' at Laodicea do seem to be so neat, tidy and uniform that I feel they were more likely part of the original installation than a later emergency repair. But I may be wrong. Some of the holes are drilled directly on the joint between two blocks, and surely no engineer would have planned it that way originally.

59. The names of the eight are the Attalos, Demophon, Madradag, Geyiklidag, Apollonius, Madradag II, Kaikos and Aksu aqueducts (Fig. 18). The first five are terracotta pipelines, the last three are in masonry conduits; the last four are of Roman date, the others Hellenistic. The most important are the Madradag and the Kaikos, the sources being respectively 40 km and 60 km distant from Pergamon, though the line of the aqueduct is much lengthened by its sinuous course. The best overall account is G. Garbrecht, 'Die Wasserversorgung des antiken Pergamon', *WAS* 2 (1987), 13-47. See also G. Garbrecht, 'L'alimentation en eau de Pergame', *Doss. Arch.* 38 (Oct.-Nov., 1979), 26-33; Fahlbusch, (n. 32 above), 65-7 (Madradag siphon); G. Garbrecht, 'Wasserwirtschaftliche Anlagen des antiken Pergamon: die Druckleitung', in *Mitt. L. Inst.* 60 (1978); G. Garbrecht & G. Holtorf, 'Wass. An. ant. Perg.: die Madradag-Leitung', *Mitt. L. Inst.* 37 (1973); G. Garbrecht & H. Fahlbusch, 'Wass. An. ant. Perg.: die Kaikos-Leitung', *Mitt. L. Inst.* 44 (1975). *SAT* 164-6. See also F. Graber, *Altertumer von Pergamon I*, 3: *Die Wasserleitungen* (Berlin, 1913). Useful summaries in English are published in the *Proceedings of the 17th Congress of the International Association of Hydraulic Research (IAHR)* (Baden-Baden, 1977): 'The development of the Pergamon water supply between 200 BC and 300 AD' (pp. 758-62: H. Fahlbusch); 'The effects of the Smyrna earthquake of 178 AD on the water supply of Pergamum' (pp. 768-9: W. Brinker).

60. Fahlbusch (n. 32 above), 66, fig. 13a, from which my Figs 20 and 21 are taken.

61. Estimates of the discharge of the siphon pipe are 45 litres per second (Garbrecht, *Doss. Arch.* (n. 59 above), 32; Garbrecht & Fahlbusch, *IAHR* (n. 59 above), 764), and 30 l/sec (Fahlbusch, (n. 32 above), 65). There are also differences in the estimate of the hydraulic gradient. The location of the header tank on Hagios Georgios at an altitude of 376 m seems to be agreed, but that of the receiving tank at the other end varies according to where it is thought to have been. Height loss runs from 41 m to 20 m (Garbrecht, *Doss. Arch.* 32; *IAHR* 764). The main point is not affected: effectively, the single pipe of the siphon carried the same quantity of water as the three gravity-feed pipes serving it, because of increased speed. Rate of flow through the siphon is estimated at 1.20 m per second.

62. Forbes: *SAT* 165. Landels 47 ('There is no clear evidence to tell us that material of which the pipeline was made ... it is highly doubtful that lead pipes could have withstood the pressure') also needs to be corrected. For the tests, see *IAHR* (n. 59 above), 764, 767, and ch. 5 n. 35 p. 418.

63. See ch. 6 n. 49 p. 429. For the pressure see e.g. Landels 48, fig. 12.

64. Les Tourillons: see p. 157 and ch. 6 n. 68 p. 431; SRA 185-7, and fig. 6, p.

191; K. Stehlin, 'Über die Colliviaria oder Colliquiaria der römischen Wasserleitungen', *Anzieger für schweizerische Altertumskunde* (1918), 167-75. Air pockets: SRA 197-200. See also p. 154 below, and Kottman (ch. 6 n. 56 below).

65. Fahlbusch (n. 32 above), 65; Vitr. VIII, 6, 8; Callebat 179-80.

66. *SAT* 165. White 162.

67. See p. 21 above.

68. The standard publication is S. Judson & A. Kahane, 'Underground drainageways in southern Etruria', *PBSR* 31 (1963), 74-99. See also J.B. Ward-Perkins, 'Veii: the historical topography of the ancient city', *PBSR* 29 (1961), 47-52, and K. Grewe, 'Etruskiche und römische Tunnelbauten in Italien', *Mitt. L. Inst.* 103 (1989), 131-52. An earlier account is Plinio Fraccaro, 'Di alcuni antichissimi lavori idraulici di Roma e della Campagna', *Opuscola* (Pavia, 1957), vol. 3, 1-36 (repr. from *Bulletino della Società Geografica Italiana* 5 (1919), 186-215); on this work, see Ward-Perkins's critical comments (op. cit., 51) who upholds instead the views of T. Ashby, *The Roman Campagna in Classical Times* (London, 1927), 239-40. See also Ward-Perkins in *Hommages à Albert Grenier*, vol. 3 (*Latomus*, vol. 58, 1962), 1636-43 ('Etruscan engineering: road-building, water-supply and drainage') and R. De La Blanchere, s.v. 'Cuniculus' in Daremberg & Saglio, *Dictionnaire des Antiquités* (Paris 1887), II, 2, 1589-94.

69. Judson & Kahane (n. 68 above), 78.

70. Ward-Perkins in *Latomus* (n. 68 above), 1643.

71. Judson & Kahane (n. 68 above), 88. The best-known *cuniculus*, at Veii, is known as the Ponte Sodo (i.e. Ponte Solido): it 'afforded a bridge of natural rock to the rich fields on the other side of the valley' (Ward-Perkins in *PBSR* (n. 68 above), 50.

72. Ward-Perkins in *Latomus* (n. 68 above), 1643. He also remarks of the Etruscan aqueduct at the Faliscan site of Corchiano (for which see *PBSR* 25 (1957), 123-7): 'The water channel (*specus*) is a *cuniculus* of characteristic form, cut along the brow first of the north and then of the south side of a narrow, cliff-girt valley and crossing from one side to the other upon a massive bridge of solid tufa masonry which rises 10 m above the silt of the valley bottom. Except that the stream in the valley is diverted through a *cuniculus*, cut in the rock-face of the cliff, instead of through an arched opening in the masonry, here are all the characteristic features of the Roman aqueduct.'

Chapter 3

1. Front. *Aq.* I, 16.

2. R. Meiggs, *Roman Ostia* (Oxford, 1973), 44.

3. See pp. 306, 330 below.

4. Anne Johnson, *Roman Forts* (London, 1983), 220.

5. See p. 157 below. Compare the complaint of Martial IX, 18.

6. K.D. White, *Roman Farming* (London, 1970), 10: 'The major industry (farming) on which this splendid edifice of culture depended for its survival has remained almost entirely inarticulate.' The point is perhaps best made by Horace, whose repeated praises of his Sabine farm make life there sound like a rest cure interrupted only by intermittent revels. Horace may for all we know have had to get up at five in the morning to see the cows got milked, but you would never think so from what he says. Either his estate was what is now called a 'hobby farm', or, if it was a serious enterprise seriously run, it was not Horace personally who did the running. Either way, he does not qualify as the voice of the true countryman.

Likewise, it has been observed by Seneca (*Ep.* 86) of Virgil's *Georgics* that, in spite of the surface air of knowledgeable and down-to-earth expertise, if any farmer tried actually to run his farm by its precepts, all he would get out of it would be a rapid bankruptcy (White, op. cit., 39-41). Both Horace and Virgil were (relatively speaking) countrymen by birth, and Horace's father was a smallholder (as well as being an auctioneer), Virgil's a farmer.

7. Napoleon III consciously identified with Augustus. The result was that during his rule Roman antiquities were accorded great respect and anything else (except Gallic) wholly neglected, even in France. The chief victim was the Greek antiquities of Provence and the Côte d'Azur, and as great a city as Marseille had to wait till as late as 1967 for its first real excavation: this is frankly incredible, but true.

8. Cherchel, 181. Local authorities are represented in the Preface, contributed by Mounir Bouchenaki, Sous-Directeur des Beaux-Arts et Antiquités at Tipasa, who concludes: 'Le mérite du présent travail est celui d'avoir posé ces différentes questions et d'inciter à un réexamen des vestiges archéologiques, à partir d'une vision décolonisée.' It must be made clear that the treatment of the archaeological evidence is in every way proper and scholarly, and the authors are consistent: Ph. Leveau, in the popular periodical *L'Histoire* 105 (Nov. 1987), 96-104, 'A quoi servaient les aqueducs romains?', urges the thesis that aqueducts in general, including the Pont du Gard, were 'un luxe inutile et couteux', and merely a 'monument à l'orgueil romain'. The reaction of the Nîmes Tourist Board remains unknown.

9. *SAT* 150; for Vitruvius on wells, see VIII, 6, 12.

10. Smith's verdict on well sinking: N.A.F. Smith, *Man and Water* (London, 1976), 70; artesian wells: ibid., 107. They were in 1126 sunk 'at Lillers, in the Artois district – hence the name', but had previously been mentioned in the writings of the Arabic geographer al-Biruni. The most familiar exponent of the Earth-floating-on-water theory was, of course, Thales. Dora Crouch (n. 22 below), 355, declares that 'Greek engineers by noting the pattern of the folded rock, could predict where water would be found below the surface'.

11. Bologna and Modena: Smith, ibid., where also the clear declaration 'Prior to the nineteenth century … artesian water was exceedingly small in quantity and highly localized in its use. In no sense did artesian water contribute to any comprehensive water-supply system; such a development was, however, to come in the nineteenth century.'

12. More specifically, 'La zone de distribution de ces puits est trés vaste: nous les retrouvons aussi bien en Angleterre, aux Pays-Bas, en Belgique, dans la vallée du Rhin, en France, en Suisse, qu'actuellement dans les plaines du Danube ou en Suède. En général, dans l'Antiquité, ils sont localisés dans les parties basses du pays et sont particulièrement nombreux dans le Nord de la Gaule et la vallée du Rhin.' (J. Mertens, 'Puits antiques à Elewijt et les puits romains en bois', *Antiquité Classique* 20 (1951), 97. Elewijt is in Belgium. For a bibliography of some 25 or so wood-lined wells and springs of various sorts, mostly from the Low Countries, see his p. 94, nn. 1-7; 95, nn. 1-5; 96, nn. 3-6. A few of these were of very primitive construction, 'puits monoxyles, consistant en un tronc d'arbre évidé' (96, n. 6; *Gallia* 1 (1943), 66; R. Dauvergne, *Sources minerales, thermes gallo-romains et occupation du sol aux Fontaines-Salées* (Paris, 1944), 46, *q.v.* also for a study of 'captages de sources'); *MAGR* 2, 452. For illustrations of the different forms of carpentry in square, wood-lined wells, see C.F.C. Hawkes & M.R. Hull, *Camulodunum* (Oxford, 1947), 126; also F. Oswald, 'The pottery of a Claudian well at Margidunum', *JRS* 13, 117-18, and fig. 10. Margidunum is a Roman fort in Nottinghamshire, half-way between Leicester and Lincoln.

13. Anne Johnson, *Roman Forts* (London, 1983), 204. See also Mertens (n. 12, above) 88; 92-3 (photographs of a particularly well-preserved example).

14. Johnson, ibid.; *SAT* 151; H. Jacobi, 'Die Be- und Ent-wasserung unserer Limeskastelle', *Saalburg Jhb. VIII* (1934), 32-60. There is a good illustration of a well changing from square to round in Peter Marsden, *Roman London* (London, 1980), 69. See also R. Merrifield, *The Roman City of London* (London, 1965), 146, and *passim* for individual wells. In Britain barrel-wells were particularly common at London (*Archaeologia* 66 (1915), 246; *Antiquaries Journal* 6 (1926), 186) and Silchester (*Arch.* 16 (1898), 19ff; 55 (1987), 413); at Silchester some of the barrels were remarkably large, almost two metres high, and may have arrived carrying imports from the Pyrenees, since their wood has been identified as silver fir (the local wood normally used in hydraulic installations was oak). See Hanson, 397-8. The same thing has been found at London (Marsden, loc. cit). Barrels seem to have been a North European (Gallic?) invention, replacing the amphorae normally used in the Mediterranean, for the storage and transport of wine; they are illustrated, being carried on carts and ships, in several well-known plaques and reliefs, notably the 'tank-wagon' from Langres (White 133, fig. 132), and the Moselle wine-barge, now in the Landesmuseum, Trier (Kretzschmer 87, fig. 143; also 77, fig. 128 (from Santander, Spain)).

15. Forbes (*SAT* 150) overstates: 'all were steined'. In Britain alone unsteined wells have been found at Dorchester, Chichester, Castleshaw, Richborough and elsewhere (Hanson, 396-7). Clay was sometimes used to line wells dug through another type of soil. We may also note that 'up until the Hellenistic period the vast majority of the wells in Athens were unlined', where the rock to be dug through was a 'soft grey-green shale' (J.M. Camp, *The Water Supply of Ancient Athens from 3000 to 86 BC* (Ph.D. thesis, Princeton, Jan. 1977; published by University Microfilms International), 176).

16. *SAT* 151.

17. Birebent 496. For a full study of N. African wells in general, see Birebent 494-9.

18. Sahara: Birebent 498. Forbes: *SAT* 150-1. On p. 183, in a table of notable dates in hydraulic history, he lists '320 BC: A 600′ well dug in the Sahara', but with no further details, reference, or evidence. This would be very impressive if only we could rely on it, but the truth is that Forbes cannot always be trusted: this item unfortunately follows immediately after he has listed two separate dates, 25-30 years apart, for the 'aqueduct of Acragas' and the 'aqueduct of Girgenti', evidently not realising that they are the same place; in the same table (p. 185) he says of the aqueducts of Lyon that 'these contain 18 syphons', the correct figure being nine. For the shafts at Laurion, see E. Ardaillon, *Les mines du Laurion* (Paris, 1897), 30-1; the deepest is the 'Puits Francisque', at 119 m. Poitiers: p. 69 and ch. 4 n. 5 p. 406.

19. At Lincoln it is possible that a force-pump was used to pump water uphill through a pipeline, but the evidence is scanty, confusing and controversial; for our purposes it is best neglected, as atypical. It also now seems that force pumps were in general much more used than hitherto supposed, and therefore may invalidate my statement. But the evidence is still fresh (J.P. Oleson, *Greek and Roman Mechanical Water-Lifting Devices* (Toronto, 1984), 300-25) and until it is more widely evaluated and studied, it is not clear just how and in what circumstances they were used. It therefore seems safer at present to assume that they played no serious part in basic urban water supply, though future studies may prove this view wrong. In rural irrigation the shaduf, noria, tympanum and other machines always had a large role.

20. Flat-bottomed buckets similar to the modern type were also known, and a very well-preserved wooden one from a well at Newstead, Scotland, is shown in White 157, fig. 162. See also my Fig. 25. It seems likely, however, that we are here dealing with a peculiarly northern type, like the barrel (which in construction it much resembles), that was not typical for Mediterranean regions. It may reflect the ready availability of wood in the North. It is hard, given the shape of wooden cask-like staves, to make a conical bucket tapering to a point, while it is quite easy with leather. F. Oswald (n. 12 above), 119, flatly declares that 'apparently buckets were not used in wells with wooden lining, but only in the later wells of Margidunum which were lined with stones. In the former, water was doubtless obtained by means of a flagon at the end of a cord.' He quotes no evidence for this, apart from a single flagon found in a wood-lined well. In London at least wells were normally wood-lined and in a number of them metal handles 'probably from wooden buckets' have been found (R. Merrifield (n. 14 above), 148). In the inventory of the Agora at Athens are three wooden buckets and thirteen of lead, all from Roman wells (Camp (n. 15 above), 279 n. 18). For the various items of equipment needed to draw water from a well, see Pollux X, 31; also Daremberg & Saglio, *Dictionnaire des Antiquités*, IV, 1, 780 nn. 16-17.

21. Shaduf: *SAT* 33; also vol. 2, 34-5; H. Hodges, *Technology in the Ancient World* (London, 1970), 119, figs. 111-13. K.D. White, *Roman Farming* (London, 1970), 157, estimates that in irrigation 'some 600 gallons a day can be hoisted to a height of six feet or more in a day' by a shaduf; the source is M.S. Drower's article in C. Singer *et al., A History of Technology* (Oxford, 1954-58), vol. 1, 525 (*q.v.*). A battery of shadufs, in series and parallel, is shown in Oleson (n. 19 above), fig. I. The shaduf is also known as the tolleno and the swipe. A Roman mosaic from Oudna, Tunisia, showing one in action in an agricultural setting, is illustrated in *Mon. Piot* III, pl. XXII, p. 200. See also Mertens (n. 12 above), 100), and Sparkes (n. 24 below), pl. XIV e.

22. Relays of men: Pliny, *NH* XXXIII, 4, 74; *SAT* 151; for a good illustration of toe-holds, see Dora P. Crouch, 'The Hellenistic water system of Morgantina, Sicily', *AJA* 88 (1984), 356, fig. 2; also *JRS* 59 (1969), 220 (toe-holds cut in chalk sides; at Dunstable). Camp (n. 15 above), 176, says they were a standard feature at Athens. Etruscan *cuniculi*: Judson & Kahane (ch. 2 n. 68 above), 85. Internal braces in wells; Johnson (n. 13 above), 204, fig. 154 (from Saalburg, Germany); Goblot (ch. 2 n. 4 above), 41, n. 4: 'Par exemple, je suis descendu dans une mine de plomb, en Sardaigne, à 180 mètres de profondeur, par des échelles de 3 mètres, avec un tout petit plancher pour passer d'une échelle à l'autre.' For ladders in mineshafts, see Ardaillon (n. 18 above), 28, who also quotes (loc. cit., n. 1) Roman mineshafts at St Laurent-le-Minier (Gard) with 'des entailles regulièrement espacées, qui servaient évidement à loger les pieds et les mains quand on montait ou que l'on descendait. Ces entailles sont disposées de telle sorte que l'ascension y est relativement commode' (Daubrée, *RA* (1881), 216). O. Davies, *Roman Mines in Europe* (Oxford, 1935), 23 (*q.v.* also for ladders and toe-holds) believes that in some Etruscan mineshafts beams were set into the side walls on such a spacing that 'porters may have sat on them and handed up baskets of ore' (compare the 'relays of men and women' in deep wells, above).

23. See *WAS* 2 (1987), 137, abb. 3.

24. Jacobi (n. 14 above). For pulleys generally, see J.W. Shaw, 'A double-sheaved pulley block from Kenchreai', *Hesp.* 36 (1967), 389-401; also Birebent 497. For a pulley and supporting beam in marble, see *Expédition Archéologique de Délos* VIII, 2, 348, fig. 213; also *ARV*[2] 1097, no. 17, and B.A. Sparkes, 'Illustrating Aristophanes', *JHS* 95 (1975), 130-1, and pl. XIV e, f. *Schol. to Ar. Lys.* 722; also Lang (ch. 2 n. 28 above), figs 8, 9, 10, 11. Compare Hor. *C.* III, 10, 10.

25. This point is reached when the total weight exceeds the tensile strength of the cross-section. As a point of interest, in modern ropes this theoretical breaking point is reached with a rope 6.7 km long (manila), 6.3 (hemp) and 5 (cotton); the figures are the same irrespective of the thickness of the rope. The maximum length recommended for actual practical use, allowing for safety factors, wear and tear of the rope (running over the pulley), etc., is 200 m. As applied to ancient wells (or mines), none of which are anything like this deep, the question is therefore purely academic, though it must be admitted that we seem to know next to nothing about the composition and hence breaking strain of ancient ropes. In modern oil wells the weight of the steel cables is often a serious practical factor. For advice on the above I am indebted to my colleague Dr G. Kardos, Professor of Engineering at Carleton University, Ottawa.

26. J. Toutain, 'Découvertes dans des puits à Alésia', *Bulletin de la Société Nationale des Antiquaires de France* (1937), 83-8, describes the remains of a wooden windlass beam, buckets, weights and associated apparatus.

27. The Tonglushan mine is a large one, on the Yangtse not far from Wukan and Huangshi. The name means Mt Verdigris, and 80,000 tons of copper is estimated to have been extracted from it over a very long period of operation, beginning around 800 BC. Several hundred vertical shafts, mostly 40-50 m deep, have been found. Water drainage was a problem. It was hoisted up the shafts by windlass and bucket, and 'implements such as wooden winches, wooden hooks, ropes, bamboo buckets, were found in these pits' – *Thonglushan: a pearl among ancient mines*, ed. Huangshi Museum, Hubei; Chinese Society of Metals, Publication Committee; and Archaeometallurgy Group, Beijing University of Iron and Steel Technology (Cultural Relics Publishing House: Beijing, 1980). Gonabad: Goblot 35-6.

28. The cerd: Oleson (n. 19 above), fig. 2.

29. Summary of noria, sakia, and the other technical terms in Oleson (n. 19 above), 10-12, with a clear and concise description of the various machines. For a full and detailed treatment, see Oleson 291-301 (the screw); 301-25 (the force pump); 325-50 (the compartmental wheel); 350-70 (the bucket-chain; for a particularly well-preserved example, see Cosa 201-4, figs 52-64); 370-85 (the sakia gear). See also Landels 58-63 (screw); 63-5 (compartment wheel); 71-5 (bucket chain); 75-83 (force pump). Rio Tinto: *SAT* 7, 221, fig. 29; Oleson fig. 115, and pp. 251-9 (w. bibliography). Useful technical details on output, pumping capacity, optimum angle of the screw, and the like, will be found in Pace 96-111 (as well as in the above works).

30. Oleson (n. 19 above) 242-6, and figs 102-8. Comapare also Herculaneum (213-14; figs 72-4), Hermoupolis (215-17, figs 75-6), Pompeii, 'Terme dell'età repubblicana' (248, fig. 109). Several other structures, particularly baths, confronted the same problem by using the treadmill/compartmented wheel, and sometimes an overlapping pair of them to attain the necessary height of lift: e.g. Ostia, Baths of Mithras (234-5, figs 95-6). R. Meiggs, *Roman Ostia* (Oxford, 1973), 144 n. 2, asserts that 'few sets of Ostian baths are without them'; he believes that the wheels raised water 'from the subsoil'.

31. Oleson (n. 19 above).

32. Saalburg: Johnson (n. 13 above), 202. Agora: Camp (n. 14 above), 183. Collapsing wells; Camp 175-6. Public wells: Hanson 395 (Queen Street, London), 'the abuse of the Queen Street wells, down which all sorts of unpleasant refuse and filth were thrown, is perhaps indicative of the treatment usually meted out to public as opposed to private property'. See ibid. for public wells at Silchester and Caerwent (*Archaeologia* 54 (1895), 19; 59 (1904), 96), located conveniently for travellers alongside the city gates. For early use of wells before the coming of aqueducts, see Paus. I, 14, 1 (Athens); Thuc. II, 48 (Peiraeus); Front. *Aq.* I, 4 (Rome).

33. 0.25-0.50 ha irrigable from one well: so Brent Shaw (n. 42 below), 133 n. 40, writing of North Africa. K.D. White, *Roman Farming* (London, 1970), 156-7. Pliny *NH* XIX, 60f.; Varro I, 1, 2; Col. XI, 3, 8.

34. 'Bottle cistern' – illustration, Crouch (n. 22 above), 356, fig. 3. Delos: *EAD* XVIII, 92-3; Laurion: Ardaillon (n. 18 above), 66-7.

35. Crouch (n. 22 above), 355, *q.v.* also for the estimate of 45 m^2 as 'about half the size of the roof of an ancient Greek or modern American tract house'. She also stresses (forthcoming study) that the co-existence of cistern and aqueduct water may often have led to a two-class system – aqueduct water for drinking, cistern water for washing, etc. This may well be right, and an important point – one thinks of the different qualities of water in the various Roman aqueduct channels carefully kept separate for different uses and purposes. But cistern water is not necessarily impure or unpleasant. In Bermuda many quite luxurious villas live off it as a matter of course.

36. See ch. 11 nn. 3 &4 p. 464.

37. Cherchel 21. In 1901 the villagers of Cannet offered a 'résistance farouche'. At Les Baux one can still see the extensive paved catchment area, with runnels cut in its surface and converging on the open cistern that they serve.

38. For this whole topic, see Camp (n. 15 above), 145-50 (full discussion). On the whole Camp is inclined to favour drought as the cause of the fall in water level, considering that the well-known chronic grain shortage in fourth-century Athens, esp. 331-324 BC (Dem. XXXIV, 37) reflects a failure by drought of local crops. The only other possibility is that underground supplies were exhausted by an increase in population, but in the fourth century at Athens the population seems rather to be in decline (A.W. Gomme, *The Population of Athens in the Fifth and Fourth Centuries BC* (Blackwell: Oxford, 1933), 26; V. Ehrenberg, *The Greek State*[2] (London, 1969), 31). Well sunk through cistern floor: Stella G. Miller, 'Menon's cistern', *Hesp.* 43 (1974), 194-9, 228-9 (F 16:8). See also *Hesp.* 23 (1954), 87-107. (Roman wells dug in cisterns: E 11:2, G 11:2, N 18:5, and N 20:3.)

39. Cherchel 39, fig. 23 (large city cistern, 1,585 m^3); 126, fig. 75 (Tipasa settling tank, with cascade for aeration of water). *Castella*: Birebent 502.

40. Cherchel 22. Tiddis baths, Cherchel 19.

41. e.g. Report (untitled) by M.H. Saladin, *Archives des Missions Scientifiques et Littéraires* (Paris) 13 (1887), 1-225. Paul Gauckler, *Enquête sur les installations hydrauliques romaines en Tunisie* (Tunis, 1897-1912); Stephane Gsell, *Enquête administrative sur les travaux hydrauliques anciens en Algérie* (Paris, 1902); P. Romanelli, *La politica romana delle acque in Tripolitania*, in *La Rinascita della Tripolitania* (Milan, 1926), 569ff.

42. Brent D. Shaw, 'Water and society in the ancient Maghreb', *Antiquités Africaines* 20 (1984), 126, largely quoting L. Carton, *Hydraulique dans l'antiquité en Barbarie*, R.T. 19 (1912), 222-3. Anyone who has looked through the *Archives* will welcome Shaw's candour. It is always refreshing to find somebody saying in print that which one has oneself not quite dared to think.

43. Saladin (n. 41 above) 106. The chief lacuna in the interpretation of Saladin and his contemporaries is the omission of a second function of these hydraulic works, that of erosion control. They were intended not only to collect and distribute water, but also to retain the silt and topsoil being carried down by the mountain torrents, and to concentrate it in areas where it formed new tracts of arable land. See p. 251 below.

44. P.L. Leveau – J.L. Paillet, *JEAR* 233: 'Une recension des citernes nous montra qu'en definitive la ville pouvait suffire à ses usages domestiques et strictement économiques par la collecte des eaux zénithales et la multiplication des captages dans l'amphithéâtre de collines gréseuses qui l'entoure. Chaque maison était pourvue de citernes recueillant les eaux de pluie.'

45. Average capacity 30-130 m^3, Cherchel 23. Compartments: 36; aisles: 39. Compare such parallels as the Piscina Mirabilis at Misenum, p. 279 below. For large public cisterns in Africa, Cherchel 41.

46. Cherchel loc. cit.

47. 'L'importance de l'évaporation doit être remise en cause. Même en plein été, sous des températures de l'air de l'ordre de 45°C et une très faible hygrométrie (de 15 à 20%), un cours d'eau perd, par évaporation, moins qu'on ne serait tenté de penser. Au dessus de celui-ci se constitue une gaine protectrice d'air dont la saturation humide passe, graduellement, de près de 100%, au niveau de l'eau, à la faible teneur de l'air ambiant. L'air saturé de cette gaine n'est entrainé ni par le cours d'eau, ni en dehors, sauf en cas de très grand vent, car les rives, même peu élevées, ont une action protectrice' (Goblot 26).

48. See, e.g., the reservoir at Henchir Baroud (Saladin (n. 41 above), 52).

Chapter 4

1. Ven Te Chow (ed.), *Handbook of Applied Hydrology* (New York, 1964), 4-1. In practice we are here dealing only with water that is 'fresh and easily available for human use'. This represents about 2-3% of all terrestrial water. Ven Te Chow is a good reference work on the whole topic of hydrogeology: see also C.O. Wisler & E.F. Brater, *Hydrology* (New York & London, 1959), esp. ch. 6, pp. 127-92, on ground water. The best and simplest account of hydrology in the ancient world is Asit K. Biswas, *History of Hydrology* (Amsterdam/New York, 1970), 1-119. The material is at an introductory level, and the chapters on the concept of water and hydrology in the Greek philosophers (53-78) and on the Nile (105-19) form a particularly convenient synthesis. See also G. Garbrecht, 'Hydrologic and hydraulic concepts in antiquity', in G. Garbrecht (ed.), *Hydraulics and Hydraulic Research* (Rotterdam/Boston, 1987), 1-22.

2. 'The soil-water reservoir supplies the largest fraction of fresh water that benefits man, that is, the water used in the production of non-irrigated crops' (Chow (n. 1 above), 4-6).

3. Porous, permeable strata are known as aquifers, impermeable ones as aquitards, aquicludes, or aquifuges (depending on technical differences). The commonest, and most permeable, aquifers are limestone, sandstone, shale and gravel. The chief aquicludes are clay, silt, and impervious (usually igneous) rocks such as granite and marble. One must note that permeability is not the same thing as softness. Limestone is a very hard rock. Students of Greek architecture will be familiar with the very much sharper edge, on things like cornice mouldings, still retained on buildings at sites that used a lot of local limestone, such as Delphi, compared with the eroded and weathered state of their equivalent at somewhere like Olympia, where there was none. When one sees, say, a classical site in which a layer of clay covers a stratum of hard bedrock limestone, one thinks of the water as seeping through the clay, not the rock, but it is the other way round.

4. The best example of this is at Corinth, where an aquiferous stratum of limestone is suddenly exposed by a break in the superimposed stratum of clay (aquiclude), and the water it is carrying bubbles forth in the springs feeding Peirene and Glauce.

5. Pimienta (n. 31 below), 70, fig. V.2.

6. Anio Vetus: Front. *Aq.* II, 90. The Anio Novus took it indirectly, from an artificial lake that Nero established on the Anio by damming it (ibid. 93). Segovia: A.R. Gallardo, *Supervivencia de una Obra Hidráulica: El Acueducto de Segovia* (Segovia, 1975), 1, and pl. 2; fig. 10, p. 3; also *WAS* 1 (1986), 19; 2 (1987), 219. Aix, St. Antonin: Fahlbusch 31; Grenier, *MAGR* 68, insists that 'il ne subsiste rien des travaux de captage antiques', but Fahlbusch, although writing 22 years later, has a photograph of the remains of the dam (his abb. 5). Aix, Traconnade: *MAGR* 72 – 'un barrage en grand appareil de fort bonne construction, "assurément romaine" '.

7. n. 14 below.

8. U. Izmirligil, 'Die Wasserversorgungsanlagen von Side', *Mitt. L. Inst.* 64 (1979); Fahlbusch 31, and his abb 7. For the Dumanli spring and springs in general, n. 14 below.

9. Front, *Aq.* I, 4.

10. Alfred Burns, 'Greek water supply and city planning', *Technology and Culture* 15, 3 (1974), 408-9.

11. Though Aristotle, *Meteor.* I, 13; II, 1-3, does give something like a description of the hydrological cycle (precipitation, rivers, sea, evaporation, precipitation). 'Descriptive rather than analytical': so J.F. Healy, *Mining and Metallurgy in the Greek and Roman World* (London, 1975), 15. See also R.J. Forbes, *Studies in Ancient Technology* (Leiden, 1966), VII, 1-94 (Ancient Geology); 'les connaissances géologiques de Vitruve, comme de manière générale celles des Anciens, apparaissent lacunaires et imprécises' – Callebat, 51; see also XLI. Vitruvius, for example, seems to live in 'une confusion permanente entre roches et sols'.

12. Plato *Leg.* 761 a-b. Compare Ar. *Meteor.* 349b30-34, 350a7-9. For a possible example of this technique in action, see William M. Murray, 'The ancient dam of the Mytikas valley', *AJA* 88 (1984), 195-203. See also Sen. *QN* III, 15.

13. Pliny *NH* II, 104, 225; XXX, 30, 55. Sen. *QN* III, 26; Paus. VIII, 14, 1; 44, 3-4; 54, 1-4; Vergil *Ecl.* X, 1-4. For similar tall tales – a spring near Delos being fed from the Nile, the Alpheus having its source in Tenedos, and the Asopos (near Sicyon) in Phrygia – see Strabo VI, 2, 4, ch. 270-1. Forbes (n. 11 above), 17-18, gives a full catalogue of these absurdities.

14. A few examples will put the question in perspective. The hot (48°C) spring feeding the baths at Bath has a daily discharge of 1,145 m^3 daily. In upstate New York, Lake Placid is a spring-fed lake, the spring delivering 45-50,000 m^3 (8-10 million gallons) daily. At this output it could have satisfied the needs of the Nîmes aqueduct (12,200 m^3) four times over, and Pompeii (6,480 m^3) eight times. One of the largest, and certainly the most celebrated, fresh-water springs in the world is La Fontaine de Vaucluse (= Lat. Vallis Clausa), near Avignon, where an entire river (and a relatively large one) suddenly appears flowing out of the base of a 230 m high cliff, which it reaches by a natural inverted siphon inside the mountain that has so far defied exploration. Its daily discharge fluctuates greatly, but is normally (winter and spring) around 1,768,000 m^3; in peak periods it runs as high as ten times this figure, for a discharge of nearly 12 million m^3 (200 m^3 per second), or over ten times the daily consumption of all aqueducts of Rome combined. The Romans tapped this spring with an aqueduct to serve Cavaillon: the excavated remains are visible 1.1 km downstream from the modern village, alongside the D24 road.

In the Mediterranean area, claims have been made for the Dumanli spring on the river Manavgat, Antalya, in S. Turkey, which has a mean discharge of 50 m^3/sec, or about 432,000 m^3 daily (J. Karanjac & G. Gunay, 'Dumanli Spring, Turkey – the largest karistic spring in the world?', *Journal of Hydrology* 45 (1980), 219-31). As noted above, at least part of its waters went to feed the aqueduct of Side. For the 'karst' phenomenon in hydrogeology, see R.G. Lafleur (ed.), *Groundwater as Geomorphic Agent* (Boston, 1984), sections 7, 8, 10, 11, 13-15. D. Crouch, who has made a study of karst, roughly describes the term as covering 'the interactions of water and limestone and other stones' (*FCAS*). For a study of this phenomenon in the Roman aqueduct system of Grand, in the Vosges region of France, see Olivier André and Valérie Bertaux, 'Les aqueducs souterrains construits par les Romains', *Doss. d. Archéologie* 162 (juil.–août 1991), 34-6 ('Le réseau karstique naturel'). Of special value we may also notice in this same issue the study of the hydrogeology of the whole catchment area (Marne/Meuse/Moselle/Rhine) by Bernard Haguenauer and Pierre Deletie, 'Mythe ou réalité de la ressource en eau du site' (pp. 67-72; with maps, showing underground karstic water movements).

A standard classification of springs by size often used by hydrologists is that enunciated by O.E. Meinzer, *U.S. Geol. Survey, Water-Supply Paper 494* (1923), 50-5 (quoted in Chow (n. 1 above), 4-35). He divides springs into eight classes according to discharge; to facilitate comparison within this book I have, roughly, converted his figures to cubic metres per 24 hours:

Class	*Discharge (m^3 per day)*	
1	302,400	and above
2	30,240	" "
3	3,024	" "
4	671	" "
5	67	" "
6	6	" "
7	0.8	" "
8	0.8	and below

15. For an excellent summary of the constituents and properties of water, see Chow (n. 1 above), ch. 19, esp. 19-23 (ground water).

16. See pp. 227-32 below.

17. For example, rivers may also contain hard water. Water acquires a calcium content by flowing over a limestone bed as well as by seeping though it, but not nearly as much or as rapidly. The river Isis, for example, is soft at its source but has already become quite calcinated by the time it reaches Oxford. And springs, already hard, do usually end up flowing into a river, which may in any case draw its source from a spring, though this will be greatly augmented along the way by surface run-off. Likewise, the rocks through which the spring has percolated may be not limestone but something like granite, from which it will acquire no calcium content, and emerge soft. Local variations abound. A spring may have a strong mineral content lacking in a well sunk only a short distance away and drawing water from the same stratum – because the mineral deposits are localised, and the water touches them before reaching the spring but after passing the well.

18. See also A. Trevor Hodge, 'Vitruvius, lead pipes and lead poisoning', *AJA* 85 (1981), 486-91.

19. Purity of rainwater: Hippocr. *Airs, Waters and Places* VIII, *'ta men oun ombria kouphotata kai glukutata esti kai leptotata kai lamprotata'*. Vitr. VIII, 2, 1 *'Itaque, quae ex imbribus aqua colligitur, salubriores habet virtutes'*. Col. I, 5, 3. Though there was agreement on its purity, it was recognised that it often became stagnant in storage (Callebat 63). 'Hard' water from rocks, Hippocr. op. cit. VIII. His terminology does not seem to include a clear linguistic expression of 'soft', and instead (above) lists 'light, sweet, fine, and clear' as the opposites of 'hard' (*sklêros*); *malakos*, the obvious word for 'soft', he does not use. I do not know whether this is the origin of our 'hard/soft' terminology.

20. Hippocr. op. cit. VII. Vitr. VIII, 1, 2. What he actually recommends is 'red rock', *rubro saxo*. The obvious translation is sandstone, particularly since this is an aquifer, but Callebat (55) opts instead for the 'tufs volcaniques' so common in Italy. Tufa is a dull reddish-brown in colour and is very porous, and hence aquiferous. For 'black earth', *terra nigra*, see Callebat 53: it seems to refer to 'une terre riche, grasse, et meuble', but the term is far from precise. 'Facing north': Vitr. VIII, 1, 6; 2, 8, *'relinquitur uti multo meliora inveniantur capita fontium quae ad septentrionem aut aquilonem spectant'*. His list of major rivers originating in the north, a thesis common in antiquity and maintained by Aristotle (*Meteor.* II, 1, 354a), does include some genuine examples, such as the Rhône, Tiber, Tigris, Euphrates and Indus, but comes badly to grief on the Rhine and the Nile; he gets round the Nile by asserting that it rises in Mauretania and therefore does flow south to start with, but his geography is in general elastic and does not inspire confidence as a practical basis for choosing the springs for a Roman aqueduct. Full exposition in Callebat 74-85. For the whole question of springs, quality and availability of water, well-digging and the like, we depend largely on literary sources. The chief ones are Pliny, Vitruvius, Palladius, Cassiodorus and the Geoponica.

21. On the connection between water purity and augury, see Vitr. I, 4, 8. Water tests, Vitr. VIII, 4, 1-2 and Callebat 133-5. The healthiness of the local inhabitants may be deceptive if in fact there is something in the water but they have acquired an immunity to it. The vegetable test is in fact a test of water hardness, which does inhibit cooking, but Vitruvius does not enlarge upon it. It is truly remarkable that the ancients, while expatiating at inordinate length upon the various qualities (often imagined) of water in various places, scarcely ever say a word about its hardness, a quality that ought to have been a paramount importance to them, in view of the havoc it wrought upon their conduits and pipes by incrustation. Test for purity of water: Pliny *NH* XXXI, 31-40.

22. The position drinking water held in the Roman menu is illustrated by the fact that in a book of 230 pages, Jacques André (*L'alimentation et la cuisine à Rome* (Paris, 1961), 163), cannot stretch out the topic for more than a half a page (which, however, see). Cato: Plutarch *Cato Maj.* 1, 7.

23. Even to the extent of their being named after it. Compare Kalonero ('good water'), a small railway junction north of Kyparissia, in Messenia, and Acqua Acetosa ('vinegar-water'), one of the eastern suburbs of Palermo.

24. High points are: water that makes your teeth fall out (Susa, Iran; ch. 23); that improves your singing (Tarsus; Magnesia; 24); that dissolves gallstones (Lyncestis; Teanum, Campania; 17); that makes you teetotal (Clitor, Arcadia; 21); that drives you mad (Ceos; 22).

25. *aquilex* (= also *indagator aquarum*), Col. II, 2, 20 (description of surface vegetation as an indication of subterranean water); Pliny *NH* XXVI, 16, 30. In the late empire, Africa enjoyed a particularly high reputation for the skill of its *aquileges* which made them much sought after, even in Italy, on the undeniably logical principle that they would have no difficulty finding water in temperate lands when they had been trained to find it in the desert (Cassiodorus *Variae* 3, 53; see Brent Shaw, 'Water and society in the ancient Maghrib', *Antiquités Africaines* 20 (1984), 125).

26. Vitr. VIII, 1, 4; 1, 1. Compare Pliny *NH* XXXI, 46; Pallad. IX, 8; Geopon. II, 4, 2. Callebat 51-9. In China, ancient miners used surface vegetation as a guide to subterranean deposits of copper.

27. The *virgulta furcata*, forked twig, used for discovering water, first appears in Agricola's *De Re Metallica* (1546). The Romans never used water-divination, and those who regularly accuse them of technological backwardness may find ironic the dictum of a modern authority: 'Divination: This superstition still persists in many sections of the United States, although no scientific basis exists for it' (Ernst Vollmer, *Enyclopedia of Hydraulics, Soil and Foundation Engineering* (Amsterdam/London/New York, 1967), 95. Just the same, I have personally seen it successfully done, at my cottage near Ottawa, Canada.

28. Juturna was in the Forum, the Camenae on the Caelian. Front. *Aq.* I, 4; Grimal 67. Both springs were supposed to have therapeutic qualities. However, channelling the spring into a built fountain-house might upset the romantically inclined, who regretted the lost sylvan setting: Juvenal's lament (III, 12-20) over the marble façade of the Camenae sounds very like Ruskin or Wordsworth bewailing the desecration of the Lake District by progress. For appreciation of a genuine rural spring, see Pliny *Ep.* VIII, 8 (the Clitumnus). For architectural decoration of springs see Franz Glaser, 'Brunnen und Nymphäen', *WAS* 2 (1987), 105-31.

29. Kallmuth: Haberey 56-63. Grüne Pütz: Haberey 64-8. See also *MAGR* 175, fig. 53 for a similar spring basin on the aqueduct of Sens (my Fig. 36). J.J. Coulton, 'Oinoanda: the water supply and aqueduct', *Anatolian Studies* 36 (1986), 19-20, has a good general discussion of the whole question.

30. Ashby 97, fig. 6.

31. Vitr. VIII, 1, 6; Callebat 60. Well-drilling: Jean Pimienta, *Le captage des eaux souterraines* (Paris, 1972), 105 (fig. VII, 3, 'Puits à drains rayonnants'). His text puts the point clearly: 'Puisque l'intérêt du puits réside dans la dimension de sa surface filtrante particulièrement utile quand l'épaisseur de la nappe captable est limitée, il est souhaitable de développer cette surface.' On tapping the spring by adit, compare Birebent 493. Speaking of Algeria: 'Les Romains savaient, en effet, que le site d'apparition d'une source n'est généralement pas un lieu favorable à l'implantation d'un captage. Ils remontaient donc le cours de l'eau jusqu'à la rencontre des couches rocheuses non délitées, non fissurées où ils étaient assurés de rassembler tout le débit possible et c'est ce qui explique que nombre de captages anciens fonctionnent encore parfaitement.'

32. Vitr. VIII, 1, 6. Callebat 60. Peirene: Bert Hodge Hill, *Corinth*, I, 6. Perachora: R.A. Tomlinson, 'Perachora: the remains outside the two sanctuaries', *BSA* 54 (1969), 204, fig. 19. See also F. Graber, 'Die Enneakrounos', *AM* 30 (1905),

60 (Athens, Megara, Aegina), and E. Ziller, 'Untersuchungen über die antiken Wasserleitungen Athens', *AM* 2 (1877), 108ff. (Athens/Periaeus). Whether such adits were in any way derived from the oriental qanat must remain uncertain.

33. The channels seem to have been up to 2 km long, and were observed by J. Auvergne ('Fontvieille inédit', in *Bulletin de la Société des amis du vieil Arles*, vol. 2, p. 133ff. quoted in Otto Stübinger, *Die Wasserleitungen von Nîmes und Arles* (Heidelberg, 1909), p. 19). The location is at Les Taillades, 2 km east of Fontvieille.

34. Apart from the artificial lakes of Nero, at Subiaco, which fed the Anio Novus (see below), the two best known examples of Roman waterworks in connection with lakes are the tunnels driven to draw off the waters of the Fucine Lake, near Avezzano, and the Alban Lake, near Rome. In both the purpose was purely drainage and land reclamation/flood control. Fucine: Tac. *An.* XII, 56-7. Suet. *Cl.* 32. The installations at the Alban Lake (*emissarium*) are still visible. There is also a less visible *emissarium* at Lake Nemi. Description and illustrations in K. Grewe (ch. 2 n. 66), 138-9; also Grewe (see Abbreviations), 71-2.

35. *MAGR* 72 (on the Traconnade aqueduct at Aix-en-Provence; the more prominent dam at Le Tholonet (p. 70) appears dubious). The only classical dams in Greece known so far are apparently the Mycenaean dam at Kophine in the hills above Tiryns, a dyke designed to divert floodwaters, and the irrigation dam in the Mytikas valley: see n. 12 above.

36. The chief works on Roman dams are: N.A.F. Smith, *A History of Dams* (London, 1971), pp. 1-50; here abbreviated to *HD*. This is easily the best and most comprehensive study of the subject, and I must here acknowledge my debt to it for this entire section on dams; the magnitude of the debt will be fully recognised only by those who consult Smith's work. N.J. Schnitter, 'A short history of dam engineering', in *Water Power* (April, 1967), 142-8; N. Schnitter Reinhardt, 'Les barrages romains' in *Doss. Arch.* 38 (Oct.-Nov., 1979), 20-5. *SAT* 160-1, surprisingly, ignores the Romans, dealing only with the Middle East. We must also add G. Garbrecht (ed.), *Historische Talsperren* (Stuttgart, 1991) (excellent illustrations).

37. *HD* 33-4; for the entire above classification and analysis I am indebted to this source, which should be consulted for further explanation.

38. *HD* loc. cit.

39. *HD* loc. cit.; Benoît, F., 'Le barrage et l'aqueduc romain de Saint-Rémy en Provence', *REA* 37 (1935), 331-9. Plan: *JEAR* 346, fig. 7.

40. A good example is the dam of Homs (Syria), for which see p. 91 below.

41. *HD* 37.

42. Kasserine: *HD* 35-6. As at Glanum, the Kasserine dam was curved but it looks as if it did not act structurally as an arched dam and the curvature was for other reasons. Proserpina, Cornalvo, Alcantarilla: *HD* 43-8; Schnitter Reinhardt, *Doss. Arch.* (n. 36 above), 22-4. Smith dates the Cornalvo dam to Hadrian.

43. Built in the third or fourth dynasty (2650-2465 BC), 30 km south of Cairo. It was 12 m high and 108 m long, consisting of 'two rubble masonry walls of 24 m base width, enclosing a 36 m thick earth core' (Schnitter (n. 36 above), 142). *HD* 1-4 gives a full account.

44. Vita-Finzi (n. 56 below), pl. II(b).

45. 'Unfortunately this basic error was to be repeated innumerable times up to the present day' (Schnitter (n. 36 above), 142).

46. It was built in AD 260 by 70,000 Roman prisoners captured along with the Emperor Valerian by Shapur I of Persia. The river was diverted during the construction work via an artificial channel still visible; *HD* 56 and pl. 17.

Schnitter Reinhardt (n. 36 above), 21 gives the same illustration.

47. *HD* 21-2. N. Glueck, *Rivers in the Desert* (New York, 1959).

48. Vita-Finzi (n. 56 below), 19.

49. Pliny *NH* III, 109; Front. *Aq* II, 93; Ashby, 253-6; *HD* 26-32.

50. The details come from the examination in 1883-84 of the then extant remains by G. Giovannoni (*L'Architettura nei Monasteri Sublacensi* (Rome, 1904)). See N.A.F. Smith, 'The Roman dams of Subiaco', *Technology and Culture* 22 (1970), 58-68.

51. *HD* 43-8; Schnitter Reinhardt (n. 36 above), 23-5; C.F. Casado, 'Las Presas Romanas en España', *Revista de Obras Publicas*, VII Congreso Internacional de Grandes Presas (1961), 361-3.

52. *HD* 48.

53. *HD* 44-6; J. de Castro Gil, 'El Pantano de Proserpina', *Revista de Obras Publicas* (1933), 449-54. Raoul Celestino Gomez, 'Chronologia des las Fabricas no Romanas del Pantano de Proserpina', *Revista de Obras Publicas* (1943), 558-61. The unusual name, Proserpina, comes from 'an inscription found nearby, which has nothing to do with the reservoir but invokes the curse of the goddess on a clothes-robber' (Paul MacKendrick, *The Iberian Stones Speak* (New York, 1969), 131-2). The inscription is published by Hubner (*CIL* II, 462), who says it was found built into the dam, though not belonging to it.

54. F.J. Wiseman, *Roman Spain* (London, 1956), 163. Compare Homs (n. 40 above) and p. 91.

55. See ch. 10 n. 11 p. 455.

56. C. Vita-Finzi, 'Roman dams in Tripolitania', *Antiquity* 35 (1961), 14-20; *HD* 37.

57. *HD* 8-12; Schnitter (n. 36 above), 142; *SAT* 161-3; see also G.A. Hathaway, 'Dams: their effect on some ancient civilisations', in *Civil Engineering* (Jan. 1958), 58-63. The water diverted by the Bavian dam was carried by a deviation canal or aqueduct to Kalatah, where it emptied into the Koshr to augment supplies available to the other three dams. For this aqueduct, see T. Jacobsen & S. Lloyd, *Sennacherib's Aqueduct at Jerwan* (Chicago, 1935). Two Minoan dams have been found at Pseira, Crete (report by R. Hope-Simpson to AIA conference, Boston, 1990; *AJA* 94 (1990), 322). They were for irrigation, of Cyclopean masonry facing and rammed earth fill, the largest being preserved to 3.5 m high, *c.* 14 m long and 3 m thick, with a Late MM *terminus post quem* dating. The only other equally early Greek dams are the drainage control dykes at Tiryns and Lake Copais (see Bibliography). See also Jost Knauss, 'Der Damm im Takka-See beim Alten Tegea', *AM* 103 (1988), 25-36. For the Nabataeans, see Glueck (n. 47 above).

58. Not to be confused with another Homs, near Leptis Magna, where there is also a Roman dam, but of more modest proportions.

59. *HD* 39-43; Schnitter Reinhardt (n. 36 above), 25. On the strength of a reference in Strabo (XVI, 1, g) 205, the dam is sometimes dated around 1300 BC and identified as Egyptian work, but this has been convincingly shown by Smith to be erroneous.

60. Gasr Khubbaz: Sir A. Stein, 'Surveys on the Roman frontier in Iran and Transjordan', *Geographical Journal* 95 (June, 1940), 430; *HD* 39.

61. Compare ch. 5 n. 12 below (the Theseion).

62. Procopius of Caesarea, *Buildings* II, 3, 16-21; *HD* 53-4. However, on the basis of his 1987-89 excavations at Dara(s), the whole idea of an arch dam is now challenged by Garbrecht (n. 36 above), 265-76.

63. *HD* 49.

Chapter 5

1. Vitr. VIII, 6,1.

2. Such as the Aqua Appia at Rome. Its channel was surface (or subterranean; i.e. not on arches like the other Roman aqueducts) throughout, except (says Front. *Aq.* 1,5) for the last 60 paces on arrival at the Porta Capena; but see Ashby 53. Isolated sections of the *specus* (i.e. the conduit) have from time to time been attributed to it, but Ashby could summarise thus: 'In fact, no trace of the *specus* of either the Appia or the Augusta [a tributary branch of the Appia] has ever been found between the springs and the City; we may note that, for good measure, the springs have never been satisfactorily identified either (Ashby 51-2).

3. 'Cut and cover': the term is a recognised one for the construction of underground railways.

4. Pont du Gard, see Esperandieu *passim*, esp. p. 16 and pl. XV, p. 49. Marcia, see Van Deman pl. XXII, 2, p. 104. For brick vaulting in an underground channel, see Esperandieu pl. III, p. 25, and pl. VII, p. 33 (at Sernhac, on the Nîmes aqueduct). Excellent underground photograph in Haberey 43 (Cologne).

5. Marcia; Van Deman 80; Brevenne; *MAGR* 127, GdM 87. These figures are naturally no more than representative and can be found to differ even for other parts of the same aqueducts. They could also be much larger. At the 'sluice-tower' of S. Cosimato the conduit of the Marcia was no less than 1.14 m wide by 3.30 m high (Ashby, 102), and the last section of the main aqueduct into Vienne, running in a tunnel, was 1.88 m wide by 2.10/2.37 m high; in 1940 a 140 m long stretch of it was cleared out and used as an air raid shelter for the population (*MAGR* 113; Pelletier, *JEAR* 299). More bizarre uses have been suggested. At Nyon (Switzerland) underground sections of the Roman aqueduct have been popularly identified as mediaeval secret passages linking convents and monasteries, to enable monks and nuns to visit each other 'bien entendu, à des fins qui n'avaient rien de réligieux'. Other sections were identified with equal confidence as ancient pipelines for wine. This belief in vinoducts, held till recently by local peasants, goes back as far as the eleventh century, when, of the Cologne aqueduct, 'on le croyait destiné à amener à la Cologne romaine les vins de Trèves'. To keep abreast of this flow would require of Roman Cologne an unremitting wine consumption of 230 litres per second, giving a whole new dimension to the Decline and Fall (Eugène Olivier, *L'alimentation d'Aventicum en eau* (Neuchâtel, 1942), pp. 7, 25 (repr. from *Bulletin de la Société Neuchâteloise de Géographie* (1942); it also appears as a section in Olivier's *Médecine et santé du Pays de Vaud*, in the Bibliothèque Historique Vaudoise 29-31 (Lausanne, 1962)). By contrast, the tunnels or galleries in mining were often very small (see ch. 6 n. 5 p. 423).

6. Recessed ledge, Ashby 272, fig. 30B (Anio Novus). For the roof profile, Ashby puts it simply: 'the roof may be flat, gabled, or vaulted, and the internal section may differ from the external' (Ashby 43, see his nn. 6, 7, 8 for numerous examples of all three from the Roman aqueducts). Flat stone slab, L. Blondel, 'L'aqueduc antique de Genève', *Genava* (1928), 52; Haberey 16-18. Channel roofed by three flat slabs, corbelled, *JEAR* 203, fig. 5 (Mont d'Or, Lyon). Pointed roof, *JEAR* 259, fig. 2.1.1.c (in a tunnel at Acquarossa, Viterbo). See also Ashby 43, nn. 6, 7, 8. For a comparison of some typical profiles and dimensions, see *JEAR* 205, fig. 7 (Lyon). We also find channels horse-shoe shaped in section, at Fréjus (*MAGR* 50), Vienne (*JEAR* 39), Poitiers (*JEAR* 44), and, exceptionally, hexagonal at Alteburg (Samesreuther 31, fig. 4).

7. Indeed, in North Africa, common practice was to use what we are here describing as the 'channel' purely as a service tunnel, with the water running in a second, narrower, channel cut in the floor of it; as Birebent puts it, 'L'*aqueduc* lui-même était construit dans la partie basse de la galerie et, d'ordinaire, n'occupait pas toute sa largeur' (Birebent 501). The same thing happened at Pompeii, where the Serino aqueduct had a vaulted conduit 0.55 × 1.30 m, but with water actually running in a 25 × 30 cm channel sunk into the floor (Eschebach, *JEAR* 88, and 103, Taf. I). In Spain the same general principle seems to have been applied to siphon pipes (SRA 191), and is already found in the sixth century BC in the tunnel of Eupalinos on Samos.

8. Overflows: GdM 105 (in the header tank of the St Genis siphon on the Gier aqueduct, Lyon). Some of the Rome aqueducts had windows cut in the side walls (Ashby 43; 103-4; 195-6) though whether intended as overflows or purely for inspection is hard to say. The water *would* of course have overflowed through them if it had risen high enough. The question is well discussed by Blackman, *PBSR* (1978), 56-7, who makes the point that the Anio Novus almost certainly ended up by running full in the section immediately after Osteriola, since the channel is almost completely blocked with incustration to within 50 cm of the top (Ashby 271-2; fig. 20 and pl. XIX b; Van Deman 296: 'solid mass of deposit'). At Osteriola there is a junction (Ashby 287) and a direct cut-off apparently of Claudian date leaves the original channel to tunnel south under the mountains and rejoin it at Fosso di Ponte Terra (Gericomio). It is the original channel (called by Van Deman 'the main line' and by Ashby 'the Tivoli loop') that is blocked, just after Osteriola junction, so the incrustation represents ancient, not mediaeval, neglect, since it must antedate the cut-off that replaced it; and, of course, a channel becomes incrusted only while the water is flowing, not after abandonment. In the Rive Droite aqueduct at Tubusuptu (Algeria), 'Dans les parcours à faible pente, les parois sont couvertes d'une épaisseur de concrétions de 2 à 3 cm et cela presque jusqu'au plafond, ce qui laisse à penser que la canalisation était généralement pleine d'eau' (Birebent 480).

9. *JEAR* 204, fig. 6 (La Brevenne); for terracotta flooring, see n. 31 below.

10. Frontinus specifies that work should be done in spring and autumn, to avoid extremes of temperature: '*quia temperamento caeli opus est ut ex commodo structura conbibat et in unitatem conroboretur*', *Aq*. II, 123. It is not clear what he has in mind as *structura*. The Loeb editor translates 'for the masonry to absorb the mortar'; Grimal (98 n. 143) feels it 'désigne plus spécialement le *blocage* de ciment formant le noyau'. *Structura* ought to mean either construction work in general or masonry. But it is hard to see how weather could affect the laying of ashlar blocks, so it must here refer to something that sets – mortar, cement or concrete. Hence the Loeb paraphrase, putting in two words (masonry, mortar) for the one in the Latin. Grimal is almost certainly right in referring it to the concrete, brick-faced core (noyau), but surely inaccurate in then describing this as cement (ciment, not béton). The only cement, as opposed to concrete, in an aqueduct was the lining inside the channel and plainly that is not what Frontinus here means by *structura*. But if the weather affected the setting of the concrete presumably it affected the cement too.

11. The standard work on the subject is now that of Roman Malinowski published in three separate articles: 'Ancient mortars and concretes: aspects of their durability', *History of Technology* (1982), 89-100; 'Concretes and mortars in ancient aqueducts', *Concrete International*, vol. 1, no. 1 (Jan., 1979), 66-76 (includes chemical analysis); and 'Einige Baustoffe Probleme der antiken Aquedukten', *JEAR* 245-74. His studies focus on the waterproof lining of the

Caesarea aqueduct, the sealant in the joints of piped aqueducts (see below), esp. Rhodes, and the waterproofing of various Roman floor mosaics. They are based on chemical analysis and experiments with modern reproductions of ancient cement (full account in the *Concrete International* article cited above, 73, 75). His comments (*History of Technology*, 90) are worth quoting on the six layers at Caesarea. They 'cause a uniaxial, normally directed shrinkage, thus restraining horizontal cracking. The lightweight, greyish layer containing coal ashes is a bond layer; the white, carbonated layer containing marble powder prevents shrinkage; and the reddish, very fine polished poszuolanie layer (of ground ceramic) assures hardening in water, impermeability, and strength.'

A further note is provided by Pascal Tissières, 'Réflexions sur quelques problèmes de l'eau à forum Claudii Vallensium', *Annales Valaisannes* 2e serie, 53e année (1978), 182 n. 27: 'Tous les mortiers romains [here = the waterproof lining of the aqueduct] sont fabriqués a l'aide d'un liant aérien – la chaux éteinte (Ca(OH)) qui durcit a l'air par carbonation (absorption de CO_2). Le mortier au tuileau résiste mieux a l'eau grace à sa forte teneur en silice du sable et la chaux, mais beaucoup plus lentement. Ainsi l'on peut dire que le temps bonifie le mortier.' It may be noted that both Malinowski and Tissières write as professional engineers. On building materials generally in Roman aqueducts, see also Heinz-Otto Lamprecht, 'Bau- und Materialtechnik bei antiken Wasserversorgungsanlagen' *WAS* 3 (1988), 129-55.

12. Vitruvius is here speaking of house walls. I have never myself seen ancient stucco, or modern either, that reflected like a mirror, but Vitruvius is quite clear that it can be achieved by polishing hard ('*politionibus crebris*'). Malinowski on removing sediment, *History of Technology* (n. 11 above), 90; on 'closed canals and tunnels', *Concrete International*, loc. cit., *q.v.* also for the orientation of the particles. Vitruvius VII 4,1 recommends crushed earthenware (= brick dust) instead of sand as an ingredient to make stucco damp-proof. There may exist another technique aimed at waterproofing the channel. The Aqua Marcia was built of ashlar blocks. In the sides of the channel, we find a curious and probably (in aqueducts) unique arrangement. At each vertical joint is drilled, vertically, a long round hole the entire depth of the course and spaced so as to be one half (i.e. of semicircular section) in each of the two adjoining blocks. This hole, of some 5-6 cm diameter, was then filled with fine cement or mortar. The process is illustrated and described by Van Deman (pl. XXII, 2, and pp. 105-6) who sees in it a substitute for clamps, a device to stop the blocks shifting. So it may be, but it would also effectively seal the joints (at least the vertical ones), and one remembers the use of a very similar technique, using lead, in the cella side walls of the Theseion in Athens. There it is normally understood as a damp-proofing technique to protect the murals the walls carried (see W.B. Dinsmoor, 'Observations on the Hephaisteion', *Hesp.* Suppl. 5 (1941), 100f. and fig. 37; A. Orlandos, *Hulika Domês Tôn Archaiôn Hellênôn* (Athens, 1959-60), vol. 2, 27-8. O. Broneer, *Hesp.* 14 (1945), 254-6, believed they were simply reinforcements for the dowels. See also G.P. Stevens, *Hesp.* 19 (1950), 158-9. My suggested extension of this interpretation to the blocks on the Marcia is of course quite conjectural. If correct, it apparently nevertheless remains as an isolated exception.

13. For a further discussion of this *sinter*, or incrustation, see pp. 227-32 below. 'The great characteristic of all the Roman waters is their extreme hardness, at least from the American point of view' (Herschel 195). He also considers the question of whether in antiquity springs had water of the same chemical composition as today, and comes to the conclusion that they mostly did. See also A. Trevor Hodge,

'Vitruvius, lead pipes and lead poisoning', *AJA* 85 (1981), 488. Among the few comparative studies of the differing hardness of the water in various aqueducts is a comparison of conduits in some twelve cities in France by M. Bailhache, 'Le débit des aqueducs gallo-romains', *JEAR* 40-1. To most of them he ascribes 'eau incrustante': among the major aqueducts, the best are at Lyon, the worst is Nîmes; Toulouse has 'eau corrosive'. He concludes (37) that 'Le cas d'eaux non incrustantes est assez rare'. The calcium carbonate content of the water at Nîmes is 227 mg per litre (Esperandieu 33). See also p. 73 above.

14. Esperandieu 34-5. He assumes an annual rate of incrustation of 0.115 cm. The extreme precision of the figure (down to one hundredth of a millimetre!) is apparently not due to uniformity in the rate of build-up, nor exactitude in the observation of it, but simply to Esperandieu dividing the total thickness among the number of years in his reconstruction, and that is what the answer happened to work out at. In its rough sense, say a millimetre a year, it may of course be quite reasonably accurate. The total existing incrustation is (maximum) 47 cm thick and Esperandieu divides it into four layers, 13, 8, 11 and 15 cm thick (= 113, 69, 95 and 130 years respectively). Taking AD 405 approx. ('les grands invasions') as the date proper maintenance ceased and the build-up began, this means that, with slight adjustments, the ends of the four layers are to be dated in AD 518 (508-510 = sieges of Nîmes by Thierry, son of Clovis, and Ibas and the Vandals); AD 579 (585 – Nîmes besieged by the Franks); AD 680 (673 = revolt of Hilderic of Nîmes against Vamba, king of the Visigoths); AD 803 (in the early ninth century the Normans 'commencèrent leur ravages dans le Midi'; territory of Nîmes totally devastated in 858). In reconstructing, Esperandieu stresses that 'On conçoit qu'une coincidence rigoureuse ne soit pas possible', but the actual layer structure of the incrustation is plainly visible, particularly around the Pont du Gard, and his basic premise does seem reasonable. So far as I know, no similar historical reconstruction has been attempted elsewhere. In the Rome aqueducts no less than 24 layers have been counted (Ashby 103).

15. Olivier (n. 5 above), 68.

16. On the Nîmes aqueduct these overflows leave marked traces as the lime content in the overflowing water itself solidifies outside the conduit. The deposits left by the overflows are often of enormous size. See also ch. 8 n. 37 p. 444. At Antioch the overflows on aqueduct bridges in time created 'saline deposits as thick as the piers themselves', leaving the whole bridge encased in this thick and amorphous mantle (D.N. Wilber, 'The plateau of Daphne', *Antioch on the Orontes*, ed. R. Stillwell (Princeton, 1938), vol. 2, 54); his fig. 5 is particularly striking. At Rome the deposit so resembled alabaster that in later ages it was dug out of the choked channels and used for altar decorations (Ashby 103). See also p. 231 below, and Figs. 162-3.

17. Nevertheless, it is an aspect of the work of his department that is passed over almost in complete silence by Frontinus: *Aq.* 122 is his only mention of it. Bailhache reasonably suggests it was too banausic to inspire interest (*JEAR* 25).

18. Birebent 466 illustrates some pipes from Djemila, Algeria, entirely blocked by calcium carbonate incrustation. We may also note Ashby's impressive and well-known illustration of the incrustation in the channel of the Anio Novus at Osteriola (Ashby pl XIX 6).

19. Vitruvius' figure has often been found impossibly low, and his text amended accordingly, from 'an *actus* between two manholes' to 'two *actus* between manholes', and Pliny (*NH* XXXI, 31) specifies two *actus*: '*in binos actus lumina esse debebunt*'. On a study of the Gier aqueduct, Jean Burdy has established that there the spacing is in fact one manhole every two *actus* (72 or 77 m, depending on the length of the foot used in calculation): Burdy, *FCAS*, 38-40; see also J. Burdy & H. Bougnol, 'Mille regards sur l'aqueduc du Gier', *Bulletin des Musées et Monuments Lyonnais* 2 and 3 (1986), 354-400. See also Burdy (loc. cit.) for an example of manholes maintaining a rigorously regular spacing even when this was realistically inadvisable – it meant putting one on a bridge, where access would be difficult and dangerous, where a few metres earlier it would have been on *terra firma*. For Vitruvius' text and its problems, Callebat 160. On the Gier aqueduct, the manholes have been published as 0.9 m square and 76 m apart (GdM 115; Jeancolas, *JEAR* 195). Carthage, 40-50 m apart (Rakob, *JEAR* 328-30). Nîmes, 80-100 m, 'tandis que dans les courbes brusques les intervalles se réduisent jusqu'à 14 mètres' (*MAGR* 96). Metz, 25-30 m (*MAGR* 202); Dougga, 50-100 m, Boulogne, 200-250 m, Bougie (Saldae), 48 m (Callebat 160). At Cologne they were provided at no set interval, but convenient to major engineering works, such as bridges, requiring servicing: an isolated group of three close together was probably 'erst und eigens für die Reparaturarbeiten angelegt' (Haberey 46). The Carthage manholes present a problem. Not only are they large and close together but midway in between each pair there is also a small square hole (45 × 45 cm) in the roof of the channel. Rakob hopefully asserts that this served for 'der Beluftung des Gerinnes', but there is apparently neither evidence nor any useful parallel. His suggestion that the large round manholes were for the removal from the channel of the wooden centering used in building its concrete vault also seems unlikely, at least to the extent that that was their exclusive or even principal purpose (Rakob, *JEAR* 315). In fact, though he stresses their size ('diese grossen Offnungen') they are no bigger than many other manholes (see n. 20 below). On manholes in general, see *MAGR* 31-2; Birebent 500. Information on manholes on the aqueducts of Rome is surprisingly sparse in view of Ashby's generally encyclopaedic treatment: *putei* are mentioned from time to time, and sometimes marked on plans where, since they are on top of a bridge, they must be inspection manholes rather than vertical shafts sunk to a channel running underground (Ashby 270, fig. 28, Anio Novus); but there is no coherent treatment of them giving details of their use and disposition, and in the example quoted above, even the dimensions go unrecorded. For Ashby, this is disappointing.

20. Round: Carthage, diameter 0.9 m (Rakob, *JEAR* 328, fig. 32). Kretzschmer 58, fig. 99 (Pompeii).

Rectangular: inspection shaft in Sernhac tunnel, Nîmes aqueduct: 1.30 × 0.60 m (Esperandieu pl. IV); Trier; 0.94 × 0.96 m (Neyses, *JEAR* 290); Gier (at Mornant): 0.95 × 0.89 m (Jeancolas, *JEAR* 202). Sens: 0.66 × 0.66 m (*MAGR* 31) (square lid on a round shaft). Metz: 1.20 × 1.20 m (*MAGR* 202). Cologne: 0.70 × 0.70 (Haberey 49, fig. 28; 80, fig. 53), also Kretzschmer 58, fig. 98.

21. 300-400 kg: Burdy, *FCAS*, 40. From Geneva is preserved a stone lid shaped like a cork, effectively plugging the manhole (*MAGR* 103, fig. 34, L. Blondel, 'L'aqueduc antique de Genève', *Genava* (1928), 53; my fig. 55). It carries recesses cut in its underside identified as handles for lifting, but though its weight is not published it is so massive that I doubt if manual power would suffice. Sens, *MAGR* 32, fig. 5. At Cologne a stone lid with a preserved iron ring for lifting it has been observed (Haberey 46; also 15, fig. 4, for photograph of a preserved square lid *in situ*). Wooden lids at Carthage: Rakob, *JEAR* 315.

22. La Brevenne: the pit is 44 cm deep, 64 cm long, and apparently ran the full width of the channel floor (preserved width 55 cm). It is at Sotizon, near Courzieu, on the early stages of the aqueduct. Jeancolas identifies it as 'un dispositif qui est très certainement un bac d'arrêt des impuretés, sables et graviers charriés par l'eau, non pas un bac de décantation comme l'indique improprement la légende du dessin', but does not enlarge further on the difference (Jeancolas, *JEAR* 191; 201, plan, fig. 3).

Carthage: unlike La Brevenne, where only one pit is known, 'sind an mehren Stellen bei Oudna und in der Ebene vor Karthago kreisrunde Vertiefungen in der Kanalsohle sichtbar' (Rakob, *JEAR* 315; illustration, abb. 40, p. 330).

23. An inscription (*CIL* XIII, 1623; *MAGR* 36) of Hadrianic date from the Gier (Lyon) aqueduct puts it clearly: '*Ex auctoritate/Imp. Caes. Traia/ni Hadriani/Aug. nemini/arandi ser/endi pang/endive ius/est intra id/spatium quod tute/lae [aquae] ductus/destinatum/est*' ('ploughing, sowing, and hoeing officially prohibited on aqueduct property'). This inscription ('La Pierre de Chagnon'), found in 1887, is, at the time of writing, preserved in a wooden case set up in the playground of the schoolhouse at Chagnon, near St Etienne. See also Front. *Aq.* II, 129, 7-9 for legal prohibitions: '*neve quis ... quid opponito, molito, obsaepito, figito, statuito, ponito, collocato, arato, serito*'. Also SRA 217 n. 119.

24. Ashby 13. The Virgo was given *cippi* by Tiberius in AD 35-7.

25. Front, *Aq.* I, 72, 3 '*ad septimum ab urbe miliarium*'; also, e.g., 70,1. A particularly important measurement was the seventh mile on the Via Latina (Front. *Aq.* I, 19, 1; Grimal 76) which essentially marks the spot where the five main aqueducts (Marcia, Tepula, Julia, Claudia and Anio Novus) left their surface channels and began the long run over elevated arcades. It is near Capanelle railway station, but more precise location is difficult: to Ashby (226 n. 1), 'The seventh milestone of the Via Latina is about 1,300 m due east of the Villa Bertone'; to Grimal (76), 'Ce point, le septième mille de la Via Latina antique, situé a 1,300 m a l'ouest de la Villa Bertone'; Van Deman (319) places the remains 'some fifty metres east of the villa'.

26. Grimal XII. The aqueduct of Nîmes also closely followed the line of two Roman roads, one from Uzès to Remoulins, whence it followed the Avignon-Nîmes road (*MAGR* 88; they are clearly shown on the map, p. 90, fig. 29).

27. The Kuttolsheim aqueduct had a 'voie de controle'; 'Elle était constituée de pierres calcaires et pourvue d'ornières espacées de 1.10 mètre' (J.J. Hatt, *JEAR* 176-7).

28. Cologne: Haberey 44, fig. 25; also pp. 42-7, fig. 27. The same solution, 'une sorte de tranchée emplie de pierres sèches destinées à éviter, en arrière, une accumulation d'eau qui pouvait entrainer la rupture de canal' was employed at Cherchel (Cherchel 142). For a statement of the problem in general terms, see Leveau, 'La construction des aqueducs', *Doss. Arch.* 38 (Oct.-Nov., 1979), 18: 'Un canal d'aqueduc est en effet la partie creuse interne d'un mur long de plusieurs dizaines de kilomètres construit à flanc de colline, sans joint de dilatation. Un tel mur fait barrage et empêche les écoulements le long du versant. Sur les versants argileux et surtout en pays mediterranéen où la sécheresse absolve de l'été alterne avec les grandes pluies d'automne, il est très exposé aux coulées de solifluxions: le mur formant barrage ne manque pas de se romper sous l'effet conjuge du conflement des argiles et du poids de la masse d'eau retenue.' At Metz an attempt was made to counter the problem by making the uphill side wall of the conduit thicker than the other (*MAGR* 202).

29. J.B. Ward-Perkins (n. 38 below), 117. He reports such leets as having been recorded 'for example, in the aqueducts of Elaeusa–Sebaste and of Korykos in Cilicia'. Even in pre-Roman Antioch, for all its splendour, 'water was brought in open channels from the suburbs and from the springs in the older part of the city'

(D.N. Wilber, in *Antioch on the Orontes*, ed. R. Stillwell (Princeton, 1938), vol. 2, 52).

30. See Dolaucothi 76. The dimensions vary widely, the estimated width running from *c.* 0.75 to 1.75 m (2 ft 6 in to 5 ft 6 in). The water ran at an estimated depth of 30-40 cm.

31. Dolaucothi 74. The aqueduct of Nyon, Switzerland (a closed concrete channel of the usual kind), was floored with ' "de grandes dalles en terre cuite" mesurant 45 × 30 × 4 cm; la face inferieure présente "trois tenons ou talons, disposés en triangle et faisant saillie d'environ demi-pouce"; ils "permettaient d'assujettir solidement les briques dans le béton". Leur poids est de 9 à 10 kg.' (Frederic Roux, 'Aqueduc romain de Divonne à Nyon', *Indicateur d'Antiquités Suisses* (1877), 720-4, quoted by Eugene Olivier, '*L'alimentation d'Aventicum en eau*' (Neuchâtel, 1942), 7). Not too much has been published on industrial or agricultural aqueducts. The best known are Dolaucothi, in Dyfed (formerly Carmarthenshire), between Lampeter and Llandovery, Wales, supplying water for washing the ore in a gold mine; the aqueduct supplying the flour-mill at Barbegal, near Arles (which seems to be a purely conventional aqueduct differing little from an urban supply); and the aqueduct serving a Spanish gold mine described in detail by Pliny *NH* XXXIII, 74-6.

32. The channel half-cut into a rock-face with the outer side built in masonry was common in aqueducts in North Africa. For vertical stone slabs, see M.H. Saladin, *Archives des missions scientifiques et littéraires* 13 (Paris, 1887), untitled report pp. 1-225, p. 62, fig. 119, detail 1 (at Foum-el-Guelta, Tunisia). Wooden planking: Haverfield on Dolaucothi, 'Sometimes channels were hewn out of the rock, sometimes a flat ledge was cut on the hillside and (as we may conjecture) wooden troughs or pipes were laid ... The wood has naturally vanished' (Dolaucothi 72). 'Stream crossings and similar obstacles were probably crossed by wooden trestling, all trace of which has disappeared' (73). 'Disposable crossings' would be on the model of the modern *falaj* (irrigation aqueducts) in Oman, which often cross gullies 'by a hollowed-out palm log', replaced if it gets carried away: Paolo M. Costa, 'Notes on traditional hydraulics and agriculture in Oman', *World Archaeology* 14.3 (1982), 280. This article offers many useful illustrations and sidelights on ancient irrigation.

33. See n. 32 above.

34. Dolaucothi and Alsietina discharge: see Table, p. 273.

35. Bronze pipes at Pergamon: so in *SAT* 161, fig. 35; see also Landels 47. White (214) queries this and rightly favours lead instead. In fact, the case has been proved by G. Garbrecht. On the principle that the corroding and decaying pipes would leave some traces of their metal impregnating the soil, soil tests for zinc, copper, and lead were carried out along the line of the siphon: 'Die in der Trasse im Mittel 56 – fach hoheren Bleiwerte bei etwa konstanten Weten der übrigen Schwermetalle lassen es als sicher erscheinen, dass die Druckleitung von Pergamon aus Blei bestanden hat' (Garbrecht, *JEAR* 154). So far as I know, bronze pipes have not been suggested for any other aqueduct.

36. See pp. 37-9 above.

37. Ephesos: ch. 2 n. 42 above. Bends: p. 118 below and my Fig. 72; 470 n. 43.

38. See ch. 2 n. 49 above. *WAS* 1 (1986), 181-3 (good photographs from Pergamon, Ephesos, Laodicea, and Palmyra). A very full list of sites where stone pipes have been found is published by J.J. Coulton, 'Oinoanda: the water supply and aqueduct', *Anatolian Studies* 26 (1986), 49. Aspendos: K.G. Lanckoronski, *Städte Pamphyliens und Pisidiens* (Vienna, 1890), vol. 1, 124, fig. 98; J.B. Ward-Perkins,

'The aqueduct of Aspendos', *PBSR* 23 (1955), 119, fig. 2; only one or two of the pierced stone blocks from the aqueduct are to be seen around the site (though the arched structure that carried them is largely intact), but a large number may easily be seen built into the piers and abutments of the Seljuk bridge over the Kopruçay (= Eurymedon) river 4 km downstream from the classical site. Other stone pipes have been recorded at Dougga (Tunisia) (illustrated by A. Baur, *Schriftenreihe der Frontinus-Gesellschaft*, Heft 10 (1987 – transactions of Symposium at Cologne, 1986), 170); Poli (near Arezzo), Lanciani/Pasqui, *NScA* (1878); Azeffoun (Algeria), S. Gsell, *Monuments antiques de l'Algérie* (Paris, 1901), vol. 1, 257; see also GdM 193 (esp. Patara), and his n. 2. Patara: illustration by Fahlbusch in *WAS* 2 (1987), 157 (abb. 24). Laodicea: Fahlbusch, abb. 48 (see also 49, 50). A Roman (? – but it does *look* convincing) stone pipe from Mainz is shown by von Kaphengst in *WAS* 3 (1988), 202 (abb. 10). Round stone pipes from the aqueduct at Padua are shown by Haberey 136, fig. 103; from Bethlehem and Jerusalem, Malinowski, *JEAR* 271. For Rome, see Herschel 162; Ashby 154. And at Avenches, Switzerland, were found 'des tambours évidés en pleine pierre, d'un demi-mètre de long, avec lumière de 30 cm et paroi de 12 à 15 cm.' (Olivier (n. 31 above), 14); the situation at Baelo (Pierre Paris & George Bonsor, *Fouilles de Belo* (Bordeaux/Paris, 1923)), 113-14 (esp. fig. 3) seems similar, and one suspects that the stone blocks are unused column drums. There was an architectural quarry close at hand.

39. For wooden pipes in general, see Hanson 419-22; Haberey 134-6, with photographs of pipes and collars (Trier); Samesreuther (153), in his index lists wooden pipes from no less than 33 different locations in Germany; general discussion: 137-9. Caerwent, Hanson 422; *Archaeologia* 61, part 2 (1909), 567. Samesreuther (137) describes them as 'Diese einfachen Leitungen, die in der Hauptsache nur örtliche Bedeutung hatten'. H.A. Wessel (*Schriftenreihe der Frontinus-Gesellschaft*, Heft 10 (symposium, Cologne, 1986), 107 reproduces a contemporary painting from the early nineteenth century showing a black work gang laying in a street water main of wooden pipes in downtown New York. 'Wood rots and splits' (White 165); does not rot (Hanson 422).

40. *Pinus, Picea* and *Alnus*. They must, he says, be buried in the earth if they are to last. The Budé editor, J. André (Pliny *NH* XVI (Paris, 1962)) is sceptical about *Picea*; it is full of knots, and after a time these fall out, leaving the pipe full of holes.

41. Collars are illustrated in Haberey 135, fig. 102; Samesreuther 135, fig. 60; E.M. Wheeler, *London in Roman Times* (London, 1930), 39 and pl. XII. See also D. Atkinson, *Wroxeter* (Oxford, 1942), 121ff.; *JRS* 29 (1939), 214 (Caister-by-Norwich: 'flanged iron rings for joining wooden water pipes'). Caerwent: Hanson 85, and 422 for lengths of sections. Internal diameter, e.g. Haberey 135, fig. 101. Packed clay and stone slabs, Samesreuther 138 (as in the aqueducts at Seulberg and Kapersburg). A. Neyses (*JEAR* 292, fig. 10; my Fig. 66) gives a drawing of a 2.4 m long length of wooden piping from Vicus Belginum (Trier), dated by dendrochronology to AD 247. It is fitted with a valve inset at its midpoint and evidently served a pump. For an access shaft to a wooden pipeline, see Samesreuther Taf. 9, 1 (facing p. 118) (Trier).

42. Samesreuther 138.

43. So Samesreuther 142, and my own observation; but for a contrary view, see Grenier, 'Les exemples de conduites en terre-cuite sont peu fréquents; on en trouve quelques-uns dans les campagnes. Elles ne sont d'un usage courant que dans les thermes ou les habitations privées' (*MAGR* 35); presumably he is referring only to France, but even there I would have thought this overstated. See also Callebat 177-9.

44. The pipeline was excavated and published in 1907 by J. Déchelette; his photograph of it is now reprinted in Danièle Bertin and Jean-Paul Guillaumet, *Bibracte* (Guides Archéologiques de la France; Ministère de la Culture et de la Communication, Paris, 1987; ISBN 2-11-080908-6), 47, fig. 14.

45. Garbrecht, *JEAR* 150.

46. For typical profiles, see Samesreuther 143 (which see also for general discussion of terracotta pipes), fig. 63; Malinowski, *JEAR* 270; Lassus, *JEAR* 229 (Antioch); Haberey 137 (Bonn). Inspection openings: Haberey 137, fig. 105, shows a fine example complete with a handle on the lid. The use of the opening for sealing the joint from the inside is well illustrated by the Enneakrounos drain in the Agora at Athens. *Hesp.* 6 (1935) 334-6, and 25 (1956) 49-52. Though not a Roman work, this offers a cogent demonstration of the principle. For an access shaft to a terracotta pipeline, analogous to that observed for a wooden one at Trier (n. 41 above), see *MAGR* 103 n. 1: 'tuyaux verticaux servant de regards' (Kuttolsheim). I wonder whether they may not have been intended rather as surge tanks to release pressure in the line (for which see A. Trevor Hodge, SRA 207-8, and fig. 11). Something similar seems to have been arranged on the terracotta pipeline of the Caesarea aqueduct in Israel (see pp. 243-4 below). See also Fig. 226.

47. Vitr. VIII, 6, 8: '*Tubuli* [V.'s normal word for terracotta pipes; lead pipes are *fistuli*] ... *fiant, sed uti hi tubuli ex una parte sint lingulati, ut alius in alium inire convenireque possint. Coagmenta autem eorum calce viva ex oleo subacta sunt inlinienda*' and see Callebat 177, ad loc. Malinowski recounts his experiments in the *Concrete International* article (n. 11 above), 75. See also Malinowski, *JEAR* 270.

48. Garbrecht calculates a discharge for the conduit of 4,000 m^3 per day, and the same for the siphon: 'Die Leistung der Zuleitung (dreistrangige Rohrleitung) und der Druckleitung ensprechen einander also' (Garbrecht, *JEAR* 152, 154; which see for illustrations of the pipes, settling tank, and siphon). One may compare Smith's anxiety over how Roman engineers could 'match the flow of an open channel to the flow capacity of a multi-piped siphon' (Smith 64).

49. A. Stieber, 'Observations concernant la conduite d'eau romaine de Kuttolsheim', *Cahiers alsaciens d'archaeologie, d'art, et d'histoire* 4 (1960), 45-52.

50. Lyon: *JEAR* 188; Caesarea: Y. Olami & Y. Peleg, 'The water supply of Caesarea Maritima', *Israel Exploration Journal* 27 (1977), 133. Surface level tank: see pp. 157-60 below for parallels with Craponne, Aspendos and further discussion.

51. The best example is at Samothrace; another has been found at Pella (J.R. McCredie, 'Samothrace: preliminary report', *Hesp.* 34 (1965), 113, and pl. 33 (c) (good illustration); Pella: Photios Petsas, *Archaeology* 27 (1964), 84, fig. 17.

52. A good example is the very sharp right-hand bend in the Arles aqueduct at Barbegal. It arrives at Barbegal running alongside the second aqueduct, which serves the mill complex, and then, in a rock-cut channel on the crest of the hill on which the mills are laid out, veers abruptly west (*MAGR* 78, fig. 27). There is no reason why this bend should not have been eased out a little.

53. Ashby, fig. 2, facing p. 69 (Anio Vetus; oblique angle, but at foot of an incline). Ashby here comments on the *specus* entering the bridge 'without the usual right-angled turn', Ashby. fig. 1, p. 51 (Marcia, at Fosso degli Arci); fig. 12 (p. 118; the Marcia at Ponte Lupo); p. 122 and fig. 13 (Marcia); fig. 14 facing p. 137 (Claudia and Anio Novus at Tor Fiscale); fig. 16, p. 197 (Claudia, S. Cosimato); fig. 18, p. 206 (Claudia, Fosso della Noce); fig. 31, p. 279 (Anio Vetus, at Grotte Sconce); fig. 32, p. 283 (Anio Novus, Ponte S. Antonio: 'A characteristic right-angled turn at the west end'). See also fig. 23, p. 238, for various sharp bends while the conduits are running on continuous arcades at Vicolo del Mandrione. See also Cherchel fig. 104A (plan, Oeud Bellah); fig. 102, (plan, Oued Soromane).

54. Cherchel fig. 99, plan, Grand pont du chabet Ilelouine. Saintes: Bailhache, *JEAR* 45, 47. There was probably a similar arrangement at Chagnon on the Gier aqueduct, where the loop was replaced by a siphon instead of a bridge, but there are no traces.

55. Aspendos: for plan, see *WAS* 2 (1987), 174; also Lanckoronski (n. 38 above), p. 122, fig. 96A; p. 123, fig. 97, repeated by Ward-Perkins (n. 38 above), p. 116, fig. 1. The identification of the buttresses as an anti-thrust device is specifically made by Kenneth D. Matthews, 'Roman aqueducts: technical aspects of their construction', *Expedition* (Journal of the University Museum, University of Pennsylvania) (Fall, 1970), p. 10 (photo caption). Segovia: plan, *WAS* 1 (1986), 164; the angle of the bend in the Aspendos aqueduct is 50°, at Segovia 55°. For other aqueduct bridges with a bend in them, see Cherchel fig. 106 (pont du Oued Nsara), and p. 118 (pile d'angle; angle of bend, 45°). All of these are well attested. A hypothetical possibility also exists at Saintes, where the alignment of the aqueduct seems to entail a 99° bend in the middle of crossing a valley. A few concrete and masonry foundations are preserved at the presumed location of the bend, but not much else. Triou suggests that not a bridge but a siphon was used, and the foundations are part of the anchoring down of the pipes needed for a 99° *geniculus*. (A. Triou, 'Les aqueducs gallo-romaines de Saintes', *Gallia* 26 (1968) 119-44. On balance, I find it improbable. The (Roman? Byzantine?) aqueduct of Kavalla crosses the city on a bridge laid out in a curve.

56. 'Uniformity of flow ...' – so Deane R. Blackman (ch. 8 n. 2 p. 439), 54.

57. The Latin for the main line, or principal element, of an aqueduct is *matrix*: see *Thesaurus Linguae Latinae* 8 (1966), 481-2 at 482.66f. (*de fistula aquaeductus*); also Brent D. Shaw (ch. 8 n. 3 p. 439), 73 n. 6.

58. Osteriola junction: Ashby 268, fig. 27. Settling tanks before junction: Kallmuth junction, on the Urft-Cologne aqueduct (Haberey 58, fig. 34; 60, fig. 36). They are quite small, one some 0.6 m square, the other 1.1 m. Grotte Sconce: Ashby 278, fig. 31. Van Deman (88) speaks of the arrangement as a 'cataract'. S. Cosimato: Ashby fig. 7, facing p. 101, and p. 102. The shaft, if it had originally been a *puteus*, presumably came right up to surface level (with the sluice from the Claudia opening into one side of it, like a door into an elevator shaft), where, equally presumably, was located the operating mechanism for the sluice. Ashby refers to the whole structure as a 'sluice-tower'; but his description is confused, and his fig. 7 makes one miss, e.g., the outstanding clarity of the illustrations in Haberey.

59. See pp. 317-19 below.

60. Grotte Sconce: Ashby 278, fig. 31. For a similar junction with an incoming branch ('the small reservoir where the waters of the two channels were once more united') on the Anio Novus at Gericomio, see Van Deman 289.

61. Haberey 25-8, inc. full plans (fig. 13) based on R. Schaltze, *Bonner Jahrb.* 135 (1930), Taf. 28. Haberey shows three sluices all in logical places, but there seems to be a possible slot for a fourth, duplicating the sluice on the intake of the branch channel. The second settling chamber is very unorthodox, being more like a vertical shaft with a manhole on top such as we have seen at S. Cosimato. The details of the operation and purpose of this installation are obscure. According to Haberey, 'die Funktion des kleineren Becks ist mir unklar', and the branch aqueduct 'gibt Ratsel auf'. He suggests it may be a relief channel into which the water could be diverted when the main line through to Cologne was closed for any reason.

62. The Eiserfey tank: Haberey 70-4, inc. plan, fig. 45. Compare with the Nîmes *castellum*, my Fig. 198, or *MAGR* 98, figs 32-3. See pp. 284-5 below, as also for an exposition of the 'public resort' hypothesis at Nîmes. My suggestion that it was used as a local fountain or watering point would be much strengthened if the coping stones bore grooves worn by ropes lowering the buckets, as we so often find on well heads; but there do not seem to be any. As restored in Haberey's fig. 45 the guard wall seems to me too low, about 70 cm, or just over knee-high, but both the height of the wall and the ancient ground level are conjectural. He also suggests that it was somehow covered over to prevent freezing, but with the water constantly swirling around in rapid motion I doubt if this would be a serious problem; possibly an ice bridge a centimetre or two thick might sometimes form on the surface. A similar form of junction with no less than three branches running into it, is conjecturally restored by Haberey (21, fig. 9) at Hurth-Hermulheim.

63. To take some examples, on the Marcia the branch to the Baths of Caracalla presumably left *via* a reservoir, and one has been located near Porta Furba in the right general area, though rather low; Winslow (43) takes it for granted this was the junction point, but Ashby (156) maintains that 'the point where the branch diverged to the Baths of Caracalla is not certain'. On the Claudia the junction point for the branch to the Villa of the Quintilii is not preserved, nor, on the Anio Novus, for the Villa of Sette Bassi (Ashby 223, 228).

64. Grewe 26-7.

65. J. Burdy, in *Préinventaire des monuments et richesses artistiques, Département du Rhône, I: l'aqueduc romain du Mont d'Or* (Lyon, 1987), 35-6. It was earlier published, but with less accuracy, by Jeancolas (*JEAR* 187). Since the entry and exit channels are not facing each other but at an angle, one wonders whether the Mont d'Or basin could perhaps have been not on the main aqueduct conduit, but alongside it, as at Siga.

66. Haberey 66, fig. 40.

67. Frontinus, 1-19; 2-67; 2-69; 2-72.

68. Ashby 226, n. 4; see also Herschel 199 (who got his information from Lanciani). Compare Monte Testaccio in Rome, composed entirely of broken fragments of empty amphorae. (E. Nash, *Pictorial Dictionary of Ancient Rome* (London, 1968), vol. 2, 411-13).

Chapter 6

1. General consideration of tunnels, GdM 230-7. The distinction between 'canal en souterrain' and 'tunnels proprement dits' (de Montauzan's terms) is not always clearly observed by modern commentators. A 'canal en souterrain', such as we have been considering in the previous chapter, was normally not tunnelled but built by 'cut and cover'. Not only was this easier and quicker, as an unlimited number of excavators could get at the job simultaneously, but there were no difficulties in orientation calling for 'des opérations délicates et une attention soutenue pour éviter les erreurs de trace' (291). See also Grewe, 69-75. We may also note the network of tunnels forming an underground urban water system at Grand, France, which, at the time of writing, is still under study: see Jean-Paul Bertaux, 'Les galeries souterraines', *Doss. d'Archéologie* 162 (juil.–août 1991), 28-33, and also Olivier André and Valérie Bertaux, 'Les aqueducs souterrains construits par les Romains', id. pp. 37-8.

2. So Ashby 271. The length was computed by Lanciani as 4.95 km and is so accepted by Van Deman (294: 'a little over four kilometres long'). The same cut-off also had a 225-250 m tunnel under Colle Castello, also not preserved. A preserved inscription of AD 88 from the area records how a certain Paquedius Festus brought the Claudia in tunnel '*sub monte Aeflano*' (*CIL* XIV, 3560; Ashby 192; see *PBSR* III, 135), but it is not at all clear where or what the tunnel was, and to Ashby (209) it 'is a mystery'. For a further example of long tunnels, though not as an aqueduct, we may compare the *emissarium* of the Fucine Lake dug by Claudius in AD 53. It was 5.5 km long and supposedly took 30,000 men 11 years to dig.

3. This has been verified in the aqueduct at Antioch, where 'Il faut supposer que des puits s'ouvraient de distance en distance, les ouvriers en partaient dans deux directions à la fois, cherchant à rejoindre ceux qui partaient des puits voisins. Parfois sans doute y parvenaient-ils parfaitement; d'autre fois une légère erreur, égale ici a 40 cm environ, faisait que les deux tronçons tout bien que mal, on a amenagé le puits de descente, au dessus du coude de la canalisation' – J. Lassus, 'Une villa de plaisance a Daphne-Yakto', *Antioch on the Orontes*, ed. R. Stillwell (Princeton, 1938), vol. 2, 100. See my Fig. 78 (Halberg).

4. A rectangular stone-cut vertical shaft in the Sernhac tunnel (about 350 m long?) is illustrated in Esperandieu pl. IV; Hauck, 118. Spoil from *putei*, Ashby xi and 44. Stone slabs and cement, also protective wall 1.5 m high. See Stübinger fig. 6-8 (Arles). The shafts are 1.2 m square, 4-5 m deep; his fig. (my Fig. 79) shows a shaft with what look like wooden planks laid across like landings, at vertical intervals of 50 cm, but there is no explanation in the text.

5. Mornant: Jeancolas, *JEAR* 195: 'Cette boucle de Mornant est une belle reussite technique: la galerie est forée sur un tracé courbe, à grand profondeur.' Crucimèle: Esperandieu 15: 'Il n'a pas été creusé en ligne droite et ses déviations, au nombre de dix a douze, pourraient tenir, non pas à un défaut de surveillance, mais au désir d'utiliser des failles ou des veines de rocher plus particulièrement tendres.'

6. *CIL* VIII 2728 = *ILS* 5795. Grewe 70; *WAS* 3 (1988), 215-18. P. MacKendrick, *The North African Stones Speak* (London 1980) 247-8; GdM 292, White 160 and 215 (for English text). R. MacMullen, *HSCP* 64 (1959), 207-35. The tunnel (dedicated *c*. AD 152) was of only medium length (500 m) and the line was marked out by stakes over the top of the mountain. (*Rigor autem depalatus erat supra montem ab orientem* (sic) *in occidentem*); apparently no *putei* were sunk and the tunnel was dug in two halves (*superior* = upstream, and *inferior* = downstream). Datus was absent from the site for four years, during which the work went ahead unsupervised (and disoriented). Above the inscription the stone slab carries representations of three female figures identified as Patientia, Virtus and Spes; in the circumstances, their invocation must be counted piquantly appropriate. At Sua (Chaouach), 9 km from Medjez-el-bab, Tunisia, there was a 665 m long tunnel without shafts, which, in the circumstances, de Montauzan considers 'assez long' (GdM 295 n. 3). For tunnelling techniques, see pp. 208-14 below.

7. But not necessarily to stand in. In the galleries in the Laurion silver mines it was 'impossible for a miner to stand upright: he had to lie on his back, or side, in cramped conditions' (J.F. Healy, *Mining and Metallurgy in the Greek and Roman World* (London, 1978), 81). The galleries were small: 'their height was never more than 1 m and often as low as 60 cm and 60-90 cm wide.' Aqueduct tunnels were usually bigger than this (p. 94 above).

8. We should be clear on terminology. A bridge is used to span something intervening, such as a river, which one could not otherwise cross. A viaduct is used to maintain height across a dip in the ground, but (usually) without it spanning anything: it is simply a cheaper substitute for an embankment. For the continuous series of arches across the plain I shall reserve the term 'arcade'.

9. The highest Roman bridges were: Pont du Gard 49-23 m; Alcantara (Spain) 50-77; Narni (Italy) 36.92; Cherchel (Algeria) 35.38; Segovia (Spain) 30.77; Mérida (Spain) 26.15. (Smith 65, which see for comparison with siphons). Siphons, see p. 155 below. The longest bridges were the Pont du l'Oued Bellah (Cherchel), 288 m, and the Pont du Gard, 275 m, if we are to exclude urban structures such as Mérida and Segovia, which are perhaps to be considered rather as continuous arcades than bridges proper.

10. Van Deman 126: Esperandieu 26, *MAGR* 91.

11. Esperandieu 26: 'Le Pont du Gard, entièrement bati de pierres de taillc.' His account is followed by Grenier, *MAGR* 93. The main arch (span 13 m) of the bridge of the Caligulan aqueduct of Antioch was also 'built of stones set without mortar' (D.N. Wilber n. 13 below), 53. Masons' marks on the Pont du Gard: Lassalle, *Doss. Arch* 58, e.g. *FRS* III (=*frons sinistra* no. 3).

12. Auguste Choisy suggested that the juxtaposed-arches technique 'a permis de n'utiliser, pour monter les voutes, qu'un cintre d'importance reduite, de la largeur d'un arc, que l'on a deplacé autant de fois qu'il était necessaire pour réaliser la voute tout entière' (quoted by Lassalle, loc. cit.). Parallels: Nîmes (Porte d'Auguste, amphitheatre, Temple of Diana), Arles (amphitheatre) Sommières (bridge).

13. 'In the Roman aqueducts, construction in concrete masonry began to replace the earlier construction above ground of cut stone masonry during the time of Agrippa and Augustus. Soon even the larger arch spans were built entirely of concrete, although the use of cut stone for the lower parts of the bridges persisted through the Claudian period.' D.N. Wilber, 'The plateau of Daphne', *Antioch on the Orontes*, ed. R. Stillwell (Princeton, 1938), vol. 2, 53 n. 9.

14. Ponte Lupo: Ashby (fig. 11, 12, and pl. V (p. 124)). The analysis of the various stages in its building and development takes Van Deman six closely argued pages (95-101). Impressively massive, it may well be 'an epitome in stone and concrete of the history of Roman construction for almost nine centuries', but it is not by chance that one so seldom sees it featuring in the glossy illustrations of handbooks of Roman civilisation. Of course, it is no help that it is usually so overgrown.

15. Antioch: Wilber (n. 13 above), p. 53; Ponte S. Pietro, Ashby 116; Pont du Gard, *MAGR* 92: the centre arch in the bottom tier and the one immediately above in the middle tier are the same size. The other arches in both these tiers, though smaller, are still larger than any other aqueduct arch, with a span of 19.2 m. Given the necessity for a wide span at one particular point in bridging the valley, the architect had a choice on either side of it: he could either use a row of the normal small arches, leaving one wide one alone in the middle, or, for the sake of uniformity, he could use wide ones all the way through. The first would probably be easier, and was done in the Ponte S. Pietro. The second meant building a quite unnecessary series of wide spans over solid ground so that all the arches would match, and thus put aesthetics before engineering simplicity and economy: this was done in the Pont du Gard. However, large spans are so uncommon in aqueduct bridges that we cannot identify either approach as normal Roman practice. The architect of the Pont du Gard does seem to have set particular store on an appearance of uniformity and regularity. 7.5 m for regular spacing of piers in most aqueducts: Cherchel 140.

16. As happened, for example, on the Pont de l'Oued Nsara, at Cherchel (Cherchel, 110). The foundations being very shallow (1 m deep) did not reach a sure footing, and the piers began individually to slide downhill. Attempts to arrest the movement by bonding together two of the piers by a 'massif de blocage' only resulted in the movement becoming collective and uniform.

17. The point is perhaps illustrated by a later version of this structure. At Maintenon, near Chartres, in France, are the imposing remains of a large aqueduct arcade 1 km long. It was built in the seventeenth century by Vauban to bring water for the fountains of Versailles across the valley of the Eure and, though based on the Pont du Gard, was never completed beyond the first storey of arches, which still stands, well illustrating the independence of the three tiers. When it became plain that this grandiose project would remain unfinished, the proposal was made that expense could be saved by turning it into the *venter* bridge of a siphon. This was not done either, though in Spain a similar project to save maintenance by turning the orthodox Roman aqueduct bridge at Valencia de Alcántra into a siphon (the bridge being truncated, with the bottom part preserved as a *venter*) was successfully executed in the nineteenth century (SRA 191). At Roquefavour, 10 km west of Aix-en-Provence, is a modern viaduct carrying the water supply for Marseille which was specifically built (in 1847) to imitate and outdo the Pont du Gard. It is a three-tier bridge, 82-65 m high and 375 m long (being the largest aqueduct bridge in the world, half as big again as the Pont du Gard; which it imitates so effectively that tourists have been known to take it for a Roman antiquity).

18. Mérida, *WAS* 1 (1986), 163, fig. 25. Cherchel: Cherchel fig. 39 (Chabet Ilelouine), and p. 74; figs. 56, 60 (Oued Bellah), and p. 97 (see also fig. 89 (Saldae)). For a general consideration of cross-bracing (*entretoisement*), see p. 138.

19. Cherchel figs 82, 100.

20. A close parallel is the celebrated and catastrophic collapse of Sir Thomas Bouch's railway bridge over the River Tay, Scotland, in 1879. The subsequent enquiry made it plain how very inadequate were the data on wind pressure to be expected, even in this modern era. A comparison with Roman practice, on the basis of even less data, is instructive. For calculations on the wind resistance of the Pont du Gard, see Hauck, 92-3; also G. Hauck, 'The Roman aqueduct of Nîmes', *Scientific American* (March 1989). The safety factor was 1:2.

21. *MAGR* 91; Esperandieu (27), 6.36 m, 4.56, and 3.06.

22. GdM 127-8, and figs 46, 48.

23. *MAGR* fig. 11 (p. 45) and fig. 19 (p. 52), GdM fig. 101, p. 247. Buttresses were also used at Mérida and, in exceptional circumstances, Aspendos: Lanckoronski (ch. 5 n. 38 p. 418), p. 121, fig. 95. They were common in the supporting masses of brick and concrete used in repairs to the bridges of the aqueducts of Rome (Ashby 118, fig. 10 (Ponte S. Pietro); 119, fig. 12 (Ponte Lupo); 216, fig. 21 (Ponte Diruto)).

24. Cherchel 138 ('Les piles d'aqueduc ont normalement la forme de pyramides à gradins'), and fig 82, 104. It was also a common technique on road bridges (Narni). See my Fig. 88 (Metz).

25. Cherchel 100 n. 95. The calculations revealed that a force of 48,379 kg acting upon the centre of gravity of Pier 25 (one of the two tallest) would be critical for the bridge and this force would be exerted by a wind of 234.6 km per hour (above). The equivalent figure for the Pont du Gard (n. 20 above) is 215 km.p.h.

26. One thinks naturally of the traditional practice of soldiers marching in column being ordered to break step when crossing a large bridge, for the same reason.

27. Winds in the Mediterranean have been very little studied in the classical context, but see A. Trevor Hodge, 'Massalia, meteorology and navigation', *The Ancient World* (published by Ares Press, Chicago), vol. 7 (1983), 67-88. The prevailing wind on the Algerian coast is the Scirocco, a southerly originating in the Sahara Desert. This would have hit the bridge on its southern side. Gales in this region, however, usually come from the north. What happens is that if the Scirocco is weak and a particularly violent Mistral comes down from the north into the Gulf of Lions, with no Scirocco to stop it, it may go right across the Mediterranean and hit the Algerian coast as a northerly gale. This is most probably what happened to the bridge. The strength of the Mistral would be weakened by the time it got to Algeria, but for its regular strength, see n. 28 below.

28. For the Mistral, see Jean Beaujeu *ad* Pliny *NH* II, 121 in the Budé edition (Paris, 1950), 201-2, and Hodge (n. 27 above), 82-4. At Marseille, a Mistral of 100 km.p.h. is registered quite often, and a force of 250 km.p.h. was recorded on 11 February 1938 at Istres airfield, between Marseille and Arles. These figures do not represent its full strength, which it attains only out at sea beyond the sight of land-bound observers, but even on land it is a force to be reckoned with, having been known to blow trains off the railway line and clothes off people (Strabo IV, 182). In antiquity it often blew down houses but was welcomed for its salubrious effects (Sen. *Q.N.* V, 17.5).

29. So A. Donnadieu, *La Pompeii de la Provence, Fréjus, Forum Julii* (Paris, 1927), 180: 'Le mistral ... avait nécessité cette précaution, la série des arcades offrant la plus grande prise à la violence du vent en raison de son orientation.' Compare Van Buren, *Wasserleitungen, RE* VII A (1955) col. 479.

30. It is not known when the cut-back took place. It is first definitely recorded by Poldo d'Abenas in 1557, but may well date back far beyond that. The project to restore the mutilated piers to something like their original form was submitted to the Etats de Languedoc in 1699 and executed in 1702. The modern road bridge alongside was added in 1743-7. Esperandieu, 38-44.

31. The Pont du Gard was as resistant to floods as to the wind. The large arches on the lower tier, though normally on dry land, did have the practical advantage of easily letting through the floodwater when the river overflowed its banks. On the upstream side all the piers, including those on land, have cutwaters to reduce erosion and damage from the current, which seems to indicate that the architect envisioned the possibility of the land-based ones being sometimes immersed. In 1958 it successfully withstood a flood that washed away the modern road bridge (at that time a clear-span suspension bridge) at Remoulins, 3 km down the river. This may partly be due to the fact that it is not entirely straight, but is, in plan, slightly bowed, with the convex side facing upstream, like a modern dam. The curvature is very slight, but, given that all the piers are founded on rock and not liable to have shifted, must be both original and intentional. Esperandieu suggests, without confidence, that 'Il se peut qu'elle ait eu pour but de permettre à l'aqueduc de mieux résister à la force du courant', and may be right, while Grenier simply records the fact of curvature (Esperandieu 30; *MAGR* 92; also accepted by Hauck 93). Roman dams were usually straight (see p. 81f. above), making it surprising if the bowed profile was here used consciously to resist the force of the current, particularly since the curvature is so slight, but I can think of no other explanation if it is deliberate. So far as I know, the Pont du Gard is the only place where this feature has been observed. N.A.F. Smith (*Transactions of the Newcomen Society* 62 (1990-91), 'The Pont du Gard and the aqueduct of Nîmes'), maintains that the curvature is accidental: 'The explanation must be that in the course of time the Pont du Gard has twisted. This might have been due to settlement, or was perhaps induced by cutting the middle piers for the one-time roadway, or conceivably Pitot's road bridge has leaned too hard from the opposite side. There is a worrying aspect to all this because the belief that what, in reality, is a defect should be applauded as clever design shows complacency over the bridge's future as well as ignorance about its construction'. This may be right.

32. S. Pietro: Ashby 118, fig. 10; Cherchel figs 104, 100. By comparison, the normal average gradient for the Marcia was 1.543 m per km, or half of the figure for S. Pietro. But figures for average gradient over a long distance are notoriously unrepresentative for any given section of the line: see p. 218f below.

33. Cherchel fig. 100; Ashby fig. 2, p. 69 and p. 70 n. 1.

34. Esperandieu 36-43 (the chapter is entitled 'Le Pont du Gard utilisé comme passage').

35. Ashby 120 and figs 11, 12.

36. Lanckoronski (ch. 5 n. 38 p. 418), 124; see also 121, fig. 95, where the excessive width of the *venter* bridge is clearly shown.

37. Lanckoronski, op. cit., 121: 'sie diente also nicht blos als Tragerin des Wasserrohres, sondern als gangbare Verbindung der Thalrander.'

38. By 'siphon' a classical archaeologist traditionally refers to what an engineer more strictly calls an 'inverted siphon', with the bend or elbow at the bottom, not the top: that is, it is like the letter U, while the true siphon is the other way up. Hydraulically the difference is important, for while an inverted siphon will commence and continue running automatically as soon as water is admitted, a true siphon has to be started artificially by some outside agency, such as a pump, though once going it will continue naturally by itself. In this book I continue the established convention, and throughout by 'siphon' I mean 'inverted siphon'. For ancient usage of the term (it is a Greek word by origin), see SRA 175. For Greek siphons, see p. 33f. above.

39. 'Did not know water would rise to its own level': so HRH the Duke of Edinburgh addressing the Royal Society (*Notes and Records of the Royal Society of London* 29, 1 (Oct. 1974), p. 18), and no doubt basing himself on standard works. Pliny (*NH* XXXI, 57: '[*aqua*] *subit altitudinem exortus sui*') clearly states the principle of water finding its own level. See also G.E. Bean, *Turkey's Southern Shore* (London, 1968), 75-6 on Aspendos. For inability to make pipes strong enough, the orthodox view is reflected by the standard modern Penguin paperback on structural engineering: 'for lack of pipes capable of conveying liquids under pressure the Romans incurred enormous expenses in building masonry aqueducts upon tall arches' – J.E. Gordon, *Structures* (Harmondsworth, 1978), 117. See n. 49 below.

40. As clearly stated by Van Buren (*RE* VIII A1, 473: 'Der Gebrauch von Druckrohrleitung war im römischen Reich weitverbreteit'). and by Ward-Perkins 117 ('[the principle of the siphon] was regularly used for local distribution ... [and] it is evident that the Republican engineers were perfectly capable of conveying water in bulk under the considerable pressure of 100 m, or approximately 10 atmospheres'). The role of siphons and the principles governing their operation are studied in detail in A. Trevor Hodge, 'Siphons in Roman aqueducts', *PSBR* 51 (1983), 174-221 (abbreviated in this book to SRA), to which the reader is referred for a fuller account: the content and argument, however, are essentially the same as are expressed here. Of special value is the study by N.A.F. Smith, 'Attitudes to Roman engineering and the question of the inverted siphon', *History of Technology*, vol. 1 (1976), 45-71 (here abbreviated to Smith), which broke completely new ground. The only ancient account of siphons is that in Vitruvius VIII, 6, 5-6; the commentary by Louis Callebat, *Vitruve, Livre VIII* (Paris, 1973; Budé series), 167-76 is excellent, but Vitruvius' text highly confusing. The indispensable source-book for a study of the archaeological remains of the best preserved and largest siphons remains Germain de Montauzan's *Les aqueducs antiques de Lyon* (Paris, 1908) (here abbreviated to GdM). Surprisingly, they are very lightly treated by Grenier, in one paragraph (*MAGR* 34). See also H. Fahlbusch, *WAS* 2 (1987), 152-6. A. Trevor Hodge, 'Siphons in Roman aqueducts', *Scientific American* 252 (June, 1985), 114-19, is a popular account.

41. No siphons at Rome: GdM, 207-8, 'Motifs de l'absence de siphons aux aqueducs de Rome'. His conclusion is that it was for economic reasons: 'Ce n'est point du tout par inexperience ou impéritie, ni par goût de vaine magnificence qu'ils ont, soit à Rome, soit ailleurs, préféré les arcades.' Vitruvius: Vitr. VIII, 6, 6 ('*per quae vis spiritus relaxetur*'), also 6.9. The whole thrust of Vitruvius' argument is that far from avoiding pressure, it must and can be handled. Avoiding pressure: Ashby, 37: 'Pressure in the conduits was created, if at all, by accident and avoided in practice.' Siphons rarely used: *SAT* 161. For a discussion of the pressure question, see pp. 232-8 below.

42. Cost of siphons: SRA 212, 220; Smith 62; Ward-Perkins 117; Cherchel 78; GdM 208.

43. It may help to list all siphons known, apart from the Hellenistic/Roman (?) ones of the East (ch. 2 n. 49 p. 396). This catalogue is based on that of Smith 54, and is not exclusive. It is almost certain that other siphons exist, even in known remains, and are only awaiting recognition.

Lyon	GdM *passim* (4 aqueducts, 9 siphons)
Rodez	*MAGR* 153
Arles	*MAGR* 85
Almuñecar	C.F. Casado, *Aquedutos Romanos en España* (Madrid, 1972) (no page ref.); SRA 190
Gades	Casado; SRA 191
Alcanadre	Casado; SRA 191
Rome	Ashby 152 (Marcia); 249-51 (Claudia); Van Deman 61 (Anio Vetus); St. *Silv.* 1, 3, 66-7 (Marcia, at Tibur). Total 4 siphons.
Angitia (near Avezanno)	M. Blake, *Roman Building Construction in Italy from Tiberius through the Flavians* (Washington, 1959), 82; Giovanni, *RPAA* 9 (1935), 70.
Alatri	GdM 194-7; SRA 192 n. 54
Saintes (?)	A. Triou, *Gallia* 26 (1968), 125-7.
Lincoln	Smith n. 37
Constantine (Algeria)	S. Gsell, *Monuments antiques de l'Algérie* (Paris, 1901), 1, 252
Aspendos	Lanckoronski (ch. 5 n. 38 p. 418), 120-4; Ward-Perkins 115-23; SRA 187-9
Hippos (Israel)	(ch. 2 n. 55 above); Jehuda Peleg, 'Die Wasserleitung nach Hippos und ihr Stein-düker', *Mitt. L. Inst.* 103 (1989), 323-36; *FCAS*, 132.

44. Before the Lyon siphons were properly published by de Montauzan, many attempts were made to reconstruct them on paper, based on the cryptic and possibly misinformed description in Vitruvius, which remains the only literary evidence for siphons. The chief studies were by Belgrand, Perrault and Delorme; they are discussed, with illustrations, by GdM 183-92.

45. Callebat 172; GdM 220. This is sometimes done in modern work.

46. True, in Fig. 102, and in other similar diagrams, the sides do usually look impressively steep but this is because of the exigencies of diagrammatic representation. At the bottom of the illustration is an elevation, partly to scale, of the whole length of the Beaunant siphon; for things like horizontal and the vertical distance it is accurate and gives a generally reliable picture of degree of slope. It will be seen that gradients on siphons are not nearly as fearsome as they often appear in diagrams.

47. Width: see SRA 183. The suggestion that the extra width was needed because across the *venter* each of the siphon pipes was divided into two smaller ones to help contain the pressure, seems unfounded. See Smith 64. Slope and air pockets: GdM 218, Smith 58, SRA 199 n. 68.

48. SRA 184.

49. M. Belgrand, *Les aqueducs romains* (Paris, 1875), 71; Smith 61; SRA 211-12, also 197 n. 63 (for the strength of pipes) and GdM 182. The statement of Forbes that 'none of the ancient materials have ever been tested against such pressures as 20 atmospheres [the pressure in the Pergamon siphon]' (*SAT* 165) is highly misleading and only the fact that Belgrand tested replicas, not originals, saves it from being quite wrong. Gordon (n. 39 above), 119-23 gives a general outline of the stresses on operating on pipes under pressure. It is worth noting that longitudinal stress on the walls of the pipe, for any given pressure, is only half circumferential: in other words, if a pipe does crack, the crack is certain to run lengthways along it, not around it.

50. The largest lead pipes I know are those that ran from the ten offtake holes in the *castellum divisorium* at Nîmes (they were not part of a siphon or under serious pressure; indeed, according to the calculations of Hauck and Novak (ch. 7 n. 11 below), they ran half-empty). The holes themselves are very large, 40 cm diameter but, allowing for a ring or collar of cement to hold it in place in the hole, the pipe was rather smaller. A tube-shaped section of the lime carbonate deposit that ringed the interior of the pipe has been found. About 8 cm thick, it was 30 cm in diameter, which gives the inner diameter of the pipe inside which it was formed. ('une espèce de moule de ces tuyaux'). Allowing for a lead pipe itself 3 cm thick, this gives a pipe of internal diameter 30 cm, external 36 cm, fitted in a hole of 40 cm, in turn giving a cement collar 2 cm thick (*Les Archives de la Commission des Monuments Historiques*, ed. Gide & J. Beaudy (Paris, 1855-72), vol. 1, 3). No fragments of these pipes now survive (but see ch. 11 n. 18 (Naumann)).

51. Solder: SRA 213 n. 107; Callebat 173, fig. 14; GdM 204, figs 80, 81. The mixture comes from analysis of some of the pipes from the Arles siphon, dredged up from the bed of the Rhône and now in the Musée Lapidaire. These are, so far as I know, the only lead siphon pipes still extant. Angitia: n. 43 above. Vitr. VIII, 8, '*coagmenta autem eorum calce viva ex oleo subacta sunt inlinienda*.' See p. 114 above. There is also slight evidence that the joints may sometimes have been lashed together by encircling cords, based on Flachat's observation of a still-operational Roman (?) aqueduct at Constantinople, where the joints were 'cordés avec des bandelettes de chanvre, comme les batons de tabac [=cigares]' – SRA 213 n. 108; GdM 190 (compare GdM 204).

52. For a discussion of the hydraulics of siphons, see pp. 232-8; *SRA* 193-202. Three forces are involved. *Friction* operates while the siphon is running and tends to retard the flow of the water rather than work on the walls of the pipe. *Static pressure* is a product of the depth of the pipe below natural water level, and is exerted whether the water is moving or not. At any given point it acts with equal force in all directions – up, down, and sideways. *Inertial thrust* is caused by the moving water encountering anything, such as a bend or obstruction, that tends to deflect it from its existing course. In siphons it is localised at bends, acts only on the outside of the curve, and only when the siphon is running.

53. Vitr. VIII, 8, '*in cavo saxi rubri haereat*'. GdM 192 identifies it as 'un porphyre ou un grès porphyroïde sans doute, roche extrêmement dure en effet'.

54. SRA 201-2. The need for prudence is fully recognised by Vitruvius (VIII, 6, 9: '*Leniter et parce a capite aqua inmittatur*').

55. For a full discussion of *colliviaria* see SRA 213-17. The MS reading in Vitruvius is doubtful so even the name, let alone the spelling, of *colliviaria* is questionable, and emendations run all the gamut from *columbaria* to *columnaria*. See GdM 187-90; Callebat 175; Van Buren, *RE* VIII A, 472-5; W.H. Plommer, *Vitruvius and Later Roman Building Manuals* (Cambridge, 1973), 28 n. 1; K. Stehlin, 'Über die Colliviaria oder Colliquiaria der römischen Wasserleitungen', *Anzeiger für schweizerische Altertumskunde* (1918), 167-75 (on which see *SRA* 188 n. 44).

56. Smith 57, writing as a recognised authority on hydraulic engineering, is clear on the point: 'In a siphon carrying water under pressure the problem of air-locking cannot arise simply because the pressure is high. There is no air pressure to relieve when the siphon is in flow.' When it is being filled up from empty it is a different matter, for the air being displaced has to find some release. The chief point is that air pockets normally form under low, not high, pressure, and are accordingly to be found at the high points of a pipeline, not in the dips. It is clearly explained by a standard modern handbook on hydraulics: 'An air valve should be located at each high point to allow the escape of air and gases and to admit air sufficiently rapidly to prevent the creation of a partial vacuum in the pipe line' – H.E. Babitt & J.J. Doland, *Water Supply Engineering* (New York, 1955), 154. In this context, low pressure means pressure below atmospheric, which accordingly will be found only where the pipeline rises *above* the natural water level (or hydraulic gradient), thus becoming a true rather than an inverted siphon. Roman siphons never did this, since they went down across valleys rather than up over peaks, and pressure in them was always atmospheric or greater. For a fuller exegesis, see SRA 197-9, 210-11. Also A. Kottmann, *Luft in Wasserleitungen-Technisches Wissen aus römischer Zeit*, Schriftenreihe der Frontinus-Gesellschaft, Heft 7 (1984), 82-5.

57. In the eighteenth century this was how piped siphons were cleaned in Spain: 'Pour nettoyer les conduits, on soulève les tubes et on les désunit: on en fait autant un peu plus loin; et au moyen d'un ringard, on parvient à introduire une corde dans la partie du conduit à nettoyer: on y attache un bouchon de paille, et en le promenant dans le conduit, on enlève les dépots qui s'y trouvent' (Andréossy (n. 77 below), 465). A similar procedure is used to clean modern siphons (ch. 2 n. 56 p. 397).

58. For *colliviaria* as drain-cocks, see GdM 187-90 (but his exposition of them (188) as releasing *vis spiritus* is not clear); Callebat 175. Pipes completely clogged: Birebent 466 illustrates some lead pipes from Djemila (Algeria) 'entièrement remplis de concrétions'.

59. See Fahlbusch 86-91. In *WAS* 2 (1987), 153, Abb. 18, a schematic drawing of a typical siphon, such a 'pressure tower' is marked as *colliviaria*. *WAS* 1 (1986),

184-5, illustrates the Aspendos towers as 'Colluviaria (Leitungsturme)' I cannot agree with the accompanying text that the identification, on the basis of modern parallels, is 'unmittelbar ersichtlich'.

60. Alatri is in Italy. Rodez (France) is attested only by topographical evidence – that is, the aqueduct did get across the valley of the Aveyron somehow, and there are no signs of a bridge (*MAGR* 153). All the others are at Lyon, Les Tourillons on the Craponne aqueduct, Ecully on the Mont d'Or, Grange Blanche on La Brevenne, and St Genis, Soucieu and Beaunant on the Gier.

61. As estimated by GdM 205. See Smith 62, SRA 212. Forbes (*SAT* 168) misquotes the figures, taking the 10,000 tons to refer to the pipes of the Beaunant siphon alone (to which he also mistakenly ascribes seven, not nine, pipes). See p. 467 n. 20.

62. Lead glut: see GdM 206, 'on sait du reste l'usage à profusion que les anciens faisaient de ce dernier métal'. Transport costs: Smith 62. Ward-Perkins 117 thinks the cost of lead was the crucial factor but believes that production and transport costs were both high. Pont du Gard quarries: Esperandieu 26; *MAGR* 91; Hauck 116-17. At Tarragona there was a quarry even closer (p. 129 above).

63. The evidence is however weak. In 1882 Bassel (*N. Sc.* (1882), 418) found small globules of spilled lead on the site which suggested an overflow from casting. But one wonders if they may not have come from some other locally conducted operation, such as sealing or joining the seams; SRA 192; GdM 194-6.

64. F. Kretzschmer 113, raises the problem and comments on the great contrast, in domestic plumbing, between the high finish of taps and their clumsy, lumpy attachment to the lead pipes. N.A.F. Smith suggests to me (private communication) that soldering irons were used.

65. An interesting parallel is the view of Alison Burford, *Greek Temple-Building at Epidauros* (Liverpool, 1969) *passim* that the governing factor in building a Greek temple was not finance but availability of skilled labour. And there is evidence that in clasical Greek construction only a small number of workmen may normally have been employed (A. Trevor Hodge, 'Bevelled joints and the direction of laying in Greek architecture', *AJA* 79 (1975), 342).

66. Cherchel 78; GdM 208.

67. Arles: *MAGR* 85. Callebat 166, 177. Statius: St. *Silv.* 1, 3, 66-7.

68. Full discussion in SRA 185-7; GdM 72ff.

69. It certainly did nothing to reduce pressure in the pipe. Pressure is a function of the vertical depth of the siphon below natural water level, and horizontal distance has nothing whatever to do with it. As soon as the second siphon ran down the ramp from the tank to ground level, therefore, pressure in it immediately became the same as if the water had never gone up to the tank at all, but stayed at ground level throughout. See SRA 195-6. 'Bringing the water up for air' is not like putting a car speedometer back to zero; the pressure does not start to build up all over again, but is already back at full strength as soon as the pipes dip to their normal level. Van Buren, in implying the contrary (*RE* VIIIA, 473: 'es war wesentlich, den Wasserdruck in gewissen Abständen zu vermindern mit Hilfe einer Erfindung, die dem Wasser die Rukkehr zu seinem Niveau gestattete, bevor es in einem weiteren Siphon eintrat'), is quite wrong. Other explanations are that it was to release air from the pipe (as we have seen, normally there was none in it) or to provide a convenient point to shut down the aqueduct when it was necessary to isolate part of it for inspection or repairs (which does not explain what then became of the diverted water, overflowing on top of piers 10 m high). Even more piquant is the suggestion recorded by GdM 73, that the piers were really part of a bridge over the River Saône, which is not only 6 km away but would have to flow uphill apparently for the express purpose of running under it.

70. See Lanckoronski (ch. 5 n. 38 p. 418) 122, 123; Ward-Perkins 116. These bends, in effect, constitute a pair of sideways *geniculi*, and it is worth remembering that a siphon can bend in plan as well as elevation. Diagrams such as Fig. 102 naturally tend to illustrate only the second, but in fact a siphon can pursue quite a complex course, bending hither and thither in three dimensions. For a fine modern example, see SRA, pl. X (a), the Siphon de l'Arc (Aix).

71. The idea that these raised tanks somehow affected pressure (whence, presumably, the term 'pressure towers') stems from the apparent (and, if so, quite erroneous) belief of Van Buren that water pressure in a siphon increases with the horizontal distance travelled (see n. 69 above). A variant proposed by G.E. Bean (*Turkey's Southern Shore* (London, 1968), 75-6) and repeated by him in the *Princeton Encyclopedia of Classical Sites* (Princeton 1976), 103, is that the purpose was to 'let the water into the open, thus allowing the air to escape from the conduit and so reducing the friction which would impede the flow'. No evidence is offered to support the claim that air in a siphon pipe causes friction, and it appears to be purely dogmatic assertion, unbased on fact of any kind. A specialist in the flow of liquids in conduits, Prof. D.R. Townsend (of the Faculty of Engineering, University of Ottawa) tells me that in such circumstances air in the siphon causes no friction. Bean also does not explain where the air came from, or how it got into the pipe, though on the basis of modern hydraulic engineering it does seem that air in the pipes is not an uncommon phenomenon. For a fuller exposition see SRA 188 n. 44. A further explanation of the whole Aspendos structure may be that it reflects a change in plan, the footings in the depression having proved unstable, necessitating a change in alignment with the two bends as a result: conversely, the irregular course may be following the only place that good footings were to be found, right from the start.

72. Cherchel 76; figs 99, 100. There is a series of four short inclined planes, of slope varying between 25° and 40°, and each ending in a circular pit now filled with sediment and rubble. GdM 175 attests to a similar system at Carthage.

73. Recret/Grézieu (both names are used independently in different publications): Jean Burdy, *L'Araire: Bulletin Périodique* no. 66 (Lyon, automne 1986), 48-9; also Jean Burdy, *WAS* 3 (1988), 192, and 196, abb. 4. Autun: H. de Fontenay, *Autun et ses Monuments* (1889), 93-101, 106 (as quoted by Burdy, *L'Araire* no. 53 (Lyon, été 1983), 42). Beaulieu: P Maurice Croquet, 'Les aqueducs romains de Beaulieu et les puits de rupture de pente', *Cahiers Ligures de Préhistoire et d'Archéologie* 15 (1966), 283-94; some of the individual falls are 6-7 m deep. Rusicade: H. Vertet, *JEAR* 358.. There the cascade shafts were evidently built externally on the hillside, like towers, rather than sunk into the ground like wells – 'Ils se présentent comme des cylindriques verticaux solidement construits en blocage comme le canal de l'aqueduc. L'extérieur porte un enduit sommaire qui laisse apparaître les pierres et sur lequel ont été tracées des lignes circulaires horizontales.' The only general discussion of this neglected topic (Autun is not even in the index of *MAGR*) seems to be GdM 173-5 ('Dénivellations par chutes'). We may also note that the two different types of cascade sometimes involve a slight confusion in terminology, being called 'cascades' in English, while French usually reserves *cascades* for the first, open, type, calling the second, closed type *chutes* or *puits de rupture de pente*.

74. The height of the arcades on the aqueducts of Rome runs to between 14 and 17 m. Given an average slope for the sides, this would mean an earth embankment 60-80 m wide at the base to attain the same height.

75. Vitr. VIII 6, 7. Callebat 176. Vitruvius suggests 200 *actus* as the interval. Callebat, finding this (7 km) unduly long, suggests emending the text from *ducentos* to *vicenos*, 20 *actus* (= 700 m). This is not unreasonable. GdM 190-1.

76. It is a question of definition. The arcades of the Rome metropolitan system are 14-27 m high, depending on location (Van Deman 126: 25 m on the Marcia). The Los Milagros arcade at Mérida, Spain, goes up to 28 m. Perhaps Segovia might more properly be called a bridge than an arcade.

77. Fuller account in Hodge, *Quinaria* (ch. 7 n. 11 p. 435), 209-10 and n. 18, based on the description of an early nineteenth-century French military engineer and diplomat, Count A.F. Andreossy, *Constantinople et le Bosphore de Thrace* (Paris, 1828), 287-91. *Suterazi* (in Turkish = 'water balance') is the name for the water tower (described, from its shape, by Andreossy as a 'pyramide hydraulique' – compare the term 'hydraulische Thurm' as applied to the 'pressure towers' at Aspendos). A *suterazi* is illustrated, with brief description, in Lanckoronski (ch. 5 n. 38 p. 418), 120, fig. 94. Cost: 'le prix d'une conduite à souterazi est estimé le cinquième de celui d'un aqueduc sur arcades' (Andreossy 389). Aqueducts on the *suterazi* principle also existed at Aleppo and Acre, in Syria (where Napoleon's forces blew it up in 1799), and at several locations in Spain, notably Puerto-Real (Cadiz), and Talavera de la Reyna (Toledo), both dating from the late eighteenth century (Andreossy 459-63): the Cadiz aqueduct drew its water from the same source as an ancient predecessor, improbably ascribed by Andreossy's informant, an army engineer, to 'les Phéniciens'. With greater verisimilitude, he also maintains that in Spain he saw 'un grand nombre d'aqueducs à *souterazi*'. A *suterazi* of more recent date, in good preservation, is still to be seen in Akko [Acre], Israel, one block west of the railway station. The existence both of a large number of *suterazis* and of the analogous arrangements at Aspendos and les Tourillons must be borne in mind when reading the strictures of Smith (57) on Choisy for proposing something very like this. Smith condemns the proposal as 'idiotic', apparently not realising that there was evidence for such a thing being actually done; but, in all honesty, I must record my doubts whether Choisy realised it either.

78. There is a parallel of a sort at Metz (my Fig. 123) where a short section of the aqueduct carries a double conduit, but divided longitudinally. The division is a median wall of bricks, slightly closer to one side than the other, but there was no question of keeping separate in two different channels water from two different sources, for the double section (which includes a large bridge over the Moselle) draws its water jointly from a common distribution tank and discharges it into another, at the other end, whence it is again conveyed by an ordinary single channel. I offer no explanation for this unusual arrangement (*MAGR* 203-5).

79. Clearly illustrated in Van Deman pl. IV, p. 13.

80. Ashby 131-2 and fig. 15, p. 139. The Tepula's channel had to be narrower because it was a smaller aqueduct with a lesser discharge, yet the height had to remain the same to permit human access for cleaning; and if the width was also the same, for the lesser volume of water, velocity of flow would have been undesirably reduced, causing silting up of the channel.

81. Juv. *Sat.* III, II, calls it '*madida*', and the Scholiast an '*arcus stillans*'. Martial III, 4, 7, 1, says the same thing. Ashby 155 and n. 6, thinks the terminal *castellum* of the *rivus Herculaneus*, a branch of the Marcia, was on top of the gateway. Van Deman 141.

82. Van Deman 17 describes his inheritance from Trajan as 'leaking channels, broken bridges, and long lines of unstable arches'. As a commentary on Roman building construction this does have some novelty.

83. Van Deman pl. VI, p. 17. Ashby fig. 25, p. 240.

84. '*Clarissima aquarum omnium in toto orbe frigoris salubritateque palma praeconis urbis Marcia est, inter reliqua deum munera urbi tributa*' (Pliny XXXI, 3, 24); '*tam candida, tam serena lucet, / Ut nullas ibi suspiceris undas,/Et credas vacuam nitere Lygdon*' (Martial IV, 42, 19-21). Pliny, *NH* 31, 24, Tac. *Ann*. XIV 22; Strabo V, 3, 13.

85. Front. I, 11, 1-2; Callebat 73 *ad loc*.

86. *JEAR* 181.

87. Front. I, 91, 4: '*Marciam ipsam, frigore et splendore gratissimam balneis ac fullonibus et relatu quoque foedis ministeriis deprehendimus servientem*.' Contamination: ibid. The chief culprit was the Anio Novus, which '*vitiabat ceteras*'.

88. For such connection in the open country, see Ashby 278 (connection from the Anio Novus to the Claudia, Marcia, and Anio Vetus), and 102 (from the Claudia to the Marcia at S. Cosimato).

89. At Tor Fiscale, near Rome, the heights above sea level of the principal aqueducts are: Anio Novus, 72.197 m; Claudia, 69.197; Julia, 65.597; Tepula, 64.097; Marcia, 62.097; Anio Vetus, 52.50 (Ashby 80 and fig. 14, p. 137). Front. I, 18, lists them in the same order. The aqueducts (except for those riding 'piggy-back') do not necessarily always maintain exactly the same vertical interval between them, as their gradients may vary.

90. The Marcia arcade was 9.5 km long, the Claudia 10.5 (Van Deman 126, 256) carried on no less than one thousand arches. The imposing effect and impressive aspect is well conveyed in the drawn reconstruction in the Deutsches Museum, Munich, illustrated by White 166, fig. 168; *WAS* 1 (1986), 89.

91. F.E. Brown, *Roman Architecture* (New York, 1961), 30.

Chapter 7

1. O.A.W. Dilke, *Greek and Roman Maps* (London, 1985), *passim*.

2. Herod. V, 49, 51: Pliny *NH* III, 17 (Augustus' map of the Roman empire, set up on public display in a colonnade in the Campus Martius).

3. O.A.W. Dilke, *The Roman Land Surveyors* (Newton Abbot, England, 1971), 112. The best known example of a city plan is the Forma Urbis Romae, a second-century city plan inscribed on marble (G. Carettoni, *S.P.Q.R.: la pianta marmorea di Roma Antica* (Roma, 1960)). For centuriation, see Dilke 133-77. The centuriated area best represented by fragments of actual surviving maps is the region around Orange (anc. Arausio): Dilke 159-77 (with illustrations); A Piganiol, *Les documents cadastraux de la colonie romaine d'Orange, Gallia* suppl. 16 (Paris, 1962). See also Dilke (n. 1 above), 'Roman stone plans'.

4. *CIL* VI 1261. See ch. 9 n. 18 p. 448.

5. '*Hinc illa contingit utilitas, ut rem statim veluti in conspectu habere possimus et deliberare tamquam adsistentes*' – Front. *Aq*. I, 17. As well as an indication of where the various features were, their size was evidently marked as well ('*ubi valles quantaeque*'), so this map presumably also had figures on it where relevant. Grimal, XI.

6. Peutinger Table: Dilke (n. 1 above) 113-20. In his n. 24 (p. 210), he expostulates against R.K. Sherk, 'Roman geographical exploration and military maps', *ANRW* II, 1 (Berlin, 1974), 534-62, who maintains that *all* Roman surveying maps show the same 'bizarre proportions' as the Peutinger Table. A fine full-size modern reproduction of the Peutinger Table is published, as a fold-out, by Konrad Miller, *Die Peutingersche Tafel* (Ravensburg, 1887/88; repr. F.A. Brockhaus GmbH., Stuttgart, 1962).

7. F. Blume, K. Lachmann & A. Rudorff, *Die Schriften der römischen Feldmesser* (Gromatici veteres), (Berlin, 1848-52, repr. 1962); bibliography in Dilke (n. 3 above), 235-6, see 45-65. See *RE* s.v. *agrimensores, gromatici*; Columella (bk. 5) also deals with the topic.

8. L. Casson, *Ships and Seamanship in the Ancient World* (Princeton, 1971), 245-6.

9. There is no island today, this area now forming the elongated peninsula on which the downtown area of modern Lyon is built, between the Saône and the Rhone before their eventual confluence to the south. In antiquity the two rivers were also joined by a connecting channel to the north, creating an island. Water supplies may have been brought to it from the west by an underwater siphon laid across the bed of the Saône, as was done across the Rhone at Arles, but this is conjecture.

10. They did of course have rainwater cisterns. For levels, *JEAR* 190. The dating of the Gier aqueduct, traditionally Hadrianic because of its association with the 'pierre de Chagnon', is now questioned (Jeancolas, *JEAR* 183).

11. The whole question of how the Romans measured aqueduct discharge is a complicated one. The theoretical capacity of any given pipe or conduit was known, but, quite apart from that, it was a matter of measuring the actual volume of water that was passing through. Frontinus was sure he could do it, evidently by installing some kind of metering device (*'mensuris positis'* – *Aq.* II, 67). But what was it? For a suggestion that it was done by measuring the head maintained by a moveable sluicegate, see A. Trevor Hodge, 'How did Frontinus measure the quinaria?', *AJA* 88 (1984), 205-16. This suggestion has been repeated and endorsed by G.F.W. Hauck & R.A. Novak, 'Water flow in the castellum at Nîmes', *AJA* 92 (1988), 393-407, and is clearly described in Hauck's article in *Scientific American* (ch. 6 n. 20 p. 425), 102-102B; see also Hauck 111.

12. Caesarea: Olami & Peleg (ch. 5 n. 50 p. 420), 135; Leptis Magna, *HD* 37. A much more enigmatic and also more serious case of making water flow uphill (?) is in the aqueduct at Lincoln, where the source is substantially lower than the delivery point in the city. The two most favoured explanations (assuming, of course, that the identification of the source is correct) seem to be that at the spring there was either a force pump, pumping the water uphill to the city through a pipeline, or some lifting device, such as a bucket chain, that raised the water, say, to the top of a tower, whence it flowed to the city downhill; Smith n. 37.

13. See *JEAR* 194 (Gier aqueduct, at Chagnon), where traces of digging for a second conduit may indicate just such a mistake ('une sorte de rattrapage d'un nivellement avorté').

14. For the material on Nîmes I have drawn largely upon Hauck. I am grateful to Dr Hauck for having kindly let me see a pre-publication MS of his book. Though a semi-fictionalised account of the building of the aqueduct, it is written by a fully qualified hydraulic engineer after careful study on the site.

15. Let us clarify. In railway or highway engineering, the 'ruling gradient' is the figure for the steepest part of the route, which thus becomes the governing factor for traffic operating over the whole of it. In aqueducts, where it is shallow, not steep, gradients that cause trouble by slowing down traffic (i.e. the water), it is reasonable to apply the term to the section where the slope is least. This is quite different from overall or average figures for the entire aqueduct, which are largely irrelevant. If one section of an aqueduct is shallow enough in slope that the water through it stops flowing (and, e.g., overflows instead), then it does not really matter what the profile is like on the rest of the route. It is the *minimum* grade that governs, and the engineer has to decide what it will be.

16. The sequence and method of construction is another neglected topic, now dealt with in Klaus Grewe, *Planung und Trassierung römischer Wasserleitungen* (Schriftenreihe der Frontinus-Gesellschaft, Supplementband I; Verlag Chmielorz GmbH., Wiesbaden, 1985) (here abbreviated to Grewe). See also Klaus Grewe, *Atlas der römischen Wasserleitungen nach Köln* (Rheinland-Verlag GmbH., Köln, 1986) (here abbreviated to Grewe, *Atlas*), 215-24 (Planung, Bemessung und Bau römischer Wasserversorgungsanlagen, by G. Garbrecht) and 225-34 (Vermessungsmethoden beim Bau von Fernwasserleitungen, by K. Grewe). It is a pleasure here to express my indebtedness to the above for much of the material in this chapter.

17. For Greek work the classic study is Alison Burford, *The Greek Temple Builders at Epidauros* (Liverpool, 1969). M.K. Thornton, 'Julio-Claudian building programs: eat, drink and be merry', *Historia* 35 (1986), 28-44, studies the various building projects of the Julio-Claudians and concludes that, in labour utilisation, they were well planned in terms of critical path programming. He fails, however, (p. 38) to take into account that some works such as the Fucine Lake tunnel must have employed only a limited workforce because of limited access to the workfaces, while in building an aqueduct there was virtually no limit to the number that could work on it simultaneously once the line was marked out. For contracts on aqueducts, the only discussion is in Grewe 24-34, and Grewe, *Atlas* 225ff., who breaks new ground in this matter.

18. The key example is at Mechernich-Lessenich, 30 km south-east of Bonn: Grewe 36; Grewe, *Atlas* 102-3 and 232. For other examples, Grewe, 38.

19. This interpretation is based on the hypothesis that the Siga aqueduct was intended to have a uniform slope throughout, and that variations in gradient must thus be the result of mistakes and efforts to correct them, rather than a deliberate and planned response to the irregularities of the terrain. The terrain is, in fact, very broken, and the aqueduct's route through it highly sinuous. Grewe 24-3; Grewe, *Atlas* 228-9.

20. The point is confirmed and clarified in a private communication from Ing. Klaus Grewe ('so liegen die Ausgleichstellen immer *vor* in der Leitung installierten Bauwerken (Aquaeduktbrücken, Sammelbecken oder Anschlussstellen) und nicht hinter diesen Punkten').

21. Grewe, *Atlas* 234. This refers only to the 70-8 km section of the aqueduct between Kallmuth and Hermuelheim, which was all built at one time: the line below Hermuelheim pre-existed, and that above Kallmuth was added later.

22. Grewe 30. There is no contradiction here with my statement above that at Siga the gradient was uniform. The gradient as planned was uniform, but the execution was imperfect, so that the real gradient, as built, was variable.

23. Grewe 36; Grewe, *Atlas* 232. The Roman foot is 0.296 m.

24. A comparison with one of the most carefully published and surveyed Roman aqueducts, Gier at Lyon, is not encouraging. A very full and detailed profile is published by GdM (folding endpapers). The most obvious stretch to compare is that between St Genis and Soucieu. The course is reasonably straight, the slope regular, and attested by frequent levels taken over a long distance (35 over 30.8 km). The commonest slope seems to be about 0.055% (within a range of around 0.052 to 0.062). This works out at *just over* half a Roman foot per 1,000, not a particularly convenient figure. On the other hand, a figure of 0.060%, which is well within the limits of possibility, does work out at a very neat and tidy result – exactly 3 Roman feet per Roman mile. This would make very good sense on this part of the Gier aqueduct and may be right. Moreover, if one measures from the end of the St Genis siphon (the obvious place the Romans would measure from?) the first change in gradient on Germain de Montauzan's profile comes at 'sur le bord du Faudangy' (3,613 km). From St Genis this is 22.112 km. 15 Roman miles comes to within 90 m of this spot (15 × 1.48 km = 22.20 km. Does this show that the aqueduct is here surveyed on a basis of Roman mileage? Optimists and cynics will draw different conclusions from these calculations, and on the value of statistics.

25. Another source of misapprehension is that sometimes archaeologists seem to publish a profile constructed by plotting all the points where the remains of the conduit could be found and a level established, and then joining them up. The resultant profile will be accurate enough for most purposes, but will automatically show gradient changes at the points observed, not necessarily where they actually occurred.

26. My acknowledgments to Prof. Ken Wilson, Faculty of Engineering, Queen's University, Kingston, Canada.

27. 'Das "Austafeln" ist im Kanalbau vielerorts heute noch gebrauchlich', Grewe 35. The 'Austafeln' is what I have called the T-board. J.J. Salinas-Pacheo, Associate Professor of Civil Engineering at Carleton University, Ottawa, confirms in conversation with me that he personally often uses something like this system in laying out the course and gradient of new roads, especially in rough country in underdeveloped nations. The simplicity of the system means it can be used in conjunction with an untrained or illiterate labour force, a point also made by Grewe (36). However, Prof. Salinas tells me that normally he uses not just three, but a whole sequence of T-boards. Provided they are carefully spaced at the same interval apart (this is important) their accuracy is quite acceptable.

28. See also full description in Dilke (n. 3 above), 73-9.

29. Vitruv. VIII, 5, 1. See also Grewe 18-21 (full discussion); K. Grewe, 'Über die Rekonstruktionversuche des Chorobates', *Allgem. Vermess.-Nachr.* 88 (1981), 206; Hultsch, *RE* III, 2, 2439-40 (s.v. *Chorobates*); Callebat 140-2; Kretzschmer 12; Adam 18-20; Pace 36-7. The *chorobates* is known to us only from the description in Vitruvius, and does not appear elsewhere.

30. As specifically noted by Vitruvius (VIII, 5, 1: '*diligentius efficitur per chorobatem, quod dioptrae libraeque fallunt*', 'the *chorobates* is more reliable because the *dioptra* and water level give misleading readings').

31. His name is also sometimes spelled 'Heron'. Hero's dates have been disputed, but he seems to have lived in the first century AD (so *OCD*). His treatise *On the Dioptra* is included in the Teubner edition of his works (ed. W. Schmidt, H. Schoene & J.L. Heiberg; Leipzig, 1899-1914). A good conveniently accessible English translation of the relevant sections (chs 3-5, 8, 14, 15) is printed in M.R. Cohen & I.E. Drabkin, *Source Book in Greek Science* (Cambridge, Mass, 1958), 336-42.

32. But one must not go overboard on this. After all, 'the designers of the Industrial Revolution saw nothing incongruous in using Classical columns, though of cast iron, in their new machines' – John Boardman *et al.*, *Greek Art and Architecture* (New York, 1967), 9: he appositely illustrates with a photograph (fig. 1) of a stationary steam engine, *c.* 1860, in which the beam is supported by a Doric column, much in the style of Hero's *dioptra*.

33. Dilke (n. 3 above), 79. The dioptra is nowhere mentioned in the *Corpus*

Agrimensorum, the chief collection of ancient texts on surveying.

34. It is often reproduced, but does seem to be a particularly elaborate model of the device. A simpler, but conjectural version is restored by Adam 9, fig. 1.

35. There are various ancient technical treatises on its use: see *OCD* s.v. *gromatici*; the standard edition of the ancient texts is F. Blume *et al.* (ed.), *Die Schriften der römischen Feldmesser* (Berlin, 1848-52); see also *Corpus Agrimensorum Romanorum*, ed. C. Thulin (Teubner, Leipzig, 1913). Dilke (n. 3 above), 66-70. *RE* VII, 2, 1881-96 (Schulten).

36. Stele illustrations in Adam 11, figs 3, 4. A variant form of the device is reconstructed by Kretzschmer 11, fig. 9, based on evidence from the Saalburg. See also Grewe 23; bibliography on the Naples *groma*, Adam 21 nn. 18-22.

37. Fully explained by Dilke (n. 3 above).

38. For modern experiments in ancient surveying techniques, see Adam 14-20; J.P. Adam, 'Groma et chorobates, exercises de topographie antique', *Mefra* 94 (1982), 1003ff.; K. Grewe, 'Über die Rekonstruktion-versuche des Chorobates', *Allgem. Vermess.-Nachr.* 88 (1981), 205.

39. Dilke (n. 3 above), 79, and 80, fig. 22 (illustration of hodometer), Vitruv. X, 9. Dilke scorns the Vitruvian version because of 'the Heath-Robinson sound of the contraption', but when a full-size model of it was built for me by one of my students, it worked very well – and still does, ten years later.

40. Dilke (n. 3 above), 73.

41. Mohammad al Karagi, *La civilisation des eaux cachées*, ed. and tr. Aly Mazaheri (Université de Nice, Institut d'Études et de Recherches Interethniques et Interculturelles; Nice, 1973), 89-96.

42. This method was observed in use in surveying Persian qanats as late as 1953: Goblot 32.

43. Goblot 33. This was observed in use in 1933.

44. Halberg: *WAS* 3 (1988), 92; 1 (1986) 152, ab. 11; F.J. Keller, 'Der römische Wasserleitungenstollen am Halberg bei Saarbrücken', *Bericht der Staatlichen Denkmalerpfliege im Saarland*, 12 (1965), 66-77. *Cuniculi*: Judson and Kahane (ch. 2 n. 68 p. 399), 88.

45. Goblot 33 (qanats).

46. The shafts in the Halberg tunnel are about 8 m apart, and 7-8 m deep. 12 of them are preserved.

47. Goblot 33.

48. Goblot 32-3, and fig. 3. The method there described is highly ingenious and, being designed for an illiterate society, functions without any measurements at all calculated or recorded in figures: everything is done by tying knots in cords to record the various distances and their relationship to each other.

49. The Bologna aqueduct is not yet widely known, but is published in a chapter by Dario Giorgetti, 'L'acquedotto Romano di Bologna: l'antico cunicolo ed i sistemi di avanzamento in cavo cieco', in *Acquedotto 2000: Bologna, l'Acqua del Duemila La Duemila Anni* (Grafis Edizioni, Casalecchio di Reno, 1985), 37-107. Casalecchio is a large suburb southeast of Bologna. It is also published in *WAS* 3 (1988), 180-5 (with colour photographs in the tunnel). The aqueduct follows the valley of the river Reno southwards from Bologna for a probable total length of 20 km, of which 17 km survive. One of its most noteworthy features, unparalleled elsewhere, is a full-scale plan of the semicircular wooden scaffolding for construction of the vault, incised on the face of the tunnel side-wall; presumably to provide guidance *in situ* to the carpenters, it is very carefully drawn though described by the excavator as a 'graffito'. There are other more orthodox graffiti, including the signature of one of the surveyors (*'librat(or) Um[---]'*; *CIL* XI, 739, a; cf. Front. *Aq*. III, 105, 4). Also of interest at the Bologna end (in the park of the Villa Ghigi) is a long sloping access gallery (for parallels to a sloping gallery instead of the more usual vertical shafts, see the high level aqueduct of Caesarea (Olami & Peleg (ch. 5 n. 50 p. 420), 129, fig. 2). It has a 318-step stairway 375 R. ft (111.0 m) long. The aqueduct was entirely underground and built by tunnelling, not cut-and-cover; it seems to have had an actual average gradient of 1 m per km (0.1%) and its discharge is estimated at 34,732 m^3 daily (assuming an average water level 1.20 m deep in the channel; this is fairly deep, but is based on deposits on the side walls). I am indebted to Prof. Michael Woloch, of McGill University, for bringing this Bologna publication to my attention.

50. Ger. 'im Gegenort'. See Grewe 69 for this whole section. It is also precisely for this reason that the Iranian *qanat* builders, although they start by sinking the deepest shaft, at the upstream or source end of the *qanat*, cut the entire tunnel beginning at the downstream end and working uphill. Goblot 31 is scathing on modern scholars who have proposed the opposite: 'Seuls des non techniciens ont pu commettre une telle erreur.'

Chapter 8

1. I have elsewhere discussed the questions more fully: A. Trevor Hodge, 'A plain man's guide to Roman plumbing', *EMC/Classical Views* (University of Calgary), 27, N.S. 2 (1983), 311-28. The question was also debated at a colloquium 'Future Currents in Aqueduct Study' in the 1987 meeting at New York of the Archaeological Institute of America, the content of which (ed. Hodge) is now published in book form (here abbreviated to *FCAS*); see esp. the Conclusion, pp. 163-72.

2. Deane R. Blackman, 'The volume of water delivered by the four great aqueducts of Rome', *PBSR* 46 (1978), 52-72, points out (53) that any attempt to calculate the depth of flow in the channel is bedevilled by constant changes in width and slope, so that the water is in effect always either recovering from the last obstruction or already starting to 'pile up' before the next one, so that 'the flow will rarely reach its uniform depth but always be in a state of change, trying to adapt to the new conditions'. The limitations of technical engineering study and its value are also discussed and clearly set out by Smith 59-61; also Smith in *FCAS*, 113-20.

3. Brent D. Shaw, 'Water and society in the ancient Maghrib', *Ant. Afric.* 20 (1984), 136. He clarifies: 'In a civilisation lacking a mechanical or engine-driven technology, the deployment of water resources on a massive scale depends on the simple principle that liquids flow downwards under the force of gravity.'

4. Haberey 37: 'Eine Kegelkugel, an irgendeiner Stelle im Kanal angestossen, würde bis ans Ende in Köln rollen.'

5. Pl. *NH* XXXI, 57. Undue emphasis on gravity flow, and on aqueducts (as opposed to other parts of the Roman supply system, which ran in closed pipes), has sometimes led modern commentators to suspect, even to declare, that the Romans did not know that water rises to its own level. I have discussed this wholly erroneous position in SRA 177 n. 8. See also p. 147 above.

6. In the Hellenistic Madradag aqueduct of Pergamon the main aqueduct was carried in three pipes of internal diameter 16-19 cm, while the last stretch, the great siphon, ran in a single pipe of internal diameter 22 cm. Because of the greater fall in the siphon section, and hence greater speed in the single pipe, the carrying capacities of the three pipes and the single one matched, at 45 l/s (4,000 cubic metres, daily). (G. Garbrecht, 'L'alimentation en eau de Pergame', *Doss. Arch.* 38 (1979), 31-2; 37 (1973), 86. See also Garbrecht, *WAS* 2 (1987), 27. A study by George Hauck & R. Novak (ch.7 n.11 p. 435) 396, 406, on the basis of calculated flow and discharge also shows that at Nîmes in the ten large offtake pipes the water probably ran so fast that they were seldom more than half full. Masonry conduits, one may generalise, never ran full (intentionally at least) and pipes always did, so this is a very unusual situation.

7. Vitr. VIII, 6, 1; Pl. *NH* XXXI, 57. Attempts have been made be reconcile the two by emending the MS reading of Vitruvius' '*semipede*' (half-foot) to '*sicilico*' (a quarter [inch]), the word used by Pliny. Normally Vitruvius would be considered the more reliable of the two and the change made the other way round, but archaeology tells us that it is Pliny's figure that corresponds to normal Roman practice. Except in exceptional circumstances, a fall of 5 m per km is impossibly steep. Pliny's figure, on the other hand, is on the whole too shallow, but nearer the truth than Vitruvius' and hence to be preferred. Callebat 146-8.

8. GdM 172. The suggestion here is that this explains why the Gier aqueduct is at half the gradient of those at Rome, the Roman water being more sediment-laden. I am not sure that this is right. A glance at the comparative gradient profiles for the other three Lyon aqueducts shows that their gradients are relatively steep (as well as being often irregular). But, Lyon or other particular cases apart, the general principle does seem reasonable. And Pace (36) reaffirms that, broadly speaking, the normal gradient of the aqueducts of Rome, 'confrontata con gli acquedotti provinciali dello stesso periodo, ha valori sensibilmente piu elevati'. The speed of flow required to prevent clogging is clearly stated in a modern handbook, Richard H. French, *Open-Channel Hydraulics* (New York, 1985), 275: 'In general, an average velocity of 2 to 3 ft/s (0.61 to 0.91 m/s) will prevent sedimentation when the silt load of the flow is low.' See n. 19 below.

9. The gradient of the water channel in the early Greek tunnel of Eupalinos on Samos, perhaps for this reason, is quite steep – 5 m in 1 km. On the Anio Novus, the gradient in the 2.25 km long tunnel under Monte Affliano (on the Osteriola-Valle Barberini cut-off) was 0.296%, or 3 m per km (Ashby 271). For a tunnel with no shafts, one would expect a steeper gradient, though this was in fact steeper than the mild gradients (around 0.1%) on the older and more circuitous route via Tivoli. Lower down the aqueduct, however, were three tunnels with very steep gradients: the 2.5 km long one between Ponte dell'Inferno and Ponte Scalino had a gradient of 1.05% (10 m per km; estimated water velocity, 3.55 m/s), followed by one 675 m long between Ponte Scalino and Ponte Amato at a gradient of 1.64% (16 m per km: velocity, 4.42), and the third, 600 m from Ponte Amato to the Fienile, 1.32% (13 m per km: velocity, 4.08). These gradients and water speeds are very high indeed; Ashby (287) suggests they were intended to avoid incrustation(?). Pace (72) erroneously attributes these figures and tunnels to the Anio Vetus, but they are correctly shown on his profile of the Anio Novus (35, fig. 14: numerazione, 21-17).

10. Examples of this are to be found on the Anio Novus (n. 9 above (the Osteriola cut-off)); the Gier (Lyon), at Chagnon, where the intervening valley was cut off by a siphon (GdM 105); at Cherchel, Oued Ilelouine (Cherchel 76): and on the Anio Vetus at Mola di S. Gregorio, where, quite exceptionally, the short (22 m long) slope required is incorporated into the bridge itself; this gives the bridge a humpbacked appearance, since the rest of it carries a channel running at the normal gradient (steep 16.35%; normal 0.76%) (Ashby 69-70, esp. 70 n. 1, and fig. 2).

11. Retarding effect of steep slope: 'The real problem was to drop, in a short cut, the height which the water previously lost in a long circuit. It was safer to do this in

a short, rapid descent than a long one in which the waters gathered momentum.' Ed.'s note to Ashby 70 n. 1. For settling basins, see above, p. 103, and *JEAR* 201, fig. 3; 330, fig. 40.

12. It is well put by Pace 36: 'in realità la *mensura declivitatis* non ha mai seguito i valori dati da Vitruvio e Plinio; vi sono tratti in cui e molto maggiore e tratti in cui e inferiore. Inoltre non sembra neppure seguire un criterio preciso.'

13. At Carthage the main line of the aqueduct runs from the source at Zaghouan to the Bordj Djedid reservoir in the city. The distance along the channel is 90.43 km (the often quoted figure of 132 km also includes a later Severan branch 33.65 km long to a second source at Ain Djoukar), and the total loss in height is 264.68 m, giving an average slope of 2.92m per km (0.28%). But in fact nearly half of this loss, no less than 137 m out of 264, is achieved in the first 6 km, between Zaghouan and Moghrane, as the channel falls precipitously down to the plains. The average gradient Zaghouan-Moghrane thus works out at 22.8 m per km (2.28%) (with sections registering figures even as high as 95.8 m per km, a gradient of 1 in 10!), while for the very much longer (84.4 km) Moghrane-Carthage section the gradient is around 1.5 m per km (0.15%), a close approximation to the Gier gradient. An aqueduct with an average gradient of 3 m per km and which yet is built through almost all its length to an actual gradient of only half that, is a potent warning against the perils of an undue faith in averages. (*JEAR* 316; the profile of the Zaghouan-Moghrane section (p. 332, abb. 43) shows that even on this stretch almost all of the height is lost in the first half of it, which is thus very steep indeed; a French translation/abridgement is published in *Doss. Arch.* 38 (Oct.-Nov., 1979) 34-42). The Segovia aqueduct may have been built in two phases, the steep mountain section being a later extension to tap further springs when the original source proved insufficient: from the present source to the city, the existing route is not a natural one if planned that way from the start. (A.R. Gallardo, *La Supervivencia de una Obra Hidraulica: el Acueducto de Segovia* (Segovia, 1975), 21-2.) The profile of the Anio Novus is conveniently to be found in Pace, 34-5, and that of the Gier at the end of GdM.

14. So Rondelet's figure of 1.543m per km for the Marcia between Arsoli and Tivoli, the correct average figure for this stretch being 0.73: 'mais, n'ayant pas tenu compte des chutes, il avait trouvé des chiffres manifestement trop forts' (GdM 170); Ashby 97 n. 1. It is instructive to compare the different attitudes of two scholars. Jean Burdy, faced on the Brevenne aqueduct at Lyon with an average gradient of over 5 m per km, finds it a 'valeur tout-à-fait inadmissible', and a consequent indication of the existence of cascades. Klaus Grewe on the Eifel aqueduct at Cologne, when confronted with much steeper gradient of 10% (over a short distance), finds it simply 'einer wahren Sturzstrecke'. (Burdy: *L'Araire, bulletin périodique* no. 66 (Autumn 1986), 'Les aqueducs romains de Lyon', 48; Grewe, *Atlas* 37). A further point should be noted. All figures published anywhere for the gradient in an aqueduct or any part of it are reached by measuring a horizontal distance between two points – perhaps kilometres, perhaps a few metres – and noting also the difference in altitude between them. Personally, I do not know of one single exception to this practice. No one, to my knowledge, has ever tried to measure directly the actual tilt or slope of the floor of the conduit. Usually, this is reasonable, for it would be very inaccurate, but it ought to be possible to establish the difference between two widely different figures: indeed, a conduit sloping at 10% ought to *look* different, to the eye, from one at 0.1%. Such a technique could sometimes be useful, for if calculations from levelling give a section of aqueduct an average gradient of 10%, and if the conduit at the beginning and end of it really does have a slope of 10%, then we may safely assume that that was the uniform gradient throughout; while if instead the height loss works out at a slope of around 10%, but the observed slope is only 0.1%, then we may assume that there was a cascade somewhere in the middle. This technique of observation may sometimes have been employed, but I cannot remember ever having seen it so stated in print. In fact, it would not surprise me if cascades were more commonly used than is generally thought. It must be re-emphasised that our knowledge of most aqueducts is not continuous: it is based on measurement at a series of separate and isolated locations, and on the assumption that between them everything was 'average'.

15. The commonest, used in this book, is to express it either in terms of so many metres height per km, or as a percentage. Sometimes this will be expressed as so many per thousand, and printed as a variation of the percentage sign: 3.54‰. Ashby seems to adopt this system but using an orthodox percentage sign, which is misleading to say the least (Ashby 38: when he speaks of 'a 10% fall', he evidently means a gradient of 1 in 100 (1%). The game is given away by the occurrence of such figures as 107.35% and 163.5%). The '1 in 100' system of reference (1 in 250, 1 in 75, etc.) is also sometimes found.

16. The normal Eifel gradient is 0.12%; Carthage, 0.15% (Grewe, *Atlas* 3; *JEAR* 316). Compare the much shallower (0.05%) gradient of qanats (Goblot 27-8). For a general summary of the gradients on the Rome city aqueducts, see Ashby 37-9; in Gaul, GdM 171; Cherchel 66 n. 77.

17. A. Trevor Hodge, 'How did Frontinus measure the quinaria?', *AJA* 88 (1984), 211.

18. Pace 71; Cherchel 87; GdM 335, 345 – on p. 172 he quotes figures for a speed of 2.30 m/sec for the Marcia but queries the result.

19. Nîmes, with its very shallow gradient, represents an extreme case. Hauck (169) assumes a speed of 0.78 m per second on the upper and steeper section (above the Pont du Gard: 16 km at the gradient of 0.67 m per km) and 0.49 m per second on the lower (34 km at an average of 0.185 per km). This gives travel times of 5.7 hours and 19.4 hours, so the water would take around 25 hours from source to *castellum*. Water speeds not exceeding 0.5/0.6 m per second were urged by Alfred Leger, *Les travaux publiques, les mines, et la metallurgie aux temps des Romains* (J. Dejey, Paris, 1875), 608. See also n. 8 above.

20. As quoted in Cherchel 66 n. 77, and 99 n. 94. J.E. Petit, *Manuel aide-mémoire des conducteurs et commis des ponts et chaussées* (Paris, 1903), lists maximum permissible gradients and water velocities for channels in three types of masonry: 'Cailloux agglomérés et schistes tendres: 2.786 m par km/1.52 m per sec.; roches tendres: 2.786/1.83; roches dures, granit: 7.342/5.05.' Though generally within these limits, higher velocities on some parts of the Roman aqueducts where the gradient was steep did exceed them. The figure to apply to the cement-lined Roman channels is probably the first one, the 'Cailloux agglomérés'.

21. GdM 172.

22. It is this surface tension which enables one, as a party trick, to float a needle in a cup of water. For the effects of friction on velocities see L.C. Urquhart, *Civil Engineering Handbook*[2] (New York, 1940), 269, 283, 316.

23. Nîmes: Hauck (103) is very clear: 'By making the conduit highly slick, the engineer could increase the water's velocity by more than 80 percent over the velocity in a rough channel.' Mont d'Or: Jean Burdy *et al.*, *Préinventaire des monuments et richesses artistiques: I, l'aqueduc romain du Mont d'Or* (Lyon, 1987), 36-7, gives the estimated discharges of the aqueduct before and after the accumulation of *sinter*, or incrustation, as 'état neuf, 10,000 a 15,000 m par jour', and 'ouvrage entartaré, 6,000 a 10,000 m par jour'; this assumes an incrustation deposit only a millimetre thick, not enough to affect the cross-section of the conduit. It is worth reprinting in summary Burdy's figures, which are calculated for three different levels in the Mont d'Or channel – half full, two-thirds, and full ('cas extrème, régime anormal de crue'):

Depth of water	*New state*		*Light incrustation*	
in channel (m)	*average velocity (m/sec.)*	*discharge (m^3 daily)*	*average velocity (m/sec.)*	*discharge (m^3 daily)*
0.30	0.84	9,400	0.535	6,000
0.40	0.91	14,000	0.59	9,000
0.60	0.985	23,000	0.65	15,000

Two points should be noted. (1) As noted, the speed of the water falls from (cm per sec.) 84 to 53 with the roughening of the channel walls. (2) Forgetting the effect of incrustation, the figures well show the effect on the speed of flow (and hence of the discharge) of the proportions and cross-sectional profile of the channel carrying it: the depth of water, and hence the amount of water, in the channel is doubled, but discharge rises from 6,000 to 15,000, because the new aqueduct profile with higher wetted sides, relative to the floor, has also led to speed rising from 0.84 m/sec. to nearly 1 m/sec. (0.985). For further consideration of this effect, see p. 225 below; for the Bazin formula, on which Burdy based his figures, see my Appendix: Facts, Figures & Formulae, p. 354.

24. Urquhart (n. 33 above), 334: 'Of all possible cross-sections for open channels, for a given area, the semicircle has the shortest perimeter and hence the highest hydraulic efficiency. The half hexagon has the highest efficiency of all trapezoidal sections. The rectangular section of highest efficiency has a depth of water equal to one half of the width.'

25. The Hauck/Novak study of the Nîmes *castellum* (ch. 7 n. 11 p. 435) suggests that since the pipes ran half-full they were, in effect, semicircular conduits, and hence reached a peak of hydraulic efficiency. I would personally suspect that, assuming its truth, this reflects coincidence rather than planning.

26. Bailhache, *JEAR* 21; Burdy (n. 23 above), 36; cf. Brinker in Grewe, *Atlas* 237.

27. So Urquhart, loc. cit. *PSBR* 80; P.S. Barna, *Fluid Mechanics for Engineers* (London, 1957), 80.

28. From Deane R. Blackman, 'The volume of water delivered by the four great aqueducts of Rome', *PBSR* 46 (1978), 67.

29. See n. 2 above.

30. Ibid.

31. For a general discussion and, in particular, its relevance to preventing lead poisoning, see A. Trevor Hodge, 'Vitruvius, lead pipes, and lead poisoning', *AJA* 85 (1981), 486-91.

32. Grewe, *Atlas* 243-7 (W. Brinker); J.C. Gilly, 'Les dépôts calcaires de l'aqueduc antique de Nîmes', *Bulletin Annuel de l'École antique de Nîmes* 6-7 (1971-2), 61ff.

33. Full discussion and review by W. Brinker in Grewe, *Atlas* 244-5. He summarises; 'Ingesamt wird die Sinterbildung also durch das Zusammenwirken

folgender Faktoren verursacht:
(1) CO_2 Gehalt der Luft
(2) CO_2 Gehalt des Regenwassers
(3) CO_2 Aufnahme in der Humus-Schicht
(4) Temperaturverhaltnisse
(a) in der Luft
(b) im Boden
(c) in der Leitung
(5) Beluftung der Leitung
(6) Turbulenz
(7) Vorhandensein basischer Baustoffe.'

34. Horst D. Schulz, in Grewe, *Atlas* 264-5.

35. Esperandieu 34; Gilly (n. 32 above), 71.

36. I say 'would expect', because there is still considerable doubt about the influence and interplay of various factors (velocity, turbulence, temperature, etc.) on *sinter* formation, some of which might negate the expected result. But, irrespective of the rate or conditions of *sinter*-formation, the argument here is that in shallow-gradient sections, more of the conduit wall was exposed to it – whatever it was, or however it worked – because the water was deeper.

37. Esperandieu 33: 'des déversements qui ont constitué de véritables blocs de rochers'. On the site they can easily be passed over as natural formations, particularly since, unlike the more modest proportions of the *sinter* layers inside the channel, they can be of great size. The largest is probably the veritable mountain formed around the terminal water tower of the aqueduct at Laodicea (see Fahlbusch, abb. 48(c)). Since *sinter* forms more readily in turbulence, it is particularly liable to accumulate in large amounts where the water leaks or overflows and cascades down the side of the aqueduct.

38. Hauck 171.

39. Grewe, *Atlas* 243-4.

40. *JEAR*, 37. His table, 40-1, shows percentage differences in discharge between 'ouvrage neuf' and 'ouvrage en fin de service' ranging from a low of 12% for Nîmes (i.e. the later discharge was 12% of the original) to a high of 119% for Toulouse (where the discharge was increased, instead of reduced, because the channel got clogged and was rebuilt larger), but there is a recognisable concentration between the 35-65% marks.

41. Very clear photograph by Horst D. Schulz in Grewe, *Atlas* 264, fig. 1.

42. The definitive account is Klaus Grewe, 'Die römische Eifel Wasserleitung als Steinbruch des Mittelalters', in Grewe, *Atlas* 269-87. See also *WAS* 3 (1988), 93-4.; *WAS* 1 (1986), 215, fig. 102. I am personally indepted to Dr Grewe for much assistance on this topic. There is also an abbreviated account in Haberey 108-12.

43. For full listings of the locations and photographs of some of the more striking specimens, see Grewe, *Atlas* loc. cit. Several are to be found at Bad Munstereifel (near Euskirchen), and in the Bonn Landesmuseum. The most distant location is the cathedral at Roskilde, outside Copenhagen, where in 1225 one episcopal and two royal tombs were closed with slabs of *sinter* imported from the Eifel aqueduct at Cologne. The material was evidently so attractive that one particular set of *sinter* columns went on a veritable odyssey: from the Abbey at Brauweiler in Pulheim, near Cologne, they were removed to Aachen, borne off thence in 1883 to St Petersburg, Russia, returned in 1897 to Berlin (as supposedly being part of the tomb of Charlemagne), and so eventually to the Schloss of Bad Homburg for incorporation in a 'romanische Halle'; they now seem to have been lost.

44. I have elsewhere considered these matters in detail and here offer a summary. See SRA 195-7 (on static pressure); and A. Trevor Hodge, 'How did Frontinus measure the quinaria?' *AJA* (184), 206 (on the speed of the water escaping from a totally submerged orifice); Hauck and Novak (ch. 7 n. 11 p. 435).

45. An interesting application of this is the clepsydra, the water-clock used in the Athenian law courts to time the speeches. Essentially it was no more than a terracotta jar with an escape hole at the bottom. Since the jar was opaque, and fitted with no dial or other indicator of the water level inside, the only way a speaker could gauge how much water remained, and hence how much more time he had, was by observing the speed and strength of the water jet escaping from the hole.

46. SRA 197ff. Smith 56-7.

47. Full discussion in SRA 203-6. It is tempting to imagine some kind of pressure release valve in the side of the siphon that, spring- or counterweight-loaded, will open to release pressure. This will not work. The basic fallacy is that such a valve is often suggested on the analogy of the safety valves on a boiler or the like. It will work to relieve excessive pressure where the pressure itself varies. But the depth of a siphon, hence the head, hence the static pressure, remains uniform; hence the valve will remain either permanently closed – in which case it is useless – or permanently open – in which case it is not a valve at all but a hole in the pipe.

48. See ch. 6 n. 77 p. 433.

49. See n. 6 above.

50. Callebat (171-2) gives the relevant formula for calculating this force.

51. See SRA 189 for an analogous situation at Aspendos. When I there suggested that the 'pressure towers' were a clumsy and complicated way of eliminating the sideways inertial thrust at the points where the aqueduct had a bend, I was far from sure of it, and now am much less so; for me, the true explanation of the Aspendos 'pressure towers' remains a puzzle.

52. One is accustomed to seeing the *geniculus* as represented in schematic diagrams illustrating the principle of the siphon (see my Fig. 102), where the angle is very acute indeed, because of the need for clarity in the drawing. In reality the angle was usually very much less (Fig. 102).

53. Vitruvius VIII, 6, 8; Callebat 179-80.

54. For a fuller discussion of these phenomena, see SRA 201-8.

55. Further discussion in SRA 207-8, esp. n. 86. A surge tank has conjecturally been restored for Saintes (A. Triou, 'Les aqueducs gallo-romains de Saintes', *Gallia* 26 (1968), 119-44; the parallel from Lincoln to which he refers is, like everything about the Lincoln aqueduct, highly dubious). At Constantinople (see GdM 190), a certain M. Flachat saw what looked to him like a surge tank (he inaccurately described it as a *ventouse*) on a Roman aqueduct, but this is not otherwise attested, and one wonders whether, even assuming the surge tank identification is correct, the aqueduct may not have been Byzantine or even Turkish: there are a lot of them around Istanbul, and it is often hard to tell them apart.

56. I cheerfully acknowledge the help of Mr Yehuda Peleg, who took me to see these remains, which are quite impossible to find without such expert local guidance.

57. For the '*colliviaria*' on the Caesarea South aqueduct, see Y. Peleg, *FCAS*, 137. I am grateful to Mr Peleg for providing the illustrations here printed. As for the larger structures, such as Aspendos, in some German publications the author feels sufficiently confident of their identification with *colliviaria* to call them that in the text, without further ado, as if the matter were conclusively settled. This I feel may be as unwise as the British habit of referring to the Aspendos structures, without inverted commas or explanation, as pressure towers, with all the attendant implications.

Chapter 9

1. C. Knapp, 'Irrigation among the Greeks and Romans', *Classical Weekly* 12 (1919), 73-4, 81-2 (and a brief note, 13 (1920), 104). Knapp's complaints on *RE* I, 1, (s.v. 'Ackerbau') 267; 287-8, written by F. Olck in 1894) are justified. Olck's account is confused, mixing together indiscriminately works of irrigation and drainage. See also W.L. Westermann, *Cl. Ph.* 12 (July, 1917), 237-43, 'Aelius Gallus and the reorganization of the irrigation system of Egypt under Augustus'.

2. On the whole question of irrigation, see *SAT* 2, 1-44. K.D. White, *Roman Farming* (London, 1970), 151-60; dry farming, 483 n. 3, also E.C. Semple, *The Geography of the Mediterranean Region in Relation to Ancient History* (London, 1932) 385-8.

3. White (n. 2 above) 283.

4. *Toronto Globe and Mail*, Saturday 18 October 1986 (page D-5).

5. M. Falkenmark & G. Lindh, *Water for a Starving World* (Boulder, Colorado, 1976; tr. from the Swedish edition, Stockholm, 1975; publ. in co-operation with the United Nations and UNESCO, and in conjunction with the UN Water Conference), pp. 8-9. For a general introduction to modern irrigation in a world context, with relevant statistics and tables, see M. & A. Biswas, *Food, Climate and Man* (John Wiley & Sons, NY, 1979), 59-65. See also ch. 11 n. 4 p. 464.

6. Compare estimated discharges of this magnitude for the Tepula, Alsietina, and Alexandrina at Rome, the Eifel at Cologne, the Gier at Lyon. Nîmes, Carthage and Trier come in the same range. It must be emphasised that estimates of aqueduct discharges are very speculative, and the list I have quoted (pp. 347-8) are not all internally consistent; but as a rough basis of comparison, they will here suffice.

7. Vitruvius X, 4-7; Strabo XVII, 1.30.C.807. The standard modern work is J.P. Oleson's monumental *Greek and Roman Mechanical Water-Lifting Devices* (Toronto, 1984).

8. For a quite uncharacteristic exception, see SRA 179, fig. 2: a rural siphon passing under a Roman road in Provence.

9. Oleson (n. 7 above) 291-301 (screw); 325-50 (compartmented wheel); 370-85 (sakia). Landels 58-75. Machines uneconomic in Italy: White (n. 2 above), 157.

10. *SAT* 7, 221-2. Oleson (n. 7 above), 251-9, and figs 113-28.

11. Strabo XVII, 1.30. See also *SAT* 2, 38-9; Oleson, 104.

12. Oleson (n. 7 above), 301-25. Landels 75-83. A possible exception is the large eight-cylinder pump (calculated discharge, 40 m^3 daily, or 1 quinaria) at St Malo, which may somehow have been used in municipal water supply (Oleson 261-3, 318-19; 321 (discharge)). This device, however, is quite unique.

13. Shaduf: *SAT* 2, 38 (Forbes does not say that this is the volume per day, but it seems a reasonable assumption). White (n. 2 above), 157 accepts this figure, though deriving it from another source (see his n. 36). Landels 63 (screw; independently confirmed by Pace 97-103, which see for all the relevant formulae and calculations), 65 (tympanum). It will be noted that these figures do not wholly correspond with the UNESCO statistics quoted above. Variation in different sources on this topic is to be expected, and the capacity of the tympanum will also vary with its individual size and how quickly it is turned (for calculations, see Oleson 348); on the basis of experiments with models, it appears that there is an optimum speed, and exceeding this does not increase discharge (Aubrey Burstall, *Simple Working Models of Historic Machines* (London, 1968), 60).

14. The inscription is *CIL* VIII 18587; *ILS* (Dessau) 5793. For the traditional view, see White (n. 2 above), 158; Birebent 403-5; W.E. Heitland, *Agricola* (Cambridge, 1921), 293; P. MacKendrick, *The North African Stones Speak* (London, 1980), 247; Felix de Patchère, 'Le règlement d'irrigation à Lamasba', in *Mélanges de l'École Française a Rome* 28 (1908), 373-405. The vital point, that the rationing scheme allows longer hours to those forced to use lifting machines, has now been successfully attacked and demolished by Brent D. Shaw, 'Lamasba: an ancient irrigation community', *Antiquités Africaines* 18 (1982), 61-103 (esp. 76-81). The explanation is, necessarily, sometimes incomplete and confusing since the inscription itself is in fragmentary state, but some points are clear. Nowhere in the text is there any explicit mention of water-lifting machines. Instead, two types of water-delivery are mentioned, *aqua ascendens* and *aqua descendens* (water rising and water descending), and from a comparison of the times allocated and the areas to be covered in this time, it is plain that the 'water rising' arrangement provided much less water: hence the assumption that it referred to properties served by water brought uphill from the main irrigation channel. Now, however, it appears that the 'water rising' and 'water descending' arrangements were locked into set dates on the calendar, on which the irrigation system switched over from one type of service to the other. The actual length of time can be reconstituted: 5 days of 'rising' (i.e. little water) is followed by 10 of 'descending', then 5, and again 10 (making 30, or one month), and so on, alternating indefinitely. Moreover, where it is possible to reconstruct the geographical layout of the plots irrigated, there is no correlation with the 'rising-descending' differentiation. Two plots alongside each other can be scheduled to receive, one, rising, the other, descending water. It does thus seem plain that the uphill/downhill interpretation, and the lifting machines that went with it, is baseless. Whatever the terminology 'rising/descending water' refers to it is some form of service that can be, so to speak, turned off and on at will, and was so at regular intervals, indefinitely repeated. It would doubtless clinch Shaw's case if he could establish what the actual procedure was – he suggests that it may have centred upon a collection reservoir at the spring that was periodically either topped up or drawn upon, introducing irregularity into the amount of water available for irrigation, but 'why this procedure was followed one can only guess' (80). Nevertheless, the evidence adduced – particularly the close linkage of the alternating 'up' and 'down' water to an orderly time schedule (and one in which the *short* time – 5 days, as opposed to 10 – is allocated to the *rising* water) – must lead us to reject this inscription as firm evidence for the use of lifting machines, even though, as Shaw puts it, the idea has 'now become enshrined in the literature'.

15. By Semple (n. 2 above), 464: here 'olives, figs, vines and all kind of fruit trees abounded where to-day flourish the famous *huertas* ... using water channels dating back to Roman times' (quoted by White (n. 2 above), 155: see also R. Dumont, *Types of Rural Economy* (London, 1957), 221-5). Martial (12, 31, 2) speaks in praise of the irrigation channels (*hoc riguae ductile flumen aquae*) at his villa.

16. At Soucieu (Gier) the overflow was in the side of the header tank, 1.30 m above the floor (SRA 180 n. 17); flow calculations, personal communication from Wilfred G. Lockett, University of Toronto. Anio Vetus: Front. II, 92 (*'in rigationem'*). For the entire Nîmes aqueduct being diverted to irrigation, see Ph. Leveau, *FCAS*, 152.

17. White (n. 2 above), 158. Those favouring the natural stream often translate Aqua Crabra as 'the Crabra brook' (so Frontinus, *Aq.* I, 9, tr. C.E. Bennett (Loeb Cl. Lib., London 1925)). See also Grimal 72 n. 26 *ad loc*. The 'brook' identification seems to centre upon a spring in the Valle della Molara that in modern times still supplied Frascati. Asbhy 163-4. The Aqua Crabra had two chief claims to fame. First, Frontinus discovered that the Rome waterworks staff had improperly diverted it into the Aqua Julia to boost the supply and give them a surplus which they could then illegally sell. He had the connection taken out, allowing the Crabra to flow again at full volume, to the uncomprehending surprise of the locals depending on it. This seems to imply that, although Frontinus specifies it was for use locally, in fact a lot of the Crabra proprietors must have been located a considerable distance downstream; if they had been anywhere near the diversion point, they would surely have known where the water was going to. Second, it is a good example of the way the irrigation water was shared out (as testified by the inscription, see below). One of the proprietors drawing water from the Crabra was Cicero (*De Leg. Agr.* III, 2, 9; *ad Fam.* XVI, 18, 3).

18. *CIL* XIV 7969; *ILS* 5793; Heitland (n. 14 above), 293. See also *CIL* VI 1261, attributed to the Crabra by Mommsen, but queried by Grimal 72 n. 26. See also Herschel 168; Shaw (n. 14 above), 74-6. Engraved on the stone is what looks like a plan of part of the aqueduct, with reservoirs and sluices (or bridges?) marked.

19. *CIL* VI, 31566 (Alsietina); *CIL* XIV, 3676 (Tivoli). Again, the text specifies the size of the *foramina* (= the service *calices*) to be used for the various farms. Lanciani (quoted, loc. cit. *ad CIL*) thinks the specifications particularly suited to conduits for agricultural irrigation. As in *CIL* VI, 1261 (above), a rudimentary plan is included on the stone. *ILS* 5771 (the Aqua Vegetiana, near Viterbo); compare also *CIL* XI, 3772 and Grimal 73 n. 33 (Tivoli).

20. Hence, probably, Hor *C.* 1, 7, 13, '*uda mobilibus pomaria rivis*'. There are three possible interpretations of why the *rivi* (channels) are *mobiles*. The word may mean simply 'swiftly moving', i.e. fleeting, fast-flowing (so *Thesaurus L. Lat.* (8.1198.45ff.). Conceivably the channels could actually themselves be movable, some sort of portable arrangement of wooden gutters to re-direct the water where needed (compare ch. 5 n. 32 p. 418). Or, it could mean that 'the water can be directed now into one channel, now into another', in rotation: so R.G.M. Nisbet & Margaret Hubbard, *A Commentary on Horace, Odes, Book I* (Oxford, 1970), 101. See also M. Nilsson, *Eranos* 43 (1945), 301ff.

21. A.K. Biswas, *History of Hydrology* (Amsterdam/London, 1970), 23-6. 'Because of the importance of water in the desert, the job of *Muqassim addayri* (the water divider) is important and prestigious.' The author refers also to C.G. Richardson, 'The measurement of flowing water', *Water and Sewage Works* 102 (1955), 379-85; Committee 861D, *Water meters: Selection, installation, testing and maintenance*: ch. 1, 'Early history of water measurement and the development of meters', *Journal of the American Water Works Association* 51 (1959), 791-9; and T.J. Abercrombie, 'Behind the veil of troubled Yemen', *National Geographic Magazine* 125 (1964), 423 (for a 'floating clock', a copper bowl with a hole in its bottom, which is set floating in the water and measures the time by the time (about 5 minutes) it takes to fill and sink). See Pliny *NH* XVIII, 51, 188. For comments on Algerian water rationing by time in more modern ages, see Shaw (n. 22 below), 169, esp. his n. 207: according to Leo Africanus, disputes over the timing were common: 'de grandes questions, dont s'ensuivent plusieurs meurtres et occisions'. In the circumstances, the French colonial administrator's 1882 order that the peasants of the Langwat oasis observe timings specified in hours, minutes and seconds (Shaw, n. 208) seems somewhat idealistic.

22. The standard authority here is the excellent and comprehensive article by Brent D. Shaw, 'Water and society in the ancient Maghrib', *Antiquités Africaines* (éditions du CNRS, Paris) 20 (1984), 121-73, upon which I have relied heavily and to which I happily express my indebtedness on this topic: in the rest of this chapter this work is abbreviated to 'Shaw'. See also H. Pavis d'Escurac, 'Irrigation et la vie paysanne dans l'Afrique du Nord Antique', *Ktema* 5 (1980), 177-91.

23. Shaw, op. cit 127 n. 19, notes parallels with Spain, especially during the Islamic period, and also with other civilisations; he appositely quotes T.F. Glick, *Irrigation and Society in Mediaeval Valencia* (Cambridge, Mass., 1972), 152: 'Simply stated, the assumption was that a great work must have been built by a great king.' The argument is often applied to the great pre-classical oriental empires: their (e.g. Nineveh's) greatness was based on extensive irrigation, which in turn could be created and administered only within the framework of a highly organised central government. On the slanted nature of archaeological findings, Shaw (p. 125) is scathing: 'In their singleminded search for hydraulic schemes which were "Roman" the surveyors were attracted almost involuntarily to ruins which were readily and obviously recognisable as such. They collated the monumental systems which were so pre-eminently suited to cataloguing in intricate and seemingly unending lists, never questioning whether or not these water systems were to be connected with the hypothetical "more prosperous" African past' (i.e. Roman period). Even scholars who have recognised that the original native farmers may have known what they were doing, still maintain that the Romans must have improved things with new techniques. (So, e.g., L. Carton, 'Hydraulique dans l'antiquité', *Revue Tunisienne* 19 (1912), 221-30.) Birebent, a modern engineer, has been criticised for publishing as Roman antiquities waterworks where there is no evidence at all for the identification or dating (Shaw 116 n. 18; see also L. Foucher, *REA* 66 (1964), 481-3; M. Leglay, *REL* 43 (1965), 666-8 and *Latomus* 25 (1966), 993-4).

24. Shaw 146ff. Negev: P. Hammond, 'Desert waterworks of the ancient Nabataeans', *Natural History* 76 (June-July 1967), 38ff.; N. Glueck, *Rivers in the Desert* (London, 1959).

25. Shaw 146. See C. Vita-Finzi, 'Roman dams in Tripolitania', *Antiquity* 35 (1961), 14-20.

26. For an appreciation of the scope of these schemes, see Y. Kedar, 'Water and soil from the desert', *Geogr. Journal* 123 (1957), recording in an area of 130 km^3 of the Negev the existence of over 17,000 wadi dams. There is no telling how many of these are ancient, if any. They are usually temporary structures of earth, and best studied from air photographs.

27. Cato, *R.R.* 155. Vivid description in White (n. 2 above) 149 n. 12. Shaw, op. cit. 137.

28. Shaw, op. cit. 155.

29. Shaw, op. cit. 160 n. 160, lists and comments on the best known ancient references. See Pliny, *NH* V, 3 (24); XVII, 3 (41); XVIII, 5 (24); 18, 21 (94); 18, 21 (95). Strabo 17.3.11 (C 831). Varro, *R.R.* 1.44-5. Sil. Ital., *Pun.* 9, 204-5. Hdt. 4, 198.

30. See p. 50 and ch. 3 n. 8 p. 400; also Leveau, *FCAS*, 159.

31. Shaw, op. cit. 138.

32. *Digest*, 8, 3, 20; 8, 3, 17. But see 43 (Ulpian) 20, 1 (39-43). Shaw, op. cit. 168.

33. Front. II, 107: '*ius impetratae aquae neque heredem neque emptorem neque ullum novum dominum praediorum sequitur*.' One guesses that rural irrigation rights, on which the farmer's livelihood depended, were seen differently from the case of urban dwellers, who, in the absence of their own personal supply, could always get water from the street fountain. The first was a necessity, hence a right, the second a convenience and hence a privilege. Grimal 95 n. 131; 96 n. 139 (b); 100 n. 155.

34. See n. 1 above. We may note Cicero, *De Off.* 2, 14; *N.D.* 2, 152; *Cato Major*, 53. Virgil, *Ecl.* 3, 111, reflects the familiarity of irrigation to the general literary reader; also *Geogr.* 1, 100-6. There are extensive references in the agronomists to what individual crops benefit from irrigation (Col. 2, 9, 17 (millet); 2, 10, 21 (hemp); 26 (clover); 18 (sesame); 23 (turnips). Varro, *R.R.* 1, 35, 1; Cato, *Agr.* 151 (cypresses).

35. For fulling, see *SAT*. 4, 81-9. At Rome, the Fullers' Guild was known as the *Conlegium Aquae* (*CIL* VI, 10298). For fullers' use of aqueducts, see Front. *Aq.* II 94 and 98; *CIL* VI, 266.

36. The 'hushing' process is described by Landels 25-6. Dolaucothi: P.R. Lewis & G.D.B. Jones, 'The Dolaucothi gold mines, I: the surface evidence', *Antiquaries Journal* 49 (1969), 244-72. See also abbreviations. Las Medulas: White 116-20; P.R. Lewis & G.D.B. Jones, 'Roman gold-mining in north-west Spain', *JRS* 60 (1970), 169-85. The mines are near León. Note also the mines at Rio Tinto (near Cadiz): White, 116-20; G.D.B. Jones, 'The Roman mines at Rio Tinto', *JRS* 70 (1980), 146-65; for water lifting see n. 10 above.

37. For illustrations of types and discussion, see *SAT* 2, 91; Landels 18-24. Pace 111; C. Singer, E.J. Holmyard *et al.*, *History of Technology* (Oxford, 1956), 2, 595; White 55 and 196-201.

38. The 'successive' theory of horizontal mill developing into vertical probably began with R. Bennett & J. Elton, *History of Corn Milling*, vol. 2 (London, 1899), 'Watermills and Windmills'. It is clearly stated and discussed by Landels 19. See also C.E. Curwen, 'The problem of early water mills', *Antiquity* 18 (1944), 130-46; L.A. Moritz, *Grain-Mills and Flour in Classical Antiquity* (Oxford, 1958), esp. chs 15-16; and pl. 14 (c); *SAT* 2, 89.

The more modern, contrary view is championed in detail by N.A.F. Smith, 'The origins of water power: a problem of evidence and expectations', *Transactions of the Newcomen Society* 55 (1983-84), 67-84, who believes not only that the vertical and horizontal wheels co-existed *pari passu* in the ancient world, but if there was any development at all from one to the other it was the other way round, the gearless horizontal wheel being a later, simplified version of the vertical one: and, in so far as the vertical wheel develops out of anything and is not purely independent, its predecessor is the water-lifting machinery of the sakia, the essential prototype being the right-angle gearing. See also H. Hodges, *Technology in the Ancient World* (Harmondsworth, 1970), 192. The whole history of the vertical water-wheel is set out and discussed by Terry S. Reynolds, *Stronger than a Hundred Men* (Baltimore/London, 1983), esp. pp. 9-46. Wikander, n. 47 below.

39. Calculations by Landels (21-2; he works on the basis of a 7 ft (2.13 m) diameter wheel turned by water flowing at 140 l/sec. As one would expect, calculations of power vary quite a lot. Using one of the Barbegal (see below) wheels, Reynolds (n. 38 above) 41 calculates its power output as 4-8 hp. C.L. Sagui, 'La meunerie de Barbegal (France) et les roues hydrauliques chez les anciens et au moyen âge', *Isis* 38 (1947), 226-7, suggests 10 hp. Stuart Fleming, 'Gallic waterpower: the mills of Barbegal', *Archaeology* 36.6 (Nov./Dec. 1983), 68, apparently opts for 8 hp, but his phrasing does not make it absolutely clear whether he is speaking of the output for one wheel or the total for the entire Barbegal system. *SAT* 2, 93, 3 hp.

40. Landels 17, N.A.F. Smith (n. 38 above), 71. *SAT* 2, 88. Full listing of all literary references, White 198-9.

41. Vitr. 10, 5; Ausonius, *Mosella* 5, 362-4. D.L. Simms, 'Water-driven saws, Ausonius, and the authenticity of the *Mosella*', *Technology and Culture* 24 (1983), 635-43. Lynn White, *Mediaeval Technology and Social Change* (Oxford, 1962), 82-3. N.A.F Smith (n. 38 above). 71. See also references in Pliny, *NH* XVII, 27; Strabo 12, 3.30 (Ch. 556 (at Cabeira, on the Black Sea; 65 BC)). Lucretius 5, 509-33; L.A. Moritz, 'Vitruvius' water-mill', *CR*, N.S. 6 (1956) 193-6. For all of this, see Landels 17-19; also Örjan Wikander, 'Water mills in ancient Rome', *Opuscula Romana* 12.2 (1979).

42. The sixteen best-known mills, as archaeologically attested (as opposed to mills in literary references), are listed in White 200-1. Agora: A.W. Parsons, 'A Roman water-mill in the Athenian Agora', *Hesp.* 5 (1936), 70-90. *SAT* 2, 95 and fig. 18. Venafrum: *SAT* 2, 93 and fig. 16. The location needs to be stressed (and has been, by N.A.F. Smith (n. 38 above), 78 n. 38), since it has been grossly mis-stated by, e.g., Terry Reynolds (n. 38 above), 36-7, who not only places Venafrum at Pompeii but then has the mill buried under the Vesuvius eruption, and uses this to date it. The 'minor correction' by M.J.T. Lewis (*Transactions of the Newcomen Society* 55 (1983-84), 82) locating Venafrum 'close to Tivoli' is itself incorrect, in the opposite direction. The true location of Venafrum is clearly shown on N.G.L. Hammond, *Atlas of the Greek and Roman World in Antiquity* (Noyes, Park Ridge, N.J., 1981), 17, Cb. It was 120 km from Tivoli, 80 km from Vesuvius, and had nothing to do with the volcanic catastrophe of AD 79.

43. *SAT* 2, 95-8. N.A.F. Smith (n. 38 above), 70. The Janiculum mills were fed by an overflow from the Aqua Traiana and are mentioned by Prud. *Contra Symm.* 2, 949-50; Procop., *Bell. Goth.* 1, 19; *Cod. Theod.* 14, 15, 4: A.W. Van Buren & G.P. Stevens, 'The Aqua Traiana and the mills on the Janiculum', *Memoirs of the American Academy in Rome* 1 (1915-16), 59-62, and 6 (1927), 137-46. Caracalla: Thorkild Schiøler & Örjan Wikander, 'A Roman water-mill in the Baths of Caracalla', *Opuscula Romana* 14.4 (Stockholm, 1983), 47-64. British mills: Smith, op. cit., 70. They are at Fullerton, Spring Valley, Leeds Castle, Ickham (Kent), Nettleton (Wiltshire) and at three locations on Hadrian's Wall (illustration Reynolds (n. 38 above), 37 (from Haltwhistle)). See also Robert J. Spain, 'The second-century Romano-British watermill at Ickham, Kent', *History of Technology* 9 (1984), 143-80.

44. This standard doctrine may be found enunciated, or at least referred to, in *SAT* 2, 98-9; Landels 16-7; Reynolds (n. 38 above), 44-5; White 56. It is convincingly attacked by M.I. Finley, *Economy and Society in Ancient Greece* (Harmondsworth, 1983), 184.

45. The orthodox view. But Ph. Leveau (*FCAS*, 152) supports the hypothesis that the Barbegal aqueduct was built to supply the city of Arles and then, when no longer suitable for urban use from sedimentation or some other cause, was demoted to industrial use in the mill, which was expressly built to take advantage of the new opportunity; the relative sequence of development is complex, based on recent (1991) excavation. Publication in *Gallia*, 1992. It is also possible that the Nîmes aqueduct, in the days of its decline, was diverted to rural irrigation, and that the massive mounds of *sinter* alongside the channel (especially on its upstream half) mark the spots where the conduit was breached to draw off water for the fields.

46. *SAT* 2, 96. Legislation: Cod. Justin. XI, 43, 10; Cassiodorus, *Var.* 3, 31, 2.

47. N.A.F. Smith (n. 38 above), *passim*. As he notes, 69-70, a vital and cogent supporting argument comes from Mark Hassall (see below). The fullest treatment is Örjan Wikander, *Exploitation of Water-Power or Technological Stagnation?* (Scripta Minora 1983-4: 3; publ. by C.K.W. Glerup, Lund, Sweden, 1984). This includes a bibliography of 112 items.

48. Mark Hassall, quoted by Smith (n. 38 above). See the very impressive map of Domesday mill locations in Reynolds (n. 38 above), 53, fig. 2-2. The author assumes (p. 52) all Domesday-listed mills were water-driven.

49. This is the obvious answer to Forbes' objection (*SAT* 2, 98) that the 'heavy investment of capital' for a mill-aqueduct would limit mills to large cities. The water re-use principle involved was of course known and exploited at Barbegal, where the same water turned eight wheels in succession (see below).

50. Thorkild Schiøler, 'Vandmollerne ved Krokodillerfloden', *Sfinx* (a journal of Aarhus University, Denmark) 8 (1985), 12-14. J.P. Oleson, 'A Roman water-mill on the Crocodilian River near Caesarea', *Zeit. des Deut. Palastina-Vereins*, vol. 100, 137-52.

51. See F. Rakob, 'Wasser als element römischer Infrastruktur', *DU Die Kunstzeitschrift* 3 (1979), 66, fig. 2.

52. The site publication is F. Benoît, 'L'usine de meunerie hydraulique de Barbegal, Arles', *RA* 15 (1940), 19-80. Discussion in C.L. Sagui, 'La meunerie de Barbegal (France) et les roues hydrauliques chez les anciens et au moyen âge', *Isis* 38 (1947), 225-31; *SAT* 2, 94; Stuart Fleming, 'Gallic waterpower: the mills of Barbegal', *Archaeology* 36.6 (1983), 68-9, 77; Robert H.J. Sellin, 'The large Roman water mill at Barbegal (France)', *History of Technology* 8 (1983), 91-109; A. Trevor Hodge, 'A Roman factory', *Scientific American* (Nov. 1990), 106-11. Also n. 45 above.

53. The usual answer is that Arles was a legionary camp, and the mills served the needs of the army. Most scholars, impressed with the scale of the installation, have been worried about it producing too much flour, rather than the reverse. Sellin (above) 101, calculates a daily output of '4.5 tonnes per day or enough to feed a population of 12,500 based on a consumption unit of 350 g per day. This corresponds closely enough to estimates of the size of population of Arles in the fourth century AD'.

54. The figures are from Sellin (n. 52 above), 98, whom I also follow, with cheerful confidence, on the absence of a reservoir. The one feature of the mill that must have existed, but which has apparently been neither found nor much discussed, is a diversion overflow at the top end, or back in the aqueduct. When the mill was closed down – probably every night – there must have been a sluice to shut off the water, and provision for it to escape elsewhere, since the aqueduct would continue flowing. But see Sellin 108.

55. *CIL* XII, 722; Benoît (n. 52 above), 72-3. The dating seems incompatible.

56. Transport difficulties: Landels 171-8; White 129-30; *SAT* 2, 161; A. Burford, 'Heavy transport in classical Antiquity', *Economic History Review* 13 (1960), 1-18. Alexandrian devices and gadgets: M.R. Cohen & I.E. Drabkin, *Source Book in Greek Science* (Cambridge, Mass., 1958), 326-9. R.S. Brumbaugh, *Ancient Greek Gadgets and Machines* (New York, 1966). My scepticism on the devices working is firmly based on factual evidence. For many years I have been teaching a course in Ancient Technology in which students often make working models of ancient devices. It is my experience that those making models of Alexandrian gadgets – automata, temple-door-openers, and similar triumphs of Hero of Alexandria – can usually only just manage to get the model to the lecture-room and persuade it to give one half-hearted performance before it bodily falls apart on the floor. Nobody who has seen and marvelled at the Heath Robinson contraptions illustrated in the relevant chapters of histories of technology will be surprised. By comparison, models of practical devices in actual use, such as lathes and bow drills, have always worked very well. Draft harness: its inefficiency may be over-estimated.

57. At Silchester (not a spa) the same thing happened with conventional baths. As Hanson (126) well puts it, 'the baths would seem to have been taken to the water rather than the other way round', and their siting determined by the existence of local springs. Silchester had no organised public water supply.

58. *Princ. Encyc.* (see ch. 6 n. 71 p. 432), 75-9; Hammond (n. 42 above), 33.

59. *MAGR* IV, 2, 401-73. This is 'stations thermales' pure and simple, not counting sanctuaries devoted to aquatic divinities, for which see 477-516. See Grenier *passim* for the whole question of 'Les monuments des eaux'.

60. For Roman baths in general, see H. Manderscheid, 'Römische Thermen: Aspekte von Architektur, Technik und Ausstattung', *WAS* 3 (1988), 101-25; Werner Heinz, *Römische Thermen* (Munich/Zurich, 1983) (with extensive bibliography and excellent illustrations).

61. B.W. Cunliffe, *Roman Bath* (London, 1970); B.W. Cunliffe, 'The Temple of Sulis Minerva at Bath', *Archaeology* 36.6 (Nov.-Dec. 1983), 16-23 (with, 20, imaginative description of the architectural/engineering treatment of the hot spring). Aquae Mattiacae: Pierre Grimal, *Roman Cities*, tr. & ed. G. Michael Woloch (Madison, Wisc., 1983), 182.

62. Oleson (n. 7 above): Pompeii: pp. 242-6, figs 102-8; Herculaneum: 213-15, 72-3; Ostia: 233-7, 93-8; Cyrene: 204, 65; Abu Mena: 181-2, 37. For showers, see F. Kretzschmer, 'La robinetterie romaine', *Rev. Arch. de l'Est et du Centre-Est* 11 (1960), 106, fig. 34, who illustrates a shower-head (*epitonium*) from the Terme al Mare, Herculaneum (my Fig. 232).

63. Oleson (n. 7 above), 384; Landels 69, 74-5.

64. Ashby 11, 14.

65. Ashby 151 (*Botte*), 158 (*Thermae Antonianae* = Baths of Carcalla).

66. 17%: *SAT* 173, based on the figures from Front. *Aq.* II, 78-87. Compare Trier, where the very large Imperial Baths were apparently fed from the Ruwer aqueduct, the only one serving the city; presumably that is where most of its water went.

67. For the principal Roman baths, see Manderscheid (n. 60 above); D. Krencker & E. Kruger, *Die Trierer Kaiserthermen (mit einer Übersicht über die wichtigsten Thermenanlagen des römischen Reiches)* (Augsburg, 1929); also plans in E. Brodner, *Untersuchungen an den Caracallathermen* (Berlin, 1951), Taf. 1, 14, 17, 18, 20, 21. G.A. Blouet, *Restauration des thermes d'Antonin Caracalla à Rome* (Paris, 1928). For general commentary, see F. Sear, *Roman Architecture* (Cornell U.P., Ithaca, 1983), 40-1, 257-62; D.S. Robertson, *Handbook of Greek and Roman Architecture* (2nd ed., Cambridge, 1943), 258-61; For baths in France, see *MAGR* 231-384.

68. See the heartfelt *cri de coeur* of W.H. Plommer, *Ancient and Classical Architecture* (London, 1956), 342-3, on how the Baths of Caracalla have now been transformed for musical performances and a full study must await 'an investigator patient enough to endure the opera'. As for the Baths of Diocletian, a large part is now occupied by the church of S. Maria degli Angeli, while the Piazza dell'Esedra is on top of the caldarium.

69. Reconstruction, with a bank of four boilers, at Lambaesis in Kretzschmer 69 (my Fig. 183). For a full exposition of the hypocaust technique and the heating of baths, see Adam 287-299. See also discussion and illustration in W.H. Plommer, *Vitruvius and Later Roman Building Manuals* (Cambridge, 1973) 12-14; W.L. Macdonald & B.M. Boyle, 'The small baths at Hadrian's villa', *J. Soc. Architect. Hist.* 39 (1980), 26 n. 55; J.B. Ward-Perkins & J.M.C. Toynbee, 'The hunting baths at Lepcis Magna', *Archaeologia* 93 (1949), 165-95 (esp. 173-7). The system of heating the water by *testudines* ('tortoises'), semicircular open-ended circulation chambers, is described by Vitr. V, 10, 1.

70. Illustrated by Kretzschmer (n. 62 above), 109, 60; this concentrates on the taps. The complete installation is shown, to smaller scale, in Kretzschmer (see list of abbreviations), 34, fig. 59; C. Merckel, *Die Ingenieurtechnik im Alterthum* (Berlin, 1889), 585, abb. 260.

71. Kretzschmer 34, fig. 58 (6); Sear (n. 67 above), 40, fig. 20.

72. The figures of 40° and 25° are as calculated by Kretzschmer 37. For air temperatures, see his interesting account (with, p. 33, fig. 57, isothermic graphs) of experiments with a reconstructed hypocaust at the Saalburg (p. 35).

73. Kretzschmer 71: 'Les conduits d'évacuation ... formaient, comme ceux des thermes imperiaux de Trèves, un labyrinthe souterrain presque inextricable.' For a plan of the basement of the Baths of St Barbara at Trier (= Trèves), showing some of the water channels, see *MAGR* fig. 121, p. 368.

74. Schiøler & Wikander (n. 43 above).

75. As Kretzschmer 38, felicitously puts it, 'on n'était pas formaliste'. It is quite possible that, as he suggests, both sexes were accommodated simultaneously. Though there was no privacy internally, it may be noteworthy that the toilet in the Forum Baths at Ostia carries on the threshold block of its entrance scratch marks showing that it was fitted with a revolving door, so that one could never look inside. For accounts of public toilets, see Kretzschmer 38, 61; *WAS* 1 (1986), 200-1; Grewe 98-101; F. Drexel (ch. 12 n. 14 p. 476); A. Scobie 415 (ch. 12 n. 14 below).

76. Illustrated in *WAS* 1 (1986), fig. 76, p. 200 (my Fig. 187). Mabel Lang's *Waterworks in the Athenian Agora* (Princeton, 1968), fig. 40, prints the plan of a 64-seater from the Roman Agora in Athens.

77. 'Peut-être ce dispositif fut-il inventé pour pallier l'absence de papier de toilette', Kretzschmer 61. The absence of paper sometimes led to the substitution of sponges.

78. So in e.g. the Baths of Djemila (Algeria): Kretzschmer 71, fig. 120 (a 23-seater); so also the Stabian Baths at Pompeii: A. Maiuri, *Pompeii: Pozzi e Condotture d'Acqua nell'Antica Città, N.Sc.* (1931), 546-76 (plans on 533, 572), *q.v.* also for plan of the layout of piping and water supply. See also Boon (n. 38 above), 117 (Silchester).

79. A. Johnson, *Roman Forts* (London, 1983), 211-14.

Chapter 10

1. Front. *Aq*, I, 11; 22.

2. Two chambers are recommended by Pliny XXXVI, 173. For the two level systems, see Alfred Neuburger, *Technical Arts of the Ancients* (New York, 1930), 432-3. He illustrates it with a multiple settling tank from Castel Gandolfo, which is not on the route of any of the Rome aqueducts and does not seem to be connected with them. See K. Schneider s.v. *piscina, RE* XX, 2, 1787-8.

3. Drawn by Callebat 190, *ad* VIII 6, 15 and GdM 309, fig. 122. The tank illustrated by Forbes, *SAT* fig. 37, p. 168, seems to be the same one, though attributed by him to the Julia. Forbes is often unreliable and I can find no trace of such a tank. Van Deman 173, fig. 19, reproduces Fabretti's plan (1680). This complex of tanks was at the foot of the Pincian, and later partly destroyed in the building of a cable railway, for which one of its chambers was used as a waiting room. It is probably Hadrianic, and certainly was added after Frontinus, who (*Aq.* I, 11) specifically states that the Virgo has no settling tanks at all. Its water was known for its purity, being ranked after the Marcia and the Claudia.

4. So in the odd-shaped settling chamber at Grüngürtel on the Cologne aqueduct (Haberey 27, fig. 13).

5. Vitr. VIII, 6, 15 suggests that for multi-chamber cisterns the water may be purified by filtering it from one into the other – '*uti percolationibus transmutari possint*', but we are here dealing with an accumulated store of rainwater to be purified by moving it around from one chamber to another. It can be allowed to seep through the filtering material in its own time, and is in any case only drawn off and used as needed. An aqueduct, constantly passing through at, say, 150 litres per second (roughly the discharge at Nîmes) and having to be processed at that rate, is a very different proposition. Neuburger (n. 2 above) quotes the use of tufa as filter material. Around AD 50, Athenaios of Attalia (a medical writer) wrote a book on the purification of water.

6. House in south-east corner of Neapolis, beside watchtower (E. Ripoll Perello, *Ampurias* (Barcelona, 1979), 27). P. MacKendrick, *The Iberian Stones Speak* (New York, 1969), 47, and fig. 3.5, p. 49.

7. André Berthier, 'Note sur un filtre romain découvert à Constantine', *Recueil des notices et mémoires de la Société Archéologique de Constantine* 69 (1955-6), 175-8. Some of the bags were in good enough condition to be removed and stored in the Gustave Mercier Museum, thanks largely to the petrified state of their contents. They had been crushed flat by being heaped up to form the barrier in the conduit, and bore on the upstream side 'un enduit formant un paroi lisse', the downstream side being clear. I am not sure whether this was a lime carbonate concretion from the water, or some kind of porous plaster deliberately applied to help the sandbag wall keep shape. I am also not sure of the purpose of the eight small channels.

8. Nîmes: Stübinger abb. 14. Pompeii: Kretzschmer 52, fig. 84, who restores two successive grilles, the second of closer mesh; also H. Eschebach, *JEAR* 103. Mabel Lang, *Waterworks in the Athenian Agora* (Princeton, 1968), fig. 22, publishes a second-century BC lead screen perforated with numerous holes.

9. Misenum, Eschebach, *JEAR* 86 n. 34; Baiae (=Bacoli), Haberey 130-1, and Daremberg & Saglio XIII, 140 ...; Constantinople, n. 11 below; Lyon, 'Citern du Grand Seminaire, GdM 321-4 and figs 126-7; J. Burdy & L. Jeancolas, 'La Grotte Berelle', in *Bulletin des Musées et Monuments Lyonnais* IV, 4 (1971), 393-413.

10. The one at Constantinople (below) was larger, but presumably Rakob does not count it as being too late. It dates to the early Byzantine era.

11. Plans and reconstruction of the Bordj Djedid cistern: Rakob, *JEAR* 320, abb. 2-4. It is now in the grounds of the Presidential Palace and inaccessible. Ph. Caillat, an engineer in charge of the reconstruction of the aqueduct in 1859, reckoned the ancient Roman discharge as 32,000 m^3 daily, but at the start of this century the discharge of the re-utilised section of the Roman aqueduct was 17,280 m^3 (Rakob 310). Compare the large cistern at Pompeii, that alongside the Forum baths, at the north-east corner of Insula VII-6. It is 5 × 15 m, and has a capacity of

430 m^3 (*JEAR* 83). The capacity of the '*piscina mirabilis*' at Misenum is 12,600 m^3 (*JEAR* 86). At Istanbul, the Cistern 'of the 1001 columns', has a capacity of 325,000 m^3. It actually has 336 columns and was built in the sixth century AD: S. Eyice, 'Byzantinische Wasserversorgungsanlagen in Istanbul', *Mitt. L. Inst.* 64 (1979), abb. 5 ('Yerebatan Istanbul').

12. So de Montauzan on cisterns generally: 'ces citernes devaient servir simplement, comme le plupart de nos réservoirs moderns, à régulariser le débit journalier, se remplissant la nuit pour donner dans le jour à chaque habitant, ainsi qu'aux fontaines publiques, un cube d'eau plus considérable que le n'aurait permis le débit naturel de l'aqueduc.' GdM 318.

13. Pompeii, Eschebach, *JEAR* 88-9 and 112-16, abbs 5-13. Nîmes, *MAGR* 48. Arles, Stübinger 27 and *MAGR* 84: it was 'une grande salle profonde dont on n'a pu mesurer que la largeur, 5m 50, couverte de dalles soutenues par des piliers'. Rome, Ashby 242 (Claudia, Anio Novus near Minerva Medica); 155 (Marcia, on top of the Porta Capena). Piranesi's engraving of the terminal reservoir of the Claudia, now destroyed, is reproduced in E. Nash, *Pictorial Dictionary of Ancient Rome* (London, 1961), vol. 1, 37.

14. Easily the most striking example is Asbhy's comments on de Montauzan's description of Nîmes. In the floor of the Nîmes *castellum* are three openings leading into the drains. This is correctly and accurately stated by de Montauzan (GdM 316), who had personally visited and studied most aqueduct sites in France. Ashby (45 n. 4), who offers no evidence of ever having gone near Nîmes himself, flatly declares him mistaken: these have to be the supply mains into the city, not the drains, for that is how Vitruvius specifies it and for Ashby that's that. In fact, he is quite wrong. Episodes such as this should engender in the reader a salutary caution towards some of the established interpretations.

15. Vitr. VIII, 6, 1-2. Full discussion in Callebat 149-56 and figs 10, 11. Another drawn reconstruction of Vitrivius' triple basin will be seen in GdM 310, fig, 123 (discussion 306ff.), reprinted in *MAGR* 33, fig. 6.

16. Size of pipes – extensive discussion in Callebat 149-50. Vitruvius' purpose 'est son désir de *protéger* les particuliers, en créant un réseau spécial, autonome, avec son budget propre' – P. Grimal, 'Vitruve et la technique des aqueducs', *Rev. Phil.* 198 (1945), 169; 'Le système de répartition que décrit Vitruve obéit apparemment aux principes fundementaux de la législation romaine: priorité imposée pour les distributions *in publico*', Callebat 154. It will be appreciated that there is much in Vitruvius' account that could be (and has been) discussed in detail, and all I have offered here is an outline.

17. 'Sa réflexion s'appuie sur la pratique contemporaine, elle n'en est pas esclave' – Grimal (n. 16 above), 174; SRA 209. For an evaluation of Vitruvius, see pp. 13ff. above. Smith (58), as usual, puts it in a nutshell: the failure of modern scholarship lies in thinking that Vitruvius is infallible, and therefore if ever he talks nonsense, it is not really nonsense, but rather our own fault for not understanding him; and this can be straightened out by interpretation, however contorted. Historians of academic thought will compare J.K. Galbraith's comments on Marxism: 'It was always possible to discuss those who believed Marx wrong as being guilty of a failure of understanding. Marx is not easy to understand. Those who thought him wrong had failed to do so.' – J.K. Galbraith, *The Affluent Society* (2nd ed., Boston, 1969), 71.

18. Mau suggested that this supply was inadequate for the needs of the known population (of around 8,000) and that there must have been a second *castellum* somewhere, but this seems to be disproved by Eschebach (*JEAR* 101 n. 101). We must also remember the cisterns.

19. Eschebach, *JEAR* 88-9; Kretzschmer 52 and figs 84-6; Callebat 152, fig. 10. The original publication is by R. Paribeni, *N.Sc.* (1903), 25ff; but see p. 321 below.

20. On the other hand, at Pompeii well water, brought up from the strata of volcanic lava, was almost undrinkable. It smelled sulphurous, was reddish in colour, and bitter to taste. For drinking water one used cisterns, directly fed with rainwater from the house-roofs. Eschebach, *JEAR* 82.

21. Originally escavated by Auguste Pelet in 1844, it is described fully in *Les Archives de la Commission des Monuments Historiques*, ed. Gide & J. Beaudry (Paris, 1855-72) vol. 1, pp. 1-4, with detailed plan. For full bibliography of the area, see *Forma Orbis Romani: carte archéologique de la Gaule romaine*, départment du Gard, ed. A. Blanchet (Paris, 1941), vii-xviii; for the *castellum*, see no. 68, pp. 60-1. A. Pelet, *Mémoires Acad. Gard.* (1845-46), 67. Stübinger 24-6 and abb. 13-14. The basic plan has often been reproduced with discussion: *MAGR* 98, fig. 32; Esperandieu 22-3; GdM 316, fig. 124; for the entry sluice see Haberey 114-15, fig. 82. Discussion only, Kretzschmer 52.

22. 1 m, GdM 316; 1.4 m, *MAGR* 97. The difference seems to depend on whether or not one counts the thickness of the surrounding paving.

23. *Archives* (n. 21 above), p. 1.

24. Pliny *NH* XXXVI, 24.

25. The height of 1.8 m given in *MAGR* 97 is erroneous, as can easily be seen from any photograph. However, the vaulted aqueduct channel supplying the *castellum* – it begins some 2.4 m behind the delivery opening – does have a height of 2.4 m from floor on top of vault (*Archives*, p. 3); this somehow may be the source of Grenier's *MAGR* figure, which I have also seen in other publications. The aqueduct channel is today inaccessible and may well have been seriously damaged during the construction on top of it of the 'ancienne citadelle' in 1685 (*MAGR*, loc. cit.), so the precision of the *Archives* figure is perhaps not above question.

26. See A. Trevor Hodge, 'How did Frontinus measure the quinaria?' *AJA* 88 (1984), 214-16, for further discussion. The existence of this very peculiar sluice seems reasonably established. The lead seal at the bottom (which is attested to by the excavator Auguste Pelet) seems to show that the bottom half did not move, the stone cover slab proves that nothing was lifted up out of the channel at the top, and the holes for the cables indicate that something did move in between. The arrangement illustrated seems the only answer, and was suggested by Pelet, *Description des Monuments Grecs et Romains* (Nîmes, 1876), 304-5. See also Stübinger 26. My own suggestion (op. cit., 216), that it was intended as a cleaning device to flush detritus out of the tunnel by creating turbulence in it, is improbable but the best I could then think of. A device such as this, in effect a sluicegate extending upwards from the bottom of the channel (instead of being lifted bodily out of the top of it) is used in modern hydraulics. It is called a variable weir. Hauck, agreeing with the existence of the double gate, and with my suggestion that it is a device for measuring the aqueduct discharge (op. cit., 215), expounds the matter fully in his articles in *AJA* (ch. 7 n. 11 p. 435) and *Scientific American* (ch. 6 n. 20 p. 425).

27. The reader may note one further complication. The offtake pipes from the *castellum* were accommodated in round holes in the side wall of the basin. These holes were of 40 cm diameter for a 30 cm internal diameter of the pipe (ch. 6 n. 50 p. 429), and were 56 cm above the floor of the *castellum*. This means that the top of the intake would be about 91 cm above the *castellum* floor (56 plus 40 less 5

(= half of the difference between 30 and 40, giving the thickness of the pipe and the cement collar fixing it in place)). Water must therefore be maintained at this minimum level in the *castellum* or the pipes will start taking in air at the top (which Hauck (op. cit.) is sure they did). This means that at the entry sluice, water on the *downstream* side must always be 91 cm high, and the entire opening is only 1.1 m high, giving only 20 cm play for the sluice to create any variation in *castellum* level. The whole question of the interrelation of water levels, volume of water delivered and its distribution in this *castellum* urgently needs to be studied.

28. *Archives* (n. 21 above), p. 3. The author quotes a parallel: 'Une disposition toute semblable se remarque dans un bassin de division de construction unique, qui fournit encore de l'eau à plusieurs quartiers de Constantinople', but does not identify it further. This interpretation is followed by V. Lassalle (*Doss. Arch.*, 38 (1979), 60): 'Un dispositif ayant sans doute pour fonction soit de servir d'obstacle au courant, soit d'arrêter des impuretés en suspension dans l'eau.'

29. For pipes, see pp. 307ff. below; for taps, pp. 322-6.

30. GdM 316: 'L'aqueduc arrivait obliquement par rapport aux ouvertures, probablement à l'effet d'atténuer les remous à l'entrée des tuyaux.' Compare the *Archives* suggestion (above) that the barrier was there to provide *agitation*. De Montauzan thinks it good hydraulic practice to cut down on the turbulence, the *Archives*, to increase it. Neither quotes any engineering authority. I would guess that the reason for the angle of the aqueduct is that that just happened to be the direction it was coming from, while the pipes were aligned as was convenient for their destination. Compare the settling tank on the Mont d'Or (ch. 5 n. 65 p. 422), which, from the 'tangential' angle of entry and exit, sounds as if it may have been intended to create a kind of whirlpool or rotary turbulence.

31. See ch. 6 n. 50 p. 429.

32. Andreossy (ch. 6 n. 77 p. 433), 412. The normal way of turning off supplies for the night is, he concedes, 'assez singulier, du moins en apparance. On jette dans le mousslouk (= *castellum*) une poignée d'herbages (du céleri sauvage) déja imprégnés d'eau. Ces herbages sont entraînés dans l'entonnoir qui se forme au dessus de l'orifice d'un lulé (= pipe), et sont arrêtés a cet orifice, qu'ils bouchent aussi exactement que s'il y avait un robinet; la simple pression d'un linge contre un orifice d'écoulement produit le même effet.' He comments: 'cette manière de boucher les lulé, qui parait un moyen grossier au premier abord, est néanmoins fondé en principe; elle est encore un preuve, en l'examinant avec attention, de la simplicité et de l'exactitude avec lesquelles on avait établi, dans l'origine, tout ce qui regardait la fourniture des eaux à Constantinople.'

33. Two similar round stone lids, complete with bronze rings for lifting them, are shown by Kretzschmer 59, fig. 100. They are on sewer manholes at Pompeii.

34. See p. 281 above; Ashby 45.

35. *Archives* (n. 21 above), 3. Following Pelet, the drain is marked 'Egout ou Naumachie' on the plan in A. Leger, *Les travaux publics, les mines et la metallurgie au temps des Romains* (Paris, 1875).

36. GdM 317; quoted from de Montauzan, *Nouvelles Archives des Missions*, vol. 25, fasc. 2; the date of excavation was 1906.

37. See also below, ch. 11 n. 1 p. 463.

38. A question of terminology arises. Sometimes modern commentators use the term '*castellum*' for devices other than the terminal *castellum* as described above, such as settling tanks on the line of the aqueduct, header tanks of siphons, or the secondary *castella* described below under the name 'water tower'. This is understandable, but the reader should be aware of the possible confusion. In this book for the sake of uniformity I will use '*castellum*' or, more precisely, 'terminal *castellum*', to refer only to the device located at the end of the aqueduct where it breaks up into branches through the city and to the secondary *castella* described below, where the branches are in turn subdivided.

39. A good example is illustrated by Ashby 160, pl. VIII b (Virgo crossing the Via Lata).

40. As sometimes happens, the clearest illustration of the principle is not always the most accurate. When built by Claudius in AD 52 the arch was outside the walls and did not form part of them, existing simply to carry the aqueduct over the two roads. In 276 it was then incorporated in the Aurelian Wall.

41. P.A. Février, *Fréjus* (*Itineraires Ligures*, Institut International d'Etudes Ligures, 1977), plan facing p. 64; *MAGR* 47. At Orange the aqueduct ran on top of the walls for part of their length, so also at Minturnae and Salona (I.A. Richmond, 'Commemorative arches and city gates in the Augustan age', *JRS* 23 (1933), 149-74, esp. 154).

42. Frontinus, *Aq*. II, 78-86. There were no *castella* on the Alsietina, which was normally used for Augustus' naumachia, though since it was also used as an emergency supplement, and even for drinking, some provision must have been made for diverting its waters into the pipes normally supplying street fountains, which implies *castella* or distribution tanks somewhere. The Claudia and the Anio Novus (once it was built) discharged into a common terminal *castellum* and their united waters were then distributed through a shared network of secondary *castella* and water towers.

43. *MAGR* 39, referring to GdM 309; 319-27.

44. Front. *Aq* II, 87.

45. Id., 106.

46. We may compare the modern parallel of a building code, established by a municipality to govern the activities of building contractors. The work they are doing is quite legal but it is a question of controlling *how* they do it, and a code with the force of law is required precisely because they are *not* municipal employees; if they were, they would simply have to do as they were told.

47. In much the same way that in modern electrical wiring in a house, building codes usually specify that wires can only be joined inside a metal junction box of approved design.

48. Callebat 94, *ad* Front. *Aq*. II, 106.

49. Herschel 206-9, with *calices* illustrated in fig. on p. 207. A *calix* from the Museo Nazionale, Rome, is shown by Kretzschmer 53, fig. 90. Grimal 84 n. 70. *CIL* XV, 7212, 7213 for inscriptions on *calices*. For sizes in *calices* see below, p. 297; Fig. 208.

50. Front. *Aq*. I, 36.

51. In scientific terms the above statement is, of course, inaccurate and oversimplified, but it conveys the general idea. The hydraulic phenomenon concerned is known as the Venturi effect and has for long been familiar. It is discussed by Herschel (206) who thinks that a Venturi adjutage (i.e. a flaring, funnel-shaped orifice with the narrow end toward the tank) attached to a long pipe of the same dimensions probably *would* increase discharge, but cannot be certain ('I should not be surprised if it did, to put it no stronger'). He maintains that the prohibition of a small *calix* followed immediately by a long length of large pipe, was a 'correct conception of the matter of head acting on orifices entertained by Frontinus, though he spoke without knowledge of the law of hydraulics as they have been developed in the eighteen hundred years following his time'. In fact, Herschel's guarded caution is unnecessary. A 'Venturi-type' adjutage would increase the discharge of water passing through it by about one-third. In a modern engineering handbook this very problem (complete with explanatory notes on Frontinus and Roman aqueducts) is set and solved, as a demonstration example for students, with all the steps in the calculation set out: R.W. Fox & A.T. McDonald, *Introduction to Fluid Mechanics* (New York, 1985), 381.

52. The standard sizes of pipe are listed in detail by Frontinus, *Aq.* I, 39-45. They are measured by diameter, and go up ¼ digit at a time (I, 25, 5). A Roman ¼ digit (*quadrans*) being 0.4625 cm, and the smallest pipe being five-quarters, it thus becomes a 'fiver', or 'quinaria', diameter 2.3125 cm. This was the basic unit of measurement as applied to volume of water (a complicated matter that will be dealt with later). Grimal (84) lists some of the commonest *calix* sizes as listed by Frontinus, with their diameter in modern measurement. I print Grimal's list, but have doubts on the reliability or practical use of measurements that go down to one-thousandth of a millimetre:

quinaria	2.3125 cm
senaria	2.775 cm
septenaria	3.2375 cm
octonaria	3.700 cm
denaria	4.625 cm
duodenaria	5.55 cm
'15-pipe'	6.9375 cm

See also n. 53 below; and p. 297 below.

53. Frontinus' entire work on the aqueducts consists of 130 chapters. Thirty-nine of them (one-third of the whole) are on the sizes of the *calices*. For a full exposition of 'The sizes of measurement nozzles (=*calices*), and Frontinus' arithmetic', see the appendix to ch. 2 in Landels 53-7. On p. 55 he tabulates 13 of the commonest *calices* with full details. They are also to be found in Kretzschmer 55; GdM 329; and Grimal 84 (above).

54. Front. *Aq.* I, 31-2. The four *calices* affected were the *duodenaria*, the *vicenaria*, the *centenaria* and the *centumvicenum*, these being pipes of 12, 20, 100 and 120 ¼-digits. The *duodenaria* was increased in diameter by 0.0775 cm, the *vicenaria* reduced by 0.9, the *centenaria* increased by 1.3 cm, and the *centumvicenum* by 6.32 cm. These changes are not to be confused with illegal and surreptitious alterations effected by the watermen with the object of cheating. They were general practice, universally known and accepted if not officially approved, though Grimal (p. 21 n. xxxi,2) does see in them evidence of the watermen, particularly under Domitian, having got out of hand to the extent of altering the official standards to suit their own convenience. The significance lies particularly in the great increases in the size of the two large pipes, which were used only as mains for supplying the *castella*, and would therefore produce a surplus in supply which the watermen could illegally retail for their own profit. Setting aside the invidious implications, one may compare a parallel from standard sizes of timber in North American building construction. The commonest size of lumber is a 'two-by-four', a term almost a household word. Frontinus would no doubt register a complaint that, though originally of two inches by four cross-section, the timbers now sold to this description are only about 1½ inches by 3½. The convention is, however, well understood and generally accepted.

55. Landels's quotation from an imaginary gang foreman has the ring of truth to it: 'You pays the money to me, guv'nor, and I squares it up at 'ead office.' As in nearly all such situations, it is important that neither side actually *says* anything illegal, though both understand perfectly what is happening. Frontinus attacks such trickery as *fraus*, fraud, but no doubt there were many householders who, when the damning oversize *calix* was pointed out to them, piously lifted their hands and 'knew nothing about it' (Landels 52). For the legal provisions on stamping *calices* and lead pipes connected to them, see Front. *Aq.* II, 105. For oversized *calices* and other forms of cheating by the watermen, II, 112-15.

56. Front. *Aq.* II, 113: '*Circa conlocandos quoque calices observari oportet ut ad lineam ordinentur nec alterius inferior calix, alterius superior ponatur. Inferior plus trahit; superior, quia cursus aquae ab inferiore rapitur, minus ducit.*' His reference to the current, '*cursus*', is of course incorrect, but his grasp of the main point is not to be faulted. Part of the trouble comes from the fact that we are here dealing with a *castellum*, in which, though the water may swirl about, there is no real current, as we normally understand the term. 'Current' implies a conduit rather than a tank. Grimal, 95 n. 133, tries to justify the term by referring it to 'le courant créé par l'ouverture de la prise inférieure. Frontinus imagine que c'est ce courant qui aspire la plus grand quantité de l'eau en n'en laissant qu'une quantité moindre par la prise supérieure.' The real reason the lower *calix* has a greater discharge is of course the greater head, its vertical distance below the natural water level in the *castellum*, under which it operates. Even with Grimal's modification, therefore, Frontinus is not right, and I would prefer to think that by current, '*cursus*', he was referring to *calices* set into the conduit, where there *was* a current. His statement is still not right – in the conduit, *calix* discharge is also governed by head, as it is everywhere – but it is closer to the natural meaning of the Latin, and is therefore to be preferred since neither interpretation will give a good hydraulic sense. One may object that Frontinus cannot be thinking of *calices* in the conduits since he has just specified that they all have to be in the *castella*, but there are two answers to this. First, there evidently *had* been a lot of *calices* implanted in the conduits and Frontinus does not say they had been removed: the tightening-up of regulations embodied in the prohibition could well apply only to new work. Second, his warning against angling the *calices* with or against the flow (below) has to apply to conduits, so he may be thinking of them here also.

57. Front. *Aq.* I, 36, 2.

58. Ibid., 2, 3. On the aims and purpose of Frontinus' book, see full discussion in Grimal XV-XVI, who stresses that despite its format as a technical treatise, 'Le *De aquae ductu* est donc, d'abord, un écrit politique'.

59. Herschel (201) sums up the opinion of many modern commentators: 'to the expert of today this seems excessively silly.'

60. See A. Trevor Hodge, 'How did Frontinus measure the quinaria?', *AJA* 88 (1984), 205-16, esp. for the present point, pp. 205-6.

61. Cl. Di Fenizio, 'Sulla portata degli acquedotti romani e determinazione della quinaria', *Giornale del Genio Civile* 14 (1916), 227-331, esp. 316. He based his argument on two factors: (1) the centre line of all adjutages, of whatever size, had to be on the same level to ensure equal speed of flow; and (2) the minimum depth of this centre line is given by the radius of the largest *calix* in use (the *centumvicenum*, radius 11.5 cm) otherwise the top of it would be above water level and taking in air. To this he adds on an extra 5 mm as a safety margin to ensure all *calices* are fully submerged (a margin that seems to me impossibly slim), bringing the total figure up to 12 cm for the very minimum distance acceptable for a standard head, and hence for a *quinaria* of uniform volume.

62. It must be emphasised that Di Fenizio's figure of 40 m^3 was intended only as a *minimum* for the *quinaria*. It could be a lot more, a fact often overlooked by subsequent writers who often quote Di Fenizio's estimate as if it were not a lower limit but a positive identification of the real figure. But in fact if we do translate Frontinus' *quinaria* figures for Roman aqueduct discharge into modern measure on the basis of Di Fenizio's 40 m^3, the result seems to come out more or less right on the rare occasions when it can be independently checked.

63. For a full discussion of Frontinus, see pp.16-18. above

64. Exposition by Eschebach, *JEAR* 92-3. Illustration in Kretzschmer 53, figs 88-9, who identifies not twelve but fourteen of them (p. 52).

65. M. Biernacka-Lubanska, 'The water-supply of Novae', *Archaeologia* 30 (1979), 57-67, p. 63, fig. 11. Novae is 4 km east of Svištov, on the south bank of the Danube which here forms the Bulgarian-Romanian border. Its modern name is Staklen, or Pametnica and it is under excavation by a joint Polish-Bulgarian expedition, the *Archaeologia* referred to being a Warsaw publication. It was served by three aqueducts running in masonry-built surface channels of U- and sometimes V-shaped cross-section, with flat stone cover slabs. One of the conduits was 'made of lead', i.e. lead pipes (p. 63), the only time that I know of this Vitruvian recommendation having been actually put into practice. I am indebted to Professor Biernaka-Lubanska for bringing this material to my attention.

66. Kretzschmer 53, fig. 88. L. Richardson, *Pompeii: an architectural history* (Baltimore, 1988), 56, suggests that the tank was kept filled by a two-man crew continually lifting up to it buckets of water on the end of a rope. This idea seems to me to be supported neither by evidence nor by common sense. And the relative levels are such that the water had plenty of head and would rise to the tank by itself (see Fig. 212).

67. This happened at Pompeii. There the main pipe ran down the Via Stabiae to the theatres, and served the Central Baths, the Stabian baths, the Temple of Isis, the Samnite Palaestra, the two theatres, the gladiators' barracks, and the Triangular Forum. A branch ran West along the Via Nolae to serve the Forum Baths, and another East along the Via dell'Abbondanza, serving the main palaestra, the amphitheatre, the Bath of Venus in the House of Julia Felix, and several private houses (Eschebach, *JEAR* 93).

68. Kretzschmer 50, fig. 83. In his drawn 'schema' he even has the towers drawn on two different levels, the private users' towers being higher than those serving the public drinking fountains. Plainly, one should not read too much into this, for a schematic diagram should not be treated as a scale plan. The drawing is repeated by Callebat (151, fig. 9). Kretzschmer is not wholly clear in his text (54) but seems to identify the water towers serving private users with the private *castella* mentioned by Frontinus, and considered above (p. 294). This seems unjustified. As he outlines it, all private users' supplies came only from private *castella*, built at private expense (identified in his diagram as UIII). But Frontinus (*Aq*. II, 106) makes it plain that private supplies were normally taken from the public *castella*, and private ones built only where no public one was convenient. Another possibility sometimes mentioned, but not yet investigated, is that a number of the water towers were served successively, in series, by the same water main, so becoming in effect something like the Turkish *suterazis* (p. 237).

69. Eschebach, *JEAR* 91-2; Kretzschmer 52; Fuller exposition in F. Kretzschmer, 'La robinetterie romaine', *RAE* 9 (1960), 110-12, upon which is based my Fig. 212; largely the same material is also to be found in F. Kretzschmer, 'Römische Wasserhähne', *Jb. Schweiz. Gesel. Urg.* 48 (1960-1), 50-62 (esp. 61). The misunderstanding and confusion is well exemplified by a comparison of two respected scholars who contradict each other flatly, neither with the slightest attempt at argument or justification: Amadeo Maiuri, *Pompeii* (Novara, 1960), 27, maintains that the water towers 'by increasing the pressure, kept the supply of water under regular control'; while Van Buren (*RE* VIII A (1955), 474) asserts that 'sie waren hauptsachlich notwendig zur Verminderung des Wasserdruckes'. As a commentary on modern exegesis of Roman hydraulics, this speaks for itself.

70. The figures are from Eschebach, *JEAR* 92, quoting the calculations of Kretzschmer in the two articles cited (n. 69 above). The figures actually given by Kretzschmer (same in both articles) are 20 m height difference and 2 kg/cm^2.

71. See n. 69 above.

72. Kretzschmer, 'Römische Wasserhähne' (n. 69 above), 61. The *castellum* at Nîmes is 50 m above ground level at the centre of the city, and would then generate a pipe pressure of 5 kg/cm^2. At Arles the *castellum*, beside the amphitheatre, is 15-20 m above the baths and street level alongside the Rhone, resulting in a pressure of 1.5-2 kg/cm^2.

Chapter 11

1. Pompeii: Kretzschmer fig. 89, p. 53. He also illustrates (fig. 91) a tank with one offtake pipe set into the side and still preserved, and maintains that, in obedience to supposed Vitruvian practice, 'Les prises d'eau de ces châteaux intermédiaires sont disposées, comme dans les principaux, à troix niveaux différents, selon l'importance des points à alimenter' (p. 52). The evidence for this has not been made clear, though Eschebach does illustrate some tanks with offtake pipes at different levels; but they come from inside a house (*JEAR* 125, fig. 32). The best-preserved secondary *castellum* was found at the north-western corner of *insula* II, 2. It is published by Maiuri (*N.Sc.* (1939), 200ff.), and was of cubic form (65 × 65 × 65 cm), of sheet lead 0.6 cm thick; it stood 1.6 m above street level on top of a pier of 0.9 m square cross-section and had both supply and offtake pipes preserved, the latter fitted with a perforated metal filter. This unique installation is no longer visible, having been destroyed by bombing in World War II. See Eschebach, *JEAR* 93-4.

Minturnae: see I.A. Richmond, 'Commemorative arches and city gates in the Augustan age', *JRS* 23 (1933), 149-74, esp. 155, fig. 5. This is the only example I know of offtake pipes being set into the *bottom* of a running conduit, but no doubt there were others. It remains a rare practice. Ashby (45), identifies this as 'the primary *castellum* of the Minturno aqueduct, which has three main discharge-pipes'. This is apparently pure guesswork. The superstructure of the gate is not preserved and there are no remains of any *castellum* or even the aqueduct itself, though it must have crossed the gate. What is preserved is the lower half of the three 'down-spouts', aligned on what should be the centre-line of the aqueduct conduit, and embedded 'in the concrete core of the south pier' of the gate. There sometimes *was* a *castellum* located above the city gate, as at the Porta Capena at Rome, but as for the one identified here by Ashby, there is no evidence either that it existed or that it did not. Of the three pipes, one served a public fountain immediately below, alongside the gateway, the other two ran off southwards to serve locations unknown. As it is not known where the pipes went nor how they were attached to the conduit, one is surprised to find Ashby seeing in the arrangement 'clear traces' of the Vitruvian triple-*castellum* theory. As noted above (pp. 280-2), it is not at all clear what Vitruvius envisaged, but it was some kind of triple division. Presumably Ashby was reminded of it by the pipes because there are three of them, but this is a poor basis for associating a confused theory with a *castellum* that is not there. Ashby's faith in Vitruvius is great. He continues (loc. cit.) by ascribing to the *castellum* at Thuburbo Minus ('another fine example') the characteristic Vitruvian triple division. This *castellum* had indeed three offtake pipes, but Ashby, quoting GdM, bases his case on the fact that the 'sluices' have three different settings, giving varying priorities. In fact only one of the three sluices has a multiple setting, and it can be set in four positions, not three (GdM 317, fig. 125), thereby invalidating much of the Vitruvian connection (see my Fig. 201(b)). We may also note Ashby's statement that 'an example of the secondary reservoir ... has yet to be sought', and giving a reference to GdM 320, on Lyon. The Pompeian tanks seem now largely to fill the gap he mentions. On the Lyon reference, see now A. Audin, 'Le réservoir terminal de l'aqueduc du Gier à Lyon', *JEAR* 13-18; also ch. 10 n. 43 p. 459.

2. Kretzschmer 52; Eschebach, *JEAR* 92, opts for twelve.

3. Eschebach, *JEAR* 90, 100-1.

4. So Eugene Olivier (ch. 5 n. 5 p. 412), 64: 'Une dizaine de litres par tête et par jour suffisent en effet pour couvrir nos besoins primordiaux.' See also his p. 67 n. 2, quoting a modern (1922) study of 'la sécheresse de 1920-21' at Lausanne and maintaining that 'en temps de pénurie, dix litres par tête suffisent indéfiniment, sans dommage pour la santé'. For a total allowance per person in modern times, Olivier (67) opts for 200 litres ('en y comprenant toutes les dépenses d'eau personelles, urbaines, et industrielles'). GdM 340ff. estimates a total discharge of 75-80,000 m^3 for Lyon, with a population of 400,000 (?). This also works out at about 200 litres per person. A modern UNESCO survey of water needs in developing countries gives what are presumably the very minimum subsistence figures: 'Each individual needs an intake of about 3 litres per day in addition to water for dishwashing, hygiene, and laundry. In the developing countries, 25 litres per person per day, i.e. 125 litres per day for a family of five, is usually enough' – M. Falkenmark & G. Lindh, *Water for a Starving World* (Boulder, Colorado, 1976; tr. from Swedish edition, Stockholm, 1975); publ. in co-operation with the United Nations and UNESCO, and in conjunction with the UN Water Conference. Eschebach's allowance of 500 litres per person is thus seen to be generous – 20 times the minimum needs. For an example of successful subsistence on 1½-2 litres daily, see ch. 2 n. 25 p. 394; also pp. 59-60 above.

It is also instructive to set this in the context of modern urban use in developed societies. In national per capita average daily water consumption the USA and Canada are far in advance of the rest of the world, with figures of 6.3 m^3 and 4.1 m^3. However, the figures for average daily *domestic* use are quite different, in Canada ranging between 0.09 and 0.32 m^3 (see also p. 327 below). The difference comes from the exclusion of agricultural and industrial consumption; it must be emphasised that the big user is not industry but agricultural irrigation (p. 247 above). The lesson is that if one is seeking illumination on ancient water use from modern figures, we must use figures for domestic use, or at least urban supply, not national averages, which are liable to be swollen and distorted by the inclusion of enormous rural use that was not part of the ancient picture (just as the ancient figures are swollen by the practice of continuous offtake, with few taps).

5. See ch. 10 n. 20 p. 457. When one is considering figures for per capita consumption based on aqueduct discharge and population one must never forget

this significant but usually indefinable supplement. Some cities lived quite happily on well water alone.

6. For a town so large and important – and close to Rome – Ostia seems to have been very poorly supplied with water. Until the early empire it had no aqueduct at all, at a time when Rome already had seven or eight, and depended on well water (which, despite the proximity of the sea, was good quality (Pliny, *Ep*. 2, 17, 25)). The aqueduct, a short one of some 7-8 km, may date from Tiberius. For details see R. Meiggs, *Roman Ostia* (Oxford, 1973), 44, 143 (map on fig. 1, p. 112). G. Becatti ascribes it to Caligula (*Ostia* (Rome, 1970; Itinerarii dei Museii), p. 11).

7. Gardens: Wilhelmina F. Jashemski, *The Gardens of Pompeii* (New York, 1979), 32-3. The Romans' love of water in their gardens is emphasised by Pierre Grimal, *Les jardins romains* (Paris, 1969), 293-9: the villas along the Via Appia at Rome profited from apparently being able to tap the aqueducts as they came to the city, and in Trastevere they used the Alsietina, the water of which was so bad that watering the garden was about all that it was good for (Front. *Aq*. I, 11). Some of these 'gardens' may, of course, have been market-gardens, vegetable plots. Ostia taverns: Gustav Hermansen, *Ostia: Aspects of Roman city life* (Edmonton, 1982), 189-90: 'water was maybe the most important commodity that bars served their customers'. The point is that practically all popular drinks, hot or cold, were long drinks, mixed with large quantities of water. I do not know whether there was anywhere that water itself was actually sold in bars, but in some of the hotter and more arid parts of the empire I would not be surprised.

8. Eschebach, *JEAR* 90: 'gleichzeitig eine willkommene Hilfe zur täglichen Reinigung des Pflasters.' See also pp. 335-6 below.

9. See ch. 12 n. 7 p. 475.

10. Plumbarii: Front. I, 25ff. Lead in Britain: see n. 13 below. In Britain, lead pipes are indeed found, but are largely concentrated in two places. One is Bath, particularly in the installations of the bath itself. The other is military sites. This includes not only forts on Hadrian's Wall, but legionary cities such as Chester and York. The military preference for the more expensive lead pipes over wooden is succinctly explained by Hanson 414: 'the "State", as opposed to private patronage, would have paid the bill.' Copper pipes: two sections of copper piping, each about 30-40 cm long and of 1.5 cm diameter, are preserved in the Schwarzenacker Freilichtmuseum, near Homburg, W. Germany (inv. nos 08-03-1966: B13). They are the only copper pipes I have seen, and were made in the usual Roman fashion with a seam along the top. Metz: the pipes are square, with a round hole up the middle (diam. 6-8 cm): *WAS* 3 (1988), 146, abb. 23; Lambrecht 180, Bild 180, 181.

11. Vitr. 8, 6, 11: '*minime fistulis plumbeis aqua duci videtur, si volumus eam habere salubrem*.' The idea that the Romans suffered from using lead pipes is surprisingly difficult to pin down. It appears to be a vague, nebulous impression, widely referred to as a generally held belief, but without anyone personally maintaining its truth or quoting anyone else who does. The origin of the view seems to be R. Kobert, *Chronische Bleivergiftung im Klassische Altertum* (in Diergard (ed.), *Beiträge aus der Geschichte der Chemie* (Leipzig, 1909)), followed by Alfred Neuburger, *Technical Arts of the Ancients* (New York, 1930), 434. The whole question has now been exhaustively discussed by Lionel and Diane Needleman, 'Lead poisoning and the decline of the Roman aristocracy', *Classical Views* (publ. by Univ. of Calgary) 29 N.S.4 (1985), 63-94. They conclude that the cause of Roman infertility was rather gonorrhea.

12. Full discussion in A. Trevor Hodge, 'Vitruvius, lead pipes and lead poisoning', *AJA* 85 (1981), 486-91. It must be emphasised that this does *not* prove that the Romans were free from lead poisoning, only that if they suffered from it, it did not come from the water pipes. There are other more probable sources, such as lead-based glaze inside wine amphoras; the wine was often in contact with it for years before being served, and the tartaric acid in wine absorbs lead much more readily than does water. I must also note a striking coincidence. At almost the same time as I published the above study, another article appeared approaching the same problem but from the viewpoint of metallurgical science; the authors came to very much the same conclusions that I did. The two studies are completely independent (in each case the authors became aware of even the existence of the other only some seven years after publication), and thus may be seen to confirm each other. See J.E. Dutrizac, J.B. O'Reilly & R.J.C. MacDonald, 'Roman lead plumbing: did it really contribute to the decline and fall of the Roman Empire?', *CIM Bulletin*, vol. 75, no. 841 (May 1982), 111-15 (*CIM* = Canadian Institute of Mining and Metallurgy). For studies on the lead content of excavated human bones, see A. Mackie, A. Townshend & H.A. Waldron, 'Lead concentrations in bones from Roman York', *Journal of Archaeological Science* 2 (1975), 235-7; and H.A. Waldron, A. Mackie & A. Townshend, 'Lead content of some Romano-British bones', *Archaeometry* 18 (1976), 221-7. The figures are conveniently summarised in Needleman (n. 11 above), table I, p. 81.

13. See ch. 6 nn. 61, 62 p. 431. Silver and lead are mined from the same ore (*galena*), which is predominantly lead sulphide but with a small proportion of silver; see C. Singer, E.J. Holmyard, A.R. Hall & T.I. Williams, *History of Technology*, vol. 1 (Oxford, 1954), 583. For ancient wastefulness of lead, see O. Davies, *Roman Mines in Europe* (Oxford, 1937) 52, 180. A classic example of the amount of lead produced, and rejected, by ancient silver mining is offered by the Athenian silver mines of Laurion, where for a century now the modern French company has been extracting large amounts of lead, partly by mining but partly by re-working the ancient Athenian slagheaps. See E. Ardaillon, *Les mines du Laurion* (Paris, 1897), 16; for ancient lead production at Laurion, 117-20; Davies, op. cit. 251. The proportions of lead to silver produced in ancient working were about 300:1 (GdM 205). The provinces most productive of lead were Spain and Britain, and Britain was so prolific that a legal limit on production was imposed (*'ut lex ultro dicatur, ne plus certo modo fiat'* – Pliny, *NH* XXXIV 164). R.F. Tylecote, *Metallurgy in Archaeology* (London, 1962), 75ff., 94ff. Weight, see p. 156 above; n. 20 below.

14. See p. 154 above.

15. The difficulty of soldering the joints was noted by Kretzschmer, 'La robinetterie romaine', *RAE* 9 (1960), 113, who refers to it to excuse the difference between the carefully machined bathroom fixtures (taps, etc.) and their 'montage ... toujours maladroit, rude et primitif'. (= also p. 62 of his 'Römische Wasserhähne', *Jahrbuch der Schweizerischen Gesellschaft für Urgeschichte* 48 (1960-61).

16. Vitr. VIII, 6, 4; Front. *Aq.* I, 25. Illustrations: Landels 43, fig. 10; Callebat 162, fig. 12; GdM 201-4, figs 74-81; *MAGR* 30, fig. 3; Pace 74, fig. 49.

17. Stone: Pace 78 ('lastra di marmo'). Sand: Hanson 415. Sand seems a lot easier and would greatly facilitate putting the inscription on the pipe (see below).

18. Vitr. VIII, 6, 4: *'fistulae ne minus longae pedum denum fundantur.'* Ten Roman feet = 2.957 m. Large pipes (38-40 cm diameter) in lengths of 11 m have been found at Nîmes, but this seems quite exceptional (R. Naumann, *Der Quelbezirk von Nîmes* (Berlin-Leipzig, 1937), 30).

19. Ostia and Rome pipes, GdM 200; medium thickness, Callebat 162 (from

Arles and Luxeuil (H. Saône, near Vesoul)); small-gauge, Callebat loc. cit. From the figures given by Vitruvius for the amount of lead used in pipes of different gauges, Landels (43) notes that they must all have been of the same thickness and finds this surprising; Callebat (164), observing the same thing, flatly refuses to accept it. It looks as if Vitruvius has got something wrong here. P. Tannery, 'Frontin et Vitruve', *R.Ph.* 21 (1897), 125, finds that the weights indicated by Vitruvius for the larger pipes are so inadequate they are 'pratiquement absurdes'. GdM 182; SRA 197 n. 63.

20. Vitruv. loc. cit. Tables in Callebat (164), Pace (76), Landels (44). Some figures may give a rough idea of the order of magnitude: a ten-foot length of *quinaria* (internal diameter, 2.5 cm), weighed 19.5 kg (60 Roman lb); of *quinquagenaria* (internal diameter, 25 cm), 196 kg (600 Roman lb). A length of *vicenaria* (internal diameter, 9 cm) weighed 69-78 kg (240 Roman lb) and about 125 ft of it weighed one ton (37.5 m, one tonne). Some of these figures are based on estimates, Vitruvius not being reliable (n. 19 above), and are therefore not always internally consistent, but the general picture is clear: for any except the smallest pipes, a standard length would take at least two men to carry it, and usually more. On the really large pipes, such as those used in siphons and the take-off pipes in the Nîmes *castellum*, Mr Richard Joy, of Ottawa, writes to me that 'With a wall thickness of 3.6 cm and an outside diameter averaging 30 cm, each length of pipe would have weighed approximately 1,100 kg (the exact weight might have been slightly less, depending on the configuration of the overlap along the seam).

'To manipulate such a weight would require a team of 24 men, with appropriate harness to distribute the load (this is based on my army experience, 6 men being told off to carry and position 600-lb units of Bailey bridging).

'If the designer of the inverted siphon had attempted to use only three pipes, instead of nine [as at Lyon], the weight per length would have been about 3,000 kg. It is difficult to envisage such a weight being manipulated into position, particularly on sloping ground.' This does seem to be a very good reason for the use of the multiple-lead-pipe siphon, and much more convincing than arguments about possible reduction of pressure.

For an alternative estimate of weights of pipes of all gauges, see pp. 348-9 below.

21. Front. *Aq.* I, 25. He expressly attributes the '*quinaria* = width of sheet' principle to 'Vitruvius and the plumbers', but adds '*Maxime probabile est quinariam dictam a diametro quinque quadrantum*'. Vitruvius generally describes the various gauges by ascribing to each a set weight of lead, not a linear measurement, but he does add '*quae lamna fuerit digitorum quinquaginta, cum fistula perficietur ex ea lamna, vocabitur quinquagenaria similiterque reliquae*' (VIII, 6, 4). It will be noted that there the implication is that the unit of measurement is the digit, not the *quadrans*, which multiplies by four the dimensions I have quoted, and gives a *quinaria* of 7.25 cm. This would *have* to refer to sheet size, since the *quinaria* is the smallest pipe, and small pipes do not have a diameter of anything like as big as 7 cm. See also Pliny *NH* XXXI, 31, 58. Further complications abound. The Romans may have used two different systems of calibration at different periods, and there may have been all kinds of overlaps between the two. An official standard was apparently established only in 11 BC, on the death of Agrippa (Front. *Aq.* II, 99). Again, these standards are only known to have applied to metropolitan Rome. Pipes in Pompeii, Lyon, or Antioch may have conformed to a completely different set of criteria. If there are chance resemblances, which could well be the case, then that could mislead us yet further. For a knowledge of the actual working of the aqueducts, our chief concern is this book, it is fortunately unnecessary to sort out in detail this bureaucrat's nightmare. A full discussion will be found in Grimal 80-1, Callebat 163-7, and GdM 327-330.

22. Pace 78. The identification of this primitive printing system was due to Lanciani (R. Lanciani, *I Commentari di Frontino Intorno le Acque e gli Acquedotti* (Rome, 1881), 416: 'ogniqualvolta occorre nella leggenda la linea di cesura la lettera o le lettere sulle quali corre la linea sono più piccole; e ciò per l'impossibilità o almeno per la difficoltà di varcare il margine della cassa').

23. The reaction of a modern technical authority is instructive: 'The thought of providing pipes today with inscriptions in relief on them giving the name of the local water board official, the plumber who installed it, and the name of the house owner, not to mention the Queen's name, is appalling' – Penn 63 (n. 29 below).

24. Conjecture. Landel 43: 'cast, or cast and rolled'. Pace (75) simply says it was 'spread' ('spendendola').

25. This is not a hypothesis, but has been mathematically calculated by Luigi Jacono, 'La misura delle antiche fistule plumbee', *Rivista di Studi Pompeiani* 1, 2 (1934-5) (Napoli, 1934), based on a comparison of some nine sections of piping from Pompeii. See Pace 88-9. The similarity of oval and circular area is graphically shown in his fig. 55. There is in any case a serious problem involved in any belief that the Romans based their standards of pipe sizes, or capacity, or discharge, upon pipe cross-section, since, with circular pipes, this would involve calculations of the area of a circle. The ancients, of course, knew how to do it (it was discovered by Archimedes), but Roman figures, and their system of fractions, made even the simplest mathematical calculations something of a tour de force, particularly when they involved π and had no concept of a zero. For example, the modern value of π is 3.1415. Frontinus based his π on the fraction 22/7 (as does Columella, *De Re Rustica* 5, 2), which, again in modern terms, computes out to 3.1428. This was a value, however, which Frontinus, lacking the decimal system, could not express. Given the fractions he was using, the closest he could get to it was (again in modern terms) either 3.1423 or 3.1458; the first is what he used. See full exposition in Pace 83-4. Herschel (221) complains that 'Working with duodecimal fractions, it would have been far easier to take $\pi = 3\frac{1}{8}$, as had been done by Vitruvius, only one hundred years before Frontinus. But he laboriously ploughs along with $\pi = 3\ 1/7$ and is always pretty near right: – except when he, or some copyist, or translator, or printer, has made a mistake; which is not without example.' No doubt ordinary working plumbers could make mistakes too, either from incomprehension or as their finger slipped on the abacus, and probably worked from some simpler, empiric rule of thumb. This may lie behind the anomaly reported by Frontinus (*Aq.* I, 24), that there were even two different digits in use, the round and the square (*rotundus* and *quadratus*), the square being three-fourteenths of its own size larger than the round. 'Digit' being a linear measure, the concept of roundness or squareness of it may seem to us anomalous, but there is nothing inherently odder about a square digit than a square foot. A square digit was like a square inch, the area formed by a square of one digit side: a round digit was the area of a circle inscribed in such a square, and therefore of one digit diameter. The use of the two digits may come from the plumber's dilemma: what was needed was a unit of circular area, pipes being round, but it was the square one that was easy to calculate.

Grimal 80; diagram in Pace 81, fig. 53, and Bennett, *Frontinus* (Heinemann, London, 1925; Loeb edition), 366 n. 2. For a general study of Frontinus' mathematics, see Herschel 220-2, 'Arithmetic, AD 97, among the Romans'; Pace 79-90 (with full table of Roman fractions and their value, p. 80); Landels 53-7.

26. 'Le tuyau, soumis à une pression de 3 atmosphères, a commencé à s'arrondir; il a pris la forme circulaire à 8 atmosphères et a supporté une pression de 18 atmosphères sans se déchirer à la commissure des lèvres' – M. Belgrand, *Les aqueducs romains* (Paris, 1875), 71, quoted by GdM 202.

27. Pliny *NH* XXXIV, 94; 158; 160. Pliny's word for 'solder' is *tertiarium*. See n. 34 below.

28. *Archaeologia* 54 (1894), 23.

29. 'It was soldered in much the same way as it is today, leaving very little surplus metal, so that a more or less round pipe without projections was obtained.' This was not much used as 'it was considered that the soldered butt joint was not strong enough to withstand the pressures involved in carrying water' – W.S. Penn, 'Roman lead pipes', *The Plumbing Trade Journal* (Great Britain) (September 1961), 62-4. It must be made clear that in his judgments Penn is apparently basing himself exclusively on Romano-British practice.

30. It is worth quoting Penn's reconstruction (n. 29 above) of the whole process, to give the views of a modern plumbing expert: 'A wooden mandrel of the correct size was used as the former for the pipe. The pipe was then bent round the mandrel, the edges being recurved outwards, although they were not allowed to touch. The mandrel was then removed and the pipe filled with a dense material, usually clay or sand. The edges of the lead to be joined would then be cleaned up by a scraping process and the lead was ready for the next stage.

'The prepared lead would be laid on its side and a mould of clay, supported by wood, would be made along the full length of the proposed joint. Molten lead was then poured into the mould. The lead was at as high a temperature as possible so that it would fuse to the flanges on either side of it. After cooling, the mould was broken away and some of the surplus lead removed. It may be observed that, considering the perfect weld obtained, far more lead was left on the joint than was really necessary. The reason for this is not known but it is possible that the plumbarii thought that a considerable amount of lead was required to achieve a strong joint.'

31. For the shape of this seam, including the pair of 'fins' projecting from its upper edges where the lead slightly overflowed the clay mould ('*bavures*'), see Callebat 162, fig. 12; also Pace 74, fig. 49.

32. By this process Belgrand produced a joint where 'l'adhérence a été complète, la solution de continuité a disparu'. The pipe was triangular. It became round at 9 atmospheres pressure, and burst at 18 (quoted in GdM 202). The similar Roman pipe is the one found at Paris, rue Gay-Lussac (my Fig. 219(c)); there the joint was so perfect that the excavation workmen refused to believe that there was one, insisting that the pipe had been cast in one piece (GdM 202). Apparently similar pipes have also been found elsewhere – Fontaines Salées, College de France, Vienne (Callebat, 163). See also J. Mahol, 'Les tuyaux de plomb. Histoire et progrès de leur fabrication', *La Nature* (1 Dec. 1937), 503-7. For live coals, Callebat 162, fig. 12. A Roman soldering iron is illustrated by Alfred Neuburger, *Technical Arts and Sciences of the Ancients* (New York, 1930), 46, fig. 53. For the process of soldering, see Neuburger, loc. cit. also *SAT* 8, 137-8, and J.T. Healey, *Mining and Metallurgy in the Greek and Roman World* (London, 1975), 239. Compare n. 30 above.

33. Ham-fisted: a modern engineer, enlarging on the difficulty of manufacturing pressure-tight pipes, maintains that you can really only say they have been properly made since around 1800, with the coming of gas lighting (and even then the earliest ones were made, in Birmingham at least, by welding musket barrels end to end). This is overstating it. The Roman pipes, though they looked clumsy, were in very extensive use and evidently worked quite well. See J.E. Gordon, *Structures* (Pelican Books, Harmondsworth, 1978), 117; he is evidently misled by the orthodox handbooks on ancient hydraulics, and their excessive, Vitruvius-based, alarm over pressure in any form, however slight. For Belgrand, see GdM loc. cit., or p. 154 above.

34. Large gauge pipes in the Museum at Arles (probably from the siphon under the Rhône) were soldered together in this way: the solder on analysis consisted of 84 parts lead to 60 tin (GdM 204 n. 1). Illustration in Haberey 141, fig. 109.

35. Paris (GdM 202, fig. 77; 204), Pompeii (Kretzschmer 54, fig. 94).

36. i.e. so that the water runs round the exposed section of nail in the middle. This remarkable procedure is found in a large (10 cm diameter) pipe from Arles, well illustrated by Haberey 140, fig. 108. Sometimes, on large pipes, joints were also strengthened by lashing with rope. Though improbable, it is attested for the piping of a supposedly Roman aqueduct siphon in Constantinople (GdM 204; 190); a certain M. Flachat observed that its pipes 'sont cordés avec des bandelettes de chanvre'. De Montauzan refers this to the joints. Another technique was, in effect, to abolish joints as such, and to run the two halves of the pipeline into opposite sides of a more or less cubical or rectangular block, of stone or cast lead, to form, in effect, a 'box-joint' (Corbridge (lead): *Archaeologia Aeliana* 4, XXXVII (1959); stone: Römisch-Germanischen Kommission, *Germania Romana*, 32, 3, p. 27, pl. 32, figs 14, 15). Compare the Greek 'marmormuffen' at Ephesos (ch. 2 n. 42 p. 395).

37. As in the warning of Vitruvius, VIII, 6, 6.

38. Gordon (n. 33 above), 121-2: 'One consequence of this must have been observed by everyone who has ever fried a sausage. When the filling inside the sausage swells and the skin bursts, the slit is almost always longitudinal. In other words, the skin has broken as a consequence of the circumferential, not the longitudinal, stress.' Roman pipes behave the same way.

39. A notable exception was the big city-main at Pompeii, which had to have its top seam frequently repaired as it was splitting under the pressure. (Eschebach, *JEAR* 91-2.) Compare Ovid, *Met.* IV, 122.

40. Haberey 132.

41. Samesreuther 29, fig. 3 (Aachen: channel dimensions 22 cm high by 21; length of sections, 60); 37, fig. 10 (Biebrich: 15 by 32; 80-90 cm. This is probably from a main aqueduct rather than an urban distribution system, but is included here for completeness, since it is fairly small). See also Haberey 134, fig. 100.

42. B. Cunliffe, *Roman Bath* (Oxford, 1969), 127, fig. 43. The duct, which has a continuous lead lid soldered on to it, is about 52 cm wide by 12 cm high. A second, slightly smaller, duct is also there illustrated, 'provenance unknown'.

43. T-junction: Samesreuther 141, fig. 62, 2 (from Cologne; also published in *Bonn. Jahrb.* 82 (1886), 75, and pl. 4). T-junctions of branch pipes running off the mains are shown by Eschebach (*JEAR* 131, fig. 43; Pompeii), and Kretzschmer (54, fig. 92; Pompeii). Y-junctions, or angled joint: Pompeii (Amadeo Maiuri (*N.Sc.* (1931), 559, fig. 6B; 558, fig. 5, photograph of joint *in situ*). This is a small-gauge pipe (apparently a *vicenaria*) taking off from a supply main. Also reproduced by Samesreuther 141, fig. 62, 1. See R. Tolle-Kastenbein, *Krümmer und Knie an antiken Wasserleitungen*, Schrift. Frontinus Gesellschaft, Heft 15, 1990 (Bochum), 13-27.

44. Oblong box: Pace 74, fig. 49; Eschebach, *JEAR* 123, fig. 27. Cylindrical, Eschebach, loc. cit., fig. 28 (all from Pompeii, Insula IV, 13).

45. Samesreuther 144, fig. 64.

46. Antioch: J. Lassus, *JEAR* 228, fig. 16. Kourion's water supply consists of two terracotta-piped aqueducts, apparently dating mostly from the second century AD. It is published by Joseph S. Last, *Proceedings of the American Philosophical Society* 119 (1975), 39-72.

47. Compare Aspendos, p. 160 above. Kourion distribution box: Last (n. 46 above), 46, fig. 4; settlement boxes, 44, fig. 1; inspection box, 44, fig. 2. The terracotta pipes are well illustrated on p. 71, pl. XI. Of particular interest is one forming a sharp, right-angled bend, or elbow, all moulded into a single section of pipe (52, fig. 7; photo, 72, pl. XII (p. 17)).

48. Hooks: Eschebach, *JEAR* 92; Samesreuther 138, fig. 60, 2, shows a wall hook, though intended for wooden pipes. Easy access: Eschebach, *JEAR* 98: 'Dies hatte den Vorteil, dass man sie leicht ausbessern konnte.' Protection: Kretzschmer 56, fig. 95. He may be right in his belief (p. 54) that they were often left exposed because it was so hard to solder the joints.

49. U-channels – Haberey 141, fig. 110. Embedded in clay, Haberey 138-9, figs 106-7 (Lamersdorf). York: R.C.H.M., *York: Eburacum* (1962), 1, 38. From the Athenian Agora come examples of lead piping actually inside a terracotta pipe (Mabel Lang, *Waterworks in the Athenian Agora* (Princeton, 1968), fig. 24); this is exceptional.

50. A modern parallel will warn us not to jump to conclusions. Englishmen still often shake their heads, in a kind of amused despair, at the number of burst pipes that occur, because their pipes are often attached to the outside of the house and exposed to the cold. They become particularly apologetic when talking to anyone from Canada where, in even the bitterest winters, burst pipes are almost unknown, because they are always inside and adequately protected. Even with a practical people like the Romans, we should not assume that they always actually did whatever it was quite plain they should do.

51. Forbes: *SAT* 167, fig. 36. Kretzschmer 50, fig. 83; he emphasises the difference by adding a 'schéma d'une distribution d'eau moderne', summarising: 'une conduite principale quittant le château d'eau constitute, en quelque sorte, un tronc qui se ramifiera en plusieurs branches vers les rues qu'il s'agit de desservir. Ces branches se subdivisent à leur tour vers les maisons des consommateurs.' This sequence cuts out the secondary *castellum*, or water tower, as a stage in the process. The 'multiplicity of small pipes in parallel' phenomenon is well illustrated by excavation at Antioch under the mediaeval church and under the paving of the Hellenistic bridge crossing the Parmenios torrent, both of which reveal about a dozen of them, side by side (of terracotta) (J. Lassus, *L'eau courante à Antioch, JEAR* 225, figs 9-10: he finds them, though not all of the same date, 'une impressionante série du tuyaux' (p.215)). Frontinus *passim*.

52. Lassus (n. 51 above, 212, 219). The distributors are known for the intersection with the main street. Their existence for intersections with the other north-south streets of the grid is surmised. Water was admitted into the secondary conduits from the main one (on the uphill side of town by 'vannes', many of which are said to have been found intact in 1933 (J. Weulersse, 'Antioche, Essai de géographie urbaine', *Bulletin des Etudes Orientales* 4 (1934), 27-9.

53. R. Meiggs, *Roman Ostia* (Oxford, 1973), 143-4. A section of the pipe, now in the Ostia Museum, is shown in Haberey 116, fig. 83.

54. So Eschebach, *JEAR* 89; Kretzschmer 94. Such is the prestige of Vitruvius that archaeologists are liable to invoke this interpretation anywhere they find three pipes together. Ashby did it for both Minturnae and Nîmes (n. 1 above). I know of no firm evidence that Vitruvius' design actually existed anywhere, or that, like so many things in his work, it was anything more than a suggestion of his own on how a *castellum should* be built: Callebat (154) calls it his 'solution personnelle'.

55. Kretzschmer 50, fig. 83; Callebat 151. This was rightly objected to by H. Fahlbusch in a paper on 'La conduite d'eau a longue distance pour l'irrigation et l'alimentation des villes, a l'epoque romaine', *International Commission on Irrigation and Drainage XI Proceedings: special session on the history of irrigation, drainage and flood control* (Grenoble, 1981), 60ff.: 'It seems senseless to instal 3 separate lead distribution systems all over a city to supply different kinds of customers' (68).

56. Eschebach, *JEAR* 106, pl. 4. In slightly amended form, the same plan is rather more clearly shown in Eschebach, 'Pompeii: la distribution des eaux dans une grande ville romaine', *Doss. Arch.* 38 (Oct.-Nov. 1979), 77. The reader may find it helpful to know that this is a condensed version, in French, of Eschebach's later (1983) *JEAR* article. In both publications there are inconsistencies (but not the same ones) between the symbols appearing on the plans and the explanatory legend. To get a good idea of the Pompeii network it is therefore best to compare and consult both works. Also shown in *WAS* 2, 202.

57. In this area, according to Eschebach's *JEAR*, pls 2, 3, and 5 (n. 56 above) there were seven shops with running water, fifteen or so private houses, four or five fountains, and three water towers, none shown as connected to the main supply in his pl. 4 (though it is not far away and they may have been connected to it). The houses include such well-known ones as the Houses of Pansa, Sallust, Apollo, Meleager, the Fountain, the Labyrinth and the Vettii. This solution for the course of the three main pipes is also maintained by Fahlbusch (n. 55 above), 68.

58. See also Eschebach, *JEAR* 131, fig. 43; id., *Doss. Arch.* (n. 56 above), 78 (illustration of junction on mains pipe under Stabian Baths).

59. Continuous flow: Front. *Aq.* II, 104. 'Roman water supply worked on the continuous offtake principle' – *SAT* 2, 172. For the impression made on a modern water engineer, see Herschel 231. Taps: 'There is one simple and obvious question which, surprisingly, cannot be answered directly from the evidence available. If a Roman householder had a piped supply of water, did he have a tap to turn it on and off?' – Landels, 52. I have elsewhere disputed this statement (Hodge, 'Vitruvius, lead pipes, and lead poisoning', *AJA* 85 (1981), 489.

60. *SAT* 176. Forbes, however, gives no reference and is in any case not always reliable. The quantity of taps found is emphasised by Kretzschmer (n. 65 below), 90, 50, who sees in them an object of industrial mass-production ('une production massive'/'billig Masenware'). A large number of such taps from Pompeii is in store in the Museo Nazionale at Naples where he inspected a crate ('la fameuse caisse/ der erwähnten Sammelkiste' – 100, 56) full of about forty of them. In an earlier age, Herschel found (and illustrated) what was probably the same set, then displayed in a glass showcase (230, 231, figs).

61. As clearly differentiated by Haberey 117: 'Im Rohrnetz innerhalb der Stadt diente der Wasserhahn als Regulier- und Absperrvorrichtung.'

62. Fr. 'robinet de débit' and 'robinet de communication'. Ger. 'Auslassehahn' and 'Durchgangshahn', also 'Auslaufhahn' and 'Absperrhahn'. See Hodge (n. 59 above), 490. According to Kretzschmer (31), ancient taps were traditionally decorated with representations of sheep and cocks, hence the name: from Old French *robin*, 'sheep', comes *robinet*, and from German *Hahn*, 'cock' (compare 'hen'), comes *Hahn*, 'tap' (compare English 'cock' (rooster) and 'cock' (tap)). In Nijmegen there is a tap with the handle shaped like a human finger (A.N. Zadoks,

Josephus Zithe and others, *Description of the Collection in the Rijksmuseum at Nijmegen* (1973), II, fig. 197).

63. The largest taps known seem to be the stopcocks fitted to the 20 cm diameter main along the Decumanus at Ostia. Two are preserved, one in the museum at Ostia, and the other *in situ*. Kretzschmer (n. 65 below), 93, 53, gives a diameter of 20 cm for the one in the museum and 30 cm for the other – 'c'est un veritable monstre!/Ein wahres Ungetum!' Haberey (116 and fig. 83) gives the smaller one an outside diameter of 16-20 cm: it is 50 cm high. We may also add the large stopcock in the Naples museum, illustrated by Herschel 198 n. 1. He gives the diameter of its water pipe as six inches (= 15 cm). Connoisseurs of the improbable will relish Herschel's statement that when, in 1886, he picked up this tap and shook it, 'I heard water splash around in it'; and his conviction that this was genuine ancient Roman water, still preserved inside, 'after having been imprisoned about eighteen hundred years'. It did not last much longer, unfortunately, and by 1897 had disappeared, 'chemically absorbed'. By contrast, the smallest taps were of diameter about 1.2 cm (Kretzschmer, loc. cit.). Stopcocks are usually larger than discharge taps, since they tend to be mounted at a division point on a main that then branches out to serve several outlets through smaller pipes: the stopcock is usually 5 cm diameter and above: so Kretzschmer 92, 52, but the figure seems high to me. Private houses often had stopcocks, and a 5 cm diameter pipe going into just one house seems impossibly large.

64. Well illustrated by Haberey 117, fig. 84, who describes the design as 'simple and reliable' ('so einfach wie zuverlässig').

65. See also SRA 202 n. 76; Haberey 116-17. The standard discussion of taps, of all kinds, is F. Kretzschmer, 'La robinetterie romaine' *RAE* (1960), 89-113, and id., 'Römische Wasserhähne', *Jahrbuch der Schweizerischen Gesellschaft für Urgeschichte* 48 (1960-61), 50-62. Essentially these are both the same article, but to facilitate research to both texts, here two page figures are quoted, as in nn. 60 and 63 above, e.g. 'Kretzschmer 93, 53', the first being the page in the French version, the second in the German. The texts are largely identical.

66. The danger from water-hammer with Roman taps was first pointed out by Haberey (117).

67. Kretzschmer (93, 52) has calculated that some of the larger taps preserved sustained an upward pressure on the top of the cylinder of around 51 kg, apparently *without* bursting.

68. McKay (n. 79 below), 242 n. 62, mentions four found at various locations in Pompeii. They were made of bronze.

69. Kretzschmer 98, 55. In 1973 the water supply to the apartment where I was living in downtown Athens was cut off and I personally had the job of turning it back on. Although the pressure in the mains was considerable, enough to carry the water up to the fifth floor, the 'tap at the mains' was of the rotary plug type just described. The friction resistance was very high. At first I tried with my fingers, and might as well not have bothered. I then borrowed from a plumber's shop the appropriate 'key', a T-shaped device with a cross-piece, or handle, at least two feet long. Even with this leverage, I was only just able to turn the tap; at first I thought I was not going to manage it.

70. But one piece of evidence indicates the opposite. The 'Pierre de Chagnon', a Hadrianic inscription found alongside the Gier aqueduct at Lyon and setting out legal prohibitions on the agricultural use of land along the right of way of the aqueduct, 'reproduit les termes même des interdictions romaines': Grenier notes that 'les règles qui s'appliquaient à Rome étaient en vigeur en Gaule' (*MAGR* 36;

GdM 108-9, 393-4). See ch. 5 n. 23 p. 417.

71. Eschebach, *JEAR* 95. Nine such tanks are marked as existing in Pompeii on his pl. III. See also his figs 32, 33.

72. 'Wohlhabende Hausbesitzer' is how Eschebach puts it (loc. cit.).

73. So Di Fenizio, followed by most modern scholars. See A. Trevor Hodge, 'How did Frontinus measure the quinaria?', *AJA* 88 (1984), 207 n. 8. See also Grimal 82; Ashby 30; *SAT* 166; GdM 337; Landels 52. Compare n. 4 above.

74. *SAT* 154. The domestic households of the rich were usually noted for their vast numbers of slaves, and since the number itself constituted a potent status symbol, finding something for them to do was often a real problem and featherbedding was rife. In villas with silver pipes, a few of the slaves were probably full-time water-pipe polishers. See J. Carcopino, *Daily Life in Ancient Rome* (Pelican ed., Harmondsworth, 1956), 83: 'The *familiae serviles* in the service of the great Rome capitalists often reached 1,000'. Such extravagances are to be found at Rome, or in country villas, not at Pompeii, a small market town. The 40 m^3 daily figure quoted applies to houses such as those at Pompeii. For the bloated establishments of Rome presumably a water supply even more gargantuan was provided.

75. For an exception see, e.g., the kitchen of the House of the Messii at Vaison-la-Romaine. Dora Crouch (forthcoming publication) rightly points out that in most studies of domestic water use, dishwashing and laundry (as opposed to fulling) are never mentioned. There is, however, not much to be said about either. The chief problem with laundry must have been getting it dry. Where to hang it? If current Mediterranean practice is anything to go by, all those balconies at Pompeii and Ostia must have been liberally festooned with clothes lines full of washing. This detail is often overlooked. There are many romantically attractive drawn restorations of the Peristyle of the House of the Vettii at Pompeii, but none show it criss-crossed with clothes lines. Where else did the Vettii hang their washing?

76. Partly because at Ostia water pressure was insufficient to serve upper floors. See R. Meiggs (n. 53 above), 239-40, and G. Hermansen (n. 7 above), 43. U.E. Paoli, *Rome* (London, 1963), 66, points out that originally 'the ancient Romans had no kitchen; they prepared their dinner in the atrium'.

77. Examples of both types, from Besançon and Pompeii, are shown by Kretzschmer 98-102, figs 29-32; 55, fig. 4.

78. From Rottweil, in S. Germany, and Studenberg, near Biel, Switzerland, Kretzschmer 103-6, fig. 33 and pls I, II; 57-60, figs 5, 6.

79. A good example of baths in a country villa is Piazza Armerina. See also, e.g., Chiragan (near St Bertrand-de-Comminges; P. MacKendrick, *Roman France* (London, 1971) 131-2). In the large number of country villas so far identified, especially in France by air photography, baths seem to come almost as standard equipment, and Horace's supposedly modest Sabine farm had a full standard set of *caldarium, tepidarium* and *frigidarium*. (A.G. McKay, *Houses, Villas and Palaces in the Roman World* (London, 1975), 112. See also 140. Bathhouses in forts need not be listed, but Saalburg, and the forts on Hadrian's Wall, may serve as examples: see Anne Johnson, *Roman Forts* (London, 1983), 33, 134. A good example of baths in a private city house is the House of the Messii (Vaison-la-Romaine): Villa of Diomede (Pompeii).

80. Boscoreale: Kretzschmer 108, fig. 37; 60, fig. 8. *Epitonium*: 106, fig. 34; 60, fig. 7 (Stabian Baths, Pompeii; also Marine Baths, Herculaneum).

81. Pliny *Ep*. V, 6, 46 (Pliny's Tuscan Villa.)

82. I hesitate to call this the muzak of the ancient world, but it may not be too bad a parallel just the same.

83. The psychological effect is real and clinically attested. In modern hospitals it is a common practice, in appropriate cases, to leave a tap dripping in the expectation that the sound will produce some desired therapeutic effect on the patient.

84. The Euripos is the narrow channel between mainland Greece and the offshore island of Euboea. It is noted for its very difficult currents, which not only run at speeds of up to 6-7 knots, but reverse direction several times daily; on Lorieus Tibertinus and the overflows, see G.C. Picard, *L'Archéologie* (1969). For 'Niles' and 'Euripi', see A. Boethius and J.B. Ward-Perkins, *Etruscan and Roman Architecture* (Harmondsworth, 1970), 162; also Frank Sear, *Roman Architecture* (London, 1982), 117.

85. The existence of a basement is common at Bulla Regia, though not elsewhere.

86. It is ironic (but very characteristically Roman) that after these elaborate measures to cool things down, the house still had a hypocaust and bath of the usual sauna type. Full publications of the hydraulics by Yvon Thebert, 'L'utilisation de l'eau dans La Maison de la Pêche', *Les Cahiers de Tunisie* (published by *La Faculté des Lettres et des Sciences Humaines de Tunis*) 19 (1971), 11-17.

Chapter 12

1. See p. 67 above.

2. Neuburger (ch. 11 n. 11 p. 465), 437. However, Kretzchmer (59) takes a contrary position: 'Un profond souci de l'hygiène jouait, ici comme ailleurs, un rôle prépondérant.' See also Lewis Mumford, *The City in History* (New York, 1961), 216.

3. Copais: *SAT* 2, 45; E.J. André Kenny, 'The ancient drainage of the Copais', *Ann. Archaeol. Anthr.* (1935), 189-206; U. Kahrstedt, 'Der Kopaissee im Altertum und die Minyschen Kanale', *JDI* 52 (1937), 1-20. Fucine Lake: tunnel length, 5,679 m: Albano, 1,425 m; Nemi, 1,650 m (Adam 267; also 22n. 46).

4. Wagon load of hay: Strabo, 5, 3, 8; Pliny *NH* XXXVI 108; 104 (Agrippa's boat). See *Pliny XXXVI*, A. Rouveret *et al.* (Budé ed., Paris, 1981), 199. For full bibliography and photographs (external only), E. Nash, *Pictorial Dictionary of Rome* (London, rev. ed. 1986) vol. 1, 258 and figs 301-6; also Garbrecht, *WAS* 1 (1986), 30, Bild 14 (internal photograph), and H.O. Lamprecht, *Caementitium Opus* (Dusseldorf, 1984), 95, figs 69-71); H. Bauer, 'Die Cloaca Maxima', *Mitt. L. Inst.* 103 (1989), 43-68, for the fullest account.

5. 'Collector': e.g. Pergamon (Garbrecht, *WAS* 1 (1986), 30, fig. 15). In Timgad there is a network of branch drains, 0.4 m × 0.8 m high, running under each street and serving the main collector drain, under the *cardo* (A. Ballu, *Les ruines de Timgad* (Paris, 1897-1911)). This is a typical arrangement. See n. 27 below (Volubilis).

6. Compare the comments of Etienne (n. 27 below), 24, and pl. XLIII, 1, on the main drain at Volubilis: 'Quand on parle d'évacuation, il faut distinguer entre l'écoulement des eaux de ruissellement et d'infiltration et l'élimination des eaux usées. On ne soucie en general que de la seconde mais la configuration du terrain et le climat de Tingitane ont posé des problèmes précis aux urbanistes volubilitains ... il s'agit donc de canaliser la majeure partie des eaux d'infiltration ou de ruissellement, amenées par les orages violents en automne et par abondantes pluies d'hiver et de printempts. Telle est la fonction du grand drain central.'

7. The ancients full recognised this: Front. *Aq.* II, 111 preserves the text of an imperial decree explicitly enunciating it: '*necesse est ex castellis aliquam partem* *aquae effluere, cum hoc pertineat non solum ad urbis nostrae salubritatem sed etiam ad utilitatem cloacarum abluendarum.*' 'An overflow from the *castella* is vital, for this is used not only to keep the city healthy but also to the practical end of scouring out the sewers.' Ashby (46), identifies the law as 'a post-Tiberian rescript'.

8. The problem is explicitly set forth by Garbrecht, *WAS* 1 (1986), 30: 'Die nicht völlig dichte Konstruktion der Kanäle sowie die offenen Einläufe in diese Mischkanalisation habe in den niederschlagsarmen Sommermonaten, wenn der Zufluss zu Städten und damit die Überschusswassermenge gering waren, sicherlich zu erheblichen Geruchsbelästigungen geführt.'

9. Neuburger (ch. 11 n. 11 p. 465), 439.

10. Compare Hanson's (92) comment on Roman Britain: 'If a public water supply in Romano-British towns should not be regarded as unusual, however, the provision of adequate drainage and sewage systems were another matter. A lack of concern for the disposal of private and public waste products of all kinds was a notable feature in many of the cantonal capitals, the exceptions generally being the large chartered towns such as Lincoln, Colchester and Verulamium.'

11. Adam 283; for 'ground surface drainage', Adam's is the only clear account existing (283-6).

12. Vitr. VII, 4. Adam 286, fig. 611 (Pompeii, Porta di Nola), figs 615-20. The double-wall principle is well illustrated by examples from Bavay (near Lille), Rheims and Bourges (anc. Avaricum): E. Will, 'Le cryptoportique de Bavay', *Revue du Nord*, 40 (1958), 493f.; 42 (1960), 403f.; 46 (1964), 207f.; E. Frezouls, *Le cryptoportique de Reims, les cryptoportiques dans l'architecture romaine, Ecole française de Rome* 14 (1974), 293f.; J.-P. Adam and Cl. Bourgeois, 'Un ensemble monumentale gallo-romain en terre dans le sous-sol de Bourges', *Gallia* 35 (1977), 1, 115f. At St Romain-en-Gal, across the Rhône from Vienne, France, we see a comparable arrangement designed to insulate the floor of a room from a soil damp with seepage or flooding from the nearby river. Under the floor is a layer of empty amphorae, buried upside down, to provide a 'vide sanitaire'. (*Gallia* 26 (1968), 580 (with illustration), 581 n. 21, records a similar system found in 1914 at Sainte-Colombe-les-Vienne, near Le Havre.) For a second example at St Romain-en-Gal, see *Gallia* 35 (1977), 490.

13. Assuming that such things existed at Pompeii. Study of the manifold complications of the range of standard pipes listed by Vitruvius and Frontinus has always been strongly slanted towards the internal, i.e. a study of these authors, illustrated, and then only rarely, by actual pipes found at Rome; there has been no real attempt to see whether the Rome specifications are relevant to other cities where pipes are often found, as at Pompeii, in large numbers.

14. For this and the whole question of Roman urban sanitation, see the comprehensive and excellent article by Alex Scobie, 'Slums, sanitation and morality in the ancient world', *Klio* 68 (1986), 399-434 (esp. 411-17); also E.J. Owens, 'The koprologoi at Athens in the fifth and fourth centuries BC', *CQ* 33 (1983), 44-50; M. Grassnick, *Gestalt und Konstruktion des Abortes im römischen Privathaus* (Munich, 1982). L. Friedlander, *Roman Life and Manners*, tr. L.A. Magnus (Barnes & Noble, New York, 1965, repr. of 1913 edition), vol. 4, 284. The urban manure trade was probably regulated by Vespasian, better known for his tax on urine ('*non olet*'). In the Orient, the contents of a town's toilets were sometimes considered sufficiently valuable to be auctioned. (Friedlander, loc. cit., who assembled the fullest bibliography of references (esp. ancient) on this understandably neglected topic; he is now replaced by Scobie. See also updated revision by F. Drexel, in Friedlander, *Darstellungen aus der Sittengeschichte Roms* (Leipzig, 1921), vol. 4, 310-11.) In relatively modern times the installation of a proper sewer system in Paris was blocked for a very long time (long after other comparable cities had one) by those contracted to clean out the city's cesspits, who formed a lobby in defence of what they evidently felt was a lucrative trade (Alain Corbin, *The Foul and the Fragrant: odor and the French social imagination* (Harvard Univ. Press, 1986), 119, 213). Any lack of sewers in Roman cities was at least, so far as we know, never due to this particular kind of obstruction. J. Carcopino, *Daily Life in Ancient Rome* (Yale, New Haven/London, 1940), 40; (Pelican ed., Harmondsworth, 1956), 52.

15. Dora Crouch, 'The Hellenistic water system of Morgantina, Sicily', *AJA* 88 (1984), 362. Although brief, this is one of the few modern considerations of the agricultural use of night soil in existence. She gives a number of useful references to recent studies in the context of ecological development. Noteworthy is the great amount of sewage that can be absorbed: 'One forested area in Pennsylvania had no trouble in accommodating 120 in. of treated sewage in a year.' See also Scobie, Owens (n. 14 above); Mumford (n. 2 above), 216. Columella (I, 6, 24) attests to the beneficial effects of farm sewage on vegetables and fruit. In nineteenth-century France there was much opposition to urban sewers on the grounds that they wasted a valuable economic and ecological resource that was much prized as a fertilizer: 'Every kilogram of urine is equivalent to a kilogram of wheat' – H. Sponi, *De la vidange au passé, au présent et au futur* (1856), 26, as quoted by Corbin (n. 14 above), 118. Various official submissions to the administration of Paris lauding the agricultural virtues of excrement are listed in Dominique Laporte's challengingly entitled *Histoire de la merde* (Paris, 1979). The relative excellence of the product reflected a class distinction, for improved sanitation, the perquisite of the rich, usually involved flushing with water and hence dilution of the vital nitrogen content; thus, studies showed that samples from the cesspool of the Grand Hotel de Paris were inferior in nitrogen by a factor of thirty to what was collected, untreated and in all its rich natural goodness, from the habitations of the poor. There is no sign that the Romans were aware of this conflict between hygiene and ecology, much less its implications for the class system.

16. J. Carcopino, *Daily Life in Ancient Rome* (Yale, New Haven/London, 1940), 40; (Pelican ed., Harmondsworth, 1956), 52. He also (39; 51), regrets 'the lack of a water system such as never existed save the imagination of too optimistic archaeologists'. Mumford (n. 2 above), 216-17, also administers a salutary correction to the conventional view of Roman sanitation: 'In the great feats of engineering where Rome stood supreme, in the aqueducts, the underground sewers, the paved ways, their total application was absurdly spotty and inefficient', and their sewage disposal 'records a low point in sanitation and hygiene that more primitive communities never descended to'. This refers to Rome itself, and is not necessarily true of provincial towns, where things were often much better, because, being new foundations, they could be properly planned right from the start.

17. The most famous, though non-Roman, example of this is the Queen's toilet at Cnossos. Though often admired as a technical and historical marvel ('a flush toilet in 1400 BC!'), I have for long had grave doubts about the drain into which it empties, for it not only has very little fall but also has a number of sharp bends as it serves *en passant* to drain rainwater from two small courtyards. It might carry off the rain, but I have fears it might prove quite ineffective as a sewage drain, even with the odd pitcher of water to help things along. Plan and section in R.W.

Hutchinson, *Prehistoric Crete* (Penguin ed., Harmondsworth 1962), 177. A. Evans, *The Palace of Minos* (London, 1921-36), vol. 1, 228-30 and fig. 172.

18. Adam 283, with illustrations (figs 607-10). Rarity of running water on upper floors, ibid. R. Meiggs, *Roman Ostia*² (Oxford, 1973), 143.

19. Fullers' urine pots: Martial VI, 93, 1; R. Meiggs, loc. cit. Vespasian taxed the practice (Suet. *Vesp.* 23). 'Neighbouring dungheap', Carcopino (n. 15 above), 42; 54.

20. '*Cacator cave Malum.*' The phrasing seems to be standard, like our 'No Parking', or 'Trespassers Prosecuted', and is found repeated in *CIL* IV, 3782, 4586, 5438. In 3832 the same warning is found actually *inside* a house, directed, it is presumed, to slaves of the domestic staff. The vague menace of '*cave malum*', 'You'll be sorry!' presumably reflects the absence of any really effective deterrent. Some proprietors sought to counter this weakness by planting nettles in places they deemed especially vulnerable, ensuring automatic retribution. The doggerel verse of *CIL* IV, 6641 approaches the problem in a more positive and constructive spirit: '*Cacator sic valeas/ut tu hoc locum transeas.*' Indelicacy, I feel, should not deprive the Latin-less reader of the translation owed him: 'Shit with comfort and good cheer/So long as you do not do it here.' This inscription is well illustrated, in full colour, by Lamprecht (n. 4 above), 99, fig. 78. See also *CIL* IV, 5242.

21. 'Traditional practice': it is traditionally accompanied by the warning cry 'gardy loo!' ('attention à l'eau'), in turn traditionally associated with eighteenth-century Edinburgh, but the universality of the practice needs no emphasis. See Mumford (n. 2 above) 292. Juvenal (3, 269-77) inveighs against it, and the jurist Ulpian (*Digest* IX, 3, 5, 1-2; 7) has a section on the legal responsibility of upper-floor tenants. Lacking any advice on what must often have been the real problem, namely how the victim would know to look upwards in time to see what was coming and where from, one suspects that Ulpian's remedies had a theoretical rather than a practical value.

22. Suet. *Nero* 26.

23. Pliny *Ep.* X, 98. In all fairness it must be noted that, in Pliny's extensive experience of municipal water schemes in Bithynia, what he usually ran up against was the opposite weakness: towns had to be restrained from plunging headlong into irresponsibly and ill-planned aquatic extravagances that would leave the treasury bankrupt and the aqueduct unfinished. *Ep.* X, 37; 39.

24. Hanson 128. Compare G.C. Boon, *Roman Silchester* (London, 1957), 161: 'Apart from the street gutters and occasional wooden drains which served to carry off waste water there was no proper drainage system at Calleva [Silchester]; nor do any of the houses seem to have been provided with drained latrines. Refuse was dumped in pits dug for the purpose, and the insulae are honeycombed with them.' The ultimate degradation must here surely be Caistor-by-Norwich, the East Anglian capital of the Iceni, which ended up being buried under its own filth. The rubbish heap around the guardrooms attached to the South Gate in the city wall became so extensive that in the third century AD access to them was impossible ('Roman Britain in 1934', *JRS* 25 (1935), 213). The entire south-eastern quarter of the town evidently turned into a collective dungheap, and one cannot honestly quibble with Hanson's (93) summing-up, that 'entry into the town from the main road from Colchester in the south must have been a memorable experience'. Sanitary horrors of this sort were of course hardly characteristic of Roman civilisation, but, while reading about the glories of the aqueducts, we should remember that the other side of the picture did also exist. The interested (and not unduly squeamish) reader will find instructive parallels from a later age fully and colourfully set out in the scholarly account by Corbin (n. 14 above) of, e.g., 'the excremental hell of Lille' (264 n. 12).

25. Livy V, 55, 2.

26. Illustrated by Kretzschmer 59, fig. 100. At this point there were two drains closely parallel, one under the sidewalk and one under the roadway, with an adjacent pair of manholes.

27. Well illustrated in Robert Etienne, *Le quartier nord-est de Volubilis* (Paris, 1960), pl. II, showing individual connections from some twenty houses either to the 'égout collecteur' or to the 'drain central' running along the decumanus. It appears as if every house in the town was connected directly to the drain network.

28. At Silchester (Hanson, 128).

29. Illustration in Adam 259, fig. 550 (House IX, 3, 5, Pompeii). The cisterns, full of drinking water collected from the roof, delivered their overflow by a conduit running under the sidewalk and opening directly on to the paving of the roadway. Adam describes the one illustrated as also carrying waste water ('évacuation du trop-plein de citerne et des eaux usées'). Another minor source of domestic waste water could be floor-washing. Mosaic floors had to be frequently washed, not as a matter of cleanliness, but to remove impurities lodging between the tesserae that could decompose and attack chemically the cement in which the mosaic was embedded (the parallel with brushing one's teeth is striking). Some mosaic-floored rooms, at least in Roman Britain, were provided with a small drain in one corner to carry off the wash-water (Boon (n. 24 above), 155). I know of no example of house sewage being discharged by pipe into the roadway (officially, as it were, as opposed to simply being thrown there), but would not be surprised if, somewhere or other, it happened.

30. Haberey 120-3; Grewe 102 and fig. 104; Lamprecht (n. 4 above), 93 and fig. 72; 96. Like the aqueduct tunnel at Vienne, it has served as an air raid shelter.

31. Lamprecht 97, fig. 73. Its appearance and proportions are similar to those at Cologne.

32. Neuburger (ch. 11 n. 11 p. 465), 443-44. He adds that a sort of plug or sluice has been found, implying that the discharge of the waters was regulated on the basis of agricultural irrigation, but gives no further source or reference.

33. Dora Crouch (n. 15 above). 'Heavier materials', she adds, 'were probably carried out to the fields on the backs of donkeys, and spread out as fertilizer, which was especially valuable for trees.' This article is almost the only study to approach ancient water use on an ecological basis.

Index

Numbers in italics (e.g. *123*) refer to illustrations or captions.